Daniel L. Hogue

D0780578

BREEDING PLANTS RESISTANT TO INSECTS

BREEDING PLANTS RESISTANT TO INSECTS

Edited by

FOWDEN G. MAXWELL
Department of Entomology
Texas A & M University

PETER R. JENNINGS
The Rockefeller Foundation

A WILEY–INTERSCIENCE PUBLICATION
JOHN WILEY & SONS
New York • Chichester • Brisbane • Toronto

Library of Congress Cataloging in Publication Data

Main entry under title:

Breeding plants resistant to insects.

(Environmental science and technology)
"A Wiley-Interscience publication."
Bibliography: p.
Includes index.
1. Plants—Disease and pest resistance. 2. Plant-
breeding. 3. Insect control—Biological control.
I. Maxwell, Fowden G. II. Jennings, Peter
Randolph, 1931-
SB750.B73   631.5'3   79-13462
ISBN 0-471-03268-9

Printed in the United States of America

10 9 8 7 6 5 4 3

CONTRIBUTORS

1. E. E. Ortman
   Department of Entomology
   Purdue University,
   West Lafayette, Indiana

   D. C. Peters
   Department of Entomology,
   Oklahoma State University,
   Stillwater, Oklahoma

2. Ernst Horber
   Department of Entomology,
   Kansas State University,
   Manhattan, Kansas

3. Dale M. Norris
   Department of Entomology,
   University of Wisconsin,
   Madison, Wisconsin

   Marcos Kogan
   Department of Entomology,
   Illinois Natural History
      Survey
      and University of Illinois,
   Urbana, Illinois

4. R. L. Gallun
   Science and Education
      Administration,
   U.S. Department of
      Agriculture and Department
      of Entomology,
   Purdue University,
   West Lafayette, Indiana

   G. S. Khush
   The International Rice
      Research Institute,
   Manila, Philippines

5. Ward M. Tingey
   Department of Entomology,
   Cornell University,
   Ithaca, New York

   S. R. Singh
   Grain Legume Improvement
      Program,
   International Institute of
      Tropical Agriculture,
   Ibadan, Nigeria

6. Stanley D. Beck
   Department of Entomology,
   University of Wisconsin,
   Madison, Wisconsin

   L. M. Schoonhoven
   Department of Animal
      Physiology,
   Agricultural University,
   Wageningen, The Netherlands

7. K. Maramorosch
   Waksman Institute of
      Microbiology,
   Rutgers—The State
      University,
   Piscataway, New Jersey

8. Raoul A. Robinson
   Consultant,
   St. Helier,
   Jersey, United Kingdom

9. D. R. MacKenzie
   Department of Plant
      Pathology,
   The Pennsylvania State
      University,
   University Park, Pennsylvania

10. Johnie N. Jenkins
    Boll Weevil Research
       Laboratory
    U.S. Department of
       Agriculture, and Mississippi
       State University,
    Mississippi State, Mississippi

11. Perry L. Adkisson
    Department of Entomology,
    Texas A & M University,
    College Station, Texas

    V. A. Dyck
    The International Rice
       Research Institute,
    Manila, Philippines

12. Jack R. Harlan
    Department of Agronomy,
    University of Illinois,
    Urbana, Illinois

    Kenneth J. Starks
    Department of Entomology,
    Oklahoma State University,
    Stillwater, Oklahoma

13. M. W. Nielson
    Forage Insects Research
       Laboratory,
    Science and Education
       Administration,
    U.S. Department of
       Agriculture,
    Tucson, Arizona

    W. F. Lehman
    Department of Agronomy and
       Range Science,
    University of California,
    Davis, California

14. Anthony Bellotti
    CIAT,
    Cali, Colombia

    Kazuo Kawano
    CIAT,
    Cali, Colombia

15. G. A. Niles
    Department of Soil and Crop
       Sciences,
    Texas A & M University,
    College Station, Texas

16. Alejandro Ortega
    International Maize and
       Wheat Improvement
       Center CIMMYT,
    Mexico D.F., Mexico

    Surinder K. Vasal
    International Maize and
       Wheat Improvement
       Center CIMMYT,
    Mexico D.F., Mexico

John Mihm
International Maize and
  Wheat Improvement
  Center CIMMYT,
Mexico D.F., Mexico

Clair Hershey
International Maize and
  Wheat Improvement
  Center CIMMYT,
Mexico D.F., Mexico

17. M. D. Pathak
The International Rice
  Research Institute,
Manila, Philippines

R. C. Saxena
The International Rice
  Research Institute,
Manila, Philippines

18. G. L. Teetes
Department of Entomology,
Texas A & M University,
College Station, Texas

19. J. W. Hanover
Department of Forestry,
Michigan State University,
East Lansing, Michigan

20. E. H. Everson
Department of Crop and Soil
  Sciences,
Michigan State University,
East Lansing, Michigan

R. L. Gallun
Science Education
  Administration,
U. S. Department of
  Agriculture and Department
  of Entomology,
Purdue University,
West Lafayette, Indiana

21. Fowden G. Maxwell
Department of Entomology,
Texas A & M University,
College Station, Texas

Dr. Reginald H. Painter
1901–1968

To Dr. Reginald H. Painter, an outstanding teacher and pioneer research worker. Many of the contributors to this volume were his students. Through his influence on them and others and in his own efforts in breeding plants resistant to insect pests, Dr. Painter has helped advance the concept of integrated pest management critical to man's ability to cope with insects today.

SERIES PREFACE

Environmental Science and Technology

The Environmental Science and Technology Series of Monographs, Textbooks, and Advances is devoted to the study of the quality of the environment and to the technology of its conservation. Environmental science therefore relates to the chemical, physical, and biological changes in the environment through contamination or modification, to the physical nature and biological behavior of air, water, soil, food, and waste as they are affected by man's agricultural, industrial, and social activities, and to the application of science and technology to the control and improvement of environmental quality.

The deterioration of environmental quality, which began when man first collected into villages and utilized fire, has existed as a serious problem under the ever-increasing impacts of exponentially increasing population and of industrializing society. Environmental contamination of air, water, soil, and food has become a threat to the continued existence of many plant and animal communities of the ecosystem and may ultimately threaten the very survival of the human race.

It seems clear that if we are to preserve for future generations some semblance of the biological order of the world of the past and hope to improve on the deteriorating standards of urban public health, environmental science and technology must quickly come to play a dominant role in designing our social and industrial structure for tomorrow. Scientifically rigorous criteria of environmental quality must be developed. Based in part on these criteria, realistic standards must be established and our technological progress must be tailored to meet them. It is obvious that civilization will continue to require increasing amounts of fuel, transportation, industrial chemicals, fertilizers, pesticides, and countless other products; and that it will continue to produce waste products of all descriptions. What is urgently needed is a total systems approach to modern civilization through which the pooled talents of scientists and engineers, in cooperation with social scientists and the medical profession, can be focused on the development of order and equilibrium in the presently disparate segments of the human environment. Most of the skills and tools that are needed are already in existence. We surely have a right to hope a technology that has created such manifold environmental problems is also capable of solving them. It is our hope that this Series in Environmental Sciences and Technology will not only serve to make this challenge more explicit to the established professionals, but that it also will help to stimulate the student toward the career opportunities in this vital area.

Robert L. Metcalf
Werner Stumm

PREFACE

The natural resistance of plants to *their* insect enemies is a common development, one that takes eons and that usually runs counter to man's current need for food and fiber of quality in abundance. This is so because the buildup of such resistance often leads to the least desirable plant material from point of view of yield and taste. To breed plants resistant to *our* insect enemies is a twentieth century phenonemon that stems from the knowledge of basic genetics and from the methodology of selecting, crossing, and hybridizing plants we have learned since Mendel's time.

No longer can we accept the view that breeding plants for good agronomic characteristics and for high yield automatically takes care of their resistance to insects. A single type of insect, as in the case of the Hessian fly, can limit yield, and we need to know precisely why this is so. Moreover, we can no longer implicitly rely on insecticides for control. Insects build up resistance to insecticides; therefore we must prepare a range of control practices in which resistance of plants to insects plays its role along with other modern methods. We need theoretical knowledge about specific insects and methodology for managing them so that we can use that resistance to our benefit for producing our food crops and fiber plants. Through his teachings and his research, Dr. R. H. Painter pioneered in these subjects; he documented them in his text in 1951. In the quarter of a century since then he, his colleagues, and his students made remarkable advances in the concepts and methods of breeding plants resistant to insects.

This book, an overview of modern concepts and methods, originated from the participation of several leading entomologists in a program, multidisciplinary and multiinstitutional in nature, which the Rockefeller Foundation sponsored on breeding cotton resistant to the bollworm and to other devastating insects of that crop in the United States. Cotton accounts for the heaviest pesticide load of any agricultural crop in the United States. Toward the close of that program, successful in producing cotton varieties that, untreated, could yield under heavy insect attack as well as standard varieties liberally treated with insecticide, the Foundation decided to review the entire topic of breeding plants for insect control. It organized an Editorial Steering Committee whose members were Dr. Fowden Maxwell, Dr. Peter Jennings, Dr. Carl Koehler, and Dr. John McKelvey, Jr.

For authorship of chapters the committee selected plant pathologists, plant breeders, and specialists in related subjects as well as entomologists. Achieving a blend of disciplines enhanced coverage of the subject in its theoretical and practical ramifications. Authors presented drafts of their chapters to their colleagues for review at a working conference at the Rockefeller Foundation Conference and Study Center at Bellagio and, subsequently, to peer reviewers.

University professors and students should find this book useful as a text; researchers beginning to explore this topic may value it as a resource. We have prepared a glossary of terms as they are commonly used, but variation in such usage may occur from chapter to chapter because of the different concepts held by the international leaders who have contributed chapters. In the interest of accuracy we have accepted the data expressed either in the metric system or in the United States equivalents as presented.

We are grateful to the Rockefeller Foundation for the interest they have taken and the support they have given in the production of this book, for use of the Foundation's Conference and Study Center at Bellagio—the venue of crucial planning sessions in organizing and developing the chapters—and for the work of the Steering Committee. We thank the authors of the chapters for their contributions and we appreciate the help of Mr. Romney, head of our Information Department, and the special editorial assistance Dr. Gilbert Tauber gave in the final stages of preparing the text for the press.

FOWDEN G. MAXWELL
PETER R. JENNINGS

College Station, Texas
New York, New York
November 1979

CONTENTS

BREEDING PLANTS RESISTANT TO INSECTS

Part One

PLANT RESISTANCE TO INSECTS
Theory and Concepts

The boll weevil, a pest whose destructiveness has won it a place in American folklore, is a principal target of resistance breeding programs. Photo by Jim Strawser, courtesy of U.S. Department of Agriculture.

1

INTRODUCTION

E. E. Ortman

Department of Entomology, Purdue University, West Lafayette, Indiana

D. C. Peters

Department of Entomology, Oklahoma State University, Stillwater, Oklahoma

1. HISTORICAL OVERVIEW

The study of plant resistance to insects dates back to the earliest days of applied entomology. The early literature contains several significant examples of differences in response of cultivars to insect attack. Havens' (1792)

report that the 'Underhill' variety of wheat was resistant to the Hessian fly is generally considered the earliest documentation of an insect resistant variety. The classic example of insect resistance is that of the American species of grapes, which were highly resistant to the grape phylloxera in comparison with susceptible European species. The grape phylloxera had disastrous effects on the French grape crop and wine industry. In the late nineteenth century the introduction of resistant vines rescued the wine industry and thereby played a major role in restoring the strength of the French Empire at a significant point in its history. Another early report (Lindley, 1831) of insect resistance was the apple variety 'Winter Majetin,' found to be resistant to the woolly apple aphid.

From those early observations, knowledge of plant resistance to insects has grown to become an important dimension of applied entomology. The development of the science of plant resistance to insects can be divided roughly into three phases: the pre-World War II era, the immediate post-World War II era, and the era of environmental awareness of recent years. Prior to World War II, initial observations by scientists had led to cooperative efforts by plant breeders and entomologists to develop improved cultivars to control insect pests. The postwar years showed a significant shift from studies of insect biology and insect–host interactions to the exploitation of the newly developed organic chemical pesticides. New pesticides were identified and synthesized with spectacular results when applied to insect populations. Research and control strategies shifted toward the new chemical-control approach. Research and development in insect biology and insect resistant plants languished during this period. Since the late 1960s, there has been a further shift toward the development of integrated systems of pest control. This change was conditioned by two major factors, the development of insect resistance to insecticides and the concern for environmental pollution stemming from the use of chemical pesticides. The realization that earlier approaches had inherent problems led to a renewed effort to study all means of insect control, including plant resistance.

A historical view of plant resistance to insects must recognize the simultaneous evolution of several sciences, principally those of genetics, plant breeding, and entomology. As insects and plants have coevolved over time, so too have the sciences that deal with their ultimate effect on man's food and fiber. Plant breeding itself can be viewed as man's manipulation of the evolutionary process.

The development of the specialty of plant resistance to insects can also be viewed from the accumulation of knowledge in its published form. The state of knowledge in this field has been surveyed periodically in review papers. The first review by Snelling (1941) listed 567 references. It showed that only 37 papers reporting resistance were published prior to 1920. Publications increased considerably after that. The first book on insect resistance in crop plants was published by Painter in 1951. It is a comprehensive review of the literature up to that time and also discusses the general principles involved in plant resistance to insects. This publication, with more than 1,000 works cited, remains the standard reference for the study of plant resistance to insects. Subsequently, significant review papers have been published by Painter (1958); Beck (1965); the National Academy of Science (1971); Maxwell et al. (1973); and Gallun et al. (1975). These reviews dealt with the development of plant resistance to insects in various crops and on special aspects of resistance, such as the characterization of the nature of resistance. Several volumes have been published on both the theoretical and practical aspects of insect/plant relationships, including works edited by Jermy (1976) and van Emden (1973).

2. ROLE OF HOST-PLANT RESISTANCE IN INSECT MANAGEMENT

The primary objective of programs on insect resistance in crop plants is to develop cultivars that are resistant to an insect pest while maintaining or improving their basic agronomic characteristic. Resistance to insects should be a basic objective of crop improvement programs conducted by plant breeders and geneticists. Similarly for entomologists, the development of crop plants resistant to insect attack should be an integral part of strategies for insect management. The role of plant resistance to insects in a breeding or insect management program varies with each crop and each insect. Its importance in a management strategy depends on the availability and utility of other control measures. Resistance may be simply a contributing feature or at other times the chief means for controlling a pest. The utility of resistant varieties and the millions of dollars saved by growers was documented by Luginbill (1969).

Insect resistance is most likely to be used as an adjunct to other control measures. Resistant cultivars may need less frequent treatment with a pesticide or pathogen, or may require lower rates of application of other control measures. The investigation of plant resistance should be an integral part of any varietal development program where insects may be a significant threat to the successful production of the crop. In the initial phases, a plant resistance program generally focuses on the key pest in a commodity. A very significant contribution would be made if all germplasm in a breeding program were evaluated for susceptibility to the key pest(s) and compared with cultivars currently grown. Though it may be quite difficult to develop resistance to several pests concurrently, it is important that studies be conducted with more than a single pest to explore the potential for multiple-pest resistance and to avoid the possibility of increased susceptibility to other pests.

In breeding and agronomic development programs, care must be taken that major crops or important production areas are not overly committed to cultivars with a common genetic base, even if that base may be the foundation for resistance to insects. Disastrous results have been observed, as with southern corn leaf blight in the United States, when genetic or cytoplasmic homogeneity was achieved in a major crop over a wide geographic area.

Insect resistance has been the principal pest control method in a number of crops. For example, resistant grapevines are still the principle means of controlling phylloxera. Furthermore, resistance to insects can serve as a means of control in unique niches where other controls are not feasible or are difficult to use. Plant resistance may afford significant advantages in situations where (1) there is a critical timing regime in which an insect is exposed for only a brief period of its life cycle; (2) the crop is of low economic value; (3) the pest is continuously present and is the single most limiting factor in successful cultivation of a crop in a wide area; or (4) other controls are not available. The Hessian fly and wheat stem sawfly are prime examples of the use of plant resistance as a primary control method. Timing is a critical factor for applying other control measures, such as pesticides, to these pests because the period of exposure is limited. Also, Hessian fly was continuously present over a wide geographic area. Infestation levels of Hessian fly in the midwestern United States, where the majority of wheat is planted to resistant varieties, have dropped from more than 90 percent to less than 10 percent as the planting of resistant cultivars has increased. Planting of highly resistant cultivars on an extensive acreage may result in the cumulative reduction of a target pest population. Holmes and Peterson (1957) demonstrated the depressive effect of continuous rearing of the wheat stem sawfly on resistant wheat. This was followed by a projection by Luginbill and Knipling (1968) which showed the suppression of the wheat stem sawfly when resistant wheats were planted. Dahms (1969) used aphid fecundity data to elucidate the theoretical effects of antibiosis on insect population dynamics.

3.   COMPONENTS OF A PROGRAM FOR PLANT RESISTANCE TO INSECTS

3.1.   Personnel

Progress in the development of plants resistant to insects, as a vital entity in pest management programs, hinges on the concept of a multidisciplinary team. It calls for a cooperative and interactive relationship between the entomologist and the plant breeders, who generally form the initial team in a plant resistance program. As the program proceeds, complex problems are likely to arise which the initial team of scientists will need help in solving. Other scientists should be added to the group as the need for their contributions is recognized. History shows that major progress in the development of resistant varieties has been made when the program of resistance has been the primary responsibility of both the entomologist and the plant breeder, not a secondary activity. For many years graduate programs in plant science have had considerable interchange between plant pathology and breeding, but much less with entomology (Russell, 1975). Positive efforts are needed to rectify this situation.

Although entomologists must lead the effort to identify sources of resistance, the plant breeder usually provides the entomologist's seed source. Once the source of resistance has been identified, the involvement of the plant breeder becomes more critical. It is important that a priority be established so that both the entomologist and breeder can work toward advancing the identified source through the many steps necessary to achieve status as a cultivar.

3.2. Insect Biology

Prior to embarking on a plant resistance program, there must be a significant pool of information on the influence of biotic and abiotic factors on the biology of the pest. This should include information on behavior, especially in relation to food habits, oviposition, and movement; definition of the parameters of growth and fecundity; and effect of the environment on pest populations. These types of information must be available to design experiments within the range of behavior and activities of the pest. It is critical to design tests that do not preclude the biological expression of important traits or characteristics of the pest or host. Generally pest and/or host responses are categorized as to their departure from the mean. The importance of detailed pest biology as it is synchronized with the development of the host plant is demonstrated by the resistance to the first and second brood European corn borer. With this pest, feeding location and feeding behavior depend on the development stage of the pest and host. Initially, in corn in the 5 to 10 leaf stage, the larva feeds on leaves in the whorl; as the corn and the insect mature, the insect becomes a stalk borer.

3.3. Pest Population

The availability of a constant and uniform insect population is essential to progress. Attention must be given to identifying the optimum pest population that will permit differentiation among genotypes. An optimum population is not necessarily a maximum population. An insect population may be obtained by (1) intensively managing existing field populations; (2) rearing populations on a natural host in an insectary, greenhouse, or growth chamber; or (3) artificially rearing, for example, many of the lepidopterous pests. Many factors dictate the method used to develop and maintain insect populations. A primary function of the entomologist on the resistance team is to understand the biology of the target species and to manipulate the population so that the infestation level will produce optimum differences among genotypes. Significant advances have been made in insect nutrition (Rodriguez, 1972; Dadd, 1973), artificial diets (Vanderzant, 1974), and mass culture (Chambers, 1977), which have led to the mass production of insects for research programs. Information on artificial diets has been compiled in several references including those by Singh (1972) and House et al. (1971). The value of culturing insect populations on a diet are given by Guthrie et al. (1965) in the case of the European corn borer. Huettel (1976) discusses criteria for maintaining a quality laboratory insect colony. Attention must be given to the biological and behavioral characteristics of the insect population, which must be comparable to those found in natural field populations. There is frequently a trade-off between greenhouse or laboratory screening with artifically reared insects and field screening with natural or augmented insect infestations. Developments in related areas such as insect dietetics, nutrition, and mass culturing have had a significant influence on progress made in breeding for plant resistance.

3.4. Genetic Sources

Success in identifying sources of resistance is directly related to the diversity of germplasm available and the probability of resistance occurring in the host populations. The search for sources of resistance is carried out in a logical sequence: first in adapted cultivars, then in plant introductions and exotic germplasm, and finally in near relatives of the cultivar. The identification of the source of resistance is followed by hybridization, selection in segregating generations, and progeny testing. Special nursery and plant propagation facilities are important for rapid advancement. Tropical nurseries are useful for seed increase and crossing during off-season. For example, an added generation of corn grown in a southern location during the winter hastens the program. Resistance is frequently found in primitive cultivars or related species. The transfer of resistance from these exotic sources may require the use of special genetic manipulations such as cell culture. An excellent example of the transfer of resistance from one species to another is the incorporation of greenbug resistance in wheat from rye. The advances made in basic science, through developments such as cell and embryo culture, have a marked impact on progress in the applied science of breeding resistance to insects.

There is now renewed interest in plant collection, and the volume published by Leon (1974) for tropical crops is a useful introduction to the general methods. The Information Services/Genetic Resources Program at the University of Colorado has developed a data bank on the characteristics of the germplasm stored in banks around the world. The International Board for Plant Genetic Resources (1976) published a priority list for crops and regions. Their criteria for priority areas were as follows: (1) the risk that genetically diverse materials will be lost owing to changes in land use; (2) the economic and social importance of the materials to be collected; (3) the recognized requirements of plant breeders for geneticially diverse materials; and (4) the size, scope, and quality of existing collections.

Plant exploration and collection is a critical activity as scientists continue to seek and utilize naturally occurring sources of resistance. Allard (1970) observed that each species contains millions or even hundreds of millions of variants, so sampling is a challenge. Unfortunately, Harlan (1972) was generally correct when he observed, "In no collection is there an adequate sampling of the spontaneous races that are most likely sources of disease and insect resistance." One frequent shortcoming of plant exploration activities is the lack of an entomologist as a member of the team.

Thus the potential for success is a function of the variation in both insect and host, coupled with the frequency of occurrence of the plant variants in the population and subsequent identification and utilization of the variants.

3.5.   Identifying Sources and Assay Techniques

The design of an assay or research method should make it possible to quantify the variations in host plants. In addition to measuring differences in host reaction, it is also important to estimate the source of variance and the heritability of the trait or traits identified. Plant resistance to insects can be described in terms of either the insect or the reaction of the plant, or as either an effect or a result. Thus plant resistance to insects is studied in two dimensions, one being variations in the host, and the other being variations in the pest population. The design of the study and the criteria and measurements used in either situation vary with the insect or plant phenomena to be measured.

In initial studies it is most important to examine a large quantity of diverse material. In these studies it may be essential only to distinguish broad differences in effect on host or pest. Rating schemes with various levels of sophistication or discriminatory power are used. A useful tool in a rating scheme is a set of pictorial standards. Later evaluation studies should permit more precise definition of the level and expression of resistance. It is important that the assay technique represent the insect–host relationship as it occurs in the field. Attention must be given to comparing plant material of the same growth stage or maturity, and to conducting the study at the growth stage when the insect generally attacks the host. Unequal seedling emergence is a major problem if plants are to be screened in the seedling stage. If vegetative plants are compared with those setting seed, the chances for regrowth and compensation are usually reduced in the latter. Strong emphasis on standardization of testing procedures must not be allowed to override the observation by a scientist of a unique event. Consideration must be given to the correlation of resistance in the seedling stage with that in more mature plants. Seedling techniques have proved very useful for mass selection for resistance to the spotted alfalfa aphid which attacks plants during all stages of development. However, a seedling screening of corn for resistance to corn leaf aphid would not be equally useful. Resistance is expressed in all growth stages in the case of alfalfa, but not in the example of the corn leaf aphid on corn.

Dahms (1972) identified 16 possible criteria used to evaluate insect resistance in plants. The slightly abridged list follows:

1. Visual evaluation of infested cultivars by observing, for example, retarded growth, lodging, cutting, and discoloration.
2. Determination of the number of surviving plants at various intervals following infestation.
3. Determination of the difference in yield between infested and noninfested plots.
4. Determination of the number of insect adults or larvae attracted to a cultivar when given a free choice.
5. Observation of the comparative effects of forced insect feeding (confinement) on plants or cultivars by measuring length of insect life cycle, mortality, reproductive rates, or molting, for example.
6. Weight of insects after definite feeding period on different cultivars.
7. Determination of the number of eggs laid.
8. Determination of the number of surviving insects and progeny produced.
9. Measurement of the amount of food insects consume.
10. Measurement of the amount of food utilized by the insect.
11. Simulation of insect damage and observation of recovery.
12. Indirect method of evaluation such as measuring root damage by amount of force required to pull a plant out of the ground.
13. Use of plant leaves or flowers in olfactometers to determine attractance.
14. Correlation of chemical factors in plants with insect response.
15. Growth and reproduction potential of insects fed various plant diets containing different plant cultivars.
16. Correlation of morphological factors with injury.

The first four are the most useful in screening a large number of entries. A relative rating scale is usually used in the initial screening process rather than counts of insects. The essential needs are to identify rapidly material worth advancing and to differentiate intermediates and susceptibles. Traditional rating scales were from 0 to 3 or 0 to 5, with the high number indicating susceptibility. However, since it is often necessary to use statistical evaluations, zeros should be avoided. Plant introductions and segregating populations must be evaluated in a manner that will identify plant-to-plant variation when it occurs; single ratings for a plot may be misleading. In studies on inheritance of resistance characteristics, it becomes more critical to quantify gradients or levels of resistance. Chesnokov (1962) published a book on methodology for studying insect resistance.

3.6. Levels

With the identification of resistance the question is generally posed, "How small an increment is usable in the development of host-plant resistance?" Statistical techniques can aid in the decision process. For several decades biologists looked on the statistical level $P = 0.05$ as the criterion for decision making. Badahur and Robbins (1950) developed mathematical logic for accepting the greater of any two means regardless of the increment between them. Nevertheless, the critical question of how small an increment of resistance is useful must be answered on an individual basis with primary consideration given to (1) the effort necessary to bring the source to commercial cultivation and (2) interaction within a total pest management program. The challenge is to provide a predictable yield over extended periods of time, consistent with total production efficiency.

3.7. Priorities, Cross Resistance

A pragmatic approach must be taken in setting priorities among pests to be considered on a specific crop. Pests may be divided into several categories. A *key pest* is one that regularly limits crop productivity. An *occasional pest* occurs at infrequent intervals but causes severe damage when present. An *incidental pest* is one constantly present but infrequently damaging. A *potential pest* is one that might occur with a change in crop and cultural practices. In developing resistance to one pest it is important to evaluate the breeding lines for resistance to other occasional, incidental, and potential pests to guard against the development of susceptibility to another insect. The frego bract condition in cotton is an excellent example of a characteristic that conditioned resistance to a primary pest but contributed to greater susceptibility to another insect. As we genetically engineer specific plant characteristics for expected performance we should routinely test lines for susceptibility to other stresses that might create an equal or worse condition.

Insect resistance is almost always species specific. However, multiple resistance can be incorporated in a single species, as has been done in alfalfa for the spotted alfalfa aphid and the pea aphid. Occasionally, a broad resistance effect has been identified, as with leaf feeding in soybeans where multiple resistance was observed to Mexican bean beetle, looper, and clover worm. A resistance factor identified in alfalfa, where the secretory glands of a trichome provide the resistance to alfalfa weevil (Shade et al., 1975), was later found to condition resistance to the pea aphid and potato leafhopper as well.

3.8. Biotypes

Breeding for resistance is a powerful tool for crop protection. It has also demonstrated the striking ability of pests to adapt to resistance in plant hosts. Culturing genetically uniform crops over a large geographical area is unnatural and has a tremendous potential to cause large-scale destruction by an adapting pest. Genetic manipulation can cause major genetic shifts over a relatively short time compared with evolutionary scale shifts, which are generally minor and are distributed over a longer period.

Biotypes were noted by Painter in the 1930s when he observed that Hessian fly from western Kansas was not damaging to certain soft wheats, whereas Hessian fly from eastern Kansas fluorished on these sources. This initial observation of the biotype potential in Hessian fly has culminated in the development of 16 recognizable biotypes based on the plant–insect interaction. Hatchett and Gallun (1970) extended the gene-for-gene concept in an insect–host system (Hessian fly–wheat) from that initially identified by Flor (1955, 1956) working with flax rust. By 1966, six examples had been noted in which insect biotypes were able to feed on formerly resistant plants, thus demonstrating the genetic versatility and plasticity of insects in relation to their host.

3.9. Causal Factors

Major contributions have been made to our understanding of insect resistance through exploration of the basic factors that condition resistance. However, biochemical information has not been useful to date in assays to identify or evaluate resistant genotypes. Plant morphological characters, such as trichomes, have been used successfully as visual surrogates for resistance factors in selection programs. Great care must be exercised when a characteristic such as pubescence is substituted for direct insect assay as the selection criterion. Presence or absence of pubescence alone does not indicate resistance. The length, diameter, rigidity, and density, of the hair must be taken into account. Also there may be resistance factors not associated with pubescence. One of the hazards in not using a direct insect assay is that different types of useful resistance will be ignored.

The identification of plants having an effect on insect behavior, feeding, and reproduction led to a series of fundamental investigations on the potential chemical basis of resistance. Painter (1951) appropriately emphasized that finding the source of resistance is the first step. Biochemical and genetic studies can only follow a successful discovery of a source of resistance. Resistance in a given plant species, as expressed in the field, is a complex phenomenon and there is generally no single chemical to condition it. Significant advances have been made in defining the chemical basis of resistance for corn borer on corn and boll weevil on cotton. Advances in defining the chemical basis of resistance are based on adequate assay techniques and development of sensitive instrumentation and analytical technology. Nomenclature has evolved, including attractants, arrestants, deterrents, stimulants, and excitants to describe chemicals that elicit particular behavioral responses. There has also been significant development of theories on plant–insect chemical ecology. Thus what in many instances was initiated as a practical search for an insect control has evolved into fundamental study of the interaction of two dynamic organisms and of the theory of host selection. However, knowledge of the chemistry of resistance and of mechanisms of host selection has not kept pace with progress in breeding for plant resistance and in the use of resistance as an insect management tool. From the initial identification of resistance as a means of combating insect problems, another whole dimension of theory and biological interrelationships has evolved. Plant resistance to insects has ignited and fed fundamental studies in entomology and plant science.

4. THE NEED

As the Committee on Genetic Vulnerability of Major Crops (1972) concluded, consumer's likes and dislikes regarding taste and appearance-influence marketability. Processors and distributors demand storability and transportability; farmers want high yields and ease of mechanical harvesting; seed companies wish to produce seed in an efficient and profitable manner. The advantages of a pest resistant variety must be compatible with these demands. Nevertheless, basic challenge in insect resistance remains that of identifying usable resistance sources and deploying them broadly in crop production.

Ears of maize are selected for testing in an international breeding program.
Photo courtesy of CIMMYT.

2

TYPES AND CLASSIFICATION OF RESISTANCE

Ernst Horber

Department of Entomology, Kansas State University, Manhattan, Kansas

Observations of insect–plant interactions reveal a wide range of plant suitabilities as hosts to insects. Variability in plants in the nature and the intensity of interaction is also reflected in the categories and definitions of resistance described in this chapter. They describe the exceptional abilities of certain plants to avoid, repel, retard, restrict, or localize insect infestation and damage, or to tolerate it by fast regrowth and recovery from injury.

Classifications of resistance phenomena may express the relative success or failure of an insect species to survive, develop, and reproduce on a plant species; or the classifications may describe the relative damage to the host plants in qualitative or quantitative terms. Snelling (1941) included in plant resistance those characteristics that enable a plant to avoid, tolerate, or recover from attacks of insects under conditions that would more severely

---

Contribution No. 1202-B, Department of Entomology, Kansas Agricultural Experiment Station, Manhattan, Kansas.

injure other plants of the same species. Painter (1951) used a more comprehensive definition than Snelling's, describing a plant's resistance as the relative amount of its heritable qualities that influence the ultimate degree of damage done by the insect. In practical agriculture, resistance represents the ability of a certain variety to produce a larger crop of good quality than would other varieties under the same insect population. Beck's definition (1965) restricts plant resistance to the collective heritable characteristics by which a plant species, race, clone, or individual may reduce the probability that an insect species, race, biotype, or individual successfully uses the plant as a host. Beck's definition narrows the spectrum of insect–plant interactions to the successful use by the insect of a plant as host, but it excludes the plant's ability to recover or repair losses after injury occurs.

1. CATEGORIES EXPRESSING VARIOUS INTENSITIES OF RESISTANCE

Interactions between insects and plants span a wide range of intensities. In terms of the insect, the interaction varies from plants being completely adequate to completely inadequate hosts. Conversely, in terms of the plant species or cultivar, the fewer insect species associated with it, and/or the lower their abundance and the less effect they exert on a plant, the more resistant the plant appears.

Resistance usually is measured by using susceptible cultivars of the same plant species as controls. Only immunity, representing complete inadequacy for insects, is an absolute term, but it is rarely encountered in plants within a host species. The terms *host plant* and *immune* exclude each other. Plants of a nonhost species would not ordinarily be classified for resistance and therefore would be considered immune. A host plant can be more or less resistant but not immune. An immune plant is a nonhost. Any degree of host reaction less than immunity is resistance; more than immune is impossible. It must be remembered, therefore, that the term immunity does not permit qualifying adjectives such as *comparatively, more, most, rather, somewhat,* or *very.* Painter (1951) used the following scale to classify degrees of decreasing resistance:

Immunity.   An immune cultivar is one that a specific insect will never consume or injure under any known condition. Thus defined, there are few, if any, cultivars immune to the attack of specific insects known to attack cultivars of the same plant species.

High resistance is demonstrated by a cultivar that has qualities that result in small damage by a specific insect under a given set of conditions.

Low resistance indicates qualities that cause a cultivar to show less damage or infestation by an insect than the average for the crop considered.

Susceptibility. A susceptible cultivar shows average or more than average damage by an insect.

High susceptibility. A cultivar shows high susceptibility when much more than average damage is caused by a specific insect.

The terms indicate the classes used by most workers in insect resistance as it is observed in the field, without analysis of the mechanisms involved. Intermediate resistance is sometimes spoken of as *moderate resistance*, which may result from one of at least three situations. A cultivar denoted as moderately resistant may consist of phenotypically similar plants, some of which have high and others low resistance because of differences in physiological characteristics. In contrast, a moderately resistant cultivar may be made up of plants derived from a single clone, which is heterozygous for incompletely dominant genes that confer high resistance when homozygous. Moderately resistant plants also may be homozygous for genes which, under given environmental conditions, produce plants that are moderately injured or infested.

2. FUNCTIONAL RESISTANCE CATEGORIES*

In describing insect–plant interactions, one often overlooks the influence of the environment. It may favor the plant or the insect unequally and unpredictably, or may alleviate or aggravate damage, and therefore affect the expression of resistance.

Certain phenomena related to resistance, but not necessarily based on heritable traits, were defined and classified by Painter (1951) as follows (italics mine):

The term *pseudoresistance* may be applied to apparent resistance which results from transitory characters in potentially susceptible host plants. Cultivars or crops showing pseudoresistance are important in economic entomology but should be distinguished from cultivars that show resistance throughout a wider range of environments. Three types may be distinguished:

(1) *Host Evasion.* Under some circumstances a host may pass through the most susceptible stage quickly or when insect numbers are reduced. Some cultivars evade insect injury by maturing early. Early maturity has been used to good advantage in

---

* Some of the definitions used in this section may vary from those in other chapters. This results from different meanings of terms between the disciplines of plant pathology and entomology. For comparison, see the definitions in Chapter 4.

economic entomology. Planting an early maturing cultivar late or other special experiments will indicate whether true resistance is present.

(2) *Induced Resistance.* This term may be used for the temporarily increased resistance resulting from some condition of plant or environment, such as a change in the amount of water or soil fertility. Such induced resistance may be of great value, especially in horticultural crops, but should not be confused with inherent differences in resistance between cultivars or individual plants.

(3) *Escape.* Escape refers to the lack of infestation of, or injury to, the host plant because of such transitory circumstances as incomplete infestation. Thus, finding an uninfested plant in a susceptible population does not necessarily mean that it is resistant. Even under very heavy infestations susceptible plants will occasionally escape so only studies of their progenies will establish their true relationship.

Attempts to classify types of resistance according to their causes may be limited by unsatisfactory analytical techniques or equipment, or by inadequate research experience. More often they are restricted by the lack of financial support for basic analytical studies. Given such limitations, priority continues to be assigned to breeding for resistance and to insect control. Lower priority is usually given to obtaining complete knowledge of the causes of resistance. Initial or superficial studies often indicated simplistic explanations that had to be abandoned when more thorough investigations revealed more complex relationships between the insect and the plant.

An empirical approach was proposed by Painter (1936, 1941, 1951). It proved a workable compromise between mere categorization of phenomena and the basic study of causative factors or processes. Painter proposed *mechanisms of resistance* which were grouped into three main categories:

1. *Nonpreference* is the insects' response to plants that lack the characteristics to serve as hosts, resulting from negative reactions or total avoidance during search for food, oviposition sites, or shelter. Nonpreference by insects is often projected as a property of the plant, which is not congruous with the process to be described. For this reason, Kogan and Ortman (1978) proposed to substitute antixenosis for the term nonpreference. It is a parallel term to "antibiosis" and conveys the idea that the plant is avoided as a "bad host."
2. *Antibiosis* includes all adverse effects exerted by the plant on the insect's biology, for example, survival, development, and reproduction.
3. *Tolerance* includes all plant responses resulting in the ability to withstand infestation and to support insect populations that would severely damage susceptible plants.

Such resistance categories are arbitrary and vaguely delineated. Not all resistance phenomena can unequivocally be assigned to one of the three categories. Nonpreference may be mistaken for antibiosis and vice versa, such as when early insect instars do not accept a plant as a host.

Tolerance, unfortunately, is often confused with low resistance. The plant reactions to insect attack collectively termed tolerance are heritable traits of great biological significance and practical value, deserving to be treated as one of the three mechanisms of resistance along with, but distinct from, antibiosis and nonpreference. These classical categories of resistance do not exclude each other but may interact and complement each other in the sense of intensifying resistance expressions, as when a nonpreferred host also exerts antibiotic effects. Similarly they may compensate each other in the sense that a moderately or highly tolerant host may not need to exhibit properties of the nonpreference or antibiosis type to any great extent to be protected from insect injury. On the other hand, a highly nonpreferred or highly antibiotic cultivar may not need much tolerance to a particular pest. Though antibiosis and nonpreference exert selection pressure on pest populations, tolerance does not.

Resistance terms more commonly found in the phytopathological than in the entomological literature include the following:

1. *Vertical* or *specific* resistance, expressed against only some biotypes of a pest species (Van der Plank, 1968).
2. *Horizontal* or *general* resistance, expressed equally against all biotypes of a pest species (Van der Plank, 1968).
3. *Hypersensitive resistance*, an intense, rapid response characterized by premature death (necrosis) of the infested tissue together with inactivation and localization of the attacking agent (Muller, 1959).

Adult plant resistance, sometimes referred to as age resistance or mature plant resistance, is manifested mainly in maturing plants and is less apparent in the seedling stage or with older plants less preferred by the insect and more difficult to damage or to kill. Adult plant resistance may involve horizontal resistance but not all horizontal resistance is concerned with the adult plant. Adult plant resistance can be demonstrated by different planting dates, as Painter (1951) showed in sorghum resistant to the chinch bug, *Blissus leucopterus leucopterus. Juvenile resistance,* often referred to as *seedling resistance,* is apparent in the seedling stage. This phenomenon is often used to identify and distinguish vertical resistance, which is most easily detected in juveniles, from horizontal resistance, which is generally most apparent in adult plants. Maize plants in the whorl stage are more resistant than in later stages to the European corn borer, *Ostrinia nubilalis.* Klun and Robinson (1969) related this to the higher DIMBOA (2,4-dihydroxy-7-methoxy-1,4-(2H)-benzoxazin-3-one) content of maize in the whorl stage. Reduced concentrations of DIMBOA in susceptible inbred lines were correlated with decreased initial resistance.

Field resistance observed under field conditions is distinct from resistance observed in the laboratory or greenhouse. It may involve seedling resistance as well as mature plant resistance and often involves resistance against all locally occurring insect biotypes. It may include all the functional categories of resistance, for example, antibiosis, nonpreference, and tolerance.

Multiple resistance protects a cultivar from different environmental hazards, for example, insects, nematodes, diseases, and injuries caused by heat, hail, drought, cold, or pollution. Genes conditioning resistance to the respective stresses may be incorporated into the same cultivar, usually in the course of several breeding cycles, by backcrossing a resistant parent with an adapted cultivar as a recurrent parent. 'Kanza' alfalfa is a synthetic population resistant to bacterial wilt, *Corynebacterium insidiosum,* the spotted alfalfa aphid, *Therioaphis maculata,* and the pea aphid, *Acyrthosiphon pisum* (Harris) (Sorensen et al. 1969).

3.   GENETIC RESISTANCE CATEGORIES

Based on the mode of inheritance, resistance phenomena may be divided into mono-, oligo-, and polygenic resistance (Van der Plank, 1968):

Monogenic resistance is governed by single genes.

Oligogenic resistance is governed by a few genes.

Polygenic resistance is governed by many genes.

One must keep in mind that resistance is the combined effect of all the genes of an individual; genes concerned primarily with resistance may also express themselves in diverse ways. Most of the resistance cases investigated would fall into the categories of oligo- and polygenic resistance. The division of resistance into these groups is popular among plant pathologists. Oligo- and polygenic resistance appear to be preferable to monogenic resistance as a strategy to safeguard against genetic vulnerability resulting from a breakdown of resistance caused by the selection of new aggressive biotypes.

The term *major gene resistance* may be applied to monogenic or oligogenic resistance, and *minor gene resistance* is used synonymously with polygenic resistance. Since all oligogenes are not necessarily major genes in the sense of being important and all polygenes are not minor genes in the sense of being unimportant, both terms must be properly defined to avoid misleading connotations. Plants may vary continuously in resistance without necessarily falling into clearly defined groups. Single-gene effects are not invariably evident enough for the gene to be identified and located. The effects of genes are usually studied by measuring the damage to segregating plant populations challenged by known insect biotypes, or by evaluating the effect of the plant on the survival, growth, and reproduction of the insect.

Multiline resistance is the resistance conveyed by mixing phenotypically similar but genotypically dissimilar pure lines. The genotypic differences between component lines usually involve vertical resistance. A multiline is grown by mixing seed of several resistant lines, which differ only in the resistance genes they carry. From the agronomic point of view, a field planted to a multiline need not appear different from a field planted to a genetically uniform cultivar, but to an insect population a multiline is a composite of different host genotypes. Resistance genes can be introduced into a multiline by adding component lines, derived from backcrossing resistant parents to an adapted standard cultivar as the recurrent parent.

Prominent pigment glands (upper photo) in the corolla of the flower on right indicate high levels of bud gossypol. The narrow, twisted bracts (lower photo) of frego genotypes of the left plant provide useful resistance to the boll weevil. Photos courtesy of G. A. Niles.

3

BIOCHEMICAL AND MORPHOLOGICAL BASES OF RESISTANCE

Dale M. Norris

Department of Entomology, University of Wisconsin, Madison, Wisconsin

Marcos Kogan

Department of Entomology, Illinois Natural History Survey and University of Illinois, Urbana, Illinois

Every green plant is inherently resistant to some herbivores. Resistance, in its broadest sense, ranges from the temporal escape mechanisms that result from phenological asynchronies to the biosynthesis of lethal complex organic molecules. Between is a vast array of phytochemical and morphological characteristics that more or less disrupt the behavior or metabolic processes involved in herbivore utilization of a plant as a host.

From the standpoint of practical use of plant resistance to insects, we are interested in characteristics that render a cultivar unsuitable, or less suitable, to an insect otherwise well adapted to feeding or ovipositing on "nonresistant" cultivars of the same plant species.

For purely didactic reasons, plant defenses with known or potential value in breeding plants resistant to insects are considered under the major groupings (1) biochemical and (2) morphological bases.

1.   BIOCHEMICAL BASES

The evolution of successful plant life by necessity carried with it a chemistry of defense. The early history of botany is largely a history of man's adaptations of phytochemical defenses for uses in human medicine. Thus we have known of chemical defenses in plants against predators and parasites since antiquity, but our use of such knowledge to breed plants more resistant to insects is still in its scientific infancy. Part of this delay in developing practical, chemically based plant resistance to insects may have resulted from inadequate early attention to the importance of understanding underlying chemical bases. The early emphases were on whether resistance was inherited, genetically manageable, and reasonably stable in practical uses.

In the last 15 years, knowledge of the chemistry of plants has mush-roomed. Several key mechanisms of chemically based resistance to insects were elucidated by the middle 1960s and provided stimuli for further research. The aglycone 2,4-dihydroxy-7-methoxy-2H-1,4-benzoxazin-3-one (DIMBOA) in *Zea mays* was shown to be a major repellent and feeding inhibitor to first-instar larvae of the European corn borer *Ostrinia nubilalis* (Klun et al., 1967). The dimeric sesquiterpene gossypol proved to be a deter-rent to certain pests of cotton, for example, *Epicauta* spp. (Maxwell et al., 1965). The aglycone 5-hydroxy-1,4-naphthoquinone (juglone), in hickory trees (*Carya* spp.) was shown to be repellent to the elm bark beetle, *Scolytus multistriatus*, but not to the closely related hickory bark beetle, *Scolytus quadrispinosus* (Gilbert et al., 1967). Deciphering the chemical bases for plant–herbivore interactions is today an important front in scien-tific inquiry.

1.1. A Submolecular Code for Chemical Ecology

A molecular code controlling the inherited nature of living systems (e.g., plants), as revealed by Watson and Crick, has given us new potentialities for using genetic techniques to "construct" plants possessing enhanced chemical defenses. Interspecific hybridization of nucleic acids is a method of this type currently being developed. However, biochemists have at least one major question to answer concerning energy exchange at the submolecular level before entomologists and plant breeders can advance rapidly with the genetic engineering of plants possessing enhanced and reasonably stable chemical defenses. The question seems to be, "Is there a submolecular code of chemical ecology?" Or, we may ask, "Is there a func-tional ordering within what currently is a gigantic, ever-increasing heap of molecular structures shown to elicit interesting responses in plants and ani-mals?" Considering the plethora of molecules implicated in chemical ecology, the comforting and probably logical answer to the above question is yes. The hypothesis that there is a submolecular code of chemical ecology is currently being tested in several research laboratories. Norris and associates have evolved a working hypothesis pertaining to some electrochemical mechanisms by which organisms communicate with one another and with their inanimate environment. They hypothesize that each species of molecule in chemical ecology readily changes its electronic excitation within a characteristic electrochemical range. By passing from a ground state to an excited state, or back to ground state, one or more electrons in a messenger take energy from, or give it to, some comparably reversible peripheral receptor macromolecules in one or more organisms (e.g., insect). The neurophysiological effect of the messenger, and possible resultant behavioral change in the organism, depend on whether there is a net receipt or donation of energy by the organism. A net receipt or donation above a threshold level each would be kairomonic to some species and allomonic to others.

Experimental findings support the hypothesis, and unifying functional electrochemical principles of chemical ecology may yet emerge. The difference between a feeding excitant and an inhibitor for *Scolytus multistriatus* was reduced, at the submolecular level, to whether a chemical messenger donated or accepted an electron (Norris, 1970). A unifying code of chemical ecology thus seemingly lies in the submolecular tendencies of chemical messengers to exchange energy with living systems.

The pioneering plant breeding efforts to make cotton more chemically resistant to the boll weevil further support the hypothesis. Gossypol, which is produced by gland cells in the leaves of cotton, is an attractant to the boll weevil (Maxwell et al., 1965). Plant breeders developed glandless cotton plants which had reduced attractiveness to the boll weevil, but which proved significantly more susceptible to the major pests *Heliothis* spp. and were damaged by blister beetles, *Epicauta* spp., that would not feed on the glandular varieties. Such experimental findings, resulting from an extensive research program, suggest that a chemical messenger is, at least, a "double agent." It attracts some species and repels others.

What does this possible principle of chemical ecology mean in the pure biological sense? Seemingly it presents a submolecular electrochemical mechanism for better understanding distributions and population densities of species among, and within, ecological habitats. A species is likely to occur where kairomones for it are dominant, and be scarce where allomones for it are abundant. Because each chemical messenger is an allomone to some species and a kairomone to others, populations of several species overlap in an orderly fashion in and among ecological habitats. If a messenger chemical is not allomonic to all insects, not even to all insect species frequenting a species or cultivar then each breeding program is likely to create plants with increased susceptibility to certain insect species and decreased susceptibility to a target pest. The history of breeding cotton resistant to the boll weevil illustrates this important point. With our limited understanding of chemical ecology, and specifically the submolecular chemical bases of plant defenses, we should not expect enduring success from a breeding effort.

1.2 Some Considerations in Breeding Programs

At this point in the infancy of breeding plants with known chemical bases of resistance, proven ingredients of such programs are few. However, at least we can stress certain apparently important considerations.

Polygenic Inheritance

Robinson and MacKenzie each emphasize elsewhere in this volume the greater merits of polygenic as compared to monogenic resistance. Chemically speaking, polygenic control should make a greater range of both qualitative and quantitative molecular defenses available through plant breeding. A polygenic-derived ability to increase incrementally the quantity of defense chemicals in cultivars may prolong the usefulness of such chemical resistance as pest insects adapt to existing chemical levels. Likewise, the "fine tuning" of multichemical resistance, which polygenic inheritance should bring, may effectively counter pest adaptations. However, the metabolic price that a plant can pay for defense without becoming noncommercial for man sets limits on such genetic engineering.

Resistance Based on Several Sites of Action

Chemical defense based largely on a limited number of closely related molecules (e.g., DIMBOA and related minor components in *Zea mays*) may appear to be quite vulnerable to the adaptive plasticity of insects, but it currently seems to be one basis of evolved chemical defense. In case such resistance proves to be commonplace, we should examine a possible basis for its perpetuation.

On first inspection, resistance based on a few similar molecules might be considered vulnerable because of possible monogenic inheritance. However, a few similar defense molecules in a plant may have a wide range of detrimental effects on the insect of concern. These chemicals may directly affect insect longevity, sensory physiology, endocrinology, and metabolism, and indirectly affect the insect through its symbiotes. Such multiple effects challenge the insect's genome at many divergent loci. The evolution of the insect species to overcome the multiple ill effects of this encountered plant chemistry may be energetically expensive, and likely to occur relatively slowly. Practical field levels of such plant resistance might not change detectably over decades. For a recent discussion of the multiple allomonic effects that single phytochemicals can have on a species of insect consult Beck and Reese (1976).

Unique Chemistry of Plant Tissues Attacked by Insects

In considering chemical resistance, it is not enough to grind up the entire plant and look for allomones against insect pests. The qualitative and quantitative differences in allomones in various parts of a plant may vary as significantly as among plant species. A study of the chemistry of defense should compare the compounds and their quantities present in tissues normally attacked versus those not attacked. Rees (1969) showed that the concentration of a secondary chemical (i.e., hypericin, hexahydroxy-dimethylnaphthodianthrone) varied as much as tenfold (i.e., 100 versus 1000 $\mu g/g$) between parts of *Hypericum hirsutum*. He did not find evidence of significant change in the concentration of hypericin during the growing season. However, seasonal changes in the concentration of 13-keto-8(14)-podocarpen-18-oic acid in needles of *Pinus banksiana* determine which foliage is consumed by the larvae of two species of *Neodiprion* sawflies (Ikeda et al., 1977). Juvenile needles contain deterrent amounts of this chemical, but older needles do not. Thus the chemistry of specific plant parts should be investigated on a seasonal basis. Other environmental changes (e.g., drought) are likely to alter the ratio between proliferation and differentiation of cells, and this also may bring marked differences in the defense chemistry of plant parts.

Physicochemistry of Involved Chemical Ecology

For phytochemicals to function as allomones against insects, some energy (i.e., electron and/or proton) exchange or sharing must occur between these compounds and the insect. This perception of the molecules has classically been viewed as involving the sensory nervous system. However, as previously stated an allomone may interact at numerous sites and with various enzymes in the insect. In all cases energy transduction (transfer) occurs. If chemical bases of plant defense are to be developed efficiently and utilized optimally, then an understanding is needed of the physicochemical modes of action of each allomone.

Research on the physicochemistry of phytochemical allomonic action against insects is in its infancy, but it offers valuable techniques (e.g., electrochemical polarography or $E_{1/2}$ analyses) for investigations of chemical defenses. The electrochemistry of *in vitro* interactions of the plant-derived repellent juglone, related 1,4-naphthoquinones, and simple *p*-benzoquinone with the lipoprotein receptors from sensory neurons of various insects indicates that (1) a minor change in the substituents on a molecule may alter its electrochemistry sufficiently to destroy its allomonic actions against some insect species while retaining those effects on other species; and (2) as predictable from (1), different species of insects, or biotypes of species, inherently have sensory receptor lipoproteins with distinct ranges of possible electrochemical interactions with chemical messengers (Norris et al., 1977; and previous publications). A specific example is that 2,3-dichloro-1,4-naphthoquinone and p-benzoquinone are significant repellents or deterrents to *Scolytus multistriatus*, but not to *Periplaneta americana*. Both radiolabeled messengers bind comparably *in vivo* or *in vitro* to the lipoprotein receptors of each insect, but the messenger-induced electrochemical shift in the receptor, essential for allomonic action through sensory physiology, occurs only in *S. multistriatus*.

The electrochemical $E_{1/2}$ analyses may be conducted routinely with a polarograph. One immediate use for this technology in plant breeding would be in monitoring for changes in the specific $E_{1/2}$ characteristics of interactions between (1) defense molecules in cultivars developed by the plant breeder and (2) receptor lipoproteins in biotypes of pest species. Once correlations are established between the $E_{1/2}$ data and the data on specific allomonic effects of involved chemicals (as derived from quantitative bioassays with individuals from several variously susceptible populations of the pest species), then insect bioassays will not be as necessary. Some plant breeding thus could be geographically or seasonally separated from the field occurrence of the pest insect, and yet chemical resistance in cultivars could be quantified electrochemically.

1.3   Chemicals that Impart Resistance

Substances conveying plant resistance to insects include inorganics (e.g., selenium), primary and intermediary metabolites (e.g., citric acid, cysteine, and certain aromatic amino acids) and secondary substances (e.g., alkaloids). A summary diagram of plant metabolism and synthesized fundamental units (i.e., building blocks) is presented in Figure 1 together with a listing of selected secondary metabolites. This scheme strives to show general biosynthetic pathways yielding major classes of secondary metabolites that include known allomones. Biosynthetically, however, the classes of secondary substances fall into at least two larger groupings. Some (e.g., isoprenoids, acetogenins, protoalkaloids, and true alkaloids) are synthesized via single major pathways; others (e.g., glycosides, flavonoids, benzophenones, certain coumarins, condensed tannins, and stilbenes) are products of more than one pathway.

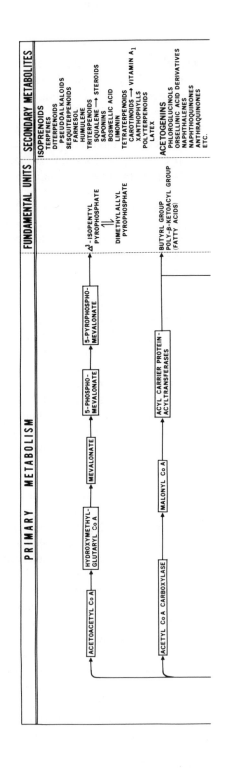

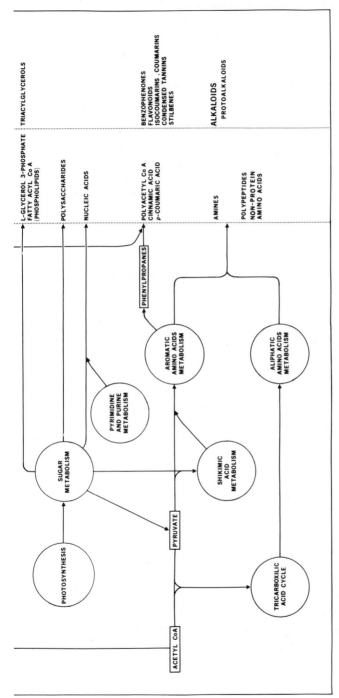

Figure 1. Schematic diagram of plant metabolism showing some fundamental units that serve as building blocks for secondary metabolites.

31

When biosynthesis involves a single pathway, knowledge of that pathway should allow breeders to monitor intermediates and to measure more efficiently their progress toward enhanced allomones in cultivars. When a defense chemical involves more than one pathway, as in the case of the benzophenones and other classes listed with them in Figure 1, an understanding of each pathway increases the number of chemical indicators by which breeders may monitor their progress.

In the case of glycosides, the aglycone may be synthesized by one or a combination of the pathways in Figure 1; the other component, though it may vary, comes via sugar metabolism. Knowledge of the biosynthesis of the specific aglycone involved in a glycoside should be especially useful in efforts to enhance such a defense chemistry because the aglycone usually is the active allomone.

Isoprenoids

This major group of phytochemicals includes the full range of terpenes (i.e., from hemiterpenes through polyterpenes), and allomones have been found especially among the monoterpenes, sesquiterpenes, triterpenoids, saponins, and various other steroids (Figure 1). Several major points emerge from our knowledge of isoprenoids as allomones in plants. The monoterpenoids are dominant components in the "volatile oils" of plant species. Research findings (e.g., α-pinene, 3-carene, gossypol, and cucurbitacin, Table 1) also demonstrate that each molecule has at least a dual messenger role. A wide range of allomonic effects has been demonstrated among these chemicals (e.g., altered insect behavior, sensory physiology, metabolism, and endocrinology). Certain plants (e.g., *Pteridium aquilinum, Achyranthes* sp., and *Abies balsamea*) contain isoprenoidal hormones that affect insect developmental rates, metamorphosis, fecundity, and longevity. The hormones may regulate the peak abundance of pest insects or the specific tissues damaged or consumed, so that the plant can escape with its vigor intact. The enumerated allomonic attributes of isoprenoids alone make these molecules extremely worthy focal points for efforts to enchance chemical resistances of plants to insects.

Acetogenins

Acetogenin means "genesis from acetate." Acetate units may form the sole building blocks of the types of chemicals listed immediately under acetogenin in Figure 1. In addition, acetate units may be a part of other plant metabolites. Acetate units have an important added role in the defense chemistry of plants as components of benzoquinones, flavonoids, certain coumarins, condensed tannins, and stilbenes, for example. The truly remarkable allelochemical, juglone, from plants in the Juglandaceae, and

Table 1.   Some Isoprenoids in Plants and Their Activities in Arthropod–Plant Interactions

Chemical Species	Source Plant	Affected Insect/Mite	Effect(s)	Reference
α-Pinene	*Pinus silvestris*	*Blastophagus piniperda*	Repellent	Oksanen et al., 1970
3-Carene	*P. silvestris*	*B. piniperda*	Repellent	Oksanen et al., 1970
α-Pinene	*P. taeda*	*Dendroctonus frontalis*	Attractant	Payne, 1970
3-Carene	*P. taeda*	*D. frontalis*	Attractant	Payne, 1970
α-Pinene	*P. ponderosa*	*D. brevicomis*	Attractant	Payne, 1970
3-Carene	*P. ponderosa*	*D. brevicomis*	Attractant	Payne, 1970
Gossypol	*Gossypium hirsutum*	*Heliothis* spp.	Deterrent	Hedin et al., 1974
Gossypol	*G. hirsutum*	*Epicauta* spp.	Feeding deterrent	Hedin et al., 1974
Gossypol	*G. hirsutum*	*Anthonomus grandis*	Feeding stimulant	Hedin et al., 1974
Cucurbitacin	Cucurbitaceae	*Tetranychus urticae*	Feeding deterrent	DaCosta and Jones, 1971
Cucurbitacin	Cucurbitaceae	*Acalymma* spp.	Arrestant, feeding excitant	DaCosta and Jones, 1971
Cucurbitacin	Cucurbitaceae	*Diabrotica* spp.	Arrestant, feeding excitant	DaCosta and Jones, 1971

DIMBOA and other benzoxazolinones from *Zea mays,* are two extensively studied types of acetogenous molecules which make plants resistant to certain insects. DIMBOA and related compounds are most abundant in young corn plants where they deter larvae of *Ostrinia nubilalis* from feeding and growing, and reduce survival. Older corn tissue does not contain such a deleterious amount of these deterrents. Thus, in this annual plant, allomonic protection of young, rapidly dividing cells versus older, differentiated cells seems evolutionarily paramount. Conversely, in the perennial deciduous dicotyledonous *Carya* juglone is at higher concentrations, in glucosidic form, in second-year or older twig growth. In such trees, current growth usually can be retarded by herbivores for more than one year without plant mortality.

Among the acetogenins we may see interesting evidence of evolution of deposition, if not also synthesis, of the allomones in those tissues important to preservation of the sporophytes of annuals and perennials.

Research on acetogenins as bases for plant defense is promising because many of these molecules possess unusual abilities to change their state of oxidation; that is, to exchange electrons and protons with their environments. Thus they possess the attributes deemed especially important in a chemical messenger. Their extraordinary abilities enable them to function at multiple sites of action in insects.

Aromatics Derived from Shikimic Acid and Acetate

These compounds are biosynthetically hybridized from the shikimic acid and acetate pathways (Figure 1). The group includes flavonoids, benzophenones, some coumarins, lignans, condensed tannins, and stilbenes. They constitute one of the more important groups of plant defense chemicals.

Flavonoids (Table 2) have received the most consideration as allomones against insects. Quercetin is an example of their allomonic ability against a series of insect pests from several orders of Insecta (Table 2). Myristicin, morin, and D-catechin are other flavonoids with proven allomonic effects on certain species; however, all four flavonoids repel or inhibit some species, or biotypes, but attract or excite others.

Lignans, sesamin and kobusin, in the nonhost *Magnolia kobus* inhibited *Bombyx mori* growth (Table 2), but α-conidendrin in host *Ulmus* stimulated *Scolytus multistriatus* feeding. The coumarins studied also have demonstrated dual roles as messengers. Coumarin-related *o*-hydroxycinnamic and *p*-hydroxycinnamic acids also functioned as allomones to *S. multistriatus*, but *trans*-cinnamic acid, without the hydroxyl substituent of the previously mentioned cinnamic acids, stimulated *S. multistriatus* feeding.

Chemicals in this major grouping merit further comparative studies of the structural properties that determine whether molecules are allomones or kairomones to particular species of insects (Norris, 1977). Specific properties that especially deserve further investigation include (1) level of oxidation in the C_3 (i.e., propane) unit; (2) functional groups (e.g., carbonyl and hydroxyl), on the propane unit or its remnant; and (3) substituents, particularly hydroxyls and methoxyls, on the aromatic rings. We have already emphasized the importance of chemical messengers in chemical ecology being readily capable of changing states of oxidation. Molecules in this group present a beautiful sequential range of oxidation states (e.g., flavonoids, Table 3) and thus seem highly evolved for serving as messengers for the intricate ordering of species distributions and population levels. In the studies with *S. multistriatus* and flavonoids, the flavan-3-ol, D-cate-

Table 2.   Some Aglycones Hybridized from Shikimic Acid and Acetate Biosynthesis Pathways of Plants and Their Activities in Insect–Plant Interactions

Chemical Species	Source Plant	Insect Species	Effect(s)	Reference
Quercetin	*Gossypium* spp.	*Anthonomus grandis*	Feeding stimulant	Hedin et al., 1974
Quercetin	*Gossypium* spp.	*A. grandis*	Stimulated larval development	Hedin et al., 1974
Quercetin	*Gossypium* spp.	*Pectinophora gossypiella*	Reduced development	Hedin et al., 1974
Quercetin	*Gossypium* spp.	*Heliothis zea*	Reduced development	Hedin et al., 1974
Quercetin	*Gossypium* spp.	*H. virescens*	Reduced development	Hedin et al., 1974
Quercetin	*Gossypium* spp.	*Schizaphis graminum*	Reduced development	Hedin et al., 1974
Quercetin	*Quercus macrocarpa*	*Scolytus multistriatus*	Feeding inhibitor	Norris, 1977
Myristicin	*Q. macrocarpa*	*Bombyx mori*	Growth inhibitor	Isogai et al., 1973
Morin	*Q. macrocarpa*	*B. mori*	Feeding excitant	Hamamura, 1970
Morin	*Q. macrocarpa*	*Heliothis virescens*	Reduced development	Hamamura, 1970
Sesamin	*Magnolia kobus*	*Bombyx mori*	Growth inhibitor	Kamikado et al., 1975
Kobusin	*M. kobus*	*B. mori*	Growth inhibitor	Kamikado et al., 1975
α-Conidendrin	*Ulmus* spp.	*Scolytus multistriatus*	Feeding stimulant	Norris, 1977

chin, was a strong feeding stimulant. The flavan-3-ols are among the flavonoids with the most reduced C_3 unit. As flavonoids with increasing levels of C_3-unit oxidation were tested on *S. multistriatus*, the effect on feeding switched to inhibition, and this increased quantitatively with the level of oxidation. Regarding functional groups in the C_3 unit, presence of a carbonyl was associated with allomonic action. When a hydroxyl was also present, allomonic effects appeared to be increased. Absence of a carbonyl, coupled

Table 3.   Oxidation States in the C_3 Units of Different Types of Flavonoids

Flavonoid	Structure of C_3 Unit[a]
Flavan-3-ols	A—CH_2—CHOH—CHOH—B
Hydrochalcones	A—CO—CH_2—CH_2—B
Chalcones	A—CO—CH=CH—B
Flavanones	A—CO—CH_2—CHOH—B
Leucoanthocyanidins	A—CHOH—CHOH—CHOH—B
Flavones	A—CO—CH_2—CO—B
Anthocyanidins	A—CH_2—CO—CO—B
Benzalcoumaranones	A—CO—CO—CH_2—B
Flavanonols	A—CO—CHOH—CHOH—B
Flavonols	A—CO—CO—CHOH—B

[a] A and B represent the corresponding rings of the flavonoid structure.

with the presence of a hydroxyl (e.g., D-catechin), was associated with kairomonic activity. Other molecular characteristics being equal, addition of a hydroxyl, either *ortho* or *para*, on *trans*-cinnamic acid changed it from a kairomone to an allomone for *S. multistriatus*. The presence of an *o*-hydroxyl on cinnamic acid makes it functionally similar to coumarins.

Some of the structural and reactive properties common to messengers in this large group of molecules, as well as in 1,4-naphthoquinones, other classical quinone-quinol couples, and other molecules capable of one-electron transfers, are discussed by Norris (1977),

Alkaloids

About 15 to 20 percent of all vascular plants contain alkaloids, which are basic, nitrogen-containing compounds. Alkaloids may be further subdivided into pseudoalkaloids (i.e., not involving amino acid and biogenic amine metabolism; e.g. diterpenoid alkaloids, Figure 1); protoalkaloids (i.e., involving biogenic amine metabolism but not having a heterocyclic ring structure, Figure 2); and true alkaloids (i.e., having the heterocyclic ring structure). Our discussion covers all three types since they involve nitrogen, a biologically active element, although the biosyntheses utilize two distinct primary pathways, the isoprenoids and amino acids-biogenic amines (Figure 1).

The toxic and medicinal powers of alkaloids have been known to pharmacologists since botanically based medications originated. Alkaloids have been especially valued for their effects on nervous systems.

The true alkaloid nicotine in *Nicotiana* functions as an allomone against many insect species. Its unusual insecticidal powers resulted in its widespread agricultural use as one of the early organic pesticides and it is still used commercially. Its range of allomonic effects to insects extends beyond the nervous system, and includes at least those additional actions that entomologists lump under "stomach poisoning."

Despite marked toxicity to many insects, the tobacco hornworm, *Manduca sexta*, copes with the nicotine in *Nicotiana* (Kogan, 1977). Thus even alkaloid-based plant resistance has discernible limits among insects.

Research efforts to understand better the roles of alkaloids in plant defense have emphasized the steroidal pseudoalkaloids. Kuhn and Gauhe (1947) conducted extensive studies on the roles of such alkaloids in deterring *Leptinotarsa decemlineata* from *Solanum* plants. Major identified compounds include solanine, tomatine, and demissine; and observed effects were deterrence of larval and adult feeding, and inhibition of the growth rate of larvae.

As another example of alkaloidal defense of plants, pilocereine and lophocereine in the cactus *Lophocereus schotti* repel or deter *Drosophila* spp. which do not use it as a host plant, but *D. pachea*, which uses it as a host, tolerates these alkaloids (Kircher et al., 1967). For more information on other specific alkaloidal allomones known in plants, and species of insects involved, see Hedin et al. (1974) and Kogan (1977).

Alkaloidal allomones usually remain effective over relatively long periods in plant evolution. However, because the actions of alkaloids tend to bring about drastic effects (e.g., death) in nonadapted insects, selection for adapted individuals may be intense. But evolution of adaptation must rely heavily on specialized methods of sequestering alkaloids and/or metabolizing them, because most insects cannot afford to expose their basic metabolic machinery to alkaloids. If such evolution does occur, then the adapted insect (e.g., *Manduca sexta*) may demonstrate a remarkable ability to utilize plant substrates richly endowed with alkaloids. Such plant resistance

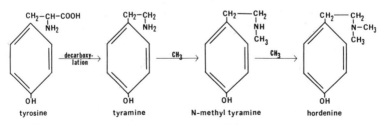

Figure 2.   Biosynthesis of the protoalkaloid hordenine from tyrosine.

may not break down often but when it begins to yield to an insect, the decline of resistance may be rapid.

Protease Inhibitors and Nonprotein Amino Acids

These chemical bases for plant resistance are lumped under one heading because of common origins in the amino acid-peptide biosynthetic pathway (Figure 1) or actions on proteinaceous enzymes.

Many plant-borne protease inhibitors are small proteins with molecular weights under 20,000 daltons, but the weight may be less than 10,000 or as large as 50,000. They may be present as dimers or tetramers (Ryan, 1973). Some phenolics and other aglycones also inhibit protease activity. The majority of these agents inhibit enzymes of animal or microbial origin with either trypsin- or chymotrypsin-like specificities. Some also inhibit proteolytic enzymes in the plant of origin.

Inhibitors have been found in many species of cultivated and wild plants, and more than one have been isolated from some plants. Liener and Kakade (1969) present an extensive table showing some plant species and families known to contain proteinaceous inhibitors, giving the common name for the inhibitor, indicating the plant part used as the source, and stating susceptibility to heat. Most proteinaceous inhibitors have been isolated from the Leguminosae and especially from their seeds.

Research on the effects of inhibitors on herbivores has concentrated on grain feeders. Lipke et al. (1954) discovered that a highly acidified extraction of raw soybeans contained an inhibitor of the proteolytic enzymes of *Tribolium confusum* and *Tenebrio molitor*. It was concluded that the natural resistance of soybeans to insect injury could be due to some toxic compound in the beans. Refinement of the extraction techniques of Lipke and co-workers allowed Applebaum and associates to demonstrate that three proteinaceous fractions from soybean strongly inhibited the growth of *T. castaneum* larvae. They have continued research on inhibitors of proteolytic enzymes in insects, and have isolated one or more from wheat and lima beans (Applebaum and Birk, 1972). The probable role of these inhibitors in plant defense was further clarified by the demonstration of Ryan and co-workers that injury to tomato and potato leaves by larvae of *Leptinotarsa decemlineata* or by a mechanical device caused the rapid accumulation of inhibitors in those leaves. Such wounding of tomato leaflets also repeatedly released a protease inhibitor inducing factor (PIIF), which is rapidly transported throughout the plant causing accumulation of a potent inhibitor of several proteases of both animals and microbes. This synthesis of inhibitors in leaves in response to wounding seems to be *de novo*, indicating the existence of a type of immune response (Walker-Simmons and Ryan, 1977). The discovery of wound-induced accumulations of protease inhibitors in many plant species may open many possibilities in terms of plant resistance. One can even speculate about the possibility of inducing "immune reactions" in plants by controlled methods not unlike the vaccination of animals.

To remind us once more that chemical bases for plant resistance always break down somewhere in the plant–herbivore interface of chemical ecology, larvae of bruchid beetles can develop on seeds of legumes despite high titers of inhibitors of proteolytic activity (Applebaum, 1964).

Among nonprotein amino acids, L-canavanine, with allomonic properties to several tested insect species (e.g., *M. sexta* and *Prodenia eridania*), was isolated from the seeds of the legume *Dioclea megacarpa* (Rosenthal et al., 1977). Of much interest is the finding that larvae of a bruchid beetle, *Caryedes brasiliensis*, which feed exclusively on these seeds, can cope metabolically with L-canavanine.

Another unusual amino acid, β-cyano-L-alanine, found in *Vicia sativa*, is allomonic to *Locusta migratoria migratorioides* (Schlesinger et al., 1976). Symptoms in the insect include a marked discharge of watery feces which results in a significant decrease in hemolymph volume within 1 day. Dehydration becomes acute and causes death within 5 days.

These two examples of allomonic nonprotein amino acids in plants suggest that further investigations of the numerous unusual amino acids in plant species are likely to provide another opportunity for plant breeders to enhance plant resistance to insect pests.

Glycosides

Glycosides are asymmetrical mixed acetals. The sugar component may be from the common plant monosaccharides; from about 10 disaccharides of which only three, rutinose, sophorose, and sambubiose, seem to occur commonly; or from about six trisaccharides. Glucose is the most commonly involved sugar moiety.

The aglycone portion may come from any, or from a combination, of the biosynthetic pathways (Figure 1). If one considers the phenolics, then there are more phenolic aglycones than any other type.

Phenolics are produced via the isoprenoid, the shikimic acid-aromatic acid, or the acetate pathway, or a combination (Figure 1). Some other, yet undescribed, types of important allelochemical aglycones found as glycosides include cyanohydrins and isothiocyanates. Both apparently are synthesized via the amino acid pathways, the latter probably from sulfur-containing amino acids.

The literature about the relative roles of the free aglycone versus the intact glycoside as active allomones for plant defense is confusing. One safe statement is that aglycones are involved wherever perception of volatiles occurs. In situations where insects probe, chew, or ingest plant tissues, the intact glycoside may play an important role in allowing the allomonic actions. Especially in cases of ingestion, hydrophilic characteristics of some glycosides may be required. However, basic allelochemical activity seems to lie with the aglycone.

The formation of glycosides in plants apparently allows the "safe" storage of otherwise highly toxic aglycones in somewhat "out of the way" places as far as basic metabolism is concerned, which may account for the plant's ability to unleash an extraordinary amount of allomonic aglycones whenever threatened or attacked by herbivores. This action commonly involves oxidative and hydrolytic enzymes which gain access to the stored glycosides especially during plant stress. Rupture of cortical plant cells by an insect is a prime example of such a situation.

Regarding examples of specific plant glycosides or their aglycones implicated as allomones against insects, Lichtenstein et al. (1962) found that the aglycone 2-phenylethyl isothiocyanate from *Brassica rapa* deterred *Drosophila melanogaster* from feeding. At higher concentrations it killed this fly. However, many insects that use crucifers as hosts utilize such isothiocyanates as kairomones (e.g., Verschaffelt, 1910). Sinigrin in crucifers is a feeding deterrent to the polyphagous aphid *Myzus persicae* (David and Gardiner, 1966). The feeding of four species of *Epicauta* was inhibited by *cis-o*-hydroxycinnamic acid glucoside, which apparently occurs in *Melilotus officinalis* and *M. alba* (Manglitz et al., 1976).

The properties of glycosides enumerated above make them strong candidates for enhancing chemically based plant resistance to insects. If plant breeders developed strategies to augment the quality and quantity of allomonic aglycones safely stored as glycosides in cultivars, then it seems that they might be participating in nature's master plan for the evolution of chemical defenses in higher plants.

1.4   Chemically Based Resistance in Future Cultivars

Present knowledge clearly indicates that chemically based resistance is a major component of the plant's total defense armament against herbivores. It is diverse in composition and extremely effective ecologically, at least if man does not interfere. It seems clear that man must achieve new levels of awareness, and practice of the derived knowledge, if genetic engineering is to be used effectively to alter both the pathways and timetables of evolution of chemically based plant defenses for our benefit.

If we hope to achieve more than sporadic and short-term success, we must further decipher the submolecular code of chemical ecology and abide by its rules in our breeding programs. We should be aware of this basic code and investigate what is happening chemically in our ecological midst.

2. MORPHOLOGICAL BASES

Morphological (physical) resistance factors interfere physically with loco-motor mechanisms, and more specifically with the mechanisms of host selection, feeding, ingestion, digestion, mating, and oviposition as opposed to those factors affecting the chemically mediated behavioral and metabolic processes discussed above. The physical barriers or deterrents to insects and other herbivores such as trichomes, surface waxes, silication, or sclerotiza-tion of tissues are, however, expressions of genetically regulated biochemical processes. In addition, allomones affecting insect behavioral and metabolic processes may occur in plant morphological structures (trichomes or bracts). Thus chemical and morphological resistance factors intertwine in a continuum of defense.

As with chemical factors, morphological defenses may act at a distance or at close range (e.g., on contact). Defenses capable of long-distance effects have been little studied among cultivated plants. Contact defenses vary from thickened nondifferentiated tissues to highly differentiated organs. Much of the existing man-enhanced practical resistance in crop plants involves morphological factors.

2.1 Remote Factors

Color and shape of plants remotely affect host selection behavior of phytophagous insects and have been associated with some resistance. In the genus *Rhagoletis* (Diptera: Tephritidae) foliage color and tree shape and size play roles in fly discrimination between nonhosts and hosts, but these cues alone do not account for host-plant specificity (Boller and Prokopy, 1976). Plant size, shape, and relative transience seem to be associated with some strategies of chemical defenses (Feeny, 1976).

Color

Experimental evidence for the involvement of color in long-range perception of plants is limited. Much of the available information comes from studies on aphids and other Sternorrhyncha (Kennedy et al., 1961; Mazokhin-Porschynakov, 1969). Most of these insects are attracted to leaves reflecting within the 500 to 600 nm range (yellow-green). Alate aphids are attracted to leaves reflecting about 500 nm, regardless of the species of plant, because they seem to be attracted to plants at a physiologically suitable stage of growth (Kennedy et al., 1961).

Little can be done by genetic manipulation to affect plant color without affecting some fundamental physiological plant processes. In many instances healthy, dark green plants are less attractive to insects than yellowing plants under stress. In a study of 13 varieties of peas in the field and in the greenhouse, Cartier (1963) observed that yellow-green plants were preferred to green plants by the pea aphid, *Acyrthosiphon pisum*. Similar reactions were observed in other insect groups.

Specific color-related resistance, however, does exist. Red cotton plants are less attractive to the boll weevil, *Anthonomus grandis,* than green plants where both grow together (Stephens, 1957). The imported cabbage worm, *Pieris rapae*, is less attracted to the red foliage of the 'Rubine' variety of Brussels sprouts than to the green ones (Dunn and Kempton, 1976). Color or intensity of light reflected from the surface of cabbage leaves affected host selection by *Brevicoryne brassicae*; red cabbages were least selected by alate aphids but presented a most favorable aspect for aphid increase after infestation (Radcliffe and Chapman, 1965, 1966). Oat cultivars with red tiller bases and pubescent shoot bases were less susceptible than other cultivars to attack by the frit fly, *Oscinella frit* (Peregrine and Catling, 1967).

Shape

Form perception probably elicits certain generalized behavioral patterns, but no resistance mechanisms among crop plants have been associated directly with plant shape. Certain morphological characteristics may, however, be linked with other resistance factors. Above-ground shape of turnip plants was not important to *Hylemya florales*, the turnip maggot, but the varieties, which generally have strong, round, and long roots, were more tolerant than the varieties with thin roots (Varis, 1958). In this case root shape could be used as a selection criterion in breeding programs.

2.2.   Close-Range or Contact Factors

The majority of recognized physical defense factors of plants operate on contact with the herbivore. The contact factors most commonly found among resistant crop plants are presented in Table 4.

Thickening of Cell Walls and Rapid Proliferation of Plant Tissues

Thickening of cell walls results from deposition of cellulose and lignin. As a consequence the tissue is tougher or more resistant to the tearing action of mandibles or to the penetration of the proboscis or ovipositor of insects.

Table 4.   Physical Resistance Factors Most Commonly Found Among Crop Plants

Plant Factors	Effect(s) on Insect
Thickening of cell walls; increased toughness of tissues	Interference with feeding and oviposition mechanisms
Proliferation of wounded tissues	Insects killed after initial injury
Solidness and other characteristics of stems	Obstruction of feeding and oviposition mechanisms; dehydration of eggs
Trichomes	Effect on feeding, digestion, oviposition, locomotion, and attachment; toxic and disruptive effects of allelochemics in glandular trichomes; provision of shelter
Accumulation of surface waxes	Effect on colonization and oviposition
Incorporation of silica	Abrasion of cuticle; feeding inhibition
Anatomical adaptations of non-specialized organs and protective structures	Various effects

Leaf toughness has been correlated with the amount of foliage consumed by the mustard beetle, *Phaedon cochleariae*. Feeding rates and larval growth were retarded when larvae fed on relatively tough turnip, kale, and Brussels sprout leaves (Tanton, 1962). Cuthbert and Davis (1972) observed that thickness of the pod wall interfered with penetration of pods of cowpea, *Vigna sinensis*, by the cowpea curculio, *Chalcodermus aeneus*. Thicker hypodermal layers were considered a resistance factor in rice to the rice stem borer, *Chilo suppressalis* (Patanakamjorn and Pathak, 1967). Resistance in varieties of sorghum to the sorghum shoot fly, *Atherigona varia soccata*, was attributed to the presence of cells with distinct lignification and thicker walls enclosing the vascular bundle sheaths within the central whorl of young leaves (Blum, 1968). Caswell and Reed (1975) showed that *Melanoplus confusus* could not completely digest the bundle sheath cells of the C_4 grasses *Panicum virgatu*, *Andropogon gerardi*, and *Schizachyrium scoporium*. These plants have leaf veins surrounded by thick-walled, organelle-rich, bundle-sheath cells.

A defense reaction consisting of proliferation of cells triggered by insect injury was observed in cotton. Larvae of young pink bollworm, *Pectinophora gossypiella*, were crushed or drowned by proliferating cells of injured tissues in certain lines; however, this character seemed to be linked to undesirable agronomic characteristics and could not be used in practical breeding programs (Adkisson et al., 1962).

Although toughness of leaf tissues is an efficient defense mechanism, breeding of improved varieties often leads to elimination of such characters, especially in crop plants consumed as leaves or fruits.

Solidness and Other Stem Characteristics

There are numerous examples of slight or profound changes in a stem feature that decrease fitness of a plant for associated herbivores. Resistance to certain stem borers is related to the nature of the stem tissues. Solid stems are responsible for the resistance of several wheat varieties to the wheat stem sawfly, *Cephus cinctus*. A certain degree of stem solidness results in damaged and desiccated eggs and impaired larval movements (Wallace et al., 1973). Mortality of eggs and first instars in the pith varied independently, however, indicating that several factors may be involved. Because light intensity affects solidness of wheat stems, this character is greatly influenced by variable weather conditions (Holmes and Peterson, 1960; Roberts and Tyrell, 1961). On sugarcane, young *Diatrea saccharalis* larvae first feed on leaf and leaf-sheath tissues; later they enter the stalk. Denticles on the midrib of leaves, number of vascular bundles, lignification of cell walls, and number of layers of sclerenchymatous cells play important roles in the resistance of sugarcane to first and second instars. As older larvae bore into the stalk, hardness of the rind and fiber content of stalks are key factors of resistance (Agarwal, 1969; Martin et al., 1975).

Hard, woody stems of *Cucurbita* spp. plants with closely packed, tough, vascular bundles are the main resistance factors against the squash vine borer, *Melittia cucurbita*. Penetration of the stems and feeding of larvae are impaired by the structural characteristics of the plant (Howe, 1949). Thick cortex in the stem of the wild tomato relative *Lycopersicon hirsutum* prevented the potato aphid, *Macrosiphum euphorbiae*, from reaching the vascular tissue (Figure 3) (Quiras et al., 1977).

Such morphological characteristics seem to vary with environmental conditions. Breeding programs using these characteristics should expose lines to a range of these conditions so that the full extent of this variability is revealed.

Trichomes

Trichomes are unicellular or pluricellular outgrowths from the epidermis of leaves, shoots, and roots (Uphof, 1962). The collective trichome cover of a

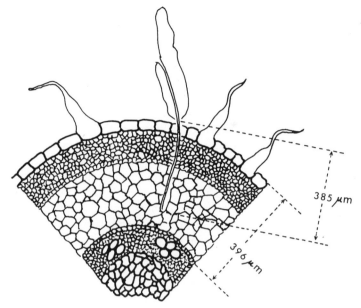

Figure 3. Thick cortex in stems of *Lycopersicon hirsutum* reported to prevent *Macrosiphum euphorbiae* from reaching vascular tissue (redrawn from Quiras et al., 1977).

plant surface is called pubescence. Several authors have attempted to classify the variety of plant trichomes; the reviews by Uphof (1962), Johnson (1975), and Hummel and Staesche (1962) present some of the most widely accepted classifications.

Trichomes serve many critical physiological and ecological functions, particularly those associated with water conservation. Levin (1973) and Johnson (1975) discussed the ecological functions of trichomes as defense against herbivores. These latter discussions relate to our interest in trichomes as resistance factors in crop plants.

Insect species respond differently to the presence of plant hairs. Pubescence as a resistance factor interferes with insect oviposition, attachment to the plant, feeding, and ingestion. However, glabrous forms of plants may be more resistant to some species. In general, the purely mechanical effects of the pubescence depend on four main characteristics of the trichomes: density, erectness, length, and shape. In some cases trichomes possess associated glands that exude secondary plant metabolites. The effect of glandular trichomes may depend on the nature of the exudate. It may be composed of allelochemics such as alkaloids or terpenes (Johnson, 1975). Such toxic substances may kill insects on contact or act as repellents. In some plants, sticky exudates glue the insects' legs and impede locomotion. The roles of pubescence as a resistance factor in crop plants, based mainly on Webster (1975), are summarized in Table 5.

Effect of Pubescence on Feeding and Digestion.   It is generally assumed that small arthropods with piercing–sucking mouth parts may be deterred from feeding on hairy plants because the tip of the proboscis cannot reach the mesophyll or the vascular bundles. Soybean pubescence "normally" consists of hairs, formed by one to three apical cell, each about 1 mm long, and one to three basal cells, at a density of about 8 hairs/mm^2 (Singh et al., 1971) (Figure 4). The angle of insertion may depart somewhat from 90°. Because the length of the mouth parts of *Empoasca fabae,* for instance, varies from 0.2 to 0.4 mm, it is conceivable that the pubescencce forms a barrier, particularly for young nymphs. From Naito's (1976a, 1976b) analyses of the feeding habits of leafhoppers attacking forage crops in

Figure 4.   Stunting of a glabrous, near-isogenic line of 'Clark' soybean caused by hopperburn. The glabrous isoline plants (foreground) are more attractive to the potato leafhopper. *Empoasca fabae,* than the normal pubescent isoline (background) plants that were planted on the same date.

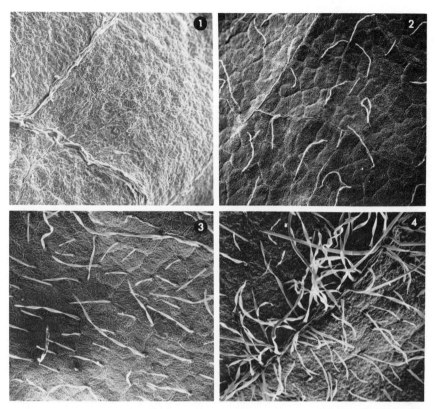

Figure 5. Scanning electron micrographs of 'Clark' soybean leaves (lower surface). Four near-isogenic lines differing in pubescence density and type: (1) glabrous isoline ($\times$17.8); (2) curly pubescent isoline ($\times$17.8); (3) normal pubescent variety; and (4) dense pubescent isoline ($\times$30.7).

Japan, one can estimate the length of the stylet sheaths of several cicadellid species feeding on sorghum and clover. The species feeding on vascular bundles had sheaths 2 to 4 mm long; the mesophyll-feeding species had sheaths only about 0.05 mm long. It seems, therefore, that the particular feeding mechanism of a leafhopper species also determines how pubescence affects feeding. Phloem or xylem feeders must insert their stylets deeper into the plant tissue; thus short trichomes may impede feeding of these species, but do not affect the mesophyll feeders. There are, however, few detailed analyses of the mechanisms of pubescence resistance to feeding. The effects of pubescent versus relatively glabrous plants on insects, however, are often dramatically different (Figure 5).

Table 5. Summary of the Role of Pubescence in the Behavioral and Physiological Response of Arthropods to Selected Crop Plants[a]

Host Plant and Insect(s)	Effects of Pubescence[b]	Behavioral and/or Physiological Process Affected[c]
COMPOSITAE		
Kuhnia eupatorioides		
Melanoplus femurrubrum, red-legged grasshopper	R	F
CUCURBITACEAE		
Bemisia tabaci, sweet potato whitefly	R?	—
EUPHORBIACEAE		
Castor bean		
Empoasca flavescens	R	F O?
FAGACEAE		
Chestnuts		
Curculio elephas	R	—
GRAMINAEAE		
Corn		
Diabrotica virgifera, western corn rootworm	R	F
Heliothis zea, corn earworm	S	O
Oats		
Oscinella frit, frit fly	R/N	O
Rice		
Chilo suppressalis, rice stem borer	R	O
Sorghum		
Atherigona varia soccata, sorghum shoot fly '	R	F
Epilachna varivestis, Mexican bean beetle	R	F L O
Heliothis zea, corn earworm	R	O
Leguminivora glycinivorella, soybean pod borer	S	O
Plathypena scabra, green cloverworm	S	O
Sericothrips variabilis, soybean thrips	N	O
Trialeurodes abutilonea, banded-wing whitefly	N	—
MALVACEAE		
Cotton		
Anthonomus grandis, boll weevil	R	O
Aphis gossypii, cotton aphid	R/S	F
Bemisia tabaci, sweet potato whitefly	S	F

Table 5. (Continued)

Host Plant and Insect(s)	Effects of Pubescence[b]	Behavioral and/or Physiological Process Affected[c]
Cicadellidae: *Empoasca* spp., *Amrasca* spp.	R	F O?
Earias fabia	S	O
Earias insulana	N	—
Heliothis zea, bollworm	S	O
Pectinophora gossypiella, pink bollworm	N	—
Pseudatomoscelis seriatus, cotton fleahopper	S	—
Spodoptera littoralis, cotton leaf worm	R	F O
Tetranychus spp., spider mites	R/S	F
Thrips tabaci, onion thrips	N	—
Trialeurodes abutilonea, banded-wing whitefly	S	—
Trichoplusia ni, cabbage looper	S	O
PASSIFLORACEAE		
Heliconius spp.	R	L
Sugarcane		
Aleurolobus barodensis, sugarcane whitefly	R	—
Melanaspis glomerata, sugarcane scale	R	—
Scirpophaga nivella, top borer	R	F
Wheat		
Mayetiola destructor, Hessian fly	S	O
Oscinella frit, frit fly	R	—
Oulema melanopus, cereal leaf beetle	R	O F
Hylemya genitalis, spring fly	S	—
LEGUMINOSAE		
Alfalfa		
Empoasca fabae, potato leafhopper	R	F O?
Beans		
Aphis craccivora, cowpea aphid	R	L
Aphis fabae, bean aphid	R	L
Empoasca fabae, potato leafhopper	R	F O? L
Etiella zinckenella, lima bean pod borer	R?	—
Thrips tabaci, onion thrips	S	Esc.

49

Table 5. (Continued)

Host Plant and Insect(s)	Effects of Pubescence[b]	Behavioral and/or Physiological Process Affected[c]
Lupine		
Acyrthosiphum pisum, pea aphid	R	—
Soybean		
Empoasca fabae, potato leafhopper	R	F O?
Deuterosminthurus yumanensis, springtail	R	—
ROSACEAE		
Strawberry		
Tetranychus urticae, two-spotted spider mite	S	F O
SOLANACEAE		
Bemisia tabaci, sweet potato whitefly	R?	—
Epitrix hirtipennis, tobacco flea beetle	R	Rep.
Heliothis virescens, tobacco budworm	S	O
Leptinotarsa decemlineata, Colorado potato beetle	R	L
Macrosiphum euphorbiae, potato aphid	R	F L
Manduca sexta, tobacco hornworm	R	Tox.
Myzus persicae, green peach aphid	R	Tox.
Tetranychus cinnabarinus, carmine spider mite	R	L Tox.
Tetranychus urticae, two-spotted spider mite	R	L Tox.
Trialeurodes vaporariorum, greenhouse whitefly	R	L
VITACEAE		
Grape		
Epitetranychus sp.	R?	—

[a] Data taken from Webster (1975) or other citations in the text.

[b] R = resistance; S = susceptibility; N = no effect; R/S or R/N = two or more references containing conflicting results; R? = resistance not clearly defined.

[c] F = feeding; O = oviposition; L = effect on locomotor activity owing to entrapment, impalement, or impediment to set a foothold; Esc. = pubescence provides shelter against predators; Rep. = repellent effect of exudates; Tox. = toxic effect of exudate from glandular trichomes.

Pubescence may also interfere with ingestion of food by small mandibulate larvae and adults. Schillinger and Gallun (1968) observed that early first-instar larvae of *Oulema melanopa,* the cereal leaf beetle, were critically affected by the pubescence of certain wheat varieties. High mortality was explained by the fact that larvae had to eat the hairs to reach the epidermis. In doing so they ingested unusually large amounts of cellulose and lignin, the basic constituents of the hairs. Death of the young larvae resulted from an unbalanced diet overly rich in fibrous materials. Larval weight was negatively correlated with increased pubescence. In addition to the dietary imbalance, Wellso (1973) observed that larvae that fed on pubescent wheat leaves were stuffed with undigested hairs, some of which pierced the gut wall.

Analyses of intake and utilization by the larvae of Mexican bean beetle, *Epilachna varivestis,* feeding on various pubescent isolines of 'Clark' soybean (Figure 5) corroborate the cereal leaf beetle data (Kogan, 1972). All consumption and utilization parameters were higher in 'Clark' "glabrous" and "curly" than in "normal" and "dense pubescent" types. Survival was 10 to 25 percent higher on the glabrous and curly pubescent types (Table 6). These results suggest that pubescence reduced the quality of the food ingested and caused greater larval mortality.

Much of the reported resistance of hairy plants to Cicadellidae, Aphididae, and other small sap-sucking insects is based on estimates of differential population buildups on hairy versus glabrous plants. It is, in most cases, impossible to assess how much of the population difference results

Table 6. Pupal Weight, Larval Weight Gain, Food Consumed, Efficiency of Conversion of Ingested Food (ECI), Developmental Time, and Survival through Pupation of the Mexican Bean Beetle, *E. varivestis,* Feeding on Four Pubescent Isolines of 'Clark' Soybean[a] (Kogan, 1972)

Clark Soybean	Mean Weight of Pupae	Total Weight Gain	Development Time (Days)	Total Weight Food Consumed	ECI[b]	Percent Survival
Normal	24.8	27.6	16.1	56.1	50.4	65.0
Dense	28.3	31.4	16.7	46.8	67.5	50.0
Glabrous	29.5	29.5	15.4	35.8	86.7	75.0
Curly	32.7	34.6	15.9	44.8	79.6	75.0

[a] Data expressed as fresh weights (mg) of larvae and pupae, and dry weight of food.
[b] ECI = (fresh weight of larvae/dry weight of food consumed) × 100.

from interference with feeding, and consequently high nymphal mortality; or from interference with oviposition.

Effect of Pubescence on Oviposition.   Much of the best data on ovipositional preference was gathered with lepidopterous species. Glabrous cotton strains of *Gossypium* spp. are less favorable substrates than pubescent strains for oviposition by *Heliothis zea* and *H. virescens* (Lukefahr et al., 1971; Stadelbacher and Scales, 1973). Table 7 summarizes data from Lukefahr et al. (1971) demonstrating the marked effect of the glabrous condition on oviposition. These data show that number of eggs was strongly correlated with hair density ($r = 0.945$, regressing number of eggs on number of trichomes/mm² on the lower surface of leaves, $y = 0.002x - 1.415$). According to Callahan (1957), however, the pubescent leaf surface provides a better foothold for the female *H. zea* and thus facilitates oviposition.

Pubescence is a trait that may be either detrimental or advantageous to insects. Although hairy cotton and pubescent soybeans are more susceptible to certain lepidopterous pests, they are more resistant to other pests of the same or other orders. Thus the glabrous types, resistant to *Heliothis* spp. and other pests of cotton, are susceptible to the cotton leafworm, *Spodoptera littoralis* (Kamel, 1965), and the boll weevil, *Anthonomus grandis* (Stephens, 1957, 1959), but resistance of pubescent cotton to the boll weevil does not persist at high population densities (Merkle and Meyer,

Table 7.   Reduction of Oviposition by *Heliothis* spp. on Glabrous Cotton Strains[a]

Strain	Type of Pubescence	Seasonal Totals/Hectare		No. Trichomes/mm² on Leaves	
		No. Eggs	No. Larvae	Upper	Lower
Normal Bayou	Hirsute	52,645a	43,330b	3.8	11.1
Bayou Sm-1	Glabrous	6,065c	10,722d	0.06	0.14
NM 1073A	Hirsute	42,412a	66,620a	2.4	10.3
NM 1073B	Glabrous	14,840b	27,180c	0	0.26
Stoneville 7A	Hirsute	75,392a	69,977a	2.7	11.9
IMAS 41-42	Glabrous	11,915b	21,122c	0	0

[a] Data from 1969 (10 counts) only, from Lukefahr et al. (1971). Original counts transformed for expression in metric system. Numbers followed by the same letter are not significantly different at 5 percent level (Duncan's multiple range test).

1963). The cotton fleahopper, *Pseudatomoscelis seriatus,* is more abundant on pubescent than on glabrous cotton, although glabrous types often sustain more damage than hairy varieties. Tolerance to fleahopper damage (but not resistance to infesting populations) seems to increase with trichome density and culminates in the extremely hairy 'Pilose' (Walker et al., 1974).

Pubescence in wheat greatly reduces cereal leaf beetle oviposition. Oviposition was reduced by 96.3 percent in the comparison of the resistant wheat 'CI 8519' with the susceptible 'Genesee' variety ('DI 12653') (Webster et al., 1973), and 99 percent in the comparison of resistant 'Vel' ('C 115890') versus susceptible ('CI 14425') varieties (Gallun et al., 1973). The susceptible wheat 'CI 12654' generally has fewer than four hairs/cm^2 and these hairs are very short. Some of the more resistant wheats may have 300 hairs/cm^2 (Schillinger and Gallun, 1968; Hoxie et al., 1975). Hoxie et al. (1975) used a system of nine wheat genotypes with six different pubescence profiles involving combinations of sparse (8 to 9 trichomes/mm^2), intermediate (10 to 12 trichomes/mm^2), and dense (23 to 33 trichomes/mm^2) pubescence consisting of short (0.05 to 0.07 mm), intermediate (0.10 to 0.12 mm), or long (0.23 to 0.28 mm) hairs. Figure 6, adapted from Howie et al. (1975), shows the varieties and lines (PI's and CI's) within each profile region with the corresponding number of eggs per plant (in preference tests) and percent survival of larvae. The data show that the varieties 'Hope' and 'CI 8519,' with dense and long pubescence, were highly resistant to both oviposition and larval feeding. Casagrande and Haynes (1976) studied the effect of pubescence on the population dynamics of the cereal leaf beetle on resistant 'Vel' and susceptible 'Genesee' wheat varieties in pure stands and in various mixtures. They observed that ovipositional preference for the glabrous 'Genesee' plants persisted even in mixed stands of the two varieties.

Pubescence as a Mechanical Barrier to Locomotion, Attachment, and Related Behavior. One of the most dramatic effects of certain nonglandular or glandular trichomes is the entrapment or impaling of arthropods that have the misfortune to alight on plants protected by these structures.

Certain varieties of bean plants possess hooked trichomes, and leafhoppers and other soft-bodied arthropods such as aphids long ago were observed impaled on these hooks (Poos and Smith, 1931; Johnson, 1953). However, the efficacy of these hair defenses was more fully revealed only by use of the electron microscope (Figure 7) (Pillemer and Tingey, 1976). Gilbert (1971), observing the effect of the hooked trichomes of *Passiflora adenopoda* on *Heliconius* spp. larvae, called them the "absolute defense."

However, each defense fails under certain conditions. Schneider (1944) observed that populations of *Aphis fabae* increased on *Phaseolus* because syrphid larvae predatory to aphids were caught by the hooked hairs and died.

Much of the work on glandular hairs was done with the Solanaceae. The purely mechanical effect of glandular trichome exudates is the entrapment and immobilization of small arthropods. The toxic effect of these exudates is

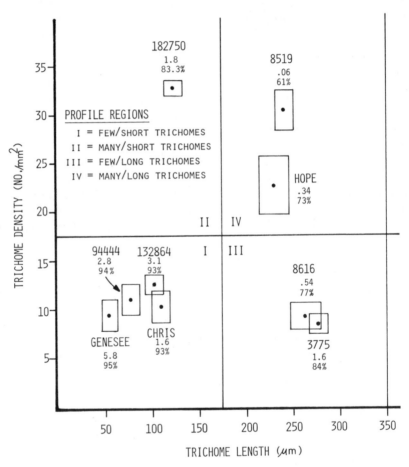

Figure 6. Varieties and lines (PI's and CI's) of wheat that fall within profile regions based on trichome density (number/mm²) and length (μm). Number of eggs per plant and percent survival of larvae of the cereal leaf beetle on each variety or line are given (adapted from Hoxie et al., 1975).

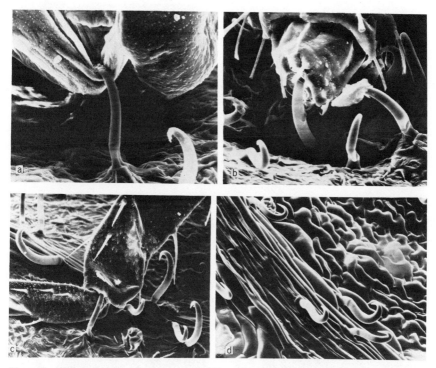

Figure 7.   Effect of hooked trichomes of field beans, *Phaseolus vulgaris,* on potato leafhopper, *Empoasca fabae,* nymphs. (*a*) trichome inserted in abdomen of a leafhopper nymph (×700); (*b*) trichome embedded in posterior of abdomen (×700); (*c*) trichome embedded in membranous tissue between leg segments (×350); (*d*) procumbent hooked trichomes of the lima bean cultivar 'Henderson Bush' (×350) (Pillemer and Tingey, 1976). Courtesy of E. A. Pillemer and W. M. Tingey, New York State College of Agriculture and Life Sciences, and the editors of *Science.*

discussed further in the next section. The wild potatoes *Solanum polyademium, S. berthaultii,* and *S. tariyense* have dense glandular trichomes. Gibson (1971) observed that when cell walls of these trichomes were ruptured by contact with the aphids *Myzus persicae* or *Macrosiphum euphorbiae,* a clear, water-soluble liquid oozed out and changed on contact with atmospheric oxygen into an insoluble black substance that hardened around the aphids' legs. The immobilized aphids quickly died. Four-lobed glandular hairs (Figure 8) on leaves and stems of *S. polyadenium* discharged a sticky substance on contact with larvae of Colorado potato beetle, *Leptinotarsa decemlineata.* This accumulated on the tarsi, immobilizing a few larvae, and causing others to fall off the plants (Gibson, 1976).

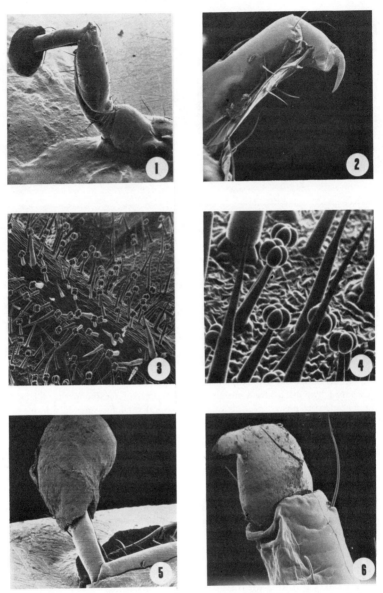

Figure 8. Four-lobed glandular hairs on leaves and stems of *Solanum polyadenium* which discharge a sticky exudate on contact: (1) tarsus of a *Leptinotarsa decemlineata* encased in a mass of the exudate, (2) terminal claw on leg of a larva (×97.5), (3) lower surface (×30), and (4) upper surface (×150) of *Solanum polyadenium* showing numerous four-lobed glandular hairs; (5) exudate-encased tibia of larva fed on an intact leaf (×56.5); and (6) tarsus of one fed on the wiped (i.e., relatively trichome-free) surface of a leaf (Gibson, 1976). Courtesy of R. W. Gibson and the editors of the *Annals of Applied Biology*.

The mechanical dislodgement of Mexican bean beetle, *Epilachna varivestis,* larvae from leaves of resistant pubescent soybean lines was observed by Van Duyn et al. (1972). In this case other resistance factors were also involved because susceptible plant lines with equally long and dense pubescence did not produce similar results. It is possible that the absence of adequate feeding stimuli or arrestants induced restlessness in the larvae, and they consequently fell off the plants.

Pubescence Associated with Allelochemical Factors.   The trichome vesture of several *Nicotiana* species exuded materials that produced toxic affects on aphids. The symptoms resembled those of nicotine poisoning— leg paralysis, loss of equilibrium, and death (Thurston and Webster, 1962). Some of these exudates were reported to contain nicotine, anabasine, and probably nornicotine (Thurston et al., 1966b). These *Nicotiana* species, as well as some *Petunia* leaf exudates, were also toxic on contact to young *Manduca sexta* larvae (Thurston et al., 1966a); toxicity was reduced when *Nicotiana* trichome exudates were removed by various washing methods (Thurston, 1970). It is interesting to note that *M. sexta* larvae are immune to *Nicotiana* alkaloids by ingestion (Self et al., 1964).

The action of glandular hairs demonstrates the close interaction of physical and biochemical plant defenses. Although repellent to many insects, glandular hairs serve as cues in host-plant finding for a few species that have evolved the mechanisms to utilize such resources. The diverse arthropod fauna associated with odorous pubescent plants is evidence of such adaptation.

Incrustation of Minerals in Cuticles

Deposits of silica are found in the epidermal walls of many plants, particularly among the Gramineae, Cyperaceae and Palmae. Deposits of calcium carbonate, in the form of cystoliths, are also found in outgrowths of some epidermal cell walls (Martin and Juniper, 1970). Calcified and silicified hairs exist on many plants (Uphof, 1962). There are several examples of resistance mechanisms related to the presence of silicate incrustations.

The mandibles of rice stem borers, *C. suppressalis,* feeding on resistant silicated rice were markedly worn. By using certain fertilizers Sasamoto (1957, 1958) promoted accelerated silication of rice. When offered a choice between silicated and normal rice, the stem borer larvae preferred the latter (Sasamoto, 1958). Higher infestations of the borer also were observed in rice grown on low-silicon than on high-silicon soils, and infestations were reduced by adding silicon to the low-silicon soils (Nakano et al., 1961). Yoshida et al. (1959) proposed that polymerized silicic acid in rice fills apertures in cellulose micelles of cell walls and forms a silicocellulose membrane. This would function as a defense against pathogens and insects. The ability to accumulate available soil silicon is characteristic of certain varieties of rice (Djamin and Pathak, 1967).

Surface Waxes

The cuticles of most vascular plants are covered with a thin layer of largely hydrophobic constituents. All substances of a "waxy" character isolated from a plant are considered under the term *wax*. However, wax chemically refers to an ester formed of a long-chain fatty acid and a high-molecular-weight aliphatic alcohol. Plant waxes vary from a fraction of a percent to several percent of the dry weight of a plant (Eglinton and Hamilton, 1963).

Cuticular waxes are rather complex mixtures including mainly *n*-alkanes (from C_7 to about C_{62}), branched or unbranched, saturated or unsaturated, as well as alcohols and acids. A detailed analysis of the surface waxes of some of the common plants is given by Martin and Juniper (1970).

The waxy coating of leaves function primarily in the mechanisms of water balance of the plant, but it also contains substances that inhibit pathogens and interfere with insect attacks. The effect of surface waxes of certain plants seems to be inhibitory to some herbivores and excitatory to others. Thus the normal waxy leaves of the sprouting broccoli, *Brassica oleracea* var. *italica,* are more resistant to attack by the cabbage flea beetle, *Phyllotreta albionica,* than a glossy-leaved mutant (Anstey and Moore, 1954). In contrast, the cabbage aphid, *Brevicoryne brassicae*, and the whitefly, *Aleurodes brassicae,* developed large colonies on normal waxy plants of narrow-stem kale, *Brassica oleracea* var. *acephala*, but did not colonize nonwaxy plants (Thompson, 1963). The difference in resistance to the aphid was attributed to the nature of the surface waxes. Twice as many alate aphids settled and produced progeny on the waxy as on the nonwaxy leaves. The surface waxes affected the rate of initial infestations, but once established on either plant type, aphids displayed no significant difference in reproductive rates (Thompson, 1967; in Martin and Juniper, 1970).

Another example of insect/plant systems influenced by the waxy nature of the cuticle is the resistance of *Rubus phoenicolasius* to the raspberry beetle, *Byturus tomentosus*. Leaves and flower buds of this *Rubus* species are heavily waxed, and the wax is richer in acidic substances than other *Rubus* species. This character was used in the breeding of raspberry plants that are unattractive to the aphid, *Amphorophora rubi* (Lupton, 1967). The efficacy of plant waxes as defenses against herbivores perhaps finds its ultimate expression in the roles they play in the mechanics of the insect-trapping plants of the genera *Byblis, Drosophyllum, Drosera,* and *Pinguicula* (see excellent summary in Martin and Juniper, 1970).

Anatomical Adaptations of Organs

Slight variations in the morphological structure of plants may result in altered fitness to herbivores. Quite often they alter the effectiveness of other factors causing mortality. Thus one of the resistance factors in sugarcane against sugarcane borer, *Diatraea saccharalis,* involves the retention of leaf sheaths. Varieties with leaf sheaths that remain intact accumulate water around the axil. Many young larvae drown in these small axillary pools. Other varieties shed lower leaves, eliminating this shelter for the young larvae (Mathes and Charpentier, 1963).

Frego is a mutant form of cotton, *Gossypium hirsutum,* in which bracts are narrow and twisted in contrast to the normally flattened bracts that more or less enclose flower buds and bolls. In this mutant the bracts leave the young flower bud or boll exposed. The presence of frego character in cotton is correlated with reduced survival of the bollworm, *Heliothis zea,* and resistance to the boll weevil, *Anthonomus grandis* (Lincoln and Waddle, 1966; Jenkins and Parrot, 1971). In Mississippi the frego character was very effective in boll weevil suppression when planted following a fall diapause program that kept overwintering populations low (Jenkins and Parrott, 1971). Stephens and Lee (1961) suggested that resistance in certain mutants of Upland cotton to the boll weevil was due partly to bracteoles of hairy plants being "sealed" during early stages of development and thus protecting temporarily the enclosed flower buds. However, Leigh et al. (1972) reported that *Lygus hesperus* was generally more abundant on frego-bract than normal-bract cotton. There was no significant effect on *Empoasca* spp., *Trichoplusia ni, Pectinophora gossypiella, Heliothis zea,* and several species of predators. They concluded that the character offered little promise under the conditions of the San Joaquin Valley, California.

Resistance in corn to *H. zea* is partially ascribed to characteristics of the husk such as length, tightness, and toughness. Some of these characters are associated with balling of the silks. Luckmann et al. (1964) showed that larvae fed heavily on silks of both susceptible and resistant varieties if pressure from the husks was reduced either by slitting the side of the husk tip or by permitting the silks to grow through a tube which reduced husk tightness.

Certain protective structures play a role in the resistance of grain in storage to stored-grain pests. The integrity of the husks of rice kernels is a key factor in the resistance of rice in storage to several coleopterous pests. Penetration into the grain occurs when small larvae find a gap between the lemma and palea of the husk (Link and Rossetto, 1972).

3. THE COMBINATION OF FACTORS

Seldom, if ever, is one factor responsible for the resistance observed in a plant. This is particularly unlikely if a complex of pests is involved. The picture is incomplete, but a few examples show that several factors interact in the process. Rice resistance to the stem borer, *Chilo suppressalis*, results from the interaction of leaf blades with a hairy upper surface, tight leaf-sheath wrapping, small stems with ridged surface, and thicker hypodermal layers (Patanakamjorn and Pathak, 1967). Resistance in cotton to the pink bollworm, *Pectinophora gossypiella*, involves absence of bracts, glabrous leaves, cell proliferation, high gossypol content, and nectariless character (Agarwal et al., 1976). Also in cotton, frego bract, red plant color, increased pubescence, and rapid fruit set provide resistance to the boll weevil, *Anthonomus grandis*; and smooth-leaved, nectariless, and high-gossypol plants offer resistance to *Heliothis* spp. (Lukefahr et al., 1966). Resistance in corn to the corn earworm, *H. zea*, has been ascribed to long husks, tight husks, blunt ear tips, flinty tip kernels, silk balling, and starchiness of kernels (Luckmann et al., 1964; Wiseman et al., 1972).

The analysis of isolated physical or chemical resistance factors provides only a partial explanation of the total syndrome manifested when an insect is exposed to a resistant plant. Resistance factors have complex complementary, synergistic, or antagonistic interactions with other plant and environmental factors. For instance, sorghum with open heads had fewer *H. zea* larvae then those with closed heads, and this probably was a consequence of larvae being more vulnerable to natural enemies (Doggett, 1964). Similarly, certain species of thrips escaped predation within the protection of dense leaf pubescence. Schuster et al. (1976) showed that smooth-leaf and frego-bract characteristics in cotton allowed *Orius insidiosus, Chrysopa* spp., and other important predators to prey effectively on pests of the boll.

There are also many possible interactions with abiotic factors. Certain plant characteristics, such as pubescence, have a profound effect on the microclimate around the plant. There are changes in reflectivity of light from pubescent surfaces which influence attractiveness to insects. Finally, the interaction among various plant species' defenses may function as an antiherbivore resource in ecological time. Monocultural cropping systems largely eliminate such defense. It is, however, conceivable that some economically practical species or varietal associations that enhance the interactions of defense processes may be developed. The use of trap crops is a present example of such man-manipulated strategies for plant resistance or protection.

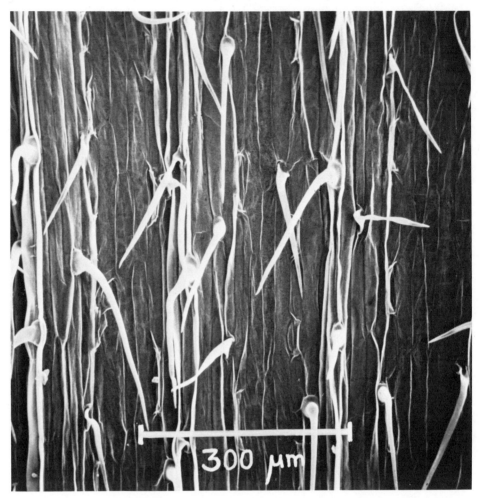

Trichomes, shown in this microphotograph, are a genetic factor responsible for resistance to oviposition on wheat by the cereal leaf beetle. Photo courtesy of USDA-SEA-AR.

4

GENETIC FACTORS AFFECTING EXPRESSION AND STABILITY OF RESISTANCE

R. L. Gallun

Science and Education Administration, U.S. Department of Agriculture, and Department of Entomology, Purdue University, West Lafayette, Indiana

G. S. Khush

The International Rice Research Institute, Manila, Philippines

1. INTRODUCTION

The expression and stability of resistance in a plant to an insect species depend on the genotype of the plant, the genotype of the insect, and the genetic interaction between the plant and insect under different environmental conditions. In this chapter we discuss only the genetic interrelationships between the plant and the insect. The environmental effects on both plant and insect are discussed in another chapter.

Since the beginning of time, man has used selection and other plant breeding methods to improve the crops on which he feeds. Moreover, man and insects have always competed for food and fiber and so have been constantly at war. However, insect resistance has only been bred into plants during the twentieth century, after man discovered that some plants have build-in protection. Even more recently, Painter (1951) and others, by their research and observations, opened the doors to crop improvements by breeding for insect resistance. Today, millions of hectares throughout the world are planted to insect-resistant cultivars.

Before reviewing present knowledge about the genetics of resistance, the genetics of parasitic ability, and the interrelationships among them, we define some terms commonly used in this chapter.

2. DEFINITION OF TERMS

2.1. Resistance

Resistance is the ability of the host plant to reduce the infestation or damage, or both, by an insect. Resistance levels vary from only slight plant defense against insects to almost total immunity. Resistance may be a result of one or more mechanisms.

1. *Mechanisms of resistance.* Mechanisms of resistance are generally classified into three categories: nonpreference, tolerance, and antibiosis (those terms are defined in Chapter 2).
2. *Types of resistance.* Two terms originally proposed by Van der Plank (1963, 1968) are used in plant disease literature to describe types of resistance. The terms are equally useful in plant insect literature. To encourage uniformity of terminology, we make a plea for their use in entomological literature.

   a. *Horizontal resistance.* This is used to describe the situation in which a series of different cultivars of the same crop infested with a series of different insect biotypes of the same species show no differential interaction. In other words, the level of resistance offered by a particular host cultivar is similar against all insect biotypes, and vice versa. Biotype nonspecific resistance or general resistance are other terms that are used. Generally, horizontal resistance is polygenically controlled and is considered to be stable and permanent.

   b. *Vertical resistance.* This term is used when a series of different cultivars of the same crop infested with a series of different insect biotypes of the same species show a differential interaction. In other words, some cultivars are classified as resistant and suffer less or no damage; others are susceptible when they are infested with the same insect biotype. Biotype specific resistance is another term. Vertical resistance is controlled by major genes or oligogenes and is considered less stable than horizontal resistance.
3. *Inheritance of resistance.* Resistance in plants can be described by three types of gene behavior.

   a. *Major genic or oligogenic resistance.* With this type of resistance genes show clear-cut and discrete segregation in the F_2 or later generations of crosses between resistant and susceptible parents. The effects are therefore qualitative. These genes are also referred to as vertical genes. If there is one oligogene for resistance, the inheritance is monogenic; if two, it is digenic.

b. *Minor genic or polygenic resistance.* With this type of resistance genes show continuous variation from susceptibility to resistance in the segregating populations of the crosses between resistant and susceptible parents. The effects are thus quantitative. Generally, a large number of genes are involved, each with a small contribution to total resistance. The level of resistance is generally low or moderate.

c. *Resistance governed by a combination of oligogenes and polygenes.* With this type of resistance, the polygenes reinforce the effect of the oligogenes and are considered "modifiers." The term multigenic is sometimes used in literature. However, multigenic can mean to 3a when two or more oligogenes are present or to 3b or 3c when polygenic resistance or combination of oligogenes and polygenes are present. Therefore it is vague and confusing and should be avoided.

4. *Gene expression in resistance.* Like other traits, resistance may be identified by variable gene expression as follows:

a. *Intra-allelic*

(1) *Recessive.* The F_1 hybrids from resistant and susceptible parents are susceptible.

(2) *Dominant.* The F_1 hybrids from resistant and susceptible parents are resistant.

(3) *Incompletely dominant.* The F_1 hybrids between resistant and susceptible parents are intermediate.

b. *Inter-allelic*

(1) *Complementary.* Two or more genes together govern the expression of the trait; one of them alone is ineffective.

(2) *Additive.* Two nonallelic genes affect the same character and enhance each other's effect.

(3) *Epistatic.* One gene inhibits the expression of another gene.

2.2. Parasitic Ability

Parasitic ability refers to the ability of an insect to survive at an expense to its host and may be vertical or horizontal. The term thus has the same meaning for insects as the term pathogenicity, which is used to denote the ability of a disease organism to attack its host. The parasitic ability is determined by the genotype of the parasite and that of its host.

1. *Genetics of parasitic ability.* The parasite has a genetic system that determines its ability to parasitize the host. It may have genes for virulence or avirulence.

a. *Virulent genes.* When a parasite is able to attack a host that has one or more genes for resistance, it has one or more virulent genes.

b. *Avirulent genes.* When the parasite is unable to attack a host with a resistance gene, it has an avirulent gene.

2. *Biotype.* The term biotype, as used in entomological literature, has the same meaning as the term race applied to pathogenic fungi. Both refer to individuals or a population of a species that is normally distinguished by criteria other than morphology including parasitic ability. In other words, one population of a species may belong to one biotype and another to a second biotype if there is a difference in the ability of the populations to parasitize a host.

3. GENETICS OF RESISTANCE IN CROP PLANTS

A great deal of literature has accumulated on inheritance of resistance to insects in crop plants. To protect a crop from the ravages of insect pests, entomologists identify the cultivars that resist insect attack; various screening techniques have been developed for this purpose. First, an insect population of significant density is required to ensure infestation. Plants can be evaluated in the field if there is high natural insect population (Figure 1); otherwise, laboratory-reared insects can be brought to the field. Plants can also be reared in the greenhouse and infested with field-collected or laboratory-reared insects. However, laboratory-reared insects are used more often because the variability for virulence within a natural insect population may minimize the value of the data for genetic analysis.

The determination of resistance may be based on plant injury or other symptoms of insect attack or on the reaction of the insects to the plant. Plant injury may be in the form of defoliated plants, as in the case of striped cucumber beetle on a squash (Nath and Hall, 1963); stunted plants, as with Hessian fly on wheat (Gallun, 1965); or death of the plant, as in "hopperburn" caused by brown planthopper in rice (Athwal et al., 1971). Insect reactions to resistant plants may be death (Hessian fly and wheat stem sawfly), reduced fecundity (spotted alfalfa aphid and greenbug), loss in weight (cereal leaf beetle), restlessness (aphids), a longer period between stages in the life cycle of the insect, or avoidance of the plant for feeding and/or ovipositing.

The interactions between the plant and insect are scored on the basis of the reaction of the plant to the insect and, less often, by the reaction of the insect to the plant. Generally, a scale is devised to classify plant reaction to the insect on the basis of damage, and hybrid populations from the crosses of resistant and susceptible parents are evaluated. The reactions of F_1, F_2,

Figure 1. Field reaction of resistant and susceptible selections to the brown planthopper. Susceptible cultivars 'IR8' and 'IR20' have been killed, but resistant selection 'IR1541-76-3-65' has suffered no visible damage (from Martinez and Khush, 1974).

and F_3 progenies are then used to determine whether the resistance is recessive, dominant, or incompletely dominant, or quantitative or qualitative (if the latter, the number of genes involved is determined). Once a gene for resistance is identified in a particular plant, that plant is crossed with other resistant plants, and the resistance of the hybrid progenies of these crosses is evaluated to determine whether the new hybrid has the same or different genes (Lakshminarayana and Khush, 1977). If two or more resistance genes are known to exist, their linkage relationships are determined as follows. First the chromosomal locations of resistance genes are determined by monosomic analysis as for Hessian fly in wheat (Gallun and Patterson, 1977); trisomic analysis, as for greenbug resistance in barley (Gardenhire et al., 1973); translocation stocks, as for earworm resistance in maize (Robertson and Walter, 1963); or substitution, as for cereal leaf beetle (Smith and Webster, 1973). Then after the resistance genes are assigned to chromosomes, they are mapped on the respective linkage groups. If the insect biotypes are available, they can be used to screen varieties for new resistance genes and to distinguish between known genes for resistance.

Because we are dealing with two biological systems in the genetic analysis for insect resistance, the following precautions are helpful to ensure the efficiency of analysis.

A genetically uniform population of the insect should be used.

The cultivars to be analyzed and the resistant and suscptible checks should be pure lines.

A suitable technique must be developed for mass rearing healthy insects for the test.

An efficient technique is essential in determining the plant–insect interactions. This technique must permit the evaluation of large volumes of segregating plant materials. Generally, a method that allows the determination of plant reaction to the insect by use of injury ratings is the simplest.

The test should be conducted under uniform environments so that resistance and susceptibility are clearly differentiated.

In studies of insect resistance, a few plants are generally misclassified. Some that are susceptible escape damage because the insect population is low and are therefore classified as resistant. Similarly, some that are resistant may be classified as susceptible because they are killed or show abnormal growth as a result of attacks of other pests, not the test insect. Such misclassifications can affect the F_2 segregation ratios, but the reactions of F_3 families can be used to determine the F_2 genotype. Thus F_3

data are more critical, and it is advisable to confirm the F_2 results from F_3 analysis.

Now that we have discussed the terms used in the literature to describe insect resistance, the procedures used to evaluate the resistance, and the techniques of genetic analysis, we are in a position to discuss specific examples of the genetics of insect resistance in crop plants. We shall attempt to summarize the present status of our knowledge about inheritance studies in major crops by citing key references. Nevertheless, this is not meant to be a comprehensive review.

3.1. Rice

More than 100 species of insects feed on the rice crop; about 20 are of major economic importance. Clear-cut cases of host resistance to at least 10 species have been recorded. Inheritance of resistance to four species has been investigated (Khush, 1977). Three of the species belong to the leaf hopper and planthopper group; the other is the rice gall midge.

1. Brown planthopper (*Nilaparvata lugens*).   This insect is one of the most serious insect pests of rice. Three biotypes are known in the Philippines and the fourth is found in southern Asia (India, Sri Lanka, and Bangladesh). Two genes for resistance, one dominant and the other recessive, were identified by Athwal et al. (1971) and named *Bph 1* and *bph 2*, respectively. Lakshminarayana and Khush (1977) analyzed 28 cultivars and identified two new genes. *Bph 3* is dominant and segregates independently of *Bph 1*. Similarly, *bph 4*, a recessive gene, segregates independently of *bph 2*. *Bph 1* and *bph 2* are closely linked (Athwal et al., 1971). Of the four genes for resistance, *Bph 1* conveys resistance to biotypes 1 and 3, and *Bph 3* and *bph 4* convey resistance to all four biotypes (Table 1).

2. Green leafhopper (*Nephotettix impicticeps*).   The inheritance of resistance to green leafhopper was first investigated by Athwal et al. (1971). Three dominant genes were identified and designated *Glh 1, Glh 2,* and *Glh 3.* These three genes segregate independently of each other. Thirteen more varieties were analyzed by Siwi and Khush (1977). All except one have single dominant genes for resistance. The cultivar Ptb 8 has a single recessive gene that was designated *glh 4.* The variety ASD 8 was found to possess a different dominant gene *Glh 5.* This gene segregates independently of the other three dominant genes.

3. White-backed planthopper (*Sogatella furcifera*).   A resistant cultivar, N22, from India was found to possess a single dominant gene for resistance (Khush and Sidhu, unpublished).

Table 1. Interrelationships between Biotypes of Brown
Planthopper and Genes for Resistance in Rice

	Plant Reaction[a]			
Gene	Biotype 1	Biotype 2	Biotype 3	Biotype 4
Bph 1	Res.	Sus.	Res.	Sus.
bph 2	Res.	Res.	Sus.	Sus.
Bph 3	Res.	Res.	Res.	Res.
bph 4	Res.	Res.	Res.	Res.

[a] Res. = resistant; Sus. = susceptible.

4. Rice gall midge *Pachydiplosis oryzae*.   Biotype variation has been
recorded in this insect. Some cultivars that are resistant in Indian are susceptible in Thailand, and vice versa. One cultivar, 'W1263,' has been the subject of several investigations but the results vary. Shastry et al. (1972) postulated two genes for resistance. Sastry and Prakasarao (1973) inferred the
presence of three genes. In another study. Satyanarayanaiah and Reddy
(1972) showed that a single dominant gene governs resistance.

3.2. Sorghum

Several insect species attack sorghum. Inheritance of resistance has been
investigated in four species: corn leaf aphid, greenbug, sorghum shoot fly,
and chinch bug.

1. Corn leaf aphid (*Rhopalosiphum maidis*).   Five biotypes of corn leaf
aphid are known (Painter and Pathak, 1962; Wilde and Feese, 1973), and
two have been used in genetic studies (Cartier and Painter, 1956). Antibiosis
suppressed the population of biotype KS1 on the resistant cultivar 'Piper
428-1' and was a dominant trait in F_2 and F_3 generations of a cross of the
resistant and a susceptible variety.

2. Greenbug (*Schizaphis graminum*).   Biotypes A and B of this insect have
been serious pests of wheat and barley. A new biotype (biotype C) feeds on
sorghum. Wood et al. (1969) identified resistant materials and Weibel et al.
(1972) and Starks et al. (1972) studied genetics of resistance. F_1 plants gave
an intermediate reaction. The reactions of F_2 populations indicated that a
single incompletely dominant gene governs resistance.

3. Shoot fly (*Atherigona varia soccata*).   Blum (1969) studied F_2 populations of eight crosses involving four parents resistant to shoot fly and two sus-

ceptible parents. The F_1's of all crosses were susceptible. Under high infestation the F_2 populations showed dominance of susceptibility. Starks et al. (1970) and Harwood et al. (1973) inferred that resistance to shoot fly was quantitative in nature.

4. Chinch bug (*Blissus leucopterus leucopterus*). Snelling and associates (Painter, 1951) investigated the inheritance of resistance to this insect. From the reaction of F_3 progenies of the cross between 'Sharon Kafir' (resistant) and dwarf yellow milo (susceptible), the presence of a single dominant gene for resistance was inferred though the result was not clear-cut. Dahms and Martin (1940) evaluated sorghum hybrids for resistance to chinch bug and found resistance to be dominant.

3.3. Wheat

Of the many insects of economic importance that attack wheat, resistance has been extensively studied in four: Hessian fly, greenbug, cereal leaf beetle, and wheat stem sawfly.

1. Hessian fly (*Mayetiola destructor*).   Eight genes for resistance to this insect have been identified; seven are dominant and one is recessive. Cartwright and Weibe (1936) and Noble and Suneson (1943) identified *H1* and *H2*; Caldwell et al. (1946) and Abdel-Malek et al. (1966) identified *H3*. The recessive gene *h4* was identified by Suneson and Noble (1950). *H5* and *H6* were identified by Shands and Cartwright (1953) and Allan et al. (1959), respectively. *H7* and *H8* were identified by Patterson and Gallun (1973).

2. Greenbug (*Schizaphis graminum*).   Painter and Peters (1956) suggested a single recessive gene for resistance in the variety 'Dickenson' selection 28A. Daniels and Porter (1958), Porter and Daniels (1963), and Abdel-Malek et al. (1966) confirmed those results. More recently E. E. Sebesta, USDA, Science and Education Administration (personal communication) successfully transferred a section of a rye chromosome into common wheat and found resistance to biotype C to be controlled by a single dominant gene. This is the only case known where a selection from an intergeneric cross has been usable for insect resistance.

3. Cereal leaf beetle (*Oulema melanopus*).   Resistance to the cereal leaf beetle in wheat occurs because of leaf pubescence, which adult insects avoid when ovipositing (Gallun et al., 1966). Ringlund and Everson (1968) studied crosses between five pubescent (resistant) and four nonpubescent (susceptible) wheat cultivars. Analysis of the pubescence in the F_1, F_2, and backcross progeny showed that this character was quantitatively inherited and that gene action was mainly additive.

4. Wheat stem sawfly (*Cephus cinctus*). Resistance in wheat to this insect is associated with solid rather than hollow stems which are common in susceptible wheats. McKenzie (1965) investigated the inheritance of resistance of solid stem in the cross of 'Red Bobs' (susceptible) and 'C.T. 715' (resistant). The data indicated that the parents differed at least in three genes that had unequal effect. The major influence was attributed to one gene. The allele of this gene for susceptibility when in homozygous condition was epistatic to the other two genes. The other two genes lacked dominance and were equal in their influence on the plant to sawfly attack.

3.4. Barley

Three common insects of wheat, Hessian fly, cereal leaf beetle, and greenbug also attack barley.

1. Hessian fly (*Mayetiola destructor*). The inheritance of resistance in the 'Delta' cultivar of barley was studied by Olembo et al. (1966) and found to be governed by a single dominant gene, *Hf*, at cooler temperature. When the plants were evaluated at higher temperature, complementary gene action was observed.

2. Cereal leaf beetle (*Oulema melanopus*). Hahn (1968) reported a recessive gene for tolerance to cereal leaf beetle in the barley variety CI66.

3. Greenbug (*Schizaphis graminum*). Gardenhire and Chada (1961) showed that a single dominant gene pair *Grb Grb* separated the resistant variety Omugi from susceptible varieties.

3.5. Corn

The genetics of resistance has been investigated for at least four insects of corn: European corn borer, corn earworm, western corn rootworm, and fall armyworm.

1. European corn borer (*Ostrinia nubilalis*). Marston (1930) first showed that the resistance of the maize variety 'Amargo' to corn borer was due to a single recessive gene. Schlosberg and Baker (1948) suggested that borer resistance was due to the cumulative effect of several genes. Singh (1953) proposed a two-gene hypothesis to explain the resistance of the corn cultivar 'A279.' Penny and Dicks (1956) studied the genetics of resistance in two crosses and postulated the presence of three genes for resistance as follows:

Cultivars 'N32' and 'MS1' both carry a major gene *A* for resistance.

'MS1' also carries two minor genes B and C that intensify resistance but have effect only in the presence of A.

'N32' carries only one of the intensifying genes.

No appreciable degree of dominance is expressed by any of these genes.

Patch et al., in 1942, reported that borer resistance was due to an undetermined number of factors and that a high degree of resistance was due to accumulation of large number of genes (Scott et al., 1964). Detailed analyses by Scott et al. (1964, 1965), Scott and Guthrie (1967), and Chiang and Hudson (1973) show beyond doubt that resistance to European corn borer is a quantitative trait.

2. Corn earworm (*Heliothes zea*).   Inheritance of resistance to corn earworm has been investigated by Robertson and Walter (1963). Widstrom and Hamm (1969), and Keaster et al. (1972). The results demonstrate that resistance is inherited quantitatively.

3. Western corn rootworm (*Diabrotica virgifera*).   Inheritance of resistance to feeding on silk by adult rootworms was investigated by Sifuentes and Painter (1964). A single recessive gene was found to govern resistance in one cultivar. Inheritance of resistance to root feeding by larvae is unknown.

4. Fall armyworm (*Spodoptera frugiferda*).   Widstrom et al. (1972) studied resistance among eight maize inbreds and their F_1 progeny. The mode of inheritance was quantitative.

3.6.   Cotton

In the United States, studies in cotton to determine the inheritance of resistance have been carried out for the boll weevil, thrips, and tobacco budworm. A comprehensive review of resistance to jassids, *Empoasca* spp., in cotton outside the United States is given by Painter (1951).

1. Boll weevil (*Anthonomus grandis*).   Resistance to boll weevil is associated with several morphological traits that are simply inherited. The best known is the monogenic recessive mutant frego bract. The mutant is characterized by relatively narrow, elongated, and twisted bracts that inhibit feeding and oviposition by weevils until population pressure becomes heavy (Hunter et al., 1965). Red plant color, imparted by the dominant gene $R1$, was one of the first morphological characters recognized as conferring boll weevil resistance; Isely (1928) worked with an intense red genotype and showed a distinct nonpreference by the weevil for red plants. More recent studies have shown that medium and light intensities of red

plant color also confer significant degrees of nonpreference (Reddy and Weaver, 1975; Glover et al., 1975). Varieties with a high degree of plant pubescence, controlled by the *H1* and *H2* genes, have long been known to possess significant resistance to boll weevil (Stephens and Lee, 1961; Hunter et al., 1965). This resistance generally is considered mechanical; that is, the enclosed flower bud is protected by interlocking of bracteoles by the trichomes. Also varieties with certain types of male sterility cause pronounced nonpreference of boll weevil. For example, Weaver (1974) observed a strong nonpreference for cytoplasmic male sterile cotton, and this association was confirmed by the later reports of Reddy and Weaver (1975) and Glover et al. (1975).

2. Thrips (*Thrips* spp.). Plant pilosity also influences resistance to thrips. Ramey (1962) studied the genetics of plant pubescence in upland cotton relative to thrip resistance. Three partially dominant genes affected pilosity (nonpreference by the insect). These genes were designated *H1* and *H2* (hairy) and *Sm* (glabrous). *H1* has been used in Africa to reduce attack by jassid.

3. Tobacco budworm (*Heliothis virescens*). Glands on cotton plants produce gossypol, which, when present in the plant, confers resistance to the tobacco budworm. Dominant alleles at *G12* and *G13* loci are responsible for the presence of glands and hence the resistance. Studies by Wilson and Lee (1971) showed that seedling damage to cotton was least and numbers of budworm larvae were lowest on plants of genotypes *G12 G12 G13 G13*, intermediate on *G12 G12 g13 g13* and *g12 g12 G13 G13*, and highest on *g12 g12 g13 g13*. Thus the two genes appear to act additively.

3.7. Trees and Fruits

The genetics of resistance to several insects that attack fruit trees has been investigated. A few examples are discussed.

1. Rosy leaf curling aphid of apple (*Dysaphis devecta*). High levels of resistance to this insect have been identified. Resistance is governed by a single dominant gene *Sd* in Cox's 'Orange Pippin.' The cultivars 'Grieve,' 'Northern Spy,' and 'Ashmeads Kernel' also possess single genes for resistance (Alston and Briggs, 1968).

2. Rosy apple aphid (*Dysaphis plantaginea*). Resistance to this aphid is governed by a single dominant gene designated Sm_h (Alston and Briggs, 1970). No biotypes have been identified.

3. The wooly apple aphid (*Eriosoma lanigerum*). This insect attacks apple, pear, hawthorn, mountain ash, and elm. Resistance in apple cultivars

has existed in heterozygous condition for more than 100 years (Painter, 1951). The variety 'Northern Spy' possesses a single dominant gene, Er, for resistance.

4. Rubus aphid ($Amphorophora\ rubi$).   This insect attacks the raspberry crop and transmits virus diseases. Numerous cases of host resistance to it have been reported. Schwartze and Huber (1937) reported that resistance of the cultivar 'Lloyd George' was controlled by two dominant genes. Existence of biotypes of the insect was first suggested by Dicker (1940) and Kronenberg and De Fluiter (1951). Knight et al. (1959) identified a dominant gene, A_1, in the resistant variety 'Baumforth A.'

Knight et al. (1960) described three strains of $A.\ rubi$ that differ in their ability to colonize on raspberries with different genetic constitutions. From that research, a number of genes were identified for resistance to rubus aphid. The American cultivar 'Chief' was shown to carry three dominant genes A_5, A_6, A_7 for resistance to strain 1. 'Chief' also carried the genes A_2, A_3, and A_4 for resistance to strain 2. A_1 is the dominant gene conferring full resistance by itself, and A_2 and A_3 are dominant complementaries since neither by itself has any effect on resistance. The gene A_1 from 'Baumforth A' that confers resistance to strains 1 and 3, when combined with gene A_3 also gives resistance to strain 2. A fourth strain of the aphid, reported by Briggs (1965), is able to develop on 'Malling Landmark,' which is resistant to strains 1 and 3. Keep et al. (1969) showed that resistance to all four strains in 'Klon 4A' from Germany was governed by a single dominant gene A_{K4A}. The resistance of $R.\ occidentalis$ L.L503 is controlled by a single dominant gene A_{L503}, which is probably identical to A_{10}.

3.8. Vegetables

The genetics of host resistance has been investigated for several insects of cucurbitaceous vegetables. Well documented cases are reviewed below.

1. The melon aphid ($Aphis\ gossypii$).   The inheritance of resistance to melon aphid in a muskmelon line was investigated by Kishaba et al. (1971) and Bohn et al. (1973). A single dominant gene for resistance was identified and designated Ag.

2. The striped cucumber beetle ($Acalymma\ vittatum$).   Genetic studies were carried out by Nath and Hall (1963) on the resistant squash cultivars 'Royal Acorn' and 'Early Golden Bush Scallop' and the susceptible 'Black Zucchini.' The F_1 hybrids showed some dominance, but all F_2 seedlings were resistant, indication that resistance is governed by several genes. Also,

backcross progenies did not segregate in a 1:1 ratio, indication of additive gene action.

3. The squash bug (*Anasa tristis*).   Studies by Benepol and Hall (1967) of this insect on the summer squash varieties 'Early Golden Bush Scallop' (resistant) and 'Black Zucchini' (susceptible) and their hybrid progeny showed that resistance is partially dominant over susceptibility. Additive variance was higher than dominance variance, which indicates quantitative inheritance.

4. Red pumpkin beetle (*Aulacophora foveicollis*).   Inheritance of resistance to this beetle in muskmelon was investigated by Vashistha and Choudhury (1974). A single dominant gene designated *Af* was found to confer resistance.

3.9.   Forages and Legumes

Many cases of host resistance to insects have been reported for forage and legume species. The genetics of resistance in a few of them has been investigated.

1. Sweetclover aphid (*Therioaphis riehmi*).   The genetics of resistance to sweetclover aphid in sweet clover was studied by Manglitz and Gorz (1968). On the basis of reactions of F_1, F_2, and backcross progeny from crosses of resistant and susceptible varieties, it was inferred that a single dominant gene conveyed resistance. An additional complementary gene appeared to be present in some resistant clovers.

2. The pea aphid of alfalfa (*Acyrthosiphon pisum*).   Inheritance of resistance to this insect was studied by Glover and Stanford (1966) in a resistant clone. One dominant gene was found to confer resistance. Several biotypes of this insect are known.

3. Spotted alfalfa aphid (*Therioaphis maculata*).   Glover and Melton (1966) studied polycrosses involving very resistant and very susceptible *Zea* clones and rated plants for resistance on the basis of the number of aphids present and amount of damage to the plants. The results showed the trait to be quantitatively inherited; similarity of parent and progeny performance indicated a reasonably high degree of heritability.

4.   GENETICS OF PARASITIC ABILITY IN INSECTS

As discussed in the preceding sections, the genetics of resistance has been investigated in numerous plant species. However, investigations of the genetics of parasitic ability in insects have been limited. Two important studies are those on Hessian fly of wheat and the rubus aphid of raspberry; only the Hessian fly–wheat study is discussed in this chapter.

When Hessian fly larvae feed on susceptible wheat plants, the plant leaves become stunted and turn dark green, and the new leaves generally fail to form. Resistant seedlings show some leaf stunting at first, but they generally recover and remain light green like uninfested plants. Larvae that feed on the resistant plants generally die. The small number that may survive remain small and do not stunt the seedlings. Thus virulent larvae (those that can survive upon and stunt the plants) can be distinguished from avirulent larvae (those that cannot) by reaction of the plant to them and by the reaction of the insect to plants of known genotype. In the case of the Hessian fly, biotype distinction is based on the virulence/avirulence of larvae to different wheats having known genes for resistance. One biotype may be virulent to a specific plant whereas another biotype may be avirulent. Thus virulence and avirulence are terms describing the insect's reaction to the plant, whereas resistance and susceptibility are terms describing the plant's reaction to the insect. A biotype's phenotype (A, B, C, etc.) can be determined by scoring plant reaction to larvae of the same progeny and the ability of the larvae to survive. Therefore, wheat seedlings of known genotype are grown and infested with a pair of flies. The female lays eggs at random, showing no host preference. The larvae can be scored as dead or alive 15 days after egg laying has ceased, when the lengths of the living larvae can be measured. The seedling reactions indicate progeny phenotypes.

When male Hessian flies form the sperm, the paternal haploid genome, that is, the chromosome set, is eliminated and only the maternal chromosomes are transmitted (Gallun and Hatchett, 1969). Consequently, F_2 and backcross progeny genotypes from crosses between adults of different biotypes vary according to the direction of the original cross between parents. Moreover, the F_1 males breed as if homozygous because they transmit only maternal chromosomes; the F_1 females breed as if heterozygous, showing normal transmission of both genomes.

Crosses between different biotypes of Hessian fly were studied by Hatchett and Gallun (1970). Table 2 shows the results of tests of F_1, F_2, and backcross progeny from intercrosses of biotypes GP and E tested on the variety 'Monon,' which carries a single dominant gene, H_3, that confers resistance to biotype GP but is ineffective against biotype E. All the F_1 flies were avirulent on 'Monon' and had the phenotype of GP; thus the virulence is recessive. Also reactions of the segregating F_2 and backcross progeny show that the difference between biotypes GP and E is caused by a single gene. As the table shows, in some crosses, no segregation occurred. This is

Table 2.   Phenotypes of Larvae from Progenies of Reciprocal Crosses between Hessian Fly Biotypes GP and E Determined on 'Monon' Wheat, which Carry the Incompletely Dominant Gene H_3 for Resistance (Hatchett and Gallun, 1970)

Cross	No. of Progenies	No. of Larvae	Percent Avirulent (GP)	Percent Virulent (E)	Ratio
F_1					
GP × E	39	3381	100	0	—
E × GP	36	2195	100	0	—
F_2					
(GP × E) × (GP × E)	35	2271	100	0	1:0
(GP × E) × (E × GP)	13	1351	50.8	49.2	1:1
(E × GP) × (E × GP)	27	1656	52.2	47.8	1:1
(E × GP) × (GP × E)	26	3066	100	0	1:0
Backcross					
(GP × E) × E	13	1329	40.2	59.8	1:1
E × (GP × E)	24	1132	100	0	1:0
(E × GP) × E	11	866	46.9	53.1	1:1
E × (E × GP)	28	2065	0	100	0:1

explained on the basis of elimination of the paternal genome of the male parent. From these genetic analyses it is concluded that biotype E of Hessian fly has one recessive gene for virulence that enables it to overcome the resistance conferred by gene *H3* in 'Monon.' This recessive gene for virulence has been designated *m*.

Five loci for virulence in Hessian fly are known. They have been designated *t, s, m, k,* and *a*. Eight biotypes of the Hessian fly have been identified (Gallun, 1977) and their genotypes have been determined by genetic analysis by using resistant varieties (Table 3).

5.   GENE-FOR-GENE CONCEPT

The gene-for-gene relationship was first expounded by Flor (1942) and then further elaborated (Flor, 1956) after it was formally defined by Person et al. (1952). It means that for every major gene for resistance in the host species, there is a corresponding matching gene for virulence in the parasite species. The host plant shows a resistant reaction if it has a resistance gene; the

Table 3. Genotypes of Eight Hessian Fly Biotypes
(Gallun, 1977)

Biotype	Wheat Varieties[a]				
	Turkey	Seneca	Monon	Knox 62	Abe
GP	*tt*	*S –*	*M –*	*K –*	*A –*
A	*tt*	*ss*	*M –*	*K –*	*A –*
B	*tt*	*ss*	*mm*	*K –*	*A –*
C	*tt*	*ss*	*M –*	*kk*	*A –*
D	*tt*	*ss*	*mm*	*kk*	*A –*
E	*tt*	*S –*	*mm*	*K –*	*A –*
F	*tt*	*S –*	*M –*	*kk*	*A –*
G	*tt*	*S –*	*mm*	*kk*	*A –*

[a] Symbols designate recessive and dominant alleles that represent virulence in the insect and susceptibility in the plant; and avirulence in the insect and resistance in the plant, respectively.

insect has an avirulent allele at the corresponding gene locus. If the insect has a virulent gene at the corresponding locus, however, the plant is susceptible. The gene-for-gene relationship has been called the matching gene theory.

This concept has been beautifully elucidated in wheat–Hessian fly. For example, four genes for virulence have been identified in the Hessian fly, and they correspond to four genes for resistance in the wheat plant. Therefore, the virulence of an insect biotype on each wheat variety is conditioned by homozygosity for a recessive virulent gene at a specific locus corresponding to a specific dominant gene of wheat that conditions resistance. A Hessian fly biotype can, therefore, be virulent to a specific wheat cultivar only if the biotype is homozygous for recessive virulent genes at all loci corresponding to loci at which the wheat plant has dominant alleles for resistance.

Specifically, the wheat variety 'Turkey' is susceptible to all fly biotypes because it has no genes for resistance. The varieties 'Seneca,' 'Monon,' 'Knox 62,' and 'Abe' are resistant to biotype GP because that biotype has at least one dominant allele for avirulence at loci corresponding to the loci in the wheat with dominant alleles that condition resistance (Table 3). Only when there is homozygosity for the recessive virulent gene at a locus in the biotype is the insect virulent to wheats that have at least one dominant allele at the corresponding locus for resistance. Thus biotype A, in addition to being virulent to 'Turkey,' is also virulent to 'Seneca' because it is homozygous for the virulent gene *s*, which corresponds to the dominant

Table 4. Five Wheat Varieties, the Resistance Genes Possessed by Them, the Eight Hessian Fly Biotypes with Virulence Genes Carried by Them, and the Insect–Plant Interactions

Wheat Variety and the Gene for Resistance	Hessian Fly Biotype and the Genes for Virulence[a]							
	GP (none)	A (*s*)	B (*s, m*)	C (*s, k*)	D (*s, m, k*)	E (*m*)	F (*k*)	G (*m, k*)
Turkey (none	Sus.	Sus.	Sus.	Sus.	Sus.	Sus.	Sus.	Sus.
Seneca (H_7 and H_8)	Res.	Sus.	Sus.	Sus.	Sus.	Res.	Res.	Res.
Monon (H_3)	Res.	Res.	Sus.	Res.	Sus.	Sus.	Res.	Sus.
Knox 62 (H_6)	Res.	Res.	Res.	Sus.	Sus.	Res.	Sus.	Sus.
Abe (H_5)	Res.	Res.	Res.	Res.	Res.	Res.	Res.	Res.

[a] Res. = resistant; Sus. = susceptible.

gene for resistance in 'Seneca.' However, it is avirulent to 'Monon,' 'Knox 62,' and 'Abe' because it has dominant avirulent alleles at the loci corresponding to the loci that carry genes for resistance in those varieties. Biotype D is homozygous at virulence loci *s, m,* and *k* and thus is virulent to varieties 'Seneca,' 'Monon,' and 'Knox 62' as well as 'Turkey.' But it is avirulent to 'Abe' because it has a dominant avirulent allele at the *a* locus, which corresponds to the resistance gene of 'Abe.'

The genes for virulence carried by eight known biotypes, the genes for resistance carried by five differential wheat varieties, and the resulting insect–plant interactions are reported in Table 4.

6. STABILITY OF RESISTANCE

During the evolution of a crop plant and the insects to which it is host, an equilibrum must be reached between the plant and the insect if both are to survive. The directions of such changes are necessarily opposed: toward greater resistance in the host and greater virulence in the parasite (Day, 1974). In other words, for a specific insect to survive on a resistant plant it must undergo a genetic change that gives it the necessary virulence to overcome this resistance. For the host plant to survive this genetic change in the insect, it must change genetically to overcome the virulencce of the insect. This coevolution has occurred in nature over the entire time the host and parasites have coexisted. Thus the host plants have accumulated genes for resistance and the insects have accumulated genes for virulence.

Today's breeders and entomologists collect varieties of crop plants from all over the world, evaluate them for resistance against local biotypes of the insects, identify resistant sources, and incorporate the resistance genes into new cultivars that are then used in commercial production. The resistant cultivars may maintain their resistance for long periods or may soon become susceptible because of the development of new biotypes of the insect.

Plant resistance governed by major or vertical genes is generally believed to be short-lived because with a single gene conditioning resistance, only a single gene mutation in the insect is required to overcome it. Thus cultivars of several crop species that were made resistant by the incorporation of major genes then influenced the development of biotypes virulent to those cultivars. We have already referred to the occurrence of biotypes in Hessian fly. The other insects in which biotypes have been identified are the greenbug, pea aphid, corn leaf aphid, spotted alfalfa aphid, cabbage aphid, grape phylloxera, chestnut gall wasp, rubus aphid, rice gall midge, and rice brown planthopper. In all these insects, the host resistance (where investigated) is under monogenic control, and biotypes arise because of the influence of host resistance. The sequence is as follows: plant breeders incorporate a major gene for resistance into cultivars. As a result, as soon as a substantial area is planted to the resistant cultivar, the insect population declines to negligible numbers. However, any individual insect with a virulent gene at the locus corresponding to the resistance gene of the host now has a selective advantage and can multiply rapidly. Thus there develops a new biotype that renders the resistant cultivar susceptible. We refer to the phenomenon as a breakdown of resistance.

The breakdown in resistance may occur after only a few years, as in brown planthopper resistance in rice, which may be an extreme example. All the rice cultivars grown in the Philippines were formerly susceptible. Then, in 1973, a serious outbreak of brown planthopper occurred, and the first brown planthopper-resistant cultivar, 'IR26,' was released. During 1975, more than a million hectares were planted to 'IR26.' However, a new planthopper biotype that was virulent to 'IR26' recently appeared in several areas of the Philippines (Khush, 1978). The same year, the new cultivars 'IR36' and 'IR38' were released, which are resistant to the new biotype because they have a different gene for resistance.

On the other hand, it is not uncommon for a resistant cultivar to maintain its resistance for 8 to 10 years, during which plant breeders can develop cultivars with different resistance genes. Moreover, when a number of vertical genes for resistance are available, they can be used at different times and locations to protect the crop. This is the case for the Hessian fly, one of the most successful programs of host resistance to control an insect through the use of vertical genes. About six vertical genes for resistance have been used in the United States to develop more than 30 resistant cultivars. In 1974 these were planted on approximately 20 million acres across the United States (Reitz and Hamlin, 1978).

Of course there are cases of major gene resistance remaining effective for relatively longer periods. The jassid-resistant cultivars of cotton developed in Africa have major resistance genes but they have remained resistant for the last 50 years. Similarly, some rice cultivars with vertical genes for resistance to green leafhopper have maintained that resistance for 40 years (Khush and Beachell, 1972).

Plant resistance governed by polygenic or horizontal resistance is considered to be more stable and longer lasting than vertical resistance. This type of resistance is biotype nonspecific and there is no gene-for-gene relationship, so there is very little danger that biotypes will develop. In fact, insect biotypes have not formed when polygenic resistance has been used, for example, European corn borer and sorghum shoot fly. Other insects for which polygenic resistance has been reported are corn earworm, fall armyworm, and squash bug. No biotypes of any of these insects have been reported.

7.   VERTICAL VERSUS HORIZONTAL RESISTANCE
IN HOST RESISTANCE PROGRAMS

7.1.   Vertical Resistance

Some authors, most notably Robinson (1976), have argued against the use of vertical resistance in the breeding of resistant crop cultivars and favor the use of horizontal resistance. However, vertical resistance has been used successfully to control insects in many programs, as we have seen. In fact, resistance to Hessian fly through the use of vertical genes has almost eradicated the insect population (Painter, 1968). Also, vertical resistance is easier to incorporate into new varieties and generally provides a high level of resistance. The only disadvantage is that it may trigger the selection of new biotypes that are able to attack the resistance cultivars. The solution is for entomologists and plant breeders to develop a dynamic program of collection and evaluation of germplasm so as to identify resistant donors and analyze them genetically. Genes that are identified must then be rapidly incorporated into breeding materials with desirable agronomic and quality characteristics. Then, when a new biotype of the insect appears, the varieties with new genes for resistance can be released.

Three strategies have been suggested for the management of vertical genes.

1. *Sequential release of cultivars with single genes for resistance.* This strategy is being followed in the wheat breeding program for Hessian fly resistance. A single gene for resistance is incorporated into a commercial variety, which is widely grown for several years. When biotypes virulent to this cultivar appear, a variety with a new gene for resistance is released.

2. *Pyramiding the vertical genes for resistance.* This strategy aims at combining two or more vertical genes for resistance into the same variety. It is being used in developing rice cultivars with resistance to brown planthopper (Khush, 1978), but there are differences of opinion about its usefulness. Nelson (1972) has hypothesized that cultivars with several vertical genes for resistance are resistant longer. Gallun (1977), on the other hand, has warned against this approach, arguing that insects will develop super biotypes to overcome the resistance so that several valuable genes will be lost at once.

3. *Development of multiline varieties.* This approach, originally suggested by Jensen (1952) and Borlaug (1958), aims at incorporating different genes into isogenic lines by backcrossing, mixing the lines in equal proportion, and releasing the result as a commercial multiline cultivar. If a component line of this cultivar becomes susceptible, it can be pulled out and replaced with another resistant line. The strategy was proposed as a way to control cereal rusts. Its use in insect control has not been explored.

7.2.   Horizontal Resistance

Horizontal resistance is generally of low level, has low heritability, and is difficult to work with. In the case of a plant such as corn, several parents with low levels of horizontal resistance are intercrossed in a diallel fashion. The progeny of those crosses are allowed to interbred with each other and mild selection pressure is applied in each generation. Outcrossing these generates new recombinants. Those recombinants with higher levels of resistance are saved and the cycle is repeated for several generations. Through this process of recurrent selection, enough polygenes for resistance are accumulated in a population to serve as a base for the development of a resistant cultivar. Therefore, genetic improvement under selection depends largely on the efficiency of the screening procedures employed and the heritability of the trait.

In the case of self-pollinated crops such as rice, on the other hand, special breeding methods are used to generate new favorable recombinants after the initial cycle of crosses between parents with low levels of horizontal resistance. For example, outcrossing can be induced in a self-pollinated crop by introducing male-sterility genes into a complete population. By recurrent selection, polygenes from different sources can be accumulated and the level of resistance can be raised. This technique is particularly useful when parents with adequate levels of host resistance are not available for a breeding program. If genes for male sterility are not available, a breeding method proposed by Jensen (1970) can be employed. This method, the diallel selective mating system, involves (1) selection of six to seven parents with low levels of resistance, (2) crossing these parents in a diallel fashion, (3) intercrossing the resulting F_1 hybrids in a diallel manner, and (4) screening the resulting double-cross progenies for resistance. Those progenies with better levels of resistance are intercrossed and the process of screening and intercrossing of selected progenies is continued until enough genes for resistance are accumulated. We are following this procedure in developing rice cultivars with resistance to stem borers (Khush, 1977).

8. SUMMARY

Resistance to insects in crop species is governed either by major (vertical) genes or by many polygenes (horizontal) that each have a small effect on the trait. Both genetic systems have been exploited in developing crop cultivars resistant to insects. Major genes generally convey high levels of resistance and are easy to incorporate into new cultivars. However, for every major gene for resistance in the host there is a corresponding gene for virulence in the insect. Therefore, where cultivars with major genes for resistance are grown widely, insect biotypes with correspond virulent genes may be selected. Thus vertical resistance may be less stable. Polygenic (or horizontal) resistance is biotype-nonspecific, is generally of moderate level, and is difficult to incorporate into improved cultivars of self-pollinated crops. However, it is considered more stable. Both types of resistance are important in the crop improvement programs, and strategies for their use are being developed.

Resistance experiments must include controls for environmental factors, which influence expression of resistance as well as growth and yield. Photo courtesy of CIMMYT.

5

ENVIRONMENTAL FACTORS INFLUENCING THE MAGNITUDE AND EXPRESSION OF RESISTANCE

Ward M. Tingey

Department of Entomology, Cornell University, Ithaca, New York

S. R. Singh

Grain Legume Improvement Program, International Institute of Tropical Agriculture, Ibadan, Nigeria

Climatic, edaphic, and cultural factors of the crop environment exert a powerful influence on arthropod–plant relationships including the relative phenomenon of plant resistance. Resistance mechanisms can be broadly placed by the degree to which they are affected by the environment (Kogan,

1975). Those mechanisms controlled primarily by inherited characters are classified in the category of "genetic resistance." They are subject to environmental influence but are not strictly controlled by the environment. Environmental factors influence both the magnitude and the expression of genetic resistance. In terms of magnitude, levels of insect and plant performance may be suppressed or enhanced, but the relative proportional relationship between resistant and susceptible germplasm is not significantly altered. In the case of expression, however, one or both classes of germplasm are affected to different degrees, resulting in significant variety–environment interaction. Analysis of such interaction is crucial in identification of environmentally stable resistance. A standard method for partitioning genetic versus environmental effects in analysis of heritability was discussed by Singh and Weaver (1972). We focus our discussion, whenever possible, on examples that demonstrate significant interactions, although such examples are relatively rare.

Other resistance phenomena are controlled to a lesser degree by gene action and are classified as "induced resistance." These phenomena are strongly influenced by the cropping environment. Cultural factors such as soil fertility, soil moisture, pesticides, and plant growth regulators affect nutritional quality of host plant tissues and appear to be particularly important in the induction of resistance.

Our principal objective in this chapter is to provide the reader with an appreciation of the importance of major edaphic, climatic, and cultural factors in the design of reliable and practical methods for assessment of resistant germplasm. Obviously, although not discussed here, numerous other variables of the physical and biotic environment including plant spatial patterns, alternate hosts, plant disease, crop rotations, competing pest species, and natural enemies, are potentially influential in the magnitude and expression of resistance.

1. TEMPERATURE

Environmental temperature is a major factor influencing fundamental plant and pest physiological processes including the magnitude and expression of resistance to insects. Either increased or reduced temperature can lead to loss of resistance. Temperature can modify the level and expression of genetic resistance in at least three ways: (1) through influence on plant physiological processes affecting host plant suitability, which indirectly alter biological performance of arthropod pests; (2) through direct influence on plant physiological and growth responses to pest feeding injury; and (3) through direct influence on behavioral and developmental biology of the pest. Most examples of temperature-induced effects on expression of genetic resistance in plants appear to involve the first two effects, either alone or in combination.

1.1. Loss of Resistance with Decreasing Temperature

Dahms and Painter (1940) were among the first to suggest that temperature modifies resistance to aphids in alfalfa. Later, Hackerott and Harvey (1959) studied the effects of several constant temperature regimes on expression of resistance in three alfalfa clones that varied in resistance to the spotted alfalfa aphid, *Therioaphis maculata.* These clones were initially selected on the basis of reduced aphid reproduction at 24°C. At 16°C, however, nymphal mortality was consistently less than at 27°C on two resistant clones compared to that on a susceptible clone. Thus clones resistant at 27°C supported limited populations at reduced temperatures. Similar results were reported by McMurtry (1962), who concluded that aphid performance on differentially resistant alfalfa clones was influenced by two principal factors: (1) an increase in reproductive rate with increasing temperature, and (2) an increase in expression of resistance with rising temperature. Through temperature mediation, the two processes acted antagonistically, although for the resistant clone 'C-84,' increased resistance was the dominant factor over all temperatures, resulting in a consistent decrease in populations. For the intermediately resistant clone 'C-902,' on the other hand, increased reproductive rate appeared to dominate up to 22°C, because rising temperature resulted in population increase. Later, Isaak et al. (1963, 1965) reported loss of resistance to both the spotted alfalfa aphid and the pea aphid, *Acyrthosiphon pisum,* at reduced temperatures in alfalfa, but they identified several clones with relatively temperature-insensitive resistance.

The influence of decreased temperature on the expression of alfalfa resistance to the spotted alfalfa aphid was further clarified by the work of Schalk et al. (1969) and Kindler and Staples (1970b). When given free choice among plants of several resistant and susceptible clones, *T. maculata* consistently selected susceptible clones at a constant temperature of 27°C. At a constant temperature of 10°C, however, the range in selection frequency between resistant and susceptible clones was much smaller. This effect was so pronounced that some clones, such as 'Nebraska 3309,' that were almost never selected at 27°C were selected as frequently as susceptible clones at 10°C (Schalk et al., 1969). Thus interpretation of plant influence on arthropod behavior should be made with consideration of possible temperature influence, as in the case of developmental resistance.

Significant loss in expression of resistance to the greenbug, *Schizaphis graminum,* in resistant sorghums also has been reported at reduced temperatures (Wood and Starks, 1972). Generally resistance to three biotypes, as measured by reproductive performance, was reduced at lower temperature, although resistance to biotype C was lost to a greater extent than to the others. In addition, reduced temperature was shown by Starks et al. (1973) to increase the selection frequency by biotype B for the resistant sorghums, 'Deer' and 'Piper,' whereas that for the susceptible sorghum 'OK-8' was relatively unaffected. By contrast, variation in environmental temperature did not appreciably alter the relative behavioral responses of biotype C to these genotypes.

1.2. Loss of Resistance with Increasing Temperature

Reduced expression of resistance to the Hessian fly, *Mayetiola destructor,* in wheat at high temperatures was first reported by Cartwright et al. (1946). Field observations by subsequent workers confirmed that resistance tended to decrease at constant temperatures greater than 18°C. Recently, Sosa and Foster (1976) described the influence of temperature on expression of resistance in four wheat cultivars ('Seneca,' 'Monon,' 'Knox 62,' 'Arthur 71'), differentially resistant to four races of the Hessian fly (Races B, C, D, 'Great Plains'). Generally, all four cultivars had less race-specific resistance with increasing temperature, as measured by percent infestation. The only exception was resistance of 'Arthur 71' to race C, which remained consistently strong at all temperatures. For several cultivars, race-specific resistance was relatively stable at all temperatures below 27°C. Significant temperature-induced loss of resistance was noted at 27°C to the 'Great Plains' race in all cultivars, and to race B in 'Knox 62.' Infestation levels in race-susceptible cultivars were generally unaffected by temperature.

Differential temperature effects also were observed after measurement of tillering in infested and noninfested genotypes. Generally, tillering of 'Arthur 71,' 'Monon,' and 'Seneca' was greater at higher temperature for all races except C, in which case tillering remained relatively constant. Tillering of 'Knox 62' was not appreciably affected by temperature, regardless of race. These observations suggest that tillering capacity of wheat may be an important component of tolerance, in compensating for temperature-induced loss of resistance.

These findings suggest complex interactions between temperature, infestation, plant growth, and races of the pest. The work reviewed above was not meant to define all interactions, but the results suggest that constant 27°C temperatures be considered for use in identifying sources of resistance stable at high temperatures.

1.3.   Effects of Alternating versus Constant Temperature

Measurements of pest resistance and assessment of resistance mechanisms are frequently made in laboratory environments at constant temperatures. However, under natural conditions plant and arthropod physiological processes are subject to diurnal, seasonal, and geographical fluctuations in temperature. Thus if significant temperature–cultivar interactions occur, results obtained at constant temperatures may vary from those observed in the field environment. An excellent example was presented by Kindler and Staples (1970b), who showed that fluctuating mean temperatures of 20, 22, and 24°C generally led to greater fecundity and nymphal survival of *Therioaphis maculata* on susceptible alfalfa clones than the same constant temperatures. By contrast, fluctuating temperatures affected aphid performance on resistant clones to a much smaller extent. Thus relative expression of resistance at each temperature regime was enhanced by use of fluctuating rather than constant temperatures. These findings suggest a practical advantage for germplasm evaluation at fluctuating temperatures, because of the somewhat broader range between aphid performance on resistant and susceptible clones. Fluctuating temperatures also simulate natural diurnal temperature changes.

1.4.   Duration of Temperature Influence and Effects on Plant versus Insect

Studies of temperature-induced changes on expression of resistance have frequently failed to differentiate direct effects on the pest from those mediated through the plant. This is an important consideration because environmental temperature has a direct influence on pest behavior and development, and may limit arthropod performance exclusive of genetic resistance factors. The effect of temperature-induced change on expression of resistance, as mediated solely through the host plant, was studied by McMurtry (1962) and Isaak et al. (1965). These workers exposed differentially resistant alfalfa clones to reversals of high–low and low–high temperatures prior to infestation by aphids. Thus the principal experimental variable subject to temperature influence was the host plant, because aphids were confined in similar physical environments during bioassays. For resistant clones, loss of resistance at 10°C and subsequent recovery of resistance at 35°C were observed within 2 to 6 days of conditioning at the opposing temperature. These results indicate that the primary effect of temperature on expression of resistance was mediated through the plant rather than directly on the insect. Furthermore, the relatively rapid shift in resistance suggests a physiological rather than morphological basis for this resistance.

2. LIGHT AND RELATIVE HUMIDITY

As with other variables of the physical environment, light quantity, quality, duration, and relative humidity can modify fundamental physiological processes of insects and their host plants. Of these four factors, low light intensity and high relative humidity have the greatest influence on expression and magnitude of plant resistance.

2.1. Loss of Resistance with Decreasing Light Intensity

One of the best examples of the influence of light on expression of resistance is that involving the wheat stem sawfly, *Cephus cinctus* in wheat. Stem solidness is the major plant factor conditioning resistance to this insect (Platt and Farstad, 1946; O'Keefe, et al. 1960). Increasing solidness of internodal tissues interferes with oviposition and with the ability of larvae to feed and tunnel, and leads to mortality of immature stages and decreased stem cutting. Platt (1941) was among the first to note a loss in resistance of plants grown in field cages or under greenhouse conditions compared to those grown in unshaded field plots. Holmes et al. (1960) and Roberts and Tyrrell (1961) showed that light intensity was a major factor conditioning expression of resistance through change in stem solidness. In the latter study the influence of light on resistance was studied using two cultivars of susceptible hollow stem wheat ('Thatcher,' 'Red Bobs'), two cultivars of susceptible durum wheat ('Melanopus,' 'Golden Ball'), and three cultivars of resistant solid stem wheat ('Rescue,' 'H4191,' 'H46146'). These workers found that stem solidness was generally reduced at decreased light intensities in the resistant solid stem wheats, while the other four varieties were unaffected. Thus the relative expression of resistance in solid stem varieties, as measured by stem cutting, and egg and larval mortality, was significantly reduced under shaded conditions. For greenhouse-grown plants of 'Rescue,' supplemental fluorescent and incandescent light prevented loss in expression of resistance at intensities of 4000 foot-candles, but not at 1500 foot-candles. Supplemental lighting in the ultraviolet wavelengths had no effect on stem solidness or expression of resistance. In addition, Holmes et al. (1960) showed that solidness of lower internodes was more important than that of upper internodes in resistance and that internodes 1 to 3 were subject to the greatest reduction in solidness by shading early in plant growth, that is, four-leaf-boot stage.

Reduced light intensity also affects expression of resistance to *Myzus persicae*, in sugar beet (Lowe, 1974). For plants grown at light intensities reduced about 75 percent from incident intensities in a greenhouse environment, fecundity and population levels of this aphid were significantly reduced on resistant and susceptible lines. However, the relative expression of resistance as measured by the ratio of aphid performance for susceptible/resistant genotypes increased with shading. Although the wider range in aphid performance on resistant and susceptible lines would appear to aid germplasm selection, the overall reduction in aphid populations had the opposite effect. That is, genotypes segregating at either end of the resistance–susceptibility continuum were readily identified, but those with intermediate levels of resistance were difficult to identify. Moreover, the reliability of the selection process was compromised at reduced light intensities, owing to the relatively greater influence of escapes and other sources of experimental error at small infestation levels.

Decreased light intensity influences tissue levels of specific chemical resistance factors in several different plant species. In potatoes, steroidal glycosides (glycoalkaloids) mediate resistance to the Colorado potato beetle, *Leptinotarsa decemlineata*, with increasing foliar concentration (Kuhn and Low, 1955; Schreiber, 1957). Pierzchalski and Werner (1958) showed that shading led to decreased concentrations of total glycoalkaloids (TGA) in foliage of several potato species differing in resistance to *L. decemlineata*. Their data suggest a genotype–light interaction because shading reduced TGA levels in a resistant accession of *Solanum chacoense* by about 59 percent on a fresh weight basis, compared to only 28 percent for a resistant accession of *S. demissum*. Although the dosage response of the Colorado potato beetle to glycoalkaloids is not well known, these results suggest a potentially serious loss in expression of resistance under shaded conditions.

A similar relationship has been reported for light intensity and levels of chemical resistance factors in maize to the European corn borer, *Ostrinia nubilalis*. Leaf-feeding resistance to this insect is conditioned by increasing tissue concentrations of benzoxazolinones and their precursor hydroxamic acids (Klun and Brindley, 1966; Klun et al., 1967). Although these compounds are present in etiolated tissue of seedlings grown in darkness, Loomis et al. (1957) and Virtanen et al. (1957) both reported greater levels of benzoxazolinones in seedling tissues as light intensity increased. These results suggest a potential loss in magnitude of leaf-feeding resistance with shading, although definitive bioassays to confirm such effects have apparently not been made.

2.2. Loss of Resistance with Increasing Relative Humidity

The influence of relative humidity on expression of resistance has been well documented for two stored-grain pests of sorghum, the maize weevil, *Sitophilus zeamais,* and the lesser rice weevil, *S. oryzae.* Rogers and Mills (1974b) studied the reaction of the former species on three differentially resistant sorghum cultivars at relative humidity regimes ranging from 43 to 71 percent. A significant cultivar–relative humidity interaction was due largely to the relatively stable resistance of 'Double Dwarf Early Shallu,' on which weevil fecundity increased to a much smaller extent with increased humidity, than that on 'Redlan' and 'Sugary Feterita.' Russell (1966) reported similar findings for *S. oryzae* on seed of four cultivars at relative humidity levels ranging from 58 to 83 percent. Adult longevity tended to increase sharply with greater relative humidity on all cultivars except 'RS-610,' on which adult longevity was consistently small and stable regardless of relative humidity.

3. SOIL FERTILITY AND MOISTURE

Management of soil fertility and moisture by application of fertilizers and irrigation water is an integral component of many cropping systems. Quantitative and qualitative variation in these environmental factors exerts dramatic influence on plant growth and development, frequently leading to alteration in the behavioral and nutritional suitability of plant tissue for phytophagous arthropods. In a few instances, the effects of soil fertility and moisture on levels and expression of genetic resistance have been studied, although most workers have tended to focus on induced effects without comparison of cultivars.

3.1. Soil Fertility

The role of mineral nutrients in fundamental plant physiological processes has been extensively studied and reviewed (Eaton, 1952; Pirson, 1955; Evans and Sorger, 1966). Several generalizations drawn from this body of knowledge bear upon the relationship between nitrogen (N), phosphorus (P), potassium (K), fertility, plant nutrition, and arthropod performance.

Depletion of soil nitrogen frequently leads to reduced concentrations of soluble amino acids and amides, whereas nitrogen excess tends to limit proteolysis, which also can lead to depleted levels of sap nitrogen. Potassium deficiency, on the other hand, has been associated with accumulation of soluble nitrogen and carbohydrates, owing to inhibition of protein synthesis and increased rates of proteolysis. Phosphorus deficiency also tends to increase soluble nitrogen levels, through inhibition of protein metabolism. Thus a common influence of variation in nitrogen, phosphorus, and potassium fertility lies with changes in levels of soluble nitrogen. Although the complex interactions associated with nitrogen, phosphorus, and potassium nutrition limit the value of generalizations, many arthropods tend to suffer from quantitative and qualitative depletion of soluble amino and amide nitrogen. This statement must be qualified, however, because phloem-feeding insects such as aphids, which depend largely on plant sap for nutrition, vary in response to nitrogen fertility of their hosts. In part, these differential responses depend on the host and aphid species, but considerable conflicting evidence still exists and remains unexplained.

Effects on Genetic Resistance

The influence of soil nutrients on resistance of alfalfa to *Therioaphis maculata* was studied by McMurtry (1962) and Kindler and Staples (1970a). In both studies, aphid performance was compared on differentially resistant clones subjected to various levels of macronutrients. McMurtry (1962) reported significantly greater survival and reproduction on resistant clones deficient in potassium, whereas phosphorus deficiency led to increased resistance. The relative expression of resistance was altered, for aphid performance on susceptible clones was unaffected. Kindler and Staples (1970a) observed, in addition, reduced expression of resistance in resistant clones treated with deficient levels of calcium or excess levels of magnesium and nitrogen.

Other studies have focused on the influence of nitrogen and phosphorus fertility on leaf-feeding resistance of maize to *Ostrinia nubilalis* (Scott et al., 1965; Cannon and Ortega, 1966). In the latter study, the relative expression of resistance was greater with incremental increases in nitrogen fertility from 10 to 300 ppm, owing to increased survival, greater leaf-feeding ratings, and more feeding lesions and tunnels on the susceptible hybrid 'WF9 × M14.' Performance on the resistant hybrid 'Oh43 × Oh51A' was unaffected. Scott et al. (1965) also observed stimulatory effects with increasing levels of nitrogen in other resistant and susceptible hybrids, although expression of resistance was not materially altered, because both classes were nearly equally affected at each fertility level. The influence of potassium fertility varied somewhat by year, but a trend for increased numbers of tunnels and greater leaf-feeding ratings was observed with increase in phosphorus from 0.5 to 10 ppm (Cannon and Ortega, 1966).

The influence of nitrogen fertility on levels and expression of genetic resistance has been studied for several other crops; results vary with the specific host and pest species. Greater feeding damage, larval survival, and larval weights of the yellow borer, *Tryporyza incertulas*, in rice were associated with increasing nitrogen levels, although resistant and susceptible cultivars were nearly equally affected at each fertility level (Manwan, 1976). Similar effects were observed by Chelliah and Subramanian (1972) for the rice gall midge, *Pachydiplosis oryzae*, on six rice cultivars. Singh and Sharma (1971), on the other hand, found no evidence of nitrogen influence on levels of resistance in sorghum to the sorghum stem fly, *Atherigona varia soccata*.

Induced Responses of Arthropods

Numerous studies of the influence of soil fertility on arthropod performance have been made without comparison of differentially resistant germplasm. In general, deficiencies in soil nitrogen frequently tend to limit growth, survival, and fecundity of insects and mites on a wide variety of crop plants including balsam fir, Brussels sprout, chrysanthemum, cotton, cucumber, peanut, lima bean, oat, potato, sorghum, strawberry, sugar beet, tobacco, and tomato (LeRoux, 1954; Coon, 1959; Morris, 1961; Henneberry, 1963; Cram, 1965; Waghray and Singh, 1965; Singh and Painter, 1965; van Emden, 1966b; Daniels et al., 1968; Legge and Palmer, 1968; Mistric, 1968; Markkula and Tittanen, 1969; Harrewijn, 1970; Rodriguez et al., 1970; Shaw and Little, 1972). Generalized statements regarding the influence of soil nitrogen on aphids must be qualified, however, because deficiencies also have been reported to enhance performance of the greenbug, pea aphid, green peach aphid, and the bean aphid, *Aphis fabae*, on various species of legumes and small grains (Arant and Jones, 1951; Barker and Tauber, 1951; Blickenstaff et al., 1954; Daniels, 1957; El-Tigani, 1962; Michel, 1963; Banks and Macaulay et al., 1964).

The influence of phosphorus and potassium fertility has been studied to a lesser extent, and varies considerably with crop and pest species. Survival, developmental rates, and fecundity of the migratory grasshopper, *Melanoplus bilituratus* were enhanced on wheat deficient in phosphorus (Smith, 1960), whereas those of the greenbug on wheat (Daniels et al., 1968) and the two-spotted spider mite, *Tetranychus urticae*, on lima bean (Henneberry, 1963), peach, cherry, and pome fruits (Harries, 1966) were adversely affected by deficient levels of phosphorus. Decreasing levels of potassium led to greater fecundity of the green peach aphid and cabbage aphid, *Brevicoryne brassicae*, on Brussels sprout (van Emden, 1966b), greater reproduction of the two-spotted spider mite on chrysanthemum, but limited fecundity of this pest on cucumber (Markkula and Tittanen, 1969).

The influence of soil nutrients other than nitrogen, phosphorus, and potassium on pests is mostly unknown, although Hagen and Anderson (1967) reported greater feeding by the western corn rootworm, *Diabrotica virgifera*, on zinc-deficient maize, whereas zinc excess in sweet corn led to decreased larval weight of the fall armyworm *Spodoptera frugiperda* (Wiseman et al., 1973).

Our knowledge of the interactions and physiological role of soil nutrients and plant nutrition for phytophagous arthropods is clearly only just beginning. Certainly, future studies focusing on plant tissue composition and nutritional responses of arthropods will contribute substantially to increased understanding. Expanded discussion of these relationships and numerous additional examples of the effect of soil fertility on arthropods can be found in papers by Rodriguez (1960), El-Tigani (1962), van Emden (1966a, 1969b), and Singh (1970).

3.2. Soil Moisture

The effects of soil moisture on the expression of genetic resistance are not well known, because most workers have restricted their studies to a single cultivar. Despite the absence of varietal comparisons, however, this work has shed considerable light on the nature of soil moisture-induced resistance and susceptibility. The most complete experimental understanding of these relationships has been developed from studies of the response of aphids to moisture stress.

Effects of Moisture Stress on Aphids

Aphid populations have been observed to increase, decrease, or remain unaffected on plants grown under moisture stress. Recent workers, particularly Kennedy et al. (1958), Kennedy and Booth (1959), and Wearing and van Emden (1967), however, have reconciled much of the ambiguity regarding the influence of moisture stress on aphids by emphasizing the action of two major interacting but opposing factors. First, moisture stress accelerates the breakdown and mobilization of leaf protein, thus enriching the amino and amide nitrogen content of phloem sap. In addition, water stress is associated with hydrolysis of starch and a subsequent rise in sucrose, particularly in older leaves (Gates, 1964). In general, levels of soluble nitrogen, sucrose, and aphid performance tend to increase together. Therefore, water stress can enhance performance if availability of nutrients has previously been limiting. Secondly, moisture stress leads to decreased cell pressure and increased viscosity of phloem sap, which can be detrimental to aphids dependent on sap pressure for optimum uptake of soluble nutrients (Weatherly et al., 1959; Auclair, 1963). Kennedy et al. (1958) and Kennedy and Booth (1959) suggested that the influence of sap enrichment would be dominant and beneficial to aphids during moderate plant moisture stress. During periods of continuous and severe water stress, on the other hand, decreased food uptake resulting from reduced turgor pressure and/or increased sap viscosity would dominate and limit aphid performance. Experimental evidence in support of these proposed mechanisms was provided by Wearing (1972), who reported increased fecundity of *Myzus persicae* and *Brevicoryne brassicae* on Brussels sprout subjected to intermittent water stress, whereas continuous moisture depletion was largely inhibitory, compared to performance on unstressed plants. These findings also support the observations of Kennedy and Booth (1959) and others that aphid outbreaks in natural environments frequently follow intermediate and periodic water stress. In practical terms, these findings suggest that maintenance of adequate soil moisture confers a relative degree of induced protection that may tend to limit or delay the buildup of aphid populations.

Effects of Moisture Stress on Other Arthropods

Water-stressed plants influence the biology of several other insects and spider mites in a manner similar to that reported for aphids. Rodriguez (1964) suggested that performance of spider mites on stressed hosts would be enhanced owing to greater accumulations of soluble carbohydrates and mineral ions. Singh (1975) reported greater infestations by *Sericothrips occipitalis,* on cowpea seedlings suffering water stress, compared to unstressed plants. Infestations of *Pachydiplosis oryzae* were also significantly greater in rice fields with inadequate water, compared to those of the same cultivars receiving optimum irrigation (Singh and Soenardi, 1973).

Larvae of some lepidopterous species appear to respond differently to plant moisture stress, depending on the relationship between plant water content and levels of nutrients and nondigestible factors. Scriber (1977), for example, showed that leaves of wild cherry low in water content limited growth and development of the cecropia moth, *Hyalophora cecropia,* compared to leaves with higher water content but similar in levels of fiber, total nitrogen, and caloric content. He suggested that leaf water affected larval growth primarily by restricting the efficiency of utilizing protein and carbohydrates. In a practical sense, these findings emphasize the need for cautious interpretation of laboratory and greenhouse studies in which plant water content is inadequately maintained or subject to extreme fluctuations.

4. PESTICIDES AND PLANT GROWTH REGULATORS

Agricultural chemicals are widely used in production of many crops, often without thorough consideration of possible influence on nontarget organisms. At rates normally used for crop protection or growth regulation these compounds are intended to be physiologically active only against target organisms. However, researchers should be aware of the possible influence of such chemicals on expression of resistance, because both stimulation and inhibition of arthropod populations have been observed following the use of many agricultural chemicals developed for purposes other than insect population regulation. The mechanisms advanced in explanation of stimulatory effects include (1) reduction in populations of natural enemies and competing species, (2) direct physiological stimulation of the arthropod, and (3) increased nutritional quality of the host plant. For inhibitory effects, the converse arguments have been used.

Though it is recognized that population shifts of pests caused by the first two mechanisms mentioned might influence the expression of genetic or induced resistance, the following discussion focuses primarily on the third, which involves a more direct relationship between the pest and its host plant. The literature contains relatively few references to the influence of agricultural chemicals on resistant and susceptible cultivars. In addition, most studies have focused on a single cultivar. With so few data available for chemical–cultivar interactions, generalizations presented in the following discussion must be interpreted with caution.

The chemical-induced effects described in this section have generally been attributed to alteration in nutritional quality of the host plant. For some examples of inhibitory effects, however, nonselective toxicity by contact or ingestion may have played a role. Only those studies in which direct chemical effects on natural enemies and competing species were excluded in the experimental design are discussed.

4.1. Herbicides

2,4-D

2,4-dichlorophenoxyacetic acid is one of the most widely used chemicals for control of broad-leafed weeds in cereal crops. One of the few studies of the influence of 2,4-D on the expression of genetic resistance was made by Gall and Dogger (1967). They described the effects on resistance of four resistant wheat cultivars ('Rescue,' 'Fortuna,' 'B50-18,' '51-3355') and one susceptible cultivar ('Selkirk') to the wheat stem sawfly, *Cephus cinctus*. Plants were treated by foliar application at a rate used for field weed control, seven days after exposure to egg laying. Larval mortality increased following treatment of 'Selkirk' and two resistant cultivars ('B50-18,' 'Rescue') but decreased on cultivar '51-3355.' No significant change was observed on 'Fortuna.' The magnitude of shift was greatest for 'Selkirk,' however, thus masking the relative expression of resistance in the other four cultivars. Generally, 2,4-D treatment did not alter the relative expression of resistance to stem infestation in the resistant cultivars, owing to the consistently large infestation of the susceptible cultivar, 'Selkirk.' Although these workers did not attempt to evaluate direct larval effects of 2,4-D versus those mediated through change in host plant condition, they failed to find any evidence of direct activity against adult sawflies or on eggs.

Maxwell and Harwood (1960) were among the first to study the influence of 2,4-D on insect development as mediated through change in host plant condition. In their studies, reproduction of *Acyrthosiphon pisum* on broad bean was consistently increased with foliar applications of 4.1 and 41.0 ppm made 24 hours before infestation. The magnitude of increase on treated compared to untreated plants ranged up to 450 percent and averaged 82 percent over both treatment rates and six dates of testing. Analyses of leaf tissues revealed large increases in levels of four amino acids (alanine, aspartic acid, glutamic acid, serine). Pooled levels of the four amino acids were 2.2-fold greater on treated compared to untreated plants. These workers suggested that the primary stimulation of pea aphid reproduction by 2,4-D was the result of quantitative increase in amino acids.

Adams and Drew (1969) likewise reported 2,4-D related increases in aphid populations for two pests of barley, *Macrosiphum avenae* and *Rhopalosiphum maidis*. For the latter species, treatment of test plants by 2,4-D lengthened the reproductive period as well. Interestingly, direct application of the herbicide to aphids tended to reduce populations of both species, indicating possible contact toxicity. Subsequently, Oka and Pimentel (1974) reported increased population levels of *R. maidis* on maize at 20 ppm, a rate equivalent to that commonly used for control of broadleaf weeds in maize.

Finally, 2,4-D has been shown to influence growth of the stem borer, *Chilo suppressalis,* on rice (Ishii and Hirano, 1963). This is one of the more complete attempts to determine the specific action of 2,4-D on insect development; larval growth responses were measured on (1) 2,4-D treated plants, and (2) sterilized diets of 2,4-D treated and untreated excised stems, with and without supplemental 2,4-D. Nitrogen and carbohydrate levels in treated and untreated plants were also assessed. In these studies, weights of larvae reared on plants treated with 0.1 percent 2,4-D were significantly greater (35 mg) than those reared on an untreated control (24 mg).

Moreover, larvae reared on excised treated stems were significantly larger (43 mg) than those reared on untreated stems with supplemental 2,4-D (38 mg) or without the chemical (36 mg). Analyses of stem tissue revealed a trend for greater protein levels in treated plants and stems compared to untreated controls. These results suggest that the growth increases of the stem borer associated with 2,4-D treatment were not due to the direct effect of the chemical on the insect, but rather to enhancement of nutritional quality through quantitative increase in protein nitrogen.

Other Herbicides

Several other herbicides, including amitrole (3-amino-1,2,4-triazole), barban (4-chloro-2-butynl N-3-chlorophenylcarbamate), dicamba (3,6-dichloroanisic acid), and MCPA (2-methyl-4-chlorophenoxyacetic acid), alter aphid populations on treated host plants (Robinson, 1960, 1961; Hintz and Schulz, 1969). The specific role of these compounds in aphid performance is not clear, although at least one, amitrole, is contact-toxic to the pea aphid (Robinson, 1961).

4.2. Chlorinated Hydrocarbon Insecticides

The indirect influence of chlorinated hydrocarbon insecticides on populations of the European red mite, *Panonychus ulmi,* and *Tetranychus urticae* through change in nutritional quality of the host plant, was studied by Rodriguez and co-workers. In a series of three papers (Rodriguez et al., 1957, 1960a, 1960b), they reported the effects of soil applications of DDT, BHC, dieldrin, and chlordane on mite populations, plant growth, and nutrient composition of apple, cotton, snap bean, and soybean. For 'Black Valentine' snap beans, 'Clark' soybeans, and 'Delta pine' cotton, the most dramatic effects on mite populations and foliar levels of nitrogen, potassium, and phosphorus were produced by BHC, followed in magnitude of effect by chlordane, dieldrin, and DDT. For example, addition of DDT to the soil of soybeans and snap beans generally tended to increase levels of total and reducing sugars, nitrogen, and population levels of *T. urticae,* while decreasing phosphorus and potassium levels. Thus populations of the two-spotted spider mite were positively correlated with sugar and nitrogen levels, and negatively correlated with phosphorus and potassium (Rodriguez et al., 1960b).

These workers did not attempt to evaluate direct chemical inhibition or stimulation of mite performance, although several of these compounds are now known to be absorbed by plant roots and translocated into aerial tissues (Finlayson and MacCarthy, 1965). Nevertheless, the striking changes in plant growth and nutrient composition of treated plants suggest a significant role for indirect mechanisms. For example, these chemicals can alter population levels of soil microorganisms (Cowley and Lichenstein, 1970) which could lead to changes in soil nutrient availability and subsequent uptake by the roots. Thus soil insecticide-associated changes in nutritional suitability of the host plant take on considerable practical significance in screening and development of genetic resistance to mites, considering the roles of carbohydrates and inorganic elements in mite nutrition (Watson, 1964; Rodriguez, 1969).

4.3. Plant Growth Retardants

Two plant growth retardants, (2-chloroethyl)trimethylammonium chloride (CCC) and (2,4-dichlorobenzyl)tributylphosphonium chloride (phosfon), used in commercial production of field and greenhouse crops to control growth and foliage color, are associated with induction of resistance to aphids and spider mites.

CCC

Interest in use of plant growth retardants for pest control was stimulated by the work of van Emden (1964), who showed that CCC treatment of Brussels sprout reduced second-generation fecundity of the cabbage aphid. Subsequently, this compound was reported to suppress fecundity or survival of four additional species (*Aphis fabae, A. nerii, A. varians, Myzus persicae*) on broad bean, oleander, black currant, and Brussels sprout, respectively (Tahori et al., 1965; Honeyborne, 1969; Smith, 1969; van Emden, 1969a).

The mechanism of action for CCC inhibition of aphid performance is not well understood. Honeyborne (1969) reported reduced survival when CCC was fed in a chemically defined diet to *A. fabae*; this suggested direct toxicity by ingestion. On the other hand, Worthing (1969), using treated diets, could not demonstrate toxicity to *M. persicae,* although fecundity and survival of aphids reared on treated chrysanthemums were significantly reduced. As for indirect action, Linser et al. (1965) reported decreased levels of amino acids in the soluble nitrogen fraction of treated wheat plants. These effects suggest that CCC-induced resistance may be due, at least in part, to quantitative changes in amino acid content of the host plant. Quantitative levels of amino acids are of great importance in the nutrition of several aphid species (Dadd and Mittler, 1965; Auclair, 1969) and in genetic resistance to aphids (Auclair et al., 1957). A somewhat different mode of action was discussed by Honeyborne (1969). Citing the findings of Shindy and Weaver (1967), who reported reduced translocation of photosynthate in CCC-treated grapevines, Honeyborne (1969) suggested that the chemical might induce resistance by decreasing the availability of nutrients in the phloem. Phloem elements and adjacent tissues are the principal feeding sites for most aphids.

Phosfon

The influence of phosfon on genetic resistance of chrysanthemum to *Myzus persicae* and *Tetranychus urticae* was reported by Worthing (1969). Spider mite populations were significantly reduced by phosfon treatment of the susceptible cultivar, 'Golden Princess Anne,' but those on the resistant cultivar, 'No. 4 Yellow Indianapolis,' were not affected, compared to untreated plants. Thus induced resistance of the former cultivar effectively masked the genetic resistance of 'No. 4 Yellow Indianapolis.' In fact, populations on treated susceptible plants were significantly smaller than those on treated plants of the resistant cultivar. Phosfon treatment also reduced nymphal survival of *M. persicae*, although both cultivars were equally affected. Survival of adults and nymphs was significantly depressed when the chemical was fed in diets at concentrations ranging from 10 to 320 ppm, suggesting action by direct ingestion toxicity. Change in host-plant nutritional suitability cannot be dismissed, however; Poole (1965) reported significant shifts in levels of amino acids and metal ions in foliage and stem tissues of treated chrysanthemum. The importance of these compounds in nutritional physiology of *M. persicae* was demonstrated by Dadd and Kreiger (1968).

Other Plant Growth Retardants

Two additional plant growth-regulating chemicals, maleic hydrazide (1,2-dihydropyridazine-3,6-dione) and SADH (succinic acid 2,2-dimethylhydrazide) alter insect populations on treated plants (Robinson, 1960; Tauber et al., 1971). The former compound is toxic to the pea aphid, as determined by feeding studies using artificial diet (Bhalla and Robinson, 1968).

4.4. Plant Growth Stimulants

Two plant growth stimulants, gibberellic acid (GA), and ethylenebisnitrourethane) (EBNU), reduce performance of *Aphis fabae, Tetranychus urticae,* and *Panonychus ulmi* on several host plants including apple, broad bean, and common bean.

Gibberellic acid

Eichmeier and Guyer (1960) observed greater than 80 percent reduction in fecundity of *T. urticae* reared on leaf disks of snap bean 'Tendergreen,' treated by foliar application, compared to that on untreated plants. Direct treatment of *T. urticae* by GA did not influence fecundity when the mites were subsequently reared on untreated plants. These results suggest that the effect of GA was mediated by ingestion toxicity or by change in host-plant suitability, as opposed to direct contact action. Subsequently, Rodriguez and Campbell (1961) reported similar results using the dry bean cultivar 'Dwarf Horticultural' and the apple cultivar 'Close.' For apple, foliar treatment with a commercial formulation, Gibrel® at 50, 100, and 500 ppm tended to decrease populations of *T. urticae*; infestation levels of *P. ulmi* were significantly depressed at 500 ppm compared to untreated trees. Levels of foliar nitrogen generally tended to decrease with increasing rates of GA. Thus populations of both mite species tended to respond in the same direction as nitrogen levels. Later, Honeyborne (1969) reported reduced fecundity of *Aphis fabae* on broad bean treated with 10, 75, and 500 ppm concentrations of GA. Moreover, adult survival was reduced when the chemical was fed to aphids using chemically defined diet, suggesting direct toxicity by ingestion. The influence of GA on levels of nutritional substances in broadbean tissues was not assessed; however, Selman and Kandiah (1971) observed changes in free amino acid levels of treated turnip leaves.

EBNU

Post-treatment effects of this plant growth stimulant on populations of *A. fabae* on broad bean were reported by Honeyborne (1969). Fecundity, adult weight, and levels of soluble plant nitrogen decreased following foliar application at 10,000 ppm, whereas the ratio of reducing sugar to soluble nitrogen increased. Because EBNU was not toxic when incorporated in diets, these results suggest indirect action through change in host nutritional quality. Increased ratio of reducing sugar to soluble nitrogen is associated with genetic resistance of peas to *Acyrthosiphon pisum* (Maltais and Auclair, 1957).

5. PHYSIOLOGICAL AGE AND OTHER PLANT VARIABLES

The suitability of plant tissues for use as food by insects and the plant's responses to insect feeding injury frequently vary with the physiological condition of plant tissues, and with specific organs or tissues. In addition, variation in physical growth characteristics such as plant height can influence host selection and subsequent population development. Because these biotic variables may be subject to manipulation in the experimental environment, their influence on levels or expression of resistance should be considered. Otherwise, results obtained under artificial conditions may not be consistent with those of the natural environment.

5.1. Physiological Age of Tissues

The use of plants and plant tissues at specific stages of development frequently is desirable because of convenience and economy. Assessment of genetic resistance at early growth stages, in particular, has been successfully used in breeding programs for several different crops, but may not be adaptable to all crops. For example, resistance of some maize genotypes to leaf feeding of *Ostrinia nubilalis* is associated with increasing concentration of 2,4-dihydroxy-7-methoxy-2H-1,4-benzoxazin-3(4H)-one (DIMBOA), a resistance factor which inhibits larval feeding and growth rates and leads to increased larval mortality (Klun et al., 1967). Most resistant and susceptible genotypes contain large levels of DIMBOA at early growth stages; however, tissue concentrations decline much more rapidly with increasing plant maturity on susceptible than on resistant germplasm (Klun and Robinson, 1969). Thus maximum expression of resistance is obtained during the mid-whorl stage of development, when the plant is normally subject to infestation by the first generation of *O. nubilalis.* Later, levels of DIMBOA in some resistant genotypes also decline significantly. Knowledge of the age-dependent expression of DIMBOA-mediated resistance is obviously critical in designing appropriate evaluation techniques. Assessment of plant material before or after the normal infestation phenology of the first generation would fail to identify some sources of resistance conditioned by DIMBOA, and possibly other resistance factors as well.

Expression of resistance in *Nicotiana* species to *Myzus persicae* also varies with plant age. Resistance is conditioned, in part, by glandular epidermal trichomes that exude several contact-toxic alkaloids (Thurston et al., 1966). Abernathy and Thurston (1969) showed that toxicity of resistant cultivars increased as the plants matured, although prior to 8 weeks of age, resistant and susceptible cultivars did not differ. From 9 to 18 weeks of age, however, aphid mortality increased from 30 to 81 percent on a resistant cultivar, 'TI 698,' whereas that on a susceptible cultivar, 'Kentucky 12,' remained the same. They attributed the change in level and expression of resistance to increased differentiation and development of glandular trichomes with plant maturity, and to the greater production of toxic secretions. Assessment of resistance at later stages of plant growth is clearly indicated in this case.

Age-related shifts in expression of resistance in alfalfa to *Therioaphis maculata* have been reported but the plant factors conditioning resistance remain unknown. Howe and Pesho (1960) observed greater expression of resistance with increasing age, as measured by plant survival. A significant cultivar–age interaction was due largely to greater survival with increasing age of two resistant cultivars, 'Moapa' and 'Lahontan,' compared to the susceptible cultivar, 'Caliverde,' which had consistently poor survival regardless of age. Even with a relatively short 9-day infestation beginning at 3 days of age, however, seedling mortality of 'Caliverde' was nearly 40 percent, compared to less than 5 percent for 'Moapa' and 'Lahontan.' Thus assessment at early growth stages would be effective in this case for identification of highly resistant germplasm, although that of intermediately resistant types might be overlooked. It should be noted that the increased survival of resistant plants at later growth stages may have been due, in part, to smaller infestation on these cultivars, although 'Lahontan' had relatively little plant mortality despite large aphid populations. Thus this work suggests that age-dependent vigor or tolerance contributed to the greater survival of 'Lahontan.' As for influence of tissue age on aphid performance, Kindler and Staples (1969) showed that the relative expression of resistance was reduced on mature compared to new growth, owing to decreased adult survival on susceptible clones. Survival on resistant clones did not vary with degree of tissue maturity. Thus impact of resistance on developmental biology of the spotted alfalfa aphid would be best assessed on relatively young tissues.

Numerous other observations of plant age-related shifts in insect performance and levels of resistance have been reported (Henneberry, 1962; Watson, 1964; Radcliffe and Chapman, 1965; Cibula et al., 1967; Morris, 1967; Kundu and Pant, 1968; van Emden and Bashford, 1969, 1971; Storms, 1969; Sutherland, 1969; de Wilde et al., 1969; Wearing, 1972a, 1972b). Generally, these phenomena have been attributed to alteration in host plant nutritional suitability. No general inferences as to influence on expression of genetic resistance can be made, however, owing to the absence of varietal comparisons or significant age–cultivar interactions.

5.2.   Plant Height

Varietal differences in plant height influence expression of resistance in cotton to oviposition by *Lygus hesperus* (Tingey and Leigh, 1974). Females preferentially selected the tallest genotypes for egg laying, but when plants were adjusted to equal height, the tallest genotypes were the least preferred. Cartier (1963) showed that tall pea cultivars had larger initial infestations of *Acyrthosiphon pisum* at early growth stages owing to greater interception of dispersing alates. At later growth stages, however, increased height was associated with reduced populations. He attributed the reversal to greater natural mortality brought about by sparser foliage and reduced sheltering. These examples emphasize the importance of plant height in free-choice assessment of resistance when plants are evaluated for insect responses conditioned by factors other than plant height.

5.3. Differential Organs and Tissues

The selection of plant parts and tissues for use in evaluation of resistance is another important consideration in the design of reliable screening methods. Kindler and Staples (1969), for example, reported greatest expression of behavioral disruption of the spotted alfalfa aphid on leaflets of resistant alfalfa clones, compared to petioles. Klun and Robinson (1969) found greatest concentrations of a European corn borer resistance factor (DIMBOA) in root, stalk, and whorl tissues of maize inbreds, although some resistant inbreds had large levels in leaves as well. Lowe (1967) reported smaller growth rates of *Aphis fabae* and *Aulacorthum solani* on basal portions of broad bean leaves compared to peripheral tissues. Growth rates and fecundity of *Myzus persicae* also differed on edges compared to more central tissues of broad bean leaves. In view of these effects, workers using plant parts other than those to which the insect is exposed during its major period of interaction with the plant should verify the consistency of their results with those obtained under natural conditions.

5.4. Excised versus Intact Tissues

Detached plant parts have been successfully used to assess resistance in a wide variety of host plants to spider mites, aphids, several lepidopterous pests, and other insects. Workers should be aware, however, that excision of plant parts can dramatically alter their chemical and structural properties through changes in transport function or induction of wound reactions. Such changes may influence the expression of resistance in detached tissues. Thomas et al. (1966), for example, observed shifts in expression of resistance associated with leaf excision in alfalfa clones previously selected for resistance to the spotted alfalfa aphid. In general, survival of adults and nymphs was significantly greater on excised than on intact trifoliolates. Moreover, clones differed in the duration of loss in resistance (Thomas and Sorensen, 1971). Similar results were reported by van Emden and Bashford (1976), who induced resistance and susceptibility of Brussels sprout to aphids by fertilization regimes of low nitrogen, high potassium and high nitrogen, low potassium, respectively. G enerally, growth rates of *Myzus persicae* and *Brevicoryne brassicae* were smaller on excised disks than on intact leaves of "susceptible" plants, whereas the reverse trend was observed on "resistant" plants. Thus use of detached tissues in these examples would tend to underestimate resistance. Workers should avoid exclusive use of detached parts in early cycles of evaluation when gene frequency and levels of resistance are sometimes low.

6. SUMMARY AND CONCLUSIONS

6.1. Temperature

The major deleterious effect of temperature on pest resistance is loss in expression at both low and high extremes depending on the crop and insect species. The best documented examples of the former involve aphids, but temperature-induced loss of resistance is generally reversible after a short period of conditioning at higher temperatures. Use of constant as opposed to fluctuating temperatures also tends to reduce the relative expression of resistance. Loss of resistance with increased temperature has been best documented for the Hessian fly in wheat, but the duration of temperature-induced loss and the effect of constant versus fluctuating temperatures are unknown. Initial assessment of alfalfa, sorghum, and wheat for resistance has frequently been made in laboratory environments using narrow temperature regimens. Since use of these methods may have contributed to the selection of temperature-sensitive resistance sources, several workers have proposed that screening and evaluation be made at either high or low temperature extremes. This strategy was suggested because most germplasm resistant to *Therioaphis maculata* and *Acyrthosiphon pisum* at low temperatures, and to *Mayetiola destructor* at high temperatures, has comparable or increased levels of resistance at the opposite temperature extreme. Workers should not discount the possibility, however, that consistent use of one temperature extreme to the exclusion of the other might lead to selection of temperature-sensitive resistance. Ideally, assessment of germplasm should be made over a range of temperatures to increase the chances for identification of temperature-stable sources of resistance. Because this strategy may not be feasible with regard to practical constraints on labor, time, and space, a reasonable compromise for laboratory or greenhouse evaluation is the use of daily temperature programs fluctuating between mean high and mean low temperatures. Such conditions more closely simulate field temperatures than the use of constant high or low temperatures, and may lead to greater success in selection of germplasm with relatively temperature-stable resistance.

6.2. Light Intensity and Relative Humidity

Reduction in light intensity of 75 percent or more than that of the field environment significantly alters the expression of insect resistance in several crop plants. The principal influence of shading on resistance appears to be through change in structural and chemical plant factors. In view of these effects, researchers should be cautious in the use of shaded environments for evaluation of resistance until the specific influence of light has been determined. Caged plants, and those grown in some types of rearing chambers or in glasshouses during periods of low insolation, are frequently subject to shading. Ideally, a range of intensities should be used in evaluation of resistance, at least in later stages of varietal development. The highest intensities selected should be consistent with those common to the cropping environment. Use of these procedures may promote identification of light-stable sources of resistance, as well as aid in determining specific effects of light on resistance. Metal halide lamps are an excellent source of supplemental light for simulation of field intensity and quality (Duke et al., 1975).

Similar cautions apply to the selection of environmental humidity regimens in evaluation of resistance in stored grains. Identification of humidity-stable sources of resistance will be aided by the use of a range in humidity. Finally, researchers should be aware that caging of plants and plant parts can markedly influence environmental temperature, as well as light intensity and relative humidity. These three variables must be alternately controlled through appropriate experimental procedures if accurate assessment of individual effects is desired.

6.3. Soil Fertility and Moisture

Variation in levels of soil nutrients can lead to altered levels and expression of genetic resistance in alfalfa and maize. In view of these effects and the widespread evidence of induced phenomena in other pest–host plant associations, it is clear that management of soil fertility is of considerable practical importance in varietal evaluation and development. Optimum crop growth and development in response to nitrogen and phosphorus fertility, in particular, occurs within relatively narrow limits, thus probably precluding practical manipulation of nitrogen and phosphorus levels for reduced pest susceptibility. In general, therefore, nutrient regimes selected for use in assessment and developmental breeding for genetic resistance should be comparable to those common to the cropping system. Certainly, excesses of nitrogen and phosphorus should be avoided in view of their association with enhanced pest performance. However, workers should consider the possibility of selecting germplasm that may be less affected by the stimulatory influence of supplemental fertilization. The phenomenon of induced resistance by manipulation of potassium fertility offers considerably more promise, in view of van Emden's (1966b) finding that aphid-inhibiting levels are frequently within the limits beneficial to crops, even in the presence of high nitrogen fertility.

Practical implications of soil moisture management on the expression of genetic resistance are less clear but use of a range in soil moisture levels consistent with those predominant in the cropping environment seems reasonable. In the case of stress-induced susceptibility, the feasibility of identifying germplasm relatively little affected by moisture deficits might be explored.

6.4. Pesticides and Plant Growth Regulators

In view of the striking and poorly defined effects of many agricultural chemicals on nontarget pests, the researcher is faced with a dilemma regarding their use in breeding for pest resistance. Because the effects vary with the chemical, the host plant, and the pest, generalizations are of questionable value. Some workers might argue that use of all chemicals be avoided in assessment of resistance until specific effects on pest performance have been determined. Certainly genetic resistance might be masked through use of stimulatory or inhibitory compounds. On the other hand, field assessment of resistance is often difficult or impossible unless nontarget pests can be reduced through use of "selective" compounds. In addition, commercial use of stimulatory products could lead to reduced field performance in genetically resistant germplasm by shifting pest population levels above economic thresholds. For crops with relatively low economic injury levels, such loss in magnitude of resistance could have serious economic impact, although compounds with less stimulatory action might be available for substitution in commercial practice.

Alternatively, other workers might argue that screening and development of genetic resistance is best accomplished through use of crop production chemicals, in the hope of identifying genotypes with relatively chemical-stable resistance. The feasibility of this strategy, however, should be determined after considering the usage patterns of popular crop production chemicals and the development of new compounds with potential commercial appeal. Ultimately, each worker must develop a strategy regarding use of crop production chemicals in breeding for genetic resistance without the benefit of general principles. Some factors for consideration in the decision include (1) previous reports of chemical-related stimulation or inhibition in specific plant–arthropod relationships, (2) evidence for chemically induced changes in plant factors involved in host selection, arthropod nutrition, or toxicity, (3) commercial longevity and extent of use of specific chemicals, and (4) availability and potential commercial acceptance of substitute compounds.

The situation with regard to chemicals involved in population inhibition must be viewed somewhat differently. Although resistance induced by nonspecific chemicals should be considered separately from genetic resistance, the former can complement the latter by reducing pest populations. The economic value of extreme population reduction is not immediately clear for crops with relatively high economic injury levels, although it may be considerable by reducing the rate of dispersal in pest populations, the magnitude of succeeding generations, and the spread of arthropod-vectored plant diseases. Thus some workers may wish to consider selection of germplasm genetically sensitive to chemical induction of resistance.

Finally, although many cases of chemically induced changes in pest performance appear to be associated with quantitative shifts in levels of plant chemicals, arthropod feeding injury to plant tissues can produce similar effects. In view of a rapidly accumulating body of evidence (Miles, 1968; Thielges, 1968; Lipetz, 1970; Green and Ryan, 1972; Bardner and Fletcher, 1974), future workers should be particularly cautious in attributing changes in plant chemistry to exogenous chemical influence, unless both pest-infested and noninfested tissues are analyzed.

In conclusion, it is clear that our knowledge of the influence of environment on genetic and induced plant resistance is just beginning. In view of the complexity of presently known interactions among a single host species, pest species, and one environmental factor, permutations of all possible environmental interactions seem almost limitless. These prospects make an appreciation of environmental influence in breeding for pest resistance essential, because such knowledge may facilitate (1) manipulation of the cropping environment for maximum levels and expression of resistance within the range of environmental variation considered optimum for other aspects of crop production, and (2) use of induced resistance as an adjunct to genetic resistance in management of arthropod pests. Induced resistance offers the option of external manipulation at desired magnitudes and intervals.

ACKNOWLEDGMENT

Thanks are due the following colleagues for their helpful reviews of the chapter: Dr. S. D. Kindler, Dr. J. G. Rodriguez, Dr. E. L. Sorensen, Dr. R. D. Sweet, and Dr. M. J. Tauber.

The cabbage looper attacks a variety of crops. Here it feeds on a soybean leaf. Photo courtesy of U.S. Department of Agriculture.

6

INSECT BEHAVIOR
AND PLANT RESISTANCE

Stanley D. Beck

Department of Entomology, University of Wisconsin, Madison, Wisconsin

L. M. Schoonhoven

Department of Animal Physiology, Agricultural University, Wageningen, The Netherlands

1. HAZARDS OF NEONATE SURVIVAL

The fecundity of any biological population tends to outstrip the long-term capacity of the habitat to support an expanding population. As a result, most neonate individuals face fearsome odds against their survival to the reproductive stage. Among insects, where the biotic potential is generally very high, the attrition rate normally exceeds 95 percent. Exceptions occur when a species is introduced into an environment that is partially devoid of the biotic and abiotic factors that normally hold the population at a stable equilibrium. These result in damaging outbreaks of the species, be it insect, vertebrate, plant, or microorganism. Agriculture tends to encourage the increase of insect populations to damaging levels by virtue of ecosystem disruption and the maintenance of artificially large host plant monocultures, frequently of genotypically uniform composition.

The neonate phytophagous insect is confronted by an array of factors inimical to its survival. Many of these factors lie outside the purview of the present discussion; these include the nonbiological density-independent influences of temperature, rainfall, soil type, and so on, as well as density-dependent biological factors such as disease, predation, and intraspecific competition. We are concerned here with the role of plant defense mechanisms, which are both biological and density-independent.

1.1. Plant Defenses

Even the most susceptible host plant of a given insect species is not defenseless, and only a small percentage of the feeding stages of the insect will survive. Numerous examples of this generalization have been reported, of which we consider only one. Many studies of the resistance of maize (*Zea mays*) to the European corn borer (*Ostrinia nubilalis*) have employed the inbred cultivar 'WF9' as the "standard susceptible" genetic line, and have compared other genetic lines to it for resistance. Under protective laboratory conditions, more than 80 percent of the newly hatched borer larvae succumbed within 6 days when reared on seedlings of 'WF9' (Beck and Lilly, 1949). These investigators also found that an age-related increase in susceptibility to the borer occurred only slightly more rapidly in 'WF9' than in the more resistant inbred lines tested. Using somewhat older plants than those employed by Beck and Lilly, Tingey et al. (1975) observed 21-day mortalities of 60 percent on 'WF9' and up to 100 percent on the most resistant maize genotypes.

From a dietetic standpoint, host plants are generally inferior to well-balanced, nutritionally complete laboratory culturing media. Several species of phytophagous Lepidoptera have been found to grow faster, to a larger body weight, and with better fecundity and longevity on artifical dietary media than on host plant tissues (Beck, 1974). Such media are devoid of physical and chemical plant defense factors, but may also lack some sensory factors, such as attractants and stimulants, that may be important to survival under natural conditions.

1.2. The Agricultural Problem

There is a distinctly precarious balance among the factors determining insect survival and successful utilization of a host plant. An agricultural cultivar that is rated as being resistant may differ from the more susceptible cultivars to only minor degrees in respect to plant defenses. Any characteristic that tips the precarious balance against host-plant utilization, even very slightly, may prove to be quite significant from the standpoint of agriculturally practicable host-plant resistance.

In this chapter we discuss the behavioral adaptations of insects, particularly as related to the utilization of their host plants. To attain a reasonable understanding of these complex phenomena, we must also consider some aspects of the evolutionary trends in the adaptive interactions between plants and insects, and the significance of these evolutionary strategies to the agricultural goal of the creation of crop cultivars that display effective, stable resistance to insect pests.

2. HOST-PLANT SPECIFICITY

2.1. Classes of Specificity

Of the million or so species of insects, approximately one-half are plant-feeding forms (Brues, 1946). Obviously, no one plant species is susceptible to attack by all phytophagous insect species, and no one insect is capable of utilizing all plants as hosts. Nevertheless, a very broad range of host-plant specificities is apparent. The most generalized feeders are species of locusts (e.g., *Schistocerca gregaria, Locusta migratoria*); the most specialized are a few species that are known to utilize only specific parts of a single plant species (e.g., northern corn rootworm, *Diabrotica longicornis*).

Traditionally, phytophagous insects have been divided into three categories, depending on the specificity of their host plant range: *monophagous* species are restricted to plants of a single species or at most a few closely related species; *oligophagous* insects feed on plants within one family or members of closely related families; and *polyphagous* insects utilize host plants from more that one botanical order. These terms have all the shortcomings attending any arbitrary divisions within a continuum. Host plant specificity can be further subdivided in terms of the plant parts being utilized (leaf miners, stalk borers, root maggots, etc.) and also by differences in the feeding specificities of the larval and adult stages of the insect. For example, the northern corn rootworm was cited above as an example of extreme monophagy, but that applies only to the larvae; the adults are polyphagous. There is also a phenological dimension involved in host-plant specificity. Different insects species utilize a given plant, or plant tissues, at different times of the year or at different developmental stages of the plant. For example, jack pine sawflies (*Neodiprion swainei* and *N. rugifrons*) feed on only the older pine needles; young needles contain a feeding deterrent (All and Benjamin, 1975; Ikeda et al., 1977). Conversely, larvae of the winter moth (*Operophtera brumata*) feed on only the very young leaves of their oak host; the mature leaves contain tannins that render them poorly digestible (Feeny, 1970).

2.2. Evolutionary Strategies

The host specificities of phytophagous insects are thus seen as a cluster of multidimensional spectra separating the insect populations in space and time. The overall pattern appears to suggest that mutational changes in the host range of many groups of insects have resulted in sufficient isolation for sympatric speciation. This diversification of host specificities has allowed partitioning of resources and exploitation of a large constellation of ecological niches while minimizing interspecific competition. These considerations might appear to support the hypothesis that the evolution of host specificity has been from polyphagy toward monophagy; that is, toward an increasing specialization of host-plant specificity (Dethier, 1954; Feeny, 1975). There is a real question, however, concerning the lengths to which this concept can be applied. That is, at what point does monophagy become disadvantageous? And what are the physiological, ecological, and evolutionary advantages and disadvantages of a high degree of polyphagy?

Broadly polyphagous insects, such as grasshoppers and cutworms, are able to move readily from plant to plant, and do not have host plants in the strict sense, but rather food plants. This may mean that the selective pressures on polyphagous species are less plant-related than are insects having greater degrees of host specificity. In the polyphagous forms, sensory discrimination of acceptable plants must be relatively unspecialized, and the metabolic machinery for degrading deleterious plant chemicals must be highly developed and broad in its range of capabilities. The energy costs of such metabolic capability should be relatively high (Feeny, 1975). A plant mutant in which a novel defense chemical—deterrent or toxin—is present would be eaten readily by the phytophagous insect only if the chemical falls within the scope of the insect's preadaptation. If it does not, the plant would be excluded from the insect's food plant range—unless the insect population has evolved the sensory and metabolic capability (with associated added metabolic costs) needed to utilize the mutant foodplant. Unless the mutant plant became dominant or highly prevalent over an extended area, it seems unlikely that there would be sufficient selective pressure to result in adaptation of the insect population. In the absence of such selective pressure, the food plant range of the polyphagous insect would become slightly more restricted than it had been previously. It is also apparent that polyphagy does not take advantage of resource partitioning strategies, and the polyphagous insects are in direct competition with all other phytophagous insects and other herbivores. By these lines of reasoning, extreme polyphagy would be difficult, perhaps impossible, to maintain over a long evolutionary period, and the trend should be toward a limited polyphagy or oligophagy.

Specialization in the form of extreme monophagy also appears to entail some evolutionary disadvantages, except perhaps in those cases where the host plant is a perennial climax species with high population densities. Otherwise, fluctuations of the plant population would have deleterious or disastrous effects on the insect population. It seems likely that the ability to utilize some additional related plant species occurring in the same general ecosystem would be of great survival value, and would be selected for in most situations. These considerations suggest greater evolutionary emphasis on host-plant specificities approaching a limited degree of oligophagy than on greater specializations in monophagy.

In view of the relative rarity of extreme monophagy and indiscriminate polyphagy, it seems reasonable to conclude that the evolutionary trend in insect host-plant specificity has been toward moderate but variable degrees of oligophagy. Such a description fits the specificities of most extant phytophagous insects (see Ehrlich and Raven, 1964; Dethier, 1970). The terms "generalist" and "specialist" in some recent literature describe the ranges of host-plant specificities observed within the very broad category traditionally designated as "oligophagous" (Feeny, 1976; Rhoades and Cates, 1976).

3. HOST-PLANT SELECTION

The term "host-plant specificity" refers to the range of plant species on which a given insect is known to occur in nature. "Host-plant selection" is the behavioral sequence by which an insect distinguishes between host and nonhost plants. A third term—host-plant preference—is also behavioral, and is used to describe the insect's predilection to select some plants in preference to others, within its host-plant range. The European corn borer, *Ostrinia nubilalis*, illustrates these concepts. The borer is not specific to maize, but is able to utilize a number of other plants. Host selection is a function of the female moth, as she will deposit eggs on maize, gladiolus, green pepper, or even potato, depending on the availability of these alternatives. However, in the presence of these several utilizable hosts, eggs will be deposited almost exclusively on maize, which is therefore the preferred host plant.

3.1   Behavior: A Sequential Activity

A functional behavior such as host selection is composed of a sequence of simpler behavioral responses. Each activity in the sequence brings the animal into a situation in which an appropriate stimulus will release the next activity. Such temporal patterns of behavioral components are associated with internal "drives," such as the urge to oviposit or the need to feed. Hungry leafhoppers, for example, locate and feed on their host plants by means of a short sequence of stimulus/reponse processes. Their flight approach to the plant is stimulated by its color; alighting is triggered by olfactory stimuli; probing the tissue with the proboscis is in response to foliage color and contact stimuli; tissue acidity stimuli guide the proboscis to the phloem; and gustatory stimuli in the phloem sap stimulate continuous feeding (Nuorteva, 1952).

The behavioral sequences involved in host selection are sometimes very complex. The cabbage root fly, *Hylemya brassicae*, has been reported to go through an elaborate series of behaviors during its host selection for oviposition (Zohren, 1968). This sequence includes (1) flight to the plant (induced by visual and probably olfactory stimuli); (2) landing on leaf surface (visual and olfactory stimuli); (3) walking on leaf surface (contact chemical stimuli); (4) walking along stem (tactile stimuli from leaf veins and stem); (5) circumventive walk around plant base and soil; (6) extrusion of ovipositor (tactile and possibly olfactory stimuli); (7) arrest of locomotion and introduction of ovipositor into soil spaces (tactile stimuli); (8) digging with hind legs; and (9) oviposition (stimuli include soil moisture, light intensity, and tactile factors).

Sequentially patterned activity is a common phenomenon, and appears to function as a checklist of characteristics identifying a host plant that is acceptable for oviposition and/or feeding. On landing on a green surface, an aphid tests the substrate with its proboscis; if this activity is not followed by the detection of appropriate chemical stimuli, the aphid withdraws its proboscis. It then resumes flight until it touches down on another green substrate, where the testing procedure is repeated. When a behavioral sequence is broken because the releasing stimulus for the next activity was not received, the insect falls back to an earlier phase of the behavioral pattern. This system of passing a number of "checkpoints," each with a different cue, minimizes the chance of making an inappropriate decision, for example, ovipositing on a nonhost plant. In this system, the nature of the adequate stimulus (the "sign stimulus") for each step can be relatively simple. Thus the insect has to make a series of simple decisions, instead of one complex decision.

3.2 Dispersal and Search

Locomotor activity may be the manifestation of two different drives, serving somewhat different functions. Dispersal may lead to a more homogeneous distribution of a given insect population and to the invasion of new areas. Search behavior, on the other hand, increases the chance of encountering stimuli initiating the behavioral concatenation culminating in either oviposition or feeding.

Dispersal activity is typical of the adult insect, but is by no means limited to the adult stage. Many neonate larvae disperse from the hatching site, and wander widely before becoming receptive to the stimuli evoking settling down and feeding. Although dispersal and search activities are often difficult to distinguish, they differ in their underlying drives and in the insect's reactions to given stimuli. For example, alate aphids may react negatively to green foliage and positively to blue sky; they then take off and fly for hours in a dispersal flight. When the behavior changes to that of search behavior, they are attracted to green surfaces, and they alight on any green plant. If the plant on which the aphid lands proves to be nonhost, the aphid reverts to the dispersal behavior and flight ensues. This alteration between dispersal and search behavior continues until an acceptable host plant is located. Similarly, an ovipositing butterfly that lays her eggs singly on different plants must be considered to alternate between dispersal and search.

Search behavior, however, is not always linked to dispersal. When a feeding insect is dislodged or runs out of food, or its feeding behavior undergoes a developmental change, it may enter a search behavior pattern that is in no way related to dispersal behavior. In ovipositional behavior, it is frequently difficult or impossible to distinguish sharply between dispersal and searching activities.

3.3. Orientation and Recognition

Both physical and chemical factors are involved in guiding the ovipositing female to potential host plants. The insect's orientation behavior is followed by recognition behavior, in which the plant is either accepted or rejected as a host. Food sources also elicit oriented locomotion and recognition behavior patterns in most insects. In some species, the eggs are not laid in the immediate vicinity of the larval food plants; for example, many locusts and grasshoppers oviposit in the soil, and do not require any plant stimuli to release ovipositional behavior. Some butterflies of the families Hesperiidae and Hepialidae eject their eggs while airborne above appropriate vegetation. In most species, however, the eggs are deposited on larval host plants, and the neonate larvae may start feeding without a difficult and hazardous search. Despite this, larvae are capable of host selection behavior, undoubtedly because they may be dislodged by rain, hail, or wind; or they may drop from the host plant to evade attack by parasites and predators. Death or depletion of the host plant may also necessitate the search for a new host.

Physical factors

Physical factors involved in host-plant orientation and recognition include both visual and tactile stimuli. Colors and shapes can be perceived by insects, especially when in locomotion. Butterflies searching for food (nectar) react positively to yellow, blue, and in some cases, ultraviolet; green induces landing when the butterfly is searching for oviposition sites. Numerous insect species have been shown to be strongly attracted by yellow; for example, the Caribbean fruit fly, *Anastrepha suspensa*, is attracted by orange and yellow, which are the colors of many fruits attacked by this polyphagous fly (Greany et al., 1977). Light reflected from green leaves contains a distinct peak in the yellow portion of the spectrum, and many whiteflies and aphids are attracted to such yellow-reflecting surfaces (Kring, 1972). These insects show a strong landing response to yellow papers and painted surfaces, and their behavior suggests that such yellow surfaces constitute supernormal sign stimuli. Larval vision is usually less well-developed than that of the imagoes, but color discrimination has been demonstrated among lepidopterous and coleopterous larvae.

Shape or contour may also play a role in insect orientation, as in the case of fruit flies orienting to a potential ovipositional site (Boller and Prokopy, 1976). Lepidopterous larvae of several species move toward vertical objects during their search for food, and locusts and grasshoppers have been shown to be attracted by vertical striped patterns, but not by horizontal contrasts (Mulkern, 1969). Nocturnal insects, which are usually active in dim light but not in complete darkness, also react to contours. Tobacco hornworm moths, *Manduca sexta*, approach any clearly delineated object such as a plant during their ovipositional flight phase; if their eyes have been coated with an opaque paint, they cannot locate objects, including host plants (Yamamoto and Jenkins, 1972). Thus visual factors play an important role in insect orientation to potential host plants in both oviposition and feeding.

Tactile factors come into play after the orientation phase of host selection, and are frequently involved in the recognition phases that immediately precede oviposition or feeding. The physical characteristics of the surface on which the female is willing to deposit eggs are frequently of paramount importance. The insect ovipositor generally bears mechanoreceptors, and in most cases tactile stimuli seem to be the only sensory information relayed by the ovipositor. The amount of foliar pubescence has been found to be of importance to oviposition in a number of instances. Some species require relatively glabrous surfaces, whereas others prefer heavily pubescent oviposition sites. Other surface characteristics are also of frequent importance. The diamondback moth, *Plutella maculipennis,* and a number of other species prefer an oviposition substrate with small crevices and cavities (Gupta and Thorsteinson, 1960). In some cases the surface shape may have tactile significance. For example, the weevil *Ceutorrhynchus maculaalba* oviposits in young seed capsules of poppies; the convexity of the seed capsule has been shown to be of decisive importance in host selection. Presumably, proprioceptors in the weevil's legs perceive the degree of convexity (Saringer, 1976). Even when the eggs are deposited in the soil near the base of the host plant, as with the cabbage root fly, the insect requires soil particles of certain size; tactile information received by the ovipositor determines whether or not oviposition will take place (Zohren, 1968). In contrast to oviposition, food selection is usually less affected by tactile factors. This does not mean, however, that the physical characteristics of plant tissue do not play a part in host selection and in the successful utilization of a host plant. In numerous cases utilization is prevented by physical barriers such as, hard spines, dense pubescence, and, sclerenchymized leaf edges. However, the acceptance of many insects of artificial diets with physical characteristics totally different from their natural host plants suggests that tactile factors may play a minor role in food recognition. The numerous mechanoreceptors located on insect mouthparts are functional in the mechanics of biting, chewing, and swallowing. Sucking insects, such as aphids, obviously need a sensitive mechanoreceptor system to monitor mechanical forces arising during probing activity with their delicate and highly specialized mouthparts.

Chemical Factors

Physical features are rarely unique to a single plant species, but chemical characteristics may show a higher degree of specificity. It is not surprising, therefore, that chemical stimuli play a major role in host-plant selection for both oviposition and feeding. Volatile chemicals are frequently involved in orientation to plants from a distance (olfactory stimuli), but are also known to stimulate biting, probing, and oviposition after the insect is in physical contact with the plant. The final recognition process leading to acceptance or rejection is usually mediated by nonvolatile chemicals acting on contact chemoreceptors.

Insects that are searching for host plants may be activated by olfactory stimuli, and as a result will show directed movements toward the odor source. The basic mechanisms underlying these oriented responses to odors are only partially understood, but apparently the role of the odor is only to trigger an orientational response to other stimuli, that is, wind direction and visual targets. When randomly moving insects enter an odor plume, they may start moving upwind (sometimes at increased speed) using optomotor stimulation and thus show positive anemotaxis (Kennedy, 1977). As long as the insect remains in the odor stream, fairly direct or weakly zigzag flight paths are maintained until it overshoots the odor source. The insect then resumes randomly directed turning, which usually brings it back into the odor stream. Such oriented behavior has been observed in both oviposition and food searching behavior.

When an insect comes into the proximity of an odor source, the concentration gradient becomes much steeper, and in principle becomes measurable, thus allowing chemotactic orientation (Kennedy, 1977). High concentrations of an attractive odor may also inhibit locomotion (arrestant effect), and consequently induce the flying insect to land (Douwes, 1968). Under close-range conditions, the insect may obtain additional olfactory information, especially with respect to differences between individual plants. For example, some individual plants in a cabbage field are more attractive to oviposition by *Pieris brassicae*, because they contain higher than average amounts of volatile allyl nitriles, which attract the cabbage butterflies (Mitchell, 1977). Discrimination between host and nonhost plants may be made at relatively short distances (5 to 20 cm). The tobacco hornworm moth lands nearly every time it approaches a tobacco plant, but almost always veers away as it approaches a nonhost (Yamamoto and Jenkins, 1972).

Orientation to potential host plants and the discrimination of host from nonhost requires a highly developed sensory system. Usually the insect must respond to the odor of a plant that is located in a stand of mixed vegetation. Specific plant "odors" are seldom single compounds, but are usually complexes of several volatile substances. The scent of potato plants, for example, is made up of a particular combination of leaf alcohols and aldehydes that are, individually, not unique to potato. The Colorado potato bettle, *Leptinotarsa decemlineata*, is attracted to the combination of compounds that is produced by the potato plant. In some cases, however, one component of the plant volatile complex is predominant with respect to insect orientation. Alpha-pinene, one of the terpenoids produced by conifers, elicits a strong response in most conifer-inhabiting insect species; most insects attacking cruciferous plants are strongly attracted by mustard oils, which are ubiquitous among members of the Cruciferae. The dominant compound in such complexes is usually called the "token stimulus," because it signals the presence of a host plant.

Once the insect has made physical contact with a plant, contact chemoreceptors located on the tarsae, antennae, mouthparts, or ovipositor receive stimuli related to the chemical characteristics of the plant surface. Special movements by the insect may intensify the chemical stimulation; examples include the tapping or scraping motions of the forelegs of many butterflies as they prepare for oviposition, and the "drumming" of maxillary and labial palps by locusts. The tarsal receptors of the cabbage butterfly, *Pieris brassicae*, have been shown to be responsive to sinigrin (Ma and Schoonhoven, 1973).

Surface testing by touching or piercing with the ovipositor, or by biting and probing with the mouthparts, is in response to chemical factors that act as "incitants"; they are occasionally the same as the odor factors that attracted the insect to the plant. If the stimuli received upon initial testing identifies the plant as an acceptable host, feeding or oviposition proceeds. These chemical factors are "stimulants." If the stimuli received on initial testing indicate an unacceptable plant the behavior pattern is interrupted, and the insect abandons the plant; such stimuli are "deterrents." Whereas attractants, repellents, and many incitants are olfactory substances, stimulants, and deterrents are usually gustatory. Many important feeding stimulants are general nutrient substances, such as sugars and amino acids, rather than host-plant-specific compounds.

3.4   Token Stimuli and Nutrients

Host specificity and host selection are governed to a large extent by the insects' responses to chemicals that are characteristic of certain plant taxa (Fraenkel, 1969). Such substances as terpenes, flavonoids, alkaloids, and nitriles act as token stimuli in orientation and host-plant recognition. In some cases, specialized receptors have been found by which the insect perceives the presence of the chemicals, for example, the sinigrin receptors on the maxillae of *Pieris brassicae* (Schoonhoven, 1967).

Insects also show behavioral reactions to a number of chemicals of very general occurrence in plants, and there has been a tendency to ignore the role of these substances in the host selection process. Several carbohydrates, most notably sucrose, glucose, and fructose, stimulate feeding by many phytophagous insects. Sucrose is one of the most potent and universal feeding stimulants known. In addition, several amino acids, sterols, phospholipids, and a few other biochemicals of general occurrence have been shown to influence feeding behavior. Many of these substances are of great nutritional importance. Through the insect's reaction to such key nutrient compounds, there is a connection between the insect's nutritional requirements and its host selection behavior. It seems likely that token stimuli act as feeding incitants, whereas general compounds, such as sucrose, act as feeding stimulants. For example, sinigrin (the principal crucifer token stimulus) may incite biting by cabbage worm larvae, but continued feeding is determined by the presence of feeding stimulants (sucrose and others) in the plant tissues being eaten (Beck, 1965). Synergistic interactions have also been demonstrated, in which a mixture of token stimuli and feeding stimulants has a much greater than additive effect on feeding (Gothilf and Beck, 1967).

In nature, an insect is never exposed to token stimuli only or to general compounds only, but always to different mixtures of these factors. Thus host-plant selection and feeding behavior are based on complex stimulus patterns in which token stimuli, general compounds, and nutrients act in concert. The insect's central nervous system receives a complex pattern of sensory input, and must "decide" which patterns are acceptable and which are not; on this basis feeding is continued or is interrupted. There is ample evidence that different insect species, even when they have overlapping host-plant specificities, are adapted to different sensory factors and therefore obtain different patterns of sensory input from the same plant (Schoonhaven, et al. 1972). Host-plant selection is therefore not based on simple sign stimuli, but rather on a complex chemical pattern, a "Gestalt," which must fit into a species-specific "innate releasing mechanism" to evoke feeding behavior.

3.5. Individual Variability

Individual differences in behavior may be due to inherited variability or to different conditions to which the insects were exposed before being tested. Genetic variability in host preferences has been reported for many species. For example, *Papilio machaon* butterflies were tested for ovipositional preference, in which the butterflies were exposed to four acceptable host species, one of which, *Peucedanum*, was known to be the preferred host. Most of the butterflies showed a marked preference for *Peucedanum*, but a small number of selected *Angelica* plant (Wiklund, 1974). Such aberrant individuals within a population might be the founders of a biotype that would be able to overcome certain types of plant resistance.

Larval feeding behavior has also been shown to subject to individual variability that might be involved in overcoming resistance. Schoonhoven (1977) offered hungry *Manduca sexta* larvae dandelion foliage (a nonhost plant). Most of the larvae refused to feed after taking an initial bite, but a few individuals fed on the dandelion foliage, usually after some delay. In lepidopterous larvae, food plant preference appears to be individually modifiable and related to previous experience. When larvae are reared on a given host plant for several days before being offered a choice of food plants, they develop a marked preference for the food plant on which they had fed previously. Such plasticity of behavior has been observed in oligophagous species, such as tobacco hornworm, as well as in polyphagous forms, such as *Heliothis zea* (Jermy et al., 1968).

3.6 Ovipositional Preference and Food Selection

In some insect orders (e.g., Lepidoptera and Diptera) the nutritional requirements of the adults are quite different from those of the larvae. Therefore, when selecting an oviposition site, the adult obtains no sensory information relating to nutritional adequacy, whereas the larvae are sensitive to such factors while feeding. Because they respond mainly to the token stimuli (incitants) the adults are usually more limited in their host-plant range than are the larvae. This difference suggests that the two types of behavior are controlled by different internal drives and different genetic factors. From the experimental finding that newly hatched larvae of *Papilio machaon* did not show preferences when offered a choice of four different host plants, whereas the adult butterflies showed distinct preferences for oviposition, Wiklund (1974) concluded that larval food plant suitability and adult oviposition preferences are determined by different gene complexes. Even in species in which the adults feed on the same plants as the immatures, there is evidence that oviposition site selection by the adult and feeding site selection by the larvae are governed by different behavioral drives and are responsive to different complexes of stimuli.

It has been suggested that in some species the adult develops a predilection to oviposit on those plants on which it fed during its larval life. This hypothesis is known as the "Hopkins host selection principle," and has been inconclusively documented by experimental results. If feeding and oviposition are behaviorally and genetically separate systems, the Hopkins host selection principle is an untenable concept.

4. PLANT DEFENSES

Most phytophagous insects live on or within their host plants. The plant provides not only food, but also shelter and essential microhabitat. The phenotypic state of the plant is therefore of great importance to the insect, which is under appreciable pressure to adapt to the environment and diet provided by the host plant. The insect does not play such an important role in the well-being of the plant, but the insect–plant interaction is not without benefits to the plant population and the ecosystem (Mattson and Addy, 1975). Plants show evolutionary adaptations in response to many selective pressures, of which insect infestation usually represents only a minor component (Jermy, 1976). Intra- and interspecific plant competition, environmental abiotic factors (water, nutrients, light), and microbial pathogens represent greater pressures than that of insect feeding, at least under most circumstances (Dethier, 1970). That is not to say that insects and other herbivores are not significant; a portion of the plant species' adaptations must be in response to the pressures exerted by plant feeding insects.

The defense mechanisms evolved by plants in response to herbivores, pathogens, and competing plants include an array of physical characteristics and a battery of chemicals that tend to render the plant repellent, toxic, or otherwise unsuited for utilization.

Physical defenses against insect depredation may involve anatomical adaptations that have adverse effects on insect behavior or that reduce the protection required by larval stages (Painter, 1951). In a number of instances, toughness of plant tissues has been shown to contribute to plant resistance, particularly to the feeding of neonate forms and tissue penetration by stem borers. Tissue toughness may be caused by high fiber content or by silica. The presence of hairs or trichomes has been shown to influence host-plant utilization, and in some cases constitutes a significant mechanism of resistance. See Chapter 3 for a more detailed discussion of the role of physical factors in plant resistance.

Tissues of the higher plants contain arrays of biochemicals that are thought to be defensive in function. They include alkaloids, steroids, phenolics, saponins, tannins, resins, essential oils, various organic acids, and other compounds. Because their metabolic roles in the plant have been obscure, they are generally known as "secondary plant chemicals," produced as metabolic by-products with defensive functions (Fraenkel, 1959, 1969). Muller (1976) has even suggested that these substances were originally "meaningless ballast" and fortuitously proved to have defensive value. Recent investigations, however, have produced evidence that many of these chemicals may have important metabolic functions in the plant (Robinson, 1974; Heftman, 1975; Seigler and Price, 1976) in addition to defense. Even if they should prove to be exclusively defensive in function, they are of primary importance to the survival of the plant and of great significance in the coevolutionary success of both plants and phytophagous insects.

In an analysis of the chemical interactions between organisms, Whittaker (1970) proposed the term "allelochemics" to replace such terms as "secondary plant chemicals" and defined an allelochemic as being a nonnutritional chemical that is produced by an individual of one species and that affects the growth, health, behavior, or population biology of another species. Allelopathy, phytoalexin production, attractants, repellents, deterrents, stimulants, inhibitors, and toxicants are examples of allelochemics or allelochemic interactions. Two classes of allelochemical effects are of particular pertinence to insect–plant interactions: (1) *allomones*, which are allelochemics tending to confer an adaptive advantage on the producing organism (host plant), and (2) kairomones, which are allelochemics tending to give an adaptive advantage to the receiving organism (phytophagous insect) (Whittaker and Feeny, 1971). Except for plants that depend on insect pollinators, or that participate in other insect–plant mutualistic relationships, allelochemics would be expected to have had a defensive function originally (allomones). Their being converted to a kairomonic function would be an evolutionary adaptation by phytophagous insects, enabling recognition and utilization of the plant as host.

Allelochemical effects on insects may be behavioral or metabolic. The latter effects are considered in greater detail in Chapter 3. It must be realized that behavior and metabolism are interdependent processes, so behavioral and metabolic effects of allelochemics are not fully separable.

Similary the "nonpreference" (behavioral) and "antibiosis" (physiological-metabolic) categories of resistance (Painter, 1951) are interdependent to varying degrees.

5. INSECT-PLANT COEVOLUTION

5.1. Adaptation and Counteradaptation

The coevolutionary relationship between insects and plants can be viewed, perhaps simplistically, as being the production of novel allomones by the plants, and their subsequent neutralization or even exploitation as kairomones by the insects. Having arisen by chance mutation or genetic recombination, a novel allomone confers a selective advantage on the plant, relieving it of some herbivore pressure and perhaps improving its competitive position in relation to other plants. Ensuing adaptive radiation might result in the allomone's becoming a characteristic of an entire population, species, or even family. Such a plant adaptation may constitute a barrier to insect utilization. The plant may thus escape from the insect's host-plant range unless a counteradaptation in the insect population restores it as an adequate host plant. A genetic mutation or recombination in the insect population might enable the mutants to utilize a previously protected plant group, with the mutant insects being able to respond positively (feeding, oviposition) to the previously prohibitive allomone, thereby converting the allomone into a functional kairomone. Such a counteradaptation will, of course, yield a distinct selective advantage to the mutant insects. They may undergo adaptive radiation, allowing new diversification in the absence of competition from nonmutant members of the parent population, as well as from other phytophagous insects and herbivores. This process may well lead to the development of host races, biotypes, and new species.

According to this interpretation, the token stimuli by which a phytophagous insect identifies suitable host plants are kairomones (attractants, stimulants) that were originally evolved as allomones (repellents, deterrents) and that have retained the original allomonic function against phytophagous species that have not evolved the required counteradaptation.

To be an effective barrier to insect attack, an allomone should have both behavioral and physiological effects, thereby severely punishing behavioral errancy and enforcing the adaptive advantage of the allomone as a deterring stimulus. Conversely, the counteradaptation by the insect must involve both behavioral and metabolic capabilities, by which the allomone becomes a positive behavioral stimulus and is readily degraded or utilized biochemically. The compilation of plant chemicals exerting effects on insect behavior by Hedin et al. (1974) includes a great many substances also represented in a compilation (Beck and Reese, 1976) of plant compounds known to exert adverse physiological effects.

The combined behavioral and physiological impact of an allelochemic is well illustrated by the results of a study of the effects of sinigrin on the feeding behavior and growth of larvae of the black swallowtail butterfly, *Papilio polyxenes* (Erickson and Feeny, 1974). Sinigrin is a mustard oil glycoside, the biologically important portion of which is the aglucone, allyl isothiocyanate, and is found in the leaves of Cruciferae. A number of crucifer-feeding insects respond to sinigrin as a token stimulus identifying an acceptable host plant. *Papilio* species do not feed on crucifers, but utilize members of the Umbelliferae, from which their token stimuli originate. When *Papilio* larvae were fed celery leaves (Umbelliferae) that had been perfused with sinigrin, they fed on the leaves in response to umbelliferous feeding stimulants but displayed greatly inhibited larval growth and high rates of mortality. Sinigrin obviously was quite toxic to the swallowtail larvae. Sinigrin is apparently one of the allomonic barriers preventing swallowtail butterfly utilization of cruciferous hosts. At least three counteradaptations would be required for these butterflies to utilize crucifers: (1) behavioral adaptation leading to oviposition on such plants; (2) behavioral adaptation permitting larval feeding on crucifers; and (3) metabolic adaptation to reduce or eliminate the toxic effect of the allylisothiocyanate component of sinigrin. Sinigrin has been shown to be a feeding stimulant in the case of the cabbage aphid, *Brevicoryne brassicae*, but a powerful deterrent to feeding by the pea aphid, *Acyrthosiphon pisum*, an insect that utilizes only Leguminosae (Nault and Styer, 1972).

A striking example of plant adaptation and insect counteradaptation is represented by recent studies of the effects of canavanine on insect growth. L-Canavanine is an analogue of L-arginine with highly toxic properties. The toxicity of canavanine has been shown to be a disruption of normal protein synthesis. Most (nonadapted) organisms incorporate L-canavanine into structural proteins in the place of L-arginine, producing faulty and physiologically incompetent proteins. However, L-canavanine is the major nitrogen-storage form found in the seeds of some legumes, and may constitute from 8 to 10 percent of the seed's dry weight. Its toxic properties are a powerful allomonic barrier, protecting the seed from being eaten by insects and higher animals. *Caryedes brasiliensis*, a beetle of the family Bruchidae, feeds almost exclusively on the canavanine-containing seeds of *Dioclea megacarpa*. The arginyl-tRNA synthetase of the larvae is able to discriminate between L-arginine and L-canavanine, and the latter is not incorporated into the insect's proteins (Rosenthal et al., 1976). The beetle larvae take the process one step further, and have been shown not only to degrade canavanine, but also to use it as a source of nitrogen for other metabolic purposes (Rosenthal et al., 1977). Thus the plant's allomonic adaptation has been exactly reversed by this insect's counteradaptation.

The use of canavanine as an allomonic barrier to herbivore utilization illustrates a simple plant defense strategy. Of alternative feasible biochemical pathways (in this case nitrogen storage), an obvious selective advantage would be gained by the adoption of a pathway in which one or more of the metabolities could contribute to plant defense. It also seems likely that in any metabolically essential biochemical reaction series in which one of the intermediate metabolites displays an allomonic potential, the reaction equilibria might be subject to adaptive selection to produce a relatively large pool of that intermediate in the plant tissues. In both instances, the allomone would be a compound with a primary physiological function as well as a valuable adaptive function in plant defense.

5.2 Plant Apparency

Evolutionary strategies in plant defense mechanisms are not clearly apparent at the present time, but recent investigations have indicated that the strategies may be based on insect host specificities as well as on the population densities and successional status of the plant species (Feeny, 1975, 1976; Rhoades and Cates, 1976). Within a given ecosystem, some perennial and climax plants (such as trees and grasses) persist from year to year, and their presence in the system is predictable in both time and space. Such plants are certain to be found by insects and other herbivores. In contrast to such readily apparent or predictable species, other plants are less predictable in both location and numbers; these plants are less apparent to their relatively host-specific enemies, and more likely to escape attack. The phytophagous insect species in the environment under discussion will differ in regard to host-plant specificities, ranging from highly specialized to rather general feeders. The specialists may be specific to one of the predictable plant species, or to one of the more ephemeral species. The generalist insects tend to feed on a range of different plants, some of which will be of either group—apparent or ephemeral.

Plant defensive chemicals can be synthesized and stored in plant structures only after some expenditure of energy; that is, defense has a metabolic price. The evolutionary strategy of plants should be to use defense mechanisms that will give the most effective protection for the smallest metabolic expenditure, in short, to optimize the cost:benefit ratio. The perennial, predictable plants may be subjected to greater feeding pressure, particularly by specialized feeders, than the more emphemeral plants, but can also afford a greater expenditure of energy for defense. The predictable plants tend to rely on high concentrations (2 to 10 percent dry tissue weight) of generalized inhibitors of digestion and assimilation, such as tannins and resins. These substances are not highly toxic, but reduce the digestibility of ingested plant material, resulting in slower insect growth and reduced general fitness. The less apparent, unpredictable plants (and the ephemeral structures of predictable plants, such as flower buds) tend to contain much lower concentrations (0.02 to 1.0 percent dry weight) of more specific highly toxic substances, to protect them against the more generalist types of herbivores. These substances do not protect effectively from specialized insects that have become adapted to the plant species, and that respond to the defense chemicals as token stimuli (kairomones) for host-plant identification. The plant relies mainly on escape as a defense against the host-specific insect; the specialized toxins and deterrents are most effective against the nonadapted generalized feeders. The deterrent and toxic substances of such plants include alkaloids, phenols, and saponins, which are both effective and relatively inexpensive from the standpoint of the expenditure of metabolic energy.

5.3. Agricultural Implications

The agricultural implications of the coevolutionary strategies of plants and insects are very important to the subject of host-plant resistance. Plants that would be relatively ephemeral and hard to locate under natural conditions, may become highly apparent under some agricultural conditions because of their high population densities and their presence in the same location year after year. Under such circumstances, the plants will certainly be discovered by the specialist insects. Such specialists are adapted to utilize the plant defense chemicals as kairomones to identify host-plant presence, and are metabolically capable of detoxifying the chemical defenses. The crop plant's defenses are directed against generalist herbivores, and the plants lack both the time and the metabolic capability to produce the chemical defenses needed by an apparent and predictable plant. Thus the agricultural difficulty of maintaining a desirable crop quality and quantity is frequently the result of elevating annual, early successional plants from a natural situation in which their obscurity was their main defense against specialized herbivora, to an artificial situation in which the plants become fully apparent to the very enemies for which they have only minimal defenses.

From the patterns observed in host-plant specificities among insects and the defense strategies among plants, an agricultural strategy for the development and use of insect-resistant crop plants should be possible. The maintenance of stable resistance must be of overriding importance in the agricultural strategy. The evolutionary history of insect–plant interactions clearly shows that no plant defense mechanism can be considered immune to counteradaptation by one or more species of phytophagous insects. This simple fact dictates that the agricultural strategy must be based in large part on a diversity of plant defense mechanisms. Such diversity must involve not only the incorporation of more than one defense system within a given agriculturally desirable genetic line, but also a diversity of defenses among the different genetic lines that may be used from one year to the next within an agricultural area. The long-term stability of commercial wheat resistance to the Hessian fly, *Mayetiola destructor*, is an excellent example of the practical importance of genetic diversity in the host plants.

The relative stability of resistance is influenced by the host-plant specificity of the insect as well as by the genetic diversity and population density of the plant. The simplest and most unstable plant resistance would be based on the introduction of a single novel defense characteristic to combat a relatively monophagous insect pest. Although insect utilization of the resistant plant might be low through several generations, the intense selective pressure would soon result in an adapted insect population. The instability of resistance to a monophagous insect may be analogous to the loss of effectiveness of an insecticide where the entire population of insects is exposed to the same insecticide year after year. The best strategy for producing stable resistance would be to screen the world's germplasm for plant population differences in defense mechanisms, and to incorporate some of those genetic factors into different agronomic cultivars. The local insect populations would likely be poorly, if at all, adapted to some of the exotic genetic factors. By alternating the genetic lines of crop plant to be employed from time to time, adaptation of the local monophagous insect population might be avoided indefinitely.

This general line of reasoning leads us to expect that plant resistance to oligophagous insects would be more stable than in the case of monophagous forms. Assuming that the agroecosystem contained some acceptable nonagricultural host plants of the insect, the introduction of a crop cultivar displaying a deterrent defensive mechanism may result in the insect's being driven to accept an alternate host that was previously less preferred. Unless the alternative host-plant density is relatively high, however, a selective advantage might still acrue to adaptation to the overwhelmingly prevalent crop cultivar. In most agricultural situations, it seems likely that the stability of resistance to the oligophagous insect would be only slightly more stable than in the case of the more highly specialized monophagous species. A diversity of introduced defense mechanisms and a diversity of resistant cultivars would be nearly as important in this case as in the case of monophagous forms.

Plant resistance to polyphagous insects should be the most stable, because the pressure for adaptation would be consistently lower than in either of the two previous cases. Because the polyphagous insects enjoy a greater range of feeding options, experience less pressure to adapt to any given plant species, and are equipped with a broader range of metabolic detoxification capabilities, it seems likely that plant resistance might be more difficult to augment through plant breeding. But if the polyphagous insect can be driven away from a resistant (deterrent) cultivar, such resistance should show a greater stability than resistance to the more specialized insects.

The pea aphid, *Acyrthosiphon pisum*, (bottom) and the spotted alfalfa aphid, *Thereoaphis maculata*, (top) are carriers of plant viruses. Photo courtesy of U.S. Department of Agriculture.

7

INSECTS AND
PLANT PATHOGENS

K. Maramorosch

*Waksman Institute of Microbiology, Rutgers—The State
University, Piscataway, New Jersey*

Breeding plants for resistance to vectors of disease agents is only one of many possible approaches to the general concept of breeding plants for resistance to insect pests. As such, it requires that consideration be given to aspects of insect resistant plants, including the various types and optimal levels of resistance, as well as genetic, environmental, and behavioral factors affecting the expression and stability of resistance. Of equal importance are the diverse interactions between vectors and plant pathogens and between insect vectors and diseased plants. The latter include the attraction of insects to color changes in diseased plants, especially changes to various shades of yellow; phytochemical changes that might benefit insect vectors nutritionally or act as attractants; and morphological changes caused by leaf curling, proliferations, bushy growth, and other abnormalities. Because of these complexities, resistance to insect vectors has seldom been the primary aim of plant breeders. When vector resistance has been obtained, it has usually been a secondary result of a breeding program aimed at an insect pest that, coincidentally, also happened to be a vector.

1. PLANT PATHOGENS

1.1. Plant-pathogenic bacteria

Most bacterial pathogens are spread by the combined action of insects, rain, wind, and man. Thus they do not depend completely on insects for dispersion. There are few instances in which the interaction between bacteria and insects is specific. More than 200 plant-pathogenic bacteria belong to the following seven genera: *Erwinia, Pseudomonas, Xanthomonas, Streptomyces, Agrobacterium, Bacterium,* and *Corynebacterium.* Of these, a few species of *Erwinia, Xanthomonas,* and *Pseudomonas* require beetles (Coleoptera) for transmission to plants, and the relationship between pathogen and the vector is specialized (Carter, 1973).

The transmission of *E. tracheiphila,* the causative agent of bacterial wilt of cucurbits, depends on two species of cucurbit beetles (*Diabrotica vittata* and *D. duodecimpunctata*) in which the bacteria overwinter. Breeding plants for resistance to cucurbit beetles has not been attempted and, to date, the control of cucurbit wilt has been based on chemical insecticides applied for beetle control.

Xanthomonas stewartii, the causative agent of corn leaf blight, is transmitted by several species of *Diabrotica* and *Chaetocneme.* Overwintering beetles retain the bacteria in their intestinal tracts. The disease has been controlled successfully by treating the flea beetles with chlorinated hydrocarbon insecticides. This suggests that breeding for resistance to beetles might also prove worthwhile, but no resistant plants have yet been reported.

Pseudomonas savastonoi, which causes the olive knot disease in the Mediterranean area and in California, is transmitted by rainwater runoff from trees in California and by the olive fly, *Dacus oleae*, in Mediterranean countries. The intestinal tract of the fly harbors the bacteria that enter the eggs of the vector via the micropyle and subsequently contaminate the larval, pupal, and finally, the adult stage. The spread of olive knot disease in the Mediterranean region can be curtailed by destroying the fly vectors with chemical insecticides. This suggests that fly-resistant olive trees could control this bacterial disease.

1.2. Fungi

Fungi represent the greatest number of pathogens responsible for plant diseases. Like bacteria, most fungal parasites of plants do not require intermediary vectors. There are a few notable exceptions (Carter, 1973), that include certain diseases of rubber trees, coffee, cacao, apple, pine, beech, and elm. Additionally, certain fungal pathogens of soybean, cotton, corn, sorghum, and cabbage also can be dispersed by insects. The Dutch elm disease fungus, *Ceratostomella ulmi*, is carried from tree to tree by several species of elm bark beetles. Spores can be retained in overwintering beetles. Annual spraying with insecticides has been the measure most commonly used to prevent dissemination and spread of the fungal spores. No attempts have been made to include resistance to beetles in breeding programs where the aim has been to produce hybrid elms resistant to the fungus (the disease agent).

1.3. Viroids

Viroids are the smallest known infectious agents. They lack a dormant form (virions) and their genomes are much smaller than those of viruses (Diener, 1971). Viroids are found associated with cell nuclei in infected plants. Half a dozen plant diseases are now known to be caused by viroids; potato spindle tuber, chrysanthemum stunt, citrus exocortis, cucumber pale fruit, and chrysanthemum chlorotic mottle have been thoroughly studied (Diener, 1977). Apparently, the coconut palm disease cadang-cadang in the Philippines is also caused by a viroid (Diener, 1977). No insect vectors of viroids are known.

1.4. Viruses

Viruses represent the main group of plant pathogens in which vectors are usually required for dissemination; other modes of virus transmission are exceptions or are infrequent. Viruses consist of either RNA or DNA and, by definition, cannot contain both types of nucleic acid, as do all other pathogens (except viroids). For proliferation, solely the nucleic acid of a virus is required. The viral protein merely serves as a protective coat, and in some cases enhances cell entry and attachment. Viruses do not multiply by binary fission and they require host cell ribosomes for multiplication. They lack high energy enzymes of the cytochrome C or Krebs cycle.

Nearly 100 years ago, when viruses were first recognized as disease agents, they were characterized by their small size, which enabled them to pass through filters that retained bacteria. However, filterability is no longer used to define viruses. Viroids and other types of pathogens, first grouped under the name "mycoplasma-like organisms" are now divided into several groups (Maramorosch, 1976), and all pass through filters that retain bacteria.

Only a few plant viruses are able to infect plants without the intervention of vectors. In this small group are viruses transmitted either through pollen or seed. No airborne transmission of viruses is known, whereby free virions would be carried from plant to plant. In nature, most viruses depend on vectors for survival. These vectors acquire viruses from a diseased plant and carry them to a healthy, susceptible one. Even viruses that can be transmitted mechanically by rubbing a leaf with virus-containing plant juice usually require a living intermediary, as exemplified by tobacco mosaic virus, where man and other animals act as "vectors" by carrying virus from plant to plant (Harris and Bradley, 1973a, 1973b).

1.5. Mycoplasma and Mycoplasma-like Organisms

The group of plant pathogens earlier confused with viruses and often termed "plant mycoplasma agents" requires description and definition. Among these agents are microorganisms that resemble bacteria, rickettsiae, or chlamydia and that possess cell walls. The second category consists of the wall-less microorganisms, now classified as Mollicutes (*Bergey's Manual*, 1974). The name "mycoplasma" is loosely used for the wall-less microorganisms, but at present no plant pathogens are known to belong to the genus *Mycoplasma*. The class Mollicutes (Table 1) comprises three families, Mycoplasmataceae, Acholeplasmataceae, and Spiroplasmataceae, and several genera that have not yet been grouped into families, such as *Thermoplasma, Ureaplasma,* and *Anaeroplasma.* The genus *Spiroplasma* is

Table 1.   Classification of Mollicutes

Class	Mollicutes
Order	Mycoplasmatales
Family I	Mycoplasmataceae
Genus I	*Mycoplasma*
Family II	Acholeplasmataceae
Genus I	*Acholeplasma*
Family III	Spiroplasmataceae
Genus I	*Spiroplasma*
Genera of uncertain affiliation	*Ureaplasma*
	Thermoplasma
	Anaeroplasma

the only one in which not only plant pathogens but also insect pathogens and pathogens of higher animals are now known to exist (Table 2). The agent of citrus stubborn disease, *S. citri,* was the first described member of this group (Saglio et al., 1973). Other spiroplasmas have been isolated and cultured on cell-free media; the corn stunt spiroplasma (Williamson and Whitcomb, 1975; Chen and Liao, 1975) a spiroplasma which causes a cactus disease of *Opuntia tuna*, changing it to *O. tuna monstrosa* (Kondo et al., 1976), and the spiroplasma responsible for potato purple top and aster yellows disease (Maramorosch, 1977). There are numerous other plant disease agents that morphologically resemble spiroplasmas but have not yet been grown in cell-free media and therefore have not been properly identified. These agents are now known under the name "mycoplasma-like organisms" (MLO's). All spiroplasmas and MLO's require insect vectors for their natural transmission to plants. Often the same vector species can also transmit viruses, sometimes making it difficult to determine whether the disease agent is a virus or a MLO.

Table 2.   Diseases of Spiroplasma Etiology

Disease	Reference
Corn stunt	Davis et al., 1972
Citrus stubborn	Saglio et al., 1973
Drosophila sex ratio	Williamson and Whitcomb, 1974
Opuntia witches' broom	Kondo et al., 1976
Suckling mouse cataracts	Tully et al., 1975
Honeybee disease	Clark, 1977
Aster yellows	Maramorosch et al., 1977

Shortly after the discovery of MLO's in the phloem of diseased plants, electron microscopists observed that certain MLO's possess cell walls and not, as all mycoplasmas and other Mollicutes, merely plasma membranes. This recognition that the agents were not Mollicutes led to antibiotic tests that further confirmed their distinction from typical MLO's. The walled microorganisms, sometimes described as rickettsialike organisms (RLO's) (Maramorosch et al., 1975) might eventually be classified as plant-pathogenic bacteria, since they possess cell walls and are usually susceptible to penicillin, which acts on cell walls. The members of the Class Mollicutes, on the other hand, having only a membrane, are not affected by penicillin but are usually susceptible to tetracycline antibiotics, which act on ribosomes. All MLO-mollicutes are restricted to the phloem of diseased plants. Some RLO's are restricted to the phloem of diseased plants, whereas others are restricted to the xylem. The latter are larger than the phloem-restricted microorganisms, and several have now been identified as bacteria (Maramorosch, 1976). The known vectors of MLO's, RLO's, and spiroplasmas belong to the leafhoppers (Cicadelloidea), planthoppers (Fulgoroidea), and psyllids (Psylloidea). It is possible that vectors of MLO's will also be found in other insect groups, but none have been incriminated to date.

2. VECTORS OF PLANT PATHOGENS

2.1. Definition of Vector

A vector is a carrier of a disease agent. Vectors of plant pathogens include fungi, nematodes, mites, and insects. In this chapter, only insect vectors will be discussed. The agricultural definition of *insect vector* differs from the medical definition. In agriculture, particularly in economic entomology and plant pathology, insect vectors are understood to carry pathogens such as viruses, MLO's, bacteria, and fungi from diseased to healthy plants. The definition includes both actual and potential vectors, that is, all insects capable of transmitting disease agents. Medical entomologists, however, define as insect vectors only those insects that under normal conditions are associated with the respective hosts and are actually transmitting disease agents. An insect not normally associated with a host, but capable of transmitting the disease agent, is not considered a vector. There are several instances in which certain insects, not normally associated with a plant species, and sometimes even unable to survive on that species, nevertheless act as transmitters of plant disease agents to and/or from that host. The broader, agricultural definition is used throughout this chapter.

2.2.   Main Groups of Insect Vectors

The order Homoptera contains 80 percent of the known insect vector species of plant pathogens (Ossiannilsson, 1966). Of these, 40 percent are transmitted by members of the suborder Auchenorrhyncha and 60 percent by the Sternorrhyncha. The Cicadellidae, with more than 100 vector species, are the most important Auchenorrhyncha (Nielson, 1962, 1979). A few vectors are known among the family Cercopidae and one in the family Flatidae (Carter, 1973). Most, but not all, leafhopper-transmitted disease agents are carried biologically by their insect vectors. More than 180 vector species of aphids occur in the suborder Sternorrhyncha (Gularostria). Other Sternorrhynchous vectors are found in the superfamilies Coccoidea, Aleyrodoidea, and Psylloidea (Carter, 1973).

The principal groups of insect vectors are listed, in order of their importance, in Table 3. Aphids and leafhoppers constitute the most numerous carriers of plant pathogens (Harris and Maramorosch, 1977; Maramorosch and Harris, 1978). Depending on climate and geography, next in importance are whiteflies and beetles, thrips, mealybugs, psyllids, membracids, and a limited number of other insect groups. Outside the class Insecta, eriophyid mites constitute an important vector group of plant viruses.

2.3.   Interactions between Insect Vectors, Disease Agents, and Plants

The interactions between disease agents and vectors can be described in terms of either the duration of pathogen retention by the vector or the more or less intimate interrelationships that exist between pathogens and vectors. The understanding of fundamental interactions between plant pathogens and insect vectors in terms of acquisitions, carryover, and inoculation into

Table 3.   Insect Vectors of Plant Pathogens

Vector Group	Disease Agent
Aphids	Viruses
Leafhoppers and planthoppers	Viruses, MLO, RLO
Whiteflies	Viruses
Beetles	Viruses, fungi, bacteria
Thrips	Viruses
Mealybugs	Viruses
Psyllids	MLO, viruses
Membracids	Viruses

susceptible plants is essential in formulating new, nonpolluting methods of plant disease control (Harris, 1977). Although nearly 100 years have passed since the discovery of the first insect vectors, knowledge of what makes an insect a vector, how to "unmake" a vector, or how to convert a vector population into a nonvector population is not yet sufficient to be of use in plant protection. We are only beginning to understand why certain plants are resistant to specific disease agents or to specific vectors, but it seems logical that plant resistance to vectors might play a role in the mechanism of field resistance to certain vector-borne disease agents. The physical and chemical characteristics of a plant can be altered by breeding so as to alter vector–plant relations. In the future, interruption of the transmission cycle could become a major goal of some breeding programs. This interruption might be effected at the acquisition, carryover, and final inoculation stages of transmission.

There are two main systems of characterizing the interactions between plant pathogens and vectors. Retention of pathogens has been termed nonpersistent, semipersistent, or persistent (Watson and Roberts, 1939; Sylvester, 1956). Harris (1977, 1978) proposes the terms noncirculative (nonpersistent and semipersistent) and circulative (persistent). The term semipersistent can better be expressed as "transitory" (Ling and Tiongco, 1978). Nonpersistent retention occurs when vectors acquire the plant pathogen during very brief probes, often as brief as 5 seconds. Similarly, the probability of plant inoculation by vectors usually maximizes after probes of 15 to 60 seconds. Vectors cease to be infective a few minutes after acquiring the pathogen, unless they are starved or chilled experimentally. Aphid vectors are among the best studied nonpersistent transmitters. No detectable latent period occurs in the case of nonpersistent transmission, and the acquisition, carryover, and inoculation phases of the transmission cycle can be completed in as little as 1 minute. For reasons that are not entirely clear, starving aphids prior to feeding on the diseased source plant increases the efficiency of transmission. Fasting for only 15 minutes greatly enhances transmission, and the proportion of insects that transmit increases with fasting periods up to 1 hour or longer. The starving effect is nullified when starved aphids are left on a source plant for more than a few minutes. Harris (1977) hypothesizes that preacquisition starvation raises the level of transmission by increasing the numbers of aphids that make intracellular probes in the epidermis. Such aphids would be more likely to acquire and later transmit virus, presumably via sap sampling and an ingestion–egestion mechanism (Harris, 1977, 1978).

Semipersistent, or transitory, transmission is characterized by the retention of plant pathogens for periods ranging from a few hours to 1 to 2 days, and in rare instances for periods longer than a week. The retention time can vary considerably with temperature and with feeding activity before, during, and after virus acquisition. In persistent transmission, a latent period occurs following acquisition during which the vector cannot transmit the pathogen. Afterward, the vectors continue to transmit the disease agent for prolonged periods, sometimes throughout their lives.

The second system in which the interaction between plant pathogen and insect vector is described is based on the type of interaction between the arthropod and the disease agent. In 1962, Kennedy et al. proposed the terms stylet-borne and circulative for the two major types of interactions. The stylet-borne disease agents include those described earlier as nonpersistent, and the circulative type comprises the persistent ones. Although the new terms gained wide acceptance, the transitory or semipersistent type of transmission remained outside the system. The semipersistent pathogens do not circulate in vectors, nor do they appear to be simply stylet-borne (Harris, 1977; Pirone and Harris, 1977). Over 80 aphid-borne viruses are noncirculative, 30 circulative, and nearly 50 remain unclassified. Four of the circulative viruses that depend on aphid vectors for their maintenance in nature, lettuce necrotic yellow, broccoli necrotic yellows, potato leaf roll, and sow thistle yellow vein viruses, are propagative; that is, they multiply in their aphid vectors. The latter two viruses are the only aphid-borne plant pathogens known to be passed transovarially to the progeny of infective female vectors (Sylvester and Richardson, 1969; Miyamoto and Miyamoto, 1966). There are no known instances of aphid-borne viruses being transmitted by a leafhopper or a whitefly. Even though the degree of specificity is less pronounced in the interactions between aphids and plant pathogens than in leafhoppers, planthoppers, or psyllids, usually only a few aphid species can transmit a given pathogen. The green peach aphid, *Myzus persicae*, can transmit a large number of viruses, but there are several viruses that cannot be transmitted by this species and that require other aphid vectors. The simple explanation that aphids transmit viruses "mechanically" no longer seems convincing (Harris, 1977; 1978).

The mechanism by which insect vectors transmit nonpersistent viruses could perhaps be explained as follows. If vectors imbibe virus-laden plant sap during brief probes, they might later transmit the virus by egesting all or a part of this material during subsequent brief probes on healthy plants. The uptake of plant sap during brief, as opposed to feeding, probes might provide the clue, because the evidence that stylet contamination is the operative mechanism is not conclusive. If aphids are stimulated to sap-sample during brief probes rather than to insert their stylets deeply for feeding in the phloem, then preaccess fasting may enchance transmission by increasing the aphid's sensitivity to the plant stimuli inducing sap sampling. Aphid behavior can be influenced by numerous factors, including day length, temperature, and host plants. The possibilities for inhibiting virus transmission by manipulating sap-sampling behavior are discussed elsewhere (Harris, 1977).

With few notable exceptions, leafhopper and planthopper vectors carry plant pathogens in the persistent manner, and numerous instances of propagative transmission are known (Maramorosch, 1963, 1969).

3. BREEDING FOR RESISTANCE TO VECTORS

3.1. Aphid vectors

There is adequate variability in plants to breed for resistance to certain aphid vectors. Resistance to aphids can depend on a variety of factors. Stimuli required by aphids have to be considered if breeding is to be directed, rather than empirical (Eastop, 1977). One of the first aspects to be considered is how to prevent aphids from alighting, to stop them from reinoculating other plants. If alighting is not prevented, transmission is more difficult to stop. Alighting can be prevented by sticky surfaces, and this feature can sometimes be incorporated into existing plant varieties. One could develop, theoretically, a plant on which aphids would probe unsuccessfully or be unable to reach the phloem, and finally starve. There is an inherent danger in this approach, however, The feeding of aphids might increase or decrease their efficiency as vectors. If they become starved, they are more likely to alight on and acquire nonpersistent viruses from source plants than when they are well fed. Nevertheless, it does not seem likely that plants could be developed that would provide a nutritional imbalance for aphid vectors.

The effect of various plant colors could be utilized. Yellow is usually most attractive to aphids. It was found that aphids would feed preferentially on orange and yellow leaves, rather than on those of various shades of green (Carter, 1966); the color of leaves may determine whether an aphid remains on a given leaf for prolonged periods. Of special interest to plant breeders is the fact that a continuous canopy of green plants is less attractive to aphids than a mosaic pattern in which the soil and the plants intermingle. Therefore breeding plants that germinate rapidly and grow fast so as to form a continuous green surface might be advantageous (Gibson and Plumb, 1977). Epidermal hairs of string beans can injure and trap aphids. Hybrids between a wild potato, *Solanum bertahultii,* and *S. tuberosum* developed aphid-trapping hairs (Gibson, 1974). Breeding for the production of nicotine has been successful in killing peach aphids on *Nicotiana tabacum* (Abernathy and Thurston, 1969), but such breeding has not been tested to determine its influence on virus transmission. Certain aphids, such as the green peach aphid, seem to prefer waxy leaves, whereas other species are repelled by them (Way and Murdie, 1965). Most aphids make several brief probes before feeding for prolonged periods. Many substances, such as flavonoids, have been linked with probing activity. Penetration into the plant phloem can sometimes be prevented or disrupted and this mechanism has been responsible, in part, for the resistance of alfalfa to *Therioaphis maculata* (Nielson and Don, 1974a). Certain species of aphids, with shorter stylets, have been prevented from feeding by the thick parenchymatous layer covering vascular bundles (Gibson, 1972). The resistance of certain apple cultivars to the wooly apple aphid, *Eriosoma lanigerum*, seems to be the result of the sclerenchymatous tissue that blocks phloem penetration (Staniland, 1924).

Certain factors are known to affect aphid vectors but cannot be used in breeding for vector resistance. For example, poor illumination decreases the suitability of plants to aphids, the water-stressed plants make aphids restless. Other factors can also be used in breeding programs. For instance, frost resistance in *Vicia faba* has enabled plants to overwinter, and the mature plants in spring are less likely to be infested by *Aphis fabae*. Early maturing potatoes, on the other hand, become more attractive to *Myzus persicae* because of early aging of leaves.

Aphid probing and penetration can be affected by the content of sinigrin, according to Nault and Styer (1972). However, quinones, alkaloids, glycosides, terpenes, and several other substances play no role in resistance because they do not occur in the phloem (Eschrich, 1970). Plants could be developed which incorporate characteristics that affect the production of winged aphids. Alatae are the main transmitters of plant viruses. According to Evans (1938) the production of alate *Brevicoryne brassicae* increases with the decrease in water-soluble carbohydrates and protein nitrogen. Differences have also been reported in alatae production on various potato varieties; the largest numbers were produced on "Katahdin" (Simpson and Shands, 1949). Changing the proportion of alatae and apterae might thus be feasible without nullifying other useful characteristics of a plant (Gibson and Plumb, 1977).

According to Maxwell and Painter (1962), susceptible plants usually contain higher amounts of free auxins than resistant plants. This indicated that aphids might feed at sites other than the phloem, thus causing a buildup of toxins in tissues where they cannot be rapidly dispersed. Brussels sprouts with open foliage have more tolerance to *B. brassicae* than plants with curled leaves, perhaps because the aphids feed in the curled areas (Dunn and Kempton, 1971).

In order to screen for resistant genotypes, one must have a thorough understanding of the interactions between aphid vectors and other types of insect vectors and the host plants. Knight and Alston (1974) pointed out that initial screening should be preceded by a survey of a wide range of potential sources of resistance to find several resistant individuals, because only a few plants might be adequate for breeding purposes. Although initial screening is often carried out in a greenhouse, field screening should always be included. And this ought to be done in diverse geographic areas so that plants are exposed to different races of the vectors, as well as to other insect pests and diseases, and to different climatic conditions.

When resistance depends on a single dominant gene, the breeding advantage is clear. At least 50 percent of the progeny of a cross with a resistant parent can be expected to be resistant. Difficulties increase when more genes are involved. Gallun (1972), Gallun et al. (1975), Maxwell et al. (1972), Dahms (1972), and Knight and Alston (1974) reviewed the genetics of resistance to several aphids. The resistance of maize to *Rhopalosiphum maidis* apparently depends on many genes. Sometimes resistance is highly specific, as in the case of the barley cultivar 'Omugi' which is resistant to the greenbug, *Schizaphis graminum*, but very susceptible to *R. maidis* (Maxwell et al., 1972). In sugar beet resistance to *A. fabae* and *M. persicae*, the same genes seemed to control both aphid settling and propagation, but not in the same measure with respect to apterae as to alatae. The mechanisms of these resistances are probably polygenically controlled (Russell, 1969).

Aphids can be harmful as vectors and as direct pests (Harris and Maramorosch, 1977). Alfalfa has been damaged directly by *Therioaphis maculata* in North America. In this instance, large populations of aphids are destructive. On the other hand, a single viruliferous aphid can cause more damage in 15 seconds, by infecting a plant, than a large number of nonviruliferous aphids feeding for prolonged periods on the same plant. Even aphids that cannot colonize a certain plant can transmit viruses during brief probing. This illustrates the difficulties in any program in which resistance to aphid vectors, rather than to aphids damaging a plant by feeding, is considered. *M. persicae* colonizes sugar beets and potatoes, but damage from feeding by these aphids is negligible compared to their spreading of viruses, especially when production of virus-free seed potatoes is at stake. Breeding of plants resistant to aphid vectors could provide a much less expensive alternative to chemicals and would seem a highly desirable goal without detrimental, environmental side effects. If aphid-vector resistant cultivars would last 10 years, as has been calculated for aphid-pest resistant ones (Horber, 1972), the return would probably exceed 300:1 compared to the 5:1 return calculated by Metcalf (1971) for chemical pesticides. Moreover, Metcalf's calculation was not aimed at vectors but at pests, where a few surviving individuals can be overlooked. Not so with vectors, however, where the aim is 100 percent kill to eliminate virus transmission. Such complete destruction by chemicals is almost impossible when aphid vectors are involved, and this is why efforts to prevent the spread of aphid-borne viruses have had only limited success.

Some forms of breeding for plant resistance have been very successful: for instance, the breeding of 'Northern Spy' apple trees resistant to *Eriosoma lanigerum* for many years, and of raspberries resistant to *Amphorophora agathonica*. Sometimes such resistant cultivars have proved less affected by viruses, but this has been accidental rather than because of planned experimental breeding.

Virus control has seldom, if ever, been the aim of a breeding program for aphid resistance.

3.2. Leafhopper Vectors

Leafhoppers have been known as vectors of viruses for nearly a century. In Japan, the first description of rice dwarf transmission dates back to the end of the nineteenth century. Until recently, it was generally assumed that all leafhopper-borne viruses are circulative in their insect vectors and retained for prolonged periods. It was also generally assumed that all plant disease agents transmitted by leafhoppers were viruses. Both of these assumptions no longer hold. The discovery by Ling (1969) that rice tungro virus is not permanently retained by leafhopper vectors and that it is transitory or semipersistent in its vector was followed by other, similar findings. Nevertheless, the total number of noncirculative leafhopper-borne viruses is very small, and this type of virus–leafhopper vector interaction is an exception rather than a rule.

Most leafhopper-borne viruses and pathogenic MLO's are transmitted by leafhoppers in the circulative manner, and many are known to multiply in their invertebrate vectors (Maramorosch and Harris, 1979). A few viruses and MLO's are transmitted transovarially to the progeny of vectors (Maramorosch, 1964; Maramorosch and Jensen, 1963; Shikata and Maramorosch, 1969; Smith, 1965).

Since 1967, it has been recognized that leafhopper and planthopper vectors carry not only plant-pathogenic viruses, but also mycoplasma-like and rickettsia-like agents, earlier confused with viruses. Sometimes the same vector species can transmit a virus and an MLO, and both agents can be transmitted separately or simultaneously (Nielson, 1979).

The interference with the spread of circulative pathogens could develop along the same principles as those for noncirculative ones. Achieving control would seem easier, since there is a closer association of the pathogens with their respective vectors. Resistance mechanisms that act after pathogen acquisition and after alighting would inhibit feeding and thus restrict further spreading of circulative disease agents. Unfortunately, large vector populations can often compensate for inefficient vector biotypes (Sohi and Swenson, 1964; Plumb, 1976). The effects of resistant cultivars on vector complexes have been studied only in a few cases.

Persistently transmitted pathogens often cause color changes in plant hosts, including yellowing, that in turn attract more vectors and thus result in a higher transmission rate (Bawden, 1964). An understanding of the interactions among vectors, pathogens, and resistant cultivars is crucial to successful breeding against vectors of plant pathogens. Among the factors that have to be considered is the increased mobility of vectors that are repelled by characteristics bred into the plants to protect them from certain pests. Though tolerance to leafhoppers may prevent direct damage, it can result in a substantial increase in viral and MLO disease transmission.

During the past 10 years, breeding for resistance to leafhoppers has resulted in the discovery of resistant varieties, described by Athwal and Pathak (1972) and Heinrichs (1979). Resistance to the brown planthopper, *Nilaparvata lugens*, and to the green leafhopper, *Nephotettix virescens*, were the first developed because these insects are important pests. It was found that one dominant and one recessive gene were responsible for resistance to the brown planthopper. Resistance to the green rice leafhopper is monogenic and dominant, and three different genes have been identified (Athwal and Pathak, 1972; Pathak, 1975). The rice variety 'IR8' is resistant to the rice delphacid, *Sogatola orizicola*, in Colombia but is susceptible to the hoja blanca virus carried by this insect. 'IR8' remains practically virus-free under field conditions because of its resistance to the vector, whereas other susceptible rice lines become infected (Jennings and Pineda, 1970). Certain rice varieties resistant to the brown planthopper, but susceptible to grassy stunt virus, have exhibited field resistance to this virus. It is not known why rice plants resistant to both vectors differ in their susceptibility to the respective viruses. The green leafhopper seldom damages rice directly and its importance is mainly as a vector of tungro virus. The brown plant-hopper, however, not only acts as a vector of grassy stunt but also causes direct feeding damage and thus constitutes a double threat (Heinrichs, 1979).

Although numerous sources of resistance to leafhoppers and planthoppers have been established in the world collection of rice germplasm, and although such resistance can be transferred to new varieties, most earlier breeding programs have been directed at virus rather than vector resistance. Recent studies have indicated that vector resistance alone may maintain the green leafhopper vector population at low levels, thus decreasing losses due to tungro virus (Pathak, 1970). It was found at IRRI that the rice variety 'Mudgo,' which is resistant to the brown planthopper but not to grassy stunt virus, often escapes virus infection (Pathak, 1970). The green leafhopper is able to feed on resistant rice varieties; however, its survival and reproduction are curtailed, probably because of lack of proper nutrients (Heinrichs, 1979). The brown planthopper does not feed on resistant varieties, apparently because of lack of feeding stimulants or the presence of feeding inhibitors.

Despite the short period during which breeding rice varieties resistant to virus vectors has been carried out, very good progress has been achieved in countries of Southeast Asia, particularly in the Phillippines, India, Thailand, Indonesia, and Sri Lanka. The national and international programs are well coordinated, and new strains, detected and released for breeding, are often resistant to vectors. Although at present the number of instances in which resistance to leafhopper vectors has been developed is limited, the existing evidence suggests that it is feasible to breed plants for leafhopper vector resistance. The study of ecological factors affecting leafhoppers and planthoppers is receiving increased attention. Leafhopper populations have been controlled by altered plant spacing and other cultural practices. If additional vector-resistant varieties could be developed through the incorporation of proper characteristics in breeding programs, such vector-resistant varieties would curtail losses due to viruses and MLO agents. If breeding for resistance to specific leafhopper vectors that act as vectors and reservoirs of plant pathogens were successful, the losses in crops such as corn, potato, rice, carrots, lettuce, spinach, and many others could be reduced considerably.

3.3. Whitefly Vectors

Whiteflies (Aleyrodidae) act as virus vectors in tropical, subtropical, and temperate zones. Their importance becomes apparent if one considers the affected crops: cassava, tobacco, cotton, and different legumes (Bird and Maramorosch, 1975). More than 30 different plant viruses have been reported transmitted by *Bemisia tabaci* (Bird and Maramorosch, 1977). The difficulties in controlling *B. tabaci* are enormous (Maramorosch, 1975) and it appears that the most feasible approach is to use plant breeding techniques to obtain varieties that would not be frequented by whiteflies. Attempts to reduce virus spread by breeding for resistance to *B. tabaci* have barely begun, and no worthwhile progress has been reported to date. Whether resistance to whiteflies can be found and incorporated into plants is not certain at this time. Since legumes are often affected by whitefly-borne viruses (Bird and Maramorosch, 1975), it might be of interest to consider this aspect in legume breeding programs.

3.4. Beetle Vectors

Beetles transmit several plant viruses (Walters, 1969; Selman, 1973). Most, if not all, viruses with beetle vectors also can be easily transmitted mechanically to plants. The viruses can be acquired during acquisition feeding periods of 24 hours or less, and they can be transmitted immediately after acquisition. Beetles can retain viruses for 48 hours as a rule, and instances of retention up to 19 days have been recorded (Walters and Henry, 1970). The mechanism of transmission is not well understood, and it is not known why some beetles can transmit one virus efficiently but not another with similar properties. Viruses have been detected in beetle hemolymph, regurgitant, and feces (Fulton et al., 1975).

The retention of viruses by beetles depends on the species of vector (Fulton et al., 1975) as well as on the type of virus (Gamez, 1972). Environmental factors and the type of host plant also play a role. The fact that the respective viruses can be transmitted mechanically led to the earlier belief that they are transmitted simply by contaminated mouthparts (Smith, 1965). Recent studies, especially on the role of beetle hemolymph, indicate that the transmission process is a complex biological phenomenon.

Among plant virus vectors are representatives of several beetle families. The majority of them are leaf beetles in the family Chrysomelidae, subfamily Galerucinae. Two species of the genus *Ceratoma* attack legumes in North and South America (Nichols et al., 1974). The cucumber beetle, *Diabrotica* spp., and members of the genera *Colaspis, Ootheca,* and *Acalymma* also act as virus vectors (Smith, 1966; Fulton et al., 1975). In addition to transmitting viruses, the various vector species of beetles also act as pests. The complex interactions between beetle vectors and viruses, the growing of multiple crops in tropical areas on small farms, and the abundance of the insects present formidable problems in attempts to control virus spread by breeding and selection for resistance to viruses and to beetle vectors.

3.5. Mealybug Vectors

Mealybugs (Coccoidea) transmit viruses affecting pineapple, sugarcane, and cacao. Chemical methods have been relied on to control these vectors. To my knowledge, breeding for resistance to mealybugs has not been attempted.

3.6. Thrips (Thysanoptera)

Only one plant virus, spotted wilt virus of tomato, also known as a pathogen of tobacco and pineapple, is carried by thrips vectors. The virus is widely distributed in Europe, Hawaii, South Africa, and Australia. Breeding plants resistant to thrips could provide an economic means for the control of spotted wilt.

4. BENEFICIAL EFFECTS OF DISEASED PLANTS ON VECTORS

It has been commonly observed that virus, MLO, and fungus-infected plants are often infested by insects, conveying the impression that such diseased plants provide a better food source than healthy plants, or otherwise attract insect pests and vectors. The reasons for this phenomenon are complex. Diseased plants, irrespective of the type of pathogen, have a different metabolism and differ chemically from healthy plants. It is conceivable that certain chemical changes in diseased plants attract insects and that their survival on such diseased plants is enhanced. In fact, survival of certain species of leafhopper vectors on aster yellows-infected plants is significantly better than on healthy ones. Diseased plants are often chlorotic, and attraction by the yellow color might explain the invasion of such diseased plants by pests and vectors.

In addition to the chemical factors, the change in physical characteristics might also favor the survival of insect vectors. Diseased plants, especially plants infected by MLO agents and certain viruses, are characterized by bushy growth, which, in turn, might provide hiding places as well as shade and moisture needed by certain vectors.

A direct beneficial effect has been described in the altered feeding and survival of *Dalbulus maidis* leafhoppers, after the insects were confirned to plants infected by the aster yellows agent (Maramorosch, 1957; Orenski, 1964). *D. maidis* is ordinarily restricted in its diet to *Zea* species and can breed only on corn and teosinte. Other grasses and all dicotyledonous plants that have been tried were unable to sustain the continuous survival and/or breeding of this species. It was found that *D. maidis* can survive surprisingly well on aster yellows-infected China aster plants. After insects were confined to such diseased plants for 10 to 14 days, their feeding habits became altered and they survived on several species of healthy dicotyledonous plants. It has been demonstrated that corn leafhoppers acquire the aster yellows agent during confinement to diseased aster plants, even though they are unable to transmit the pathogen to healthy plants. The aster yellows MLO is retained by the corn leafhopper for several weeks and it is conceivable that it influences the sterol metabolism of the insects.

Fortunately, neither the corn stunt spiroplasma, transmitted by *D. maidis*, nor the royado fino virus, carried by the same vector, infects dicotyledonous plants in nature. Otherwise, the altered vector metabolism could become disastrous to new host plants.

Alteration by breeding of a plant species that normally provides the food for a vector species so that it becomes unsuitable for the vector might force the insect population to search for new food sources. The described changes in *D. maidis* are a possible means of altering the metabolism of other vectors in a way that might be quite common but that has not been studied in detail in the past. It is obvious that effects that benefit insect vectors are undesirable insofar as crop production and protection is concerned.

5. CONCLUSIONS

The control of insect vectors that transmit plant pathogens requires either the drastic reduction and/or elimination of the vectors, or the efficient interruption of the transmission cycle. To prevent vector transmission, each newly arriving vector would have to be killed before it had a chance to feed or even to probe. Conventional pesticides have not been totally effective in controlling the dissemination of vector-borne pathogens. Among the most difficult problems are the attempts to interrupt or suppress the disease cycle in the transmission of nonpersistent viruses. When brief probing of the epidermis is sufficient to inoculate a plant, insecticides can rarely influence virus dissemination. Induced transitory repellency has been achieved in certain instances by using oil sprays to prevent the feeding of vectors. Coating of leaf surfaces with thin layers of repellent substances has merely a transitory effect. It is hoped that future breeding programs will incorporate breeding for resistance to specific vector feeding to achieve permanent, rather than transitory, resistance to vectors.

In order to breed for resistance to vectors, particularly to aphid and leafhopper vectors, there is a need for adequate knowledge of virus transmission and of host selection, as well as of vector behavior, a subject that has only recently been investigated. Plant pathologists have paid little attention to the appearance of vector biotypes that break resistance. An exception is the study of biotypes of *R. maidis* by Saksena et al. (1964), in which transmission was taken into account. Other workers have mainly made the observation that there were differences between pea aphids transmitting pea enation mosaic virus (Bath and Chapman, 1967) or *R. padi* transmitting barley yellow dwarf virus (Rochow and Eastop, 1966). These differences were due to biotypes. It is as important to test what occurs when plants become diseased following virus transmission. Diseased plants often attract large numbers of vectors, thus increasing the chances for further dissemination of pathogens. It is of special importance to assure that resistance to a vector does not cease when the plant becomes diseased. The ideal solution would be to obtain vector resistance that would, at the same time, restrict the dissemination of virus, in instances in which the vector is also a pest. Most breeding for resistance to virus diseases has been directed against the viruses, although it might sometimes be easier to breed against the vector. This is especially so when the same vector species acts as carrier for several viruses.

In most instances, breeding must be directed against several insect pests, not against a single species, if the eventual goal is to replace the old variety with a new one that will not require chemical pesticides. In the case of vectors, often a single species or even a race of a species is involved in virus transmission, and the breeding should take this into account. If each resistance trait is controlled by a single dominant gene, it is sometimes possible to combine resistance against several insect pests in a single host plant. Knight and Alston (1974) described the use of a multiple approach, in which resistance of apples to a leaf-curling aphid, apple mildew, collar rot, and the woolly aphid were successfully combined.

Resistance obtained to a single or several vectors ought to be durable, each under a variety of conditions. Though the climatic differences can be anticipated, the appearance of new races of the vector are more difficult to anticipate and cope with. Numerous biotypes have been described among aphid vectors (Rochow and Estop, 1966) and their appearance illustrates well the tremendous difficulties in breeding cultivars for resistance to aphid vectors. The development of *A. pisum* biotypes has been shown to be affected by environmental factors (Kilian and Nielson, 1971). Selective pressures for producing new biotypes, combined with the rapid multiplication of aphids, resulted in the large number of recognized aphid biotypes. As early as 1933 Storey (1933) recognized so-called races among leafhopper vectors, some able and others unable to act as virus vectors. Breeding against both types could, in this case, be the solution. But where a large number of biotypes is likely to appear, as with aphid vectors, the problem is very difficult.

The success already achieved in breeding for resistance to certain aphid and leafhopper pests that also act as virus vectors clearly indicates that vector resistance can be incorporated in breeding programs. Such programs present challenging and exciting research problems for the future. If successful, they would play an important role in protecting crops from diseases in today's hungry world. The infection of a crop by a virus is a dynamic process that can be interrupted if the proper knowledge is available and effort is made in the right direction. The balance can be tipped in favor of the plant host, but this requires collaboration of virologists, entomologists, plant pathologists, and breeders.

A researcher rates and records stalk damage to maize caused by the sugar cane borer. Photo courtesy of CIMMYT.

8

THE PATHOSYSTEM CONCEPT

Raoul A. Robinson

Consultant, St. Helier, Jersey, United Kingdom

1. THE PATHOSYSTEM CONCEPT

The most important fundamental in science is the *pattern*. A pattern is an arrangement of units. A word is a pattern; so is a molecule or a mathematical formula. A *system* is a series of patterns of patterns which are called systems *levels*. In a dynamic system, *balance* is crucial. Systems balance is the maintenance of a dynamic equilibrium, and a balanced system is thus a stable system. Systems balance must obviously occur at every systems level if stability is to be maintained. Balance is achieved and maintained by systems *control* which may be autonomous, deterministic (i.e., imposed by man), or both.

A *pathosystem* is a subsystem of an ecosystem and is defined by the phenomenon of parasitism. As with an ecosystem, the geographical, conceptual, biological, and other boundaries of a pathosystem may be defined according to convenience. A plant pathosystem is one in which the hosts are plants; the parasites are normally considered to include insects, mites, nematodes, fungi, bacteria, mycoplasmas, and viruses, but not avian or mammalian herbivores. (By convention, entomologists often restrict the use of the term "parasite" to the hyperparasites of crop pests; this usage is not employed here.)

There are two categories of plant pathosystem. The wild pathosystem is a balanced system; were it not, it could not have survived evolutionary competition. It is also a completely autonomous system. The crop pathosystem differs in that it has an element of deterministic control imposed by man. Cultivars differ from wild plants and cultivation differs from a wild ecosystem. The crop pathosystem is usually (but not necessarily) an unbalanced system.

The pathosystem concept can thus be defined as the analysis and management of systems balance at all systems levels within the pathosystem.

2. THE TERMS "VERTICAL" AND "HORIZONTAL"

The terms *vertical* and *horizontal,* in reference to resistance, were coined by Van der Plank (1963) as abstract, conceptual terms without any literal or descriptive function. Each term has an absolute definition and, in addition, can be conditionally defined for use in special contexts. Obviously, such conditional definitions must be accurate and they must fit the facts. In practice, this means that they must conform with the original diagrams of Van der Plank (1963) from which they are derived.

The absolute definition of vertical resistance is that it involves a gene-for-gene relationship (Flor, 1942); that is, for each resistance gene in the host, there is a corresponding, matching gene in the parasite. Like antigens and antibodies, the presence of both is necessary for the demonstration of either. If a host has several resistance genes, it is resistant to all races of the parasite which lack one or more of the corresponding genes; and it is susceptible to all races of the parasite which possess all (or more than all) of them. Because the parasite can eventually match any combination of genes in the host, vertical resistance is *within* the capacity for microevolutionary change of the parasite, just as the insecticide DDT is within the capacity for change of houseflies. For this reason, vertical resistance is usually temporary resistance in agriculture.

The absolute definition of horizontal resistance is that it does not involve a gene-for-gene relationship. Like the fungicide Bordeaux mixture, it is *beyond* the capacity for microevolutionary change of the parasite. For all practical purposes, it is thus permanent resistance in agriculture.

3. THE VERTICAL PATHOSYSTEM

When discussing the vertical and horizontal pathosystems, it is convenient to restrict the conceptual boundaries of the pathosystem to the interaction of one species of parasite with one species of host. The term "interaction" has two distinct meanings depending on whether it is used in a host–parasite or a statistical context. In this chapter, a statistical usage is always qualified by "differential."

Because the vertical pathosystem is defined by the gene-for-gene relationship, vertical resistance and vertical parasitic ability depend on each other. The vertical pathosystem thus exhibits a differential interaction. That is, a series of parasite differentials is necessary for the identification of any one resistance, and a series of host differentials is necessary for the identification of any one parasitic ability. Person (1959) first showed that a gene-for-gene relationship can be demonstrated without genetic studies in either the host or the parasite. This is because the various combinations of vertical genes (vertical genomes) can be arranged in the same sequence in both the host and the parasite. Such an arrangement is only possible if there is a gene-for-gene relationship, and it results in a special category of differential interaction now known as the "Person differential interaction" (Robinson, 1976).

Habgood (1970) devised a system of nomenclature that can be applied to vertical genomes, regardless of whether the genes in question are in the host or the parasite. Each pair of matching genes is named as one of the geometric series 2^0, 2^1, 2^2, 2^3, etc., with arithmetic values 1, 2, 4, 8, etc., each value being double that of its predecessor. This system has the advantage that the sum (Habgood name) of any combination of gene values is unique. Thus the Habgood name "15" can only mean a combination of genes 8, 4, 2, and 1. If a series of vertical genomes is arranged in the numerical order of their Habgood names, the individual genes are distributed in a regular pattern (see Figure 1, vertical genomes). The gene with arithmetic value one occurs every alternate unit; that with value two, every alternate pair; that with value four, every alternate foursome; and so on to infinity.

In Figure 1, the Habgood nomenclature and the Person differential interaction have been combined to produce the Person/Habgood differential interaction (Robinson, 1976), which can be taken as the model of a vertical pathosystem. The mathematical properties of this model are so simple and elegant that they have the ring of scientific truth; they are also extremely important. If a new gene is added to the Person/Habgood differential interaction, irrespective of the number of genes already present, five changes occur:

1. The number of vertical genomes (i.e., Habgood names) is doubled and is 2^n (where n = number of vertical genes).
2. The number of matching host–parasite interactions (i.e., black dots in Figure 1) is trebled and is 3^n.
3. The total number of interactions, both matching (black dots) and non-matching (blanks), is quadrupled and is 4^n.
4. The proportion of matching to total interactions is reduced by a factor of 0.75 and is $(\frac{3}{4})^n$; this is a new biological constant which is discussed further below.
5. The pattern of matching interactions (black dots) is replicated three times in the form of a bottom-left triangle, and there is a mirror-image symmetry each side of the diagonal from bottom-left to top-right.

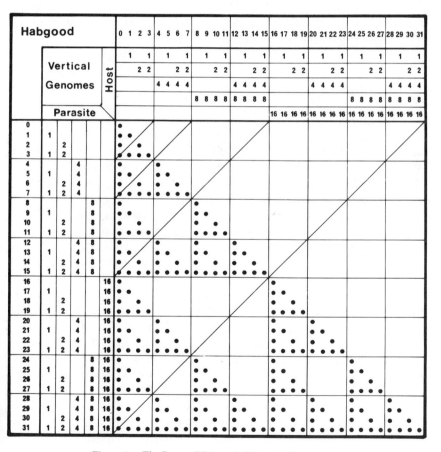

Figure 1. The Person/Habgood differential interaction.

The significance of these mathematical relationships will become apparent during the discussion of the wild pathosystem below. It should be noted that the Person/Habgood differential interaction is proof of a gene-for-gene relationship and hence of a vertical pathosystem. This proof does not require genetic studies in either the host of the parasite. All that is necessary is that all the known host–parasite interactions (both matching and nonmatching) of a series of host and parasite differentials can be accommodated to a Person/Habgood differential interaction.

Finally, vertical resistance is the principal cause of "physiological specialization" in which resistance operates against some populations of the parasite but not others. Such resistance is temporary because it fails to operate following an appropriate change in the parasite population. There are other forms of physiological specialization that are not due to vertical resistance. The most important of these is the polyphyletic pathosystem (Robinson, 1976); However, the details are beyond the scope of this discussion and are published elsewhere (Robinson, 1979).

4. THE HORIZONTAL PATHOSYSTEM

4.1. The Horizontal Pathosystem Model

Because all vertical resistance involves a gene-for-gene relationship, it follows that the inheritance of all vertical resistance is controlled by various combinations of major genes. It follows also that all polygenically inherited resistance is horizontal, and that any major gene resistance that does not involve a gene-for-gene relationship is also horizontal. The model in Figure 2 represents a horizontal pathosystem in which the inheritance is polygenic.

It may be assumed that resistance and parasitic ability are each controlled by many alleles which may be + or −. In the host, each 1 percent of + alleles increases resistance by 1 percent, and each 1 percent of − alleles decreases resistance by 1 percent. Similarly, in the parasite, each 1 percent of + alleles and each 1 percent of − alleles increase and decrease parasitic ability by 1 percent, respectively.

If a pathotype (i.e., a parasite population in which all individuals have a given parasitic ability in common) has 100 percent − alleles, it has no parasitic ability and consequently there is no parasitism, irrespective of the percentage of − alleles in the host.

Equally, if a pathodeme (i.e., a host population in which all individuals have a given resistance in common) has 100 percent + alleles, it has no susceptibility, and consequently there is no parasitism, irrespective of the percentage of + alleles in the parasite.

Pathodemes; percent (−) alleles											mean
	100	90	80	70	60	50	40	30	20	10	0	
100	100	90	80	70	60	50	40	30	20	10	0	50
90	90	81	72	63	54	45	36	27	18	9	0	45
80	80	72	64	56	48	40	32	24	16	8	0	40
70	70	63	56	49	42	35	28	21	14	7	0	35
60	60	54	48	42	36	30	24	18	12	6	0	30
50	50	45	40	35	30	25	20	15	10	5	0	25
40	40	36	32	28	24	20	16	12	8	4	0	20
30	30	27	24	21	18	15	12	9	6	3	0	15
20	20	18	16	14	12	10	8	6	4	2	0	10
10	10	9	8	7	6	5	4	3	2	1	0	5
0	0	0	0	0	0	0	0	0	0	0	0	0
mean	50	45	40	35	30	25	20	15	10	5	0	25

(Row labels at left, top to bottom: Pathotypes; percent (+) alleles — 100, 90, 80, 70, 60, 50, 40, 30, 20, 10, 0)

Figure 2. The horizontal pathosystem model.

The 100 percent level of parasitism is obtained with the interaction of 100 percent + alleles in the parasite with 100 percent − alleles in the host. These relationships are fixed, regardless of how many alleles may in fact be controlling the inheritance of either the resistance or the parasitic ability. If x is the percentage of − alleles in the host (i.e., percentage susceptibility) and y is the percentage of + alleles in the parasite (i.e., percentage parasitic ability), then the percentage parasitism is $xy/100$.

4.2. Properties of the Horizontal Pathosystem

The model in Figure 2 can now be used to illustrate the properties of the horizontal pathosystem. It should be noted, however, that the percentage parasitism shown in the model is a theoretical scale which is not easily equated with the percentage of host tissue actually destroyed by the parasite. It is presumed also that in the wild pathosystm there is an equilibrium gene frequency in both the host and the parasite population. Consequently, the extremes shown in the model are not normally manifested.

Independence and Constant Ranking

Because there is no gene-for-gene relationship, horizontal resistance and parasitic ability are independent of each other; either can be increased or decreased regardless of the other. In statistical terms, this means that there is no differential interaction. The ranking of the pathodemes, according to their resistance, is constant, irrespective of which pathotype they are tested against; and the ranking of the pathotypes, according to their parasitic ability, is also constant, irrespective of which pathodeme they are tested against. Provided that it has some susceptibility, and that this is known, one pathodeme thus identifies any pathotype; similarly, one pathotype identifies any pathodeme.

Permanence

For any pathodeme, there is a maximum level of parasite damage which is due to the interaction with the pathotype possessing 100 percent + alleles; a subsequent, microevolutionary change in the parasite population can only lead to a reduction in this level of damage. Horizontal resistance is consequently permanent, at least in the foreseeable, microevolutionary future. However, uncontrolled genetic changes in the host may reduce the level of resistance under agricultural conditions. Similarly, screening for horizontal resistance must be conducted with a pathotype possessing a high percentage of + alleles; otherwise subsequent microevolutionary changes in the parasite population could lead to significant increases in the level of parasite damage.

Domestication

Horizontal resistance, being variable in the wild state, can be domesticated, that is, increased and stabilized by artificial selection. This is done in the same way as the levels of sucrose in sugarcane and sugar beet were increased above the natural levels in their wild progenitors. As the breeder raises the percentage of + alleles for resistance in the host until it approaches 100 percent, so the level of resistance increases until it approaches immunity. The levels of parasite damage are then extremely low, irrespective of high percentages of + alleles in the parasite.

Complete Control of Parasites

The maximum level of horizontal resistance can only be determined experimentally. It is possible, for example, that 100 percent of + alleles in the host might confer less than absolute resistance; or conversely, that absolute resistance may be conferred by less than 100 percent of + alleles. Absolute resistance is probably unnecessary in agriculture since, below a certain level of intensity, parasite damage becomes negligible. Furthermore, as Van der Plank (1975) has pointed out, it is possible to have "population immunity" when the host individuals are less than immune. This occurs when each parasite unit produces less than one daughter unit; the parasite population then shrinks and lacks epidemiological competence. Whatever practical demonstrations may reveal, therefore, it is at least theoretically possible that horizontal resistance will provide not only a permanent control, but also a complete control of some, and possibly many, crop parasites.

Natural Levels

In this context, the ecological phenomenon of "spotty distribution" is of major entomological significance. Because it is a variable phenomenon, horizontal resistance declines in the absence of the parasite owing to negative selection pressure. For genetic reasons, this decline is usually completed in some 10 to 15 generations of crossing in the host population, when the parasite is totally absent, and it is then called the "unselected level" of horizontal resistance (Robinson, 1976). A spotty distribution may be sequential or spatial. For example, under natural conditions, a locust swarm normally invades about once in 10 to 15 generations of an annual host, which then has minimal resistance. The Rossetto hypothesis states that a spatial spotty distribution confers a positive survival value or the parasite (Robinson, 1976). It occurs typically with *Cicadulina* spp., the vectors of maize streak virus in Africa. Only a small portion of the total maize population is infested each season and, as a consequence of this negative selection pressure, the resistance levels to both the virus and its vector are normally low. Many wild hosts thus exhibit relatively low levels of horizontal resistance to some of their parasites; but even low levels of a variable survival value can be raised by artificial selection.

No Good Source of Resistance

For similar reasons, most cultivars have inadequate horizontal resistance to their major parasites; indeed, were this not so, the parasites would be of minor importance. This loss of horizontal resistance is one of the failings of conventional plant breeding techniques and is discussed further below. It is clear also that high levels of horizontal resistance can be accumulated from very susceptible parent cultivars possessing low percentages of + alleles, provided that each parent possesses more or less different + alleles; that is, provided that a reasonably wide genetic base is used. Moreover, the conventional "good source" of resistance is unnecessary when one breeds for horizontal resistance.

Comprehensive Horizontal Resistance

Finally, horizontal resistance is ubiquitous; it occurs in every host against each of its parasites, even if it is currently inadequate in most cultivars. This postulation is justified below (Section 5) in the discussion of the esodemic. The horizontal resistance to any one species of parasite should be regarded as a single, variable survival value which is independent of all the other horizontal resistances to all other species of parasite. The new approach to plant breeding (Section 8) emphasizes the *simultaneous* domestication of all desirable, variable survival values. This means that all horizontal resistances are increased equally and progressively during some 10 to 15 generations of screening a genetically flexible gene pool. The target is resistance that is permanent, complete, and comprehensive; that is, a crop husbandry that is virtually and permanently free from parasites.

However, each geographical pathosystem is different from all others. The distribution and epidemiological competence of the various parasite species differ from one geographical pathosystem to another. Therefore, comprehensive horizontal resistance must be defined as resistance to all *locally* important parasites. This means that "on-site screening" and a multiplicity of breeding programs are essential.

5.   THE WILD PATHOSYSTEM

At this point it becomes useful to borrow some terms and ideas from plant pathology in which a crucial distinction is made between two kinds of infection.

When a host individual is infected by a fungal spore, that spore causes *auto-infection* if it was produced on the same host individual, and it causes *allo-infection* if it was not produced on the same host individual. The entomological equivalent of infection is oviposition by a parasite that reaches the adult stage on or in that same host individual. (Feeding by an adult does not correspond to infection.) That part of the epidemic which is due solely to auto-infection is called the *esodemic*; and that part of the epidemic which is due solely to allo-infection is called the *exodemic*. There are thus two subdivisions of the epidemic based on two kinds of infection. There are also two kinds of resistance and parasitic ability, based on the presence or absence of the gene-for-gene relationship. It now transpires that there are also two kinds of wild pathosystem, the continuous and the discontinuous.

5.1.   The Discontinuous Pathosystem

The discontinuous pathosystem is illustrated in Figure 3 which concerns a leaf parasite of a deciduous tree species. The host population has spatial discontinuity owing to genetic heterogeneity, and sequential discontinuity of leaf tissue owing to its deciduous habit. The parasite population is similarly discontinuous. It may be a fungus or an insect such as an aphid; the principles are identical.

There are 2^n vertical resistances in the host population and 2^n vertical parasitic abilities in the parasite population. If there are 16 vertical genes in

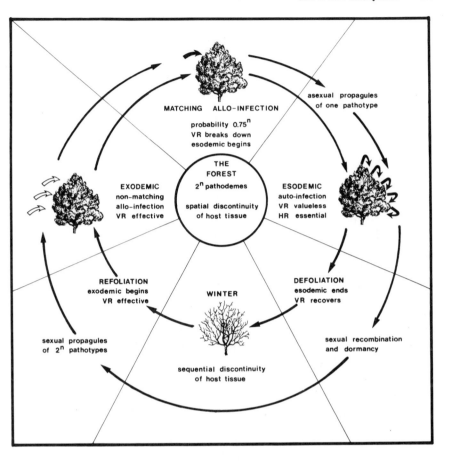

Figure 3.   The discontinuous wild pathosystem.

the pathosystem, there will be 65,536 vertical genomes and it may be assumed (see below) that all of them occur with a nearly equal frequency in both the host and the parasite populations. Thus when one parasite individual makes contact with one host individual, the probability of this being a matching allo-infection is the same as the proportion of matching interactions in the Person/Habgood differential interaction, that is, $(\frac{3}{4})^n$. Table 1 shows that, for every eight genes added to the vertical pathosystem, this probability is reduced to one tenth With 16 genes, it is only 0.01. Clearly the function of vertical resistance is to reduce allo-infection; because some matching always occurs, vertical resistance cannot prevent all allo-infection, it can only reduce the exodemic.

Once a matching all-infection does occur, the vertical resistance of the individual tree breaks down and the esodemic begins. In most pathogens

Table 1. Proportion of Matching Allo-infection

Number of Vertical Genes (n)	Proportion of Matching Allo-Infection $(^3\!/_4)^n$
0	1.0
8	0.1
16	0.01
24	0.001
80	1×10^{-10}

and some insect parasites the esodemic is characterized by asexual reproduction. Every leaf of that tree has the same vertical resistance and there is no further need for population heterogeneity in the parasite. All auto-infection is matching infection; vertical resistance is valueless in the esodemic, and auto-infection can be reduced only by horizontal resistance. Because every epidemic (or infestation) has an esodemic, it follows that horizontal resistance is essential and universal. (With systemic diseases and one-generation insect parasites the esodemic does not involve auto-infection as such, but the horizontal resistance is still essential following the breakdown of vertical resistance.)

The esodemic ends with leaf fall and the vertical resistance then "recovers." This is the converse of a breakdown. The following spring, there will be new leaves and the vertical resistance will again be effective because the tree will be parasite-free and can only be allo-infected. In the discontinuous wild pathosystem, therefore, there is a seasonal cycle of the breakdown and recovery of vertical resistance in each host individual. So long as the host population heterogeneity is maintained, the reduction of the exodemic by vertical resistance is a stable and permanent phenomenon.

It must be remembered that this is a wild pathosystem in which the systems control is autonomous, and in which a good systems balance is essential at every systems level. The evolutionary trend can only be toward systems balance; otherwise the system would be unstable and could not survive evolutionary competition. For the purposes of this discussion, systems balance need be considered only at three levels of the pathosystem. First, and highest, is the balance between the host and parasite populations; then balance between the vertical and horizontal pathosystems; and, finally, balance within the vertical pathosystem and within the horizontal pathosystem.

In the simplest terms, balance is the point at which the positive and negative selection pressures for a given variable are equal. In detail, of course, balance means more than this. The requirements of balance vary from one geographical pathosystem to another, and from one period to another within one pathosystem. The concept of balance embraces the phenomenon of systems *resilience*. The system can tolerate wide swings away from the optimum and still recover. This is homeostasis; it ensures that the system is well buffered.

Balance between the host and parasite populations is clearly essential. If the parasitic ability of the parasite is too great, the evolutionary survival of the host is impaired; if the host becomes extinct, the parasite becomes extinct also. It follows that there must be an absolute limit to the parasitic ability of parasites. Equally, if the resistance of the host is too low, evolutionary survival is again threatened. But so long as host survival is not threatened, absolute resistance is an unnecessary survival value which tends to be lost owing to negative selection pressure.

Balance between the vertical and horizontal pathosystems results from the selection pressure that these two subsystems exert on each other. If the reduction of the exodemic by vertical resistance is excessive, there is inadequate selection pressure for horizontal resistance. A matching allo-infection may then be very rare but it would still occur; the host individual would then have inadequate protection against auto-infection. Similarly, if the horizontal resistance is too great the frequency of vertical genes tends to decline owing to negative selection pressure.

Balance within the vertical pathosystem means that all vertical genomes must occur with approximately equal frequency. If one vertical genome predominated, the pathosystem would tend to the "boom or bust" situation of agriculture, which represents an extreme loss of systems balance and stability.

Balance within the horizontal pathosystem is reflected in the reproductive rates, and hence the evolutionary survival, of both the host and the parasite; neither must be impaired by the other. The positive and negative selection pressures are equal, both for horizontal resistance in the host, and for horizontal parasitic ability in the parasite.

5.2. The Continuous Pathosystem

The second category of wild pathosystem is the continuous pathosystem in which there is spatial and sequential continuity of host tissue and, as a consequence, the esodemic never ends. It is found typically in evergreen perennial host populations which have genetic uniformity because of a natural vegetative reproduction, and sequential continuity because of the absence of a closed season. It occurs typically in the wild progenitors of crops such as sugarcane, cassava, banana, and sisal. Vertical resistance cannot occur in a continuous pathosystem because, though it could break down, it would never recover and consequently would have no evolutionary survival value. The esodemic never stops and, though it may fluctuate, even to the point of virtual dormancy, the infection is auto-infection, and hence matching infection. A continuous pathosystem is protected exclusively by horizontal resistance.

Crops derived from a continuous wild pathosystem are the easiest to breed for resistance because they possess only horizontal resistance. Although physiologic specialization that is not due to vertical resistance may occur, this can be easily avoided during breeding (Robinson, 1979).

6. THE CROP PATHOSYSTEM

In the course of cultivating, domesticating, and breeding his crops, man has altered the systems balance, often disastrously.

At the highest systems level, domestication is a single survival value. Some cultivars are better domesticated than others, and inferior cultivars do not survive the domestic competition. Domestication is thus the agricultural equivalent of Darwinian fitness in the wild ecosystem. At the next systems level, domestication has the three components of yield, quality, and resistance; and a balanced domestication means that none of these components is deficient. At a still lower level, each of these components has many subcomponents. Resistance, for example, is made up of many different resistances to many different species of parasite. If only one resistance is inadequate, the cultivar is classified as susceptible and hence inferior. Most modern cultivars are unbalanced in this respect, having a major susceptibility to one or more species of parasite.

However, the main loss of balance has occurred in still lower systems levels in the crop pathosystem. In the traditional approach to breeding crops for vertical resistance, balance has been lost between the vertical and horizontal pathosystems, and within both the vertical and horizontal pathosystems.

During breeding for vertical resistance, horizontal resistance tends to decline because of negative selection pressure. This phenomenon was first recognized by Van der Plank (1963), who called it the "vertifolia effect" after a potato cultivar in which it was particularly prominent. The negative selection pressure for horizontal resistance results from the fact that all breeding for vertical resistance is conducted during the exodemic. All infection is nonmatching allo-infection. If a matching infection does occur, the line in question is discarded as valueless. However, the vertical resistance does not normally break down until the final selection has become a widely grown cultivar. It is then found to be very susceptible because of the vertifolia effect. It is this general low level of horizontal resistance in vertically resistant cultivars which makes the "bust" of the boom-and-bust cycle so severe.

Balance within the vertical pathosystem has been almost totally destroyed. In place of the 2^n vertical pathodemes of the wild pathosystem, a crop pathosystem has only one pathodeme cultivated as a pure line or clone. Each plant within such a crop is the epidemiological equivalent of one leaf of one tree in Figure 3. Once one matching allo-infection occurs, the vertical resistance of the entire crop breaks down and this is why vertical resistance is temporary in agriculture.

The apparently endless repetition of resistance failures since the beginning of the century has led to a corresponding loss of balance in scientific opinion. Most crop scientists now believe that all resistance is impermanent and is bound to fail sooner or later. Indeed, the crop scientist who is optimistic, positive, and confident is now a very rare individual. Possibly the most important feature of horizontal resistance is that it can be expected to restore confidence in the ultimate potential of crop science.

The loss of balance within the horizontal pathosystem has been even more important. It has already been shown (Figure 2) that the horizontal resistance to one species of parasite is a variable survival value that can be increased or decreased with positive or negative selection pressure. This means that horizontal resistance can and should be domesticated to levels above that of the wild pathosystem. However, cultivars are notoriously more susceptible than wild plants; their levels of horizontal resistance are usually less than those of the wild pathosystem, except, perhaps, when there is a spotty distribution of the parasite. High levels of horizontal resistance are necessary in the crop pathosystem for two reasons. Firstly, many features of agriculture, such as unnaturally high host-population densities and crop uniformity, tend to intensify the epidemics and infestations. Secondly, the natural level of horizontal resistance is enough—but only enough—to ensure the evolutionary survival of the host. Evolutionary survival is not a significant criterion in the crop pathosystem; what matters is that neither the yield nor the quality of the crop product is damaged by parasites. In practice, this is likely to mean that the domesticated level of horizontal resistance must be so high that the crop is virtually parasite-free. It remains to be seen whether such levels of horizontal resistance can be achieved to all locally important parasites. The sugarcane breeders of Hawaii have come very close to this ideal and there are good grounds for thinking that a similar achievement is possible in many other crops in many other areas.

7. THE DOMESTICATION OF RESISTANCE

Domestication means that a natural, variable survival value desired by man is increased and stabilized by artificial selection to levels above the natural optimum of the wild ecosystem. Both vertical and horizontal resistance can be domesticated.

7.1. Vertical Resistance

Van der Plank (1968) first formulated the concept of *strength* of vertical genomes. Strength refers to the relative rarity of the matching vertical pathotype, and it can be demonstrated in two ways. When a matching vertical pathotype is common, the resistance breaks down quickly and is described as weak; conversely, when the pathotype is rare, the resistance remains effective for a longer period and is described as strong. Similarly, a vertical resistance may be abandoned after its breakdown; the matching pathotype, having been common, may become rare again slowly or quickly, and the resistance is then described as weak or strong, respectively. This phenomenon is analogous to that of insecticide-resistant insects. DDT-resistant houseflies appear quickly and remain abundant for a long time following the abandonment of this insecticide, which thus confers a "weak" protection against houseflies. But it confers a "strong" protection against malarial mosquitoes.

It was postulated above that, in a wild pathosystem, all vertical genomes occur with equal frequency and this can only mean that they are of equal strength. The crop pathosystem, however, is very different. When a single vertical resistance gene is transferred, by man, from a wild host to a cultivar, which may even be a different species, and which is grown in a different environment, the strength of that gene is liable to change; some genes become stronger and others weaker. Evidence for this comes mainly from late blight of potato in Europe, incited by *Phytophthora infestans,* where the strength of vertical resistance genes derived from *Solanum demissum* in Mexico has been greatly altered in the new host and the new environment.

Obviously, gene strength is a valuable characteristic. A strong vertical resistance remains effective for a longer period and, following its break-down, it can be reemployed after a shorter interval. Indeed, if vertical resistance is to be properly and effectively managed in the crop patho-system, strong genes are essential. Effective management employs spatial or sequential patterns of different vertical resistances. A spatial pattern can involve a mixture of different vertical pathodemes in a single cultivar, which is then known as a multiline. Or the pattern can involve a regional mixture of crops so deployed that a migratory insect is faced with a series of different vertical resistances in the course of its migration. A sequential pattern involves regular, controlled changes from one vertical resistance to another in such a way that the matching vertical pathotypes never have a chance to become common.

Breeding for vertical resistance thus involves a search for the strongest vertical genes. Even more valuable, various combinations of vertical genes (vertical genomes) can lead to further increases in strength. It may even be possible to increase the strength to the point where the matching vertical pathotype lacks epidemiological competence entirely. The vertical resistance is then effectively permanent. This possibility was recognized by Van der Plank (1975) and has been called *frozen* vertical resistance (Robinson, 1976).

7.2. Horizontal Resistance

Breeding for horizontal resistance has a far greater potential than breeding for vertical resistance. It is a polygenically inherited, variable survival value which can be raised by artificial selection to maximum levels that may well be the equivalent of immunity. Given suitable breeding techniques, all desirable variables can be raised equally and simultaneously to provide both a comprehensive horizontal resistance and a balanced domestication. Indeed, at the highest systems level, the one character of "balanced domestication" is a single variable and a single selection criterion.

8.   BREEDING CROPS FOR A BALANCED DOMESTICATION

8.1. Traditional Plant Breeding

Traditional plant breeding has always accepted the need for a "good source" of resistance; it has depended on *gene transfer* methodologies. This approach usually (but not necessarily) leads to physiologic specialization and resistance which is impermanent. The new concepts and methodologies rely on *changes of gene frequency* within existing, susceptible populations of the host. These changes in gene frequency must occur when vertical resistance is either absent or inoperative, that is, during the esodemic. These are the two basic differences which promise to produce permanent resistance and a balanced domestication.

There are two reasons why the new approach is unlikely to be popular with traditional plant breeders. Firstly, and perhaps not unexpectedly, almost every step in the new approach is the exact opposite of the conventional approach. Secondly, the new approach is based almost entirely on theoretical science; indeed, its validity still requires practical demonstration. But there is a strong implication that the theoretical basis of the traditional approach is unsound, and this implication will, no doubt, be resented. That the traditional approach has led to extreme loss of balance in the crop pathosystem is indisputable, however.

It is, of course, impossible to predict the results of research. The research targets can be precisely defined, but only time and experiment will reveal their feasibility.

8.2. Targets

The first target is to achieve a level of comprehensive horizontal resistance to all locally important parasites such that there is a crop husbandry which is virtually and permanently free from parasites. The second target is cumulative crop improvement in the sense that, with comprehensive horizontal resistance, a good cultivar need never be replaced except with a better cultivar. The third target is to approach the ultimate potential of crop production.

8.3. The Multidisciplinary Approach

A man who studies an ecosystem cannot be either a botanist or a zoologist; he must be both and more besides. Similarly, a man who studies a crop pathosystem must be prepared to develop a working knowledge of all the crop science disciplines. The first step, therefore, is to form a multidisciplinary team whose aim is to achieve a balanced pathosystem. Each member of that team must regard himself primarily as a "pathosystem manager" and only secondarily as a specialist in one of the various crop science disciplines. The main contributions of each discipline are as follows.

8.4. Breeding

The breeder should aim to work at the highest systems level and to reduce deterministic control to the minimum necessary to achieve a balanced domestication. He relies as much as possible on the autonomous control of the system itself to achieve systems balance for him. His work is based on the model of the horizontal pathosystem (Figure 2) and his function is to manipulate a gene pool derived from a number of susceptible cultivars, each with a relatively low percentage of + alleles, but each with more or less different + alleles. There is no "good source" of resistance, just as there was originally no "good source" of sucrose in the domestication of sugarcane or sugar beet. The gene pool manipulation requires a large host population in which there is a high proportion of random cross-pollination. There must be high selection coefficients for the three main components of domestication:

yield, quality, and resistance to all locally important parasites. The breeder aims to change gene frequencies in favor of these selection criteria by mass selection methods during as many generations as may be required; 10 to 15 generations are probably necessary.

The techniques for achieving a large-scale random polycross in self-pollinated crops vary according to the species. When a single fruit produces a large number of seeds (e.g., tomato, potato, tobacco), hand emasculation and pollination with mixed pollen from selected male parents is feasible. Male sterility is available in some species and male gametocides are particularly useful in cereal crops. These chemicals are easier to use than the technique of genetic male sterility; they can be applied to any individual or population, their effects are not inherited, and they have no undesirable, linked characters. At present, no effective male gametocides are known for dicotyledonous crops (e.g., grain legumes) but natural cross-pollination, which is usually 1 to 5 percent in autogamous species, can often be relied on, the crosses being identified by a suitable marker gene. The Stoetzer strategy (Robinson, 1976) is a suitable approach.

The breeding strategy is essentially one of population genetics, and the various techniques used by maize population breeders are appropriate. Unlike pedigree breeding, which tends to emphasize the parents of a successful cross, this approach emphasizes progenies. The selected individuals of each screening generation are randomly polycrossed to become the parents of the next screening generation. All selection criteria are assumed to be polygenically inherited (exceptions must be treated as special cases) and the breeding method aims to shift the mode of each variable toward the desired extreme. All major variables are taken into account and, collectively, constitute the single variable of "balanced domestication" which is raised by a small degree in each screening generation.

One of the disadvantages of pedigree breeding is that it tends to be labor-intensive. Both the need for controlled cross-fertilization and the consequent intensity of labeling and recording severely limit the number of individuals and the number of screening criteria that can be handled. In the new approach, all these labor-intensive procedures are swept aside. In wheat, for example, the traditional approach produces 1,000 to 2,000 labeled and recorded crosses each year. The new approach with male gametocides produces several million unlabeled, unrecorded crosses each year. The only scoring is by eye for the one character of balanced domestication; and the only labeling and recording is made by the simple process of harvesting the selected individuals. In detail, each selected individual may have been scored several times, with the easiest scores made first and the most difficult last; and there is a drastic reduction in the number of selections following each scoring. The final selections should provide only enough seed to sow the next generation (see Section 8.12).

8.5. Entomology

The entomologist must ensure that selection pressure is exerted for resistance to all locally important pests. Clearly, these pests must all be present regularly, both spatially and sequentially. All spotty distribution must be eliminated. The best general technique is to grow borders of susceptible hosts which are then artificially infested if necessary. With screening populations that occupy a large area, because of the size of the host individuals (e.g., cassava), it may be necessary to interplant suitably sized screening blocks with susceptibles. The susceptibles should be planted earlier than the screening population to build up the infestations, and to prevent any cross-pollination with undesirable pollen. If the flowering periods overlap, the borders may be either deflowered or slashed.

Various techniques can be devised for the mechanical redistribution of pests with a spatial spotty distribution. Failing all else, the screening can be confined to heavily infested areas of the screening population; or it can avoid individuals which are entirely pest-free on the grounds that they are more likely to be escapes than immune. The possibilities of screening for resistance to post-harvest pests should be considered.

8.6. Pathology

The pathologist has a closely similar role. Spotty distribution is rare in pathology except with insect-borne viruses and soil-borne pathogens. The latter can be reduced by the avoidance of rotation. However, the main task of the pathologist is to ensure that all screening is conducted in the esodemic. This means that all vertical resistance must be eliminated or, if this is not possible, inoperative. The genetic elimination of all vertical resistance genes is possible in relatively few crops such as potatoes. More commonly, the vertical resistance has to be made inoperative. This is achieved epidemiologically by first selecting one vertical pathotype, and then finding as many good parent cultivars as possible with full vertical susceptibility to it. By employing only *one* pathotype, the vertical resistance is eliminated in all the parents, and in all their progenies, in all subsequent generations. However, in crops with vertical resistance to several different species of parasite (e.g., wheat, rice, tomato, field beans), it is necessary to select a single vertical pathotype of each parasite species. Obviously, the pathotype with the widest host range should be selected first, and so on to the one with the narrowest range of host cultivars. The borders must be susceptible to all the nominated vertical pathotypes and they are inoculated with them prior to each screening generation.

This technique for making vertical resistance inoperative during the screening process means that new cultivars produced from the program will possess vertical resistance which may be operative in farmers' fields. This is unimporant as the horizontal resistance should be so high that, when the vertical resistance breaks down, no one will even notice.

Apart from these measures to ensure selection pressure for comprehensive horizontal resistance, the epidemics should be as natural as possible.

8.7. Physiology

The physiologist's main task is to ensure that there are no undue susceptibilities to environmental stresses, other than those due to parasites. Variable stresses (e.g., short-term drought) are best screened by a staggered planting of subdivisions of the screening population; only the subdivision most critically exposed to the stress is used for screening. The physiologist should also concern himself with the various components of yield and quality, working always at the higher systems levels. He can also exert selection pressure for characters such as time of maturation, daylength response, and so forth.

8.8. Agronomy

The agronomist must ensure that the final selections are suited to the farmers' requirements. *On-site* screening must be conducted under field conditions, and in the area, the time of year, and according to the farming system of future cultivation. All major agonomic factors such as spacing and fertilizer use must be taken into account and the screening must conform with them.

8.9. Other Disciplines

Specialists in other disciplines, such as soil science, weed control, and food technology, can be consulted as necessary.

8.10. The Nonspecialist

In practice, the actual work of screening is best conducted by one man, preferably a nonspecialist, who has been suitably instructed by specialists in the various disciplines. Ideally, he should have that rare gift of a "good eye" for a plant which is healthy, vigorous, and superior to the screening population as a whole.

8.11.   Interplot Interference

The overriding importance of interplot (or interplant) interference (Van der Plank, 1963; James 1973, 1976) has only recently been appreciated. Parlevliet and van Ommeren (1975) have shown that plants grown in isolation may have less than 1 percent of the disease in plants suffering from interplot interference. Interplot interference involves allo-infection, and this is perhaps the greatest weakness of traditional approach to plant breeding. Vertical resistance prevents allo-infection and, as a result, it looks superb in small plots although its impermanence is not apparent. In comparison, the effects of horizontal resistance are greatly reduced by interplot interference; it then looks poor and neither its permanence nor its effectiveness in widespread cultivation is obvious. Traditionally, when the vertical resistance in a small breeders' plot breaks down, the line in question is discarded as valueless. This is a mistake because this is the point at which the esodemic begins and selection pressure for horizontal resistance can be exerted.

Interplot interference is important in two ways when screening for horizontal resistance. The parasite densities may be so high in the screening population that the work appears to be futile; it could be abandoned if the real situation were not appreciated. Secondly, during screening, all measurements of resistance must be relative, not absolute. The *least* parasitzed individuals are selected; they may have much more parasite damage than would occur if they were cultivated as pure lines without interplot interference.

In the earliest screening generations, the parasite damage may be so great that total destruction is threatened. The screening population should then be protected with pesticides to ensure that the least susceptible individuals can produce seed. Conversely, as the work progresses, the overall level of parasite damage declines owing to the gradual accumulation of horizontal resistance. The artificial intensification of the parasite populations may then have to be increased if the selection pressures are to be maintained.

8.12.   Categories of Population

Most annual crops are cultivated only once each year. Because screening must be conducted during that same season, it is possible to breed only at the rate of one screening generation per year. However, various categories of supplementary generation can be exploited in the intervals between screening generations. To obtain a high selection coefficient, only enough of the best individuals of the screening generation are selected to provide seed for the next generation. The selection coefficient can be raised by a supplementary "multiplication generation" of a correspondingly smaller number of selections. It may also be desirable to have a separate "crossing generation" for the random polycross, although crossing can, if necessary, be combined with either the multiplication or the screening generation. In the latter event, it is necessary to have a negative screening to eliminate the worst individuals prior to flowering. A few "special-purpose" generations may be interposed to correct one or two serious imbalances (e.g., extreme susceptibilities) in an otherwise reasonably balanced, if low level, of domestication. Obviously, favorable imbalances (e.g., very high yield) are not detrimental provided they are regarded as secondary selection criteria which are subordinate to the primary aim of a balanced domestication. In some crops, post-harvest screening may be desirable to accumulate resistance to storage parasites. Finally, the screening population may be proliferated to provide separate screening programs for a variety of different geographical pathosystems, farming systems, soil types, and so forth. If necessary, it can also be modified by the addition of new genetic material, subject to the epidemiological elimination of vertical resistance, described above.

8.13.   Formation of New Cultivars

A suitable number of the best-looking individuals can be chosen from each screening generation for multiplication and testing as potential new cultivars. Vegetatively propagated crops (e.g., sugarcane, potato, cassava) present no problems, but seed-propagated corps must be selfed for five to six generations to form pure lines, and clearly all selection pressures must be maintained during this process. A promising possible shortcut is pollen cell culture; the resulting haploid plant is then doubled and is homozygous at all loci. However, the pollen cells are also segregating, and a relatively large number of them are required to permit a final selection of the differing homozygous lines. This subject has been recently reviewed by Reinert and Bajaj (1977).

Both the probability and the quality of such new cultivars increase with successive screening generations. The earlier cultivars will be quickly replaced with newer and superior cultivars; this process can be continued until the law of diminishing returns begins to be severely limiting. At that point modification of the screening population by the addition of new genetic material will become desirable. In many crops, however, particularly in the developing world, these earlier cultivars may well be superior to existing cultivars and would be valuable as "stopgaps" pending their replacement with even better cultivars.

8.14. Measurement of Horizontal Resistance

Absolute measurements of horizontal resistance are best obtained with statistically controlled field trials designed to assess the crop loss due to parasites (James et al., 1971; James, 1974). The trials compare either the current screening population or potential new cultivars with a mixture of the original parent cultivars, under the same "on-site" conditions and parasite selection pressures with all vertical resistances and spotty distributions eliminated. The treatments include appropriate combinations of insecticides and fungicides to assess the losses caused by various categories of parasites. In addition, there must be a reference trial, grown in isolation, to assess interplot interference, and the main trial results must be corrected accordingly.

8.15. Demonstration of Horizontal Resistance

The horizontal nature (i.e., permanence) of the resistance of potential new cultivars must be demonstrated prior to their release to farmers. The best test is genetic; the resistant line is crossed experimentally with a susceptible line to produce a segregating progeny. A normal distribution of susceptibility indicates a polygenic inheritance which is proof of horizontal resistance. The progenies must be inoculated with a pathotype taken from the susceptible parent, and the test must be repeated with each major species of parasite in the pathosystem.

8.16. Reinforcement of Horizontal Resistance

The ultimate levels of comprehensive horizontal resistance that can be achieved by this new approach can only be conjectured. If some species of parasite cannot be completely controlled in this way, the horizontal resistance will have to be reinforced. This can be done either with domesticated vertical resistance (if it occurs) or with an integrated control. Such an integrated control will be greatly facilitated by even modest improvements in the current levels of horizontal resistance.

8.17. Perennial Crops

Although the principles are identical, horizontal resistance breeding in perennials differs mainly in the size of the individual plants, and hence the impractical area of the screening population, and in the generation time which makes a series of screening generations a very long-term process. With crops of ancient origin (e.g., stone and pome fruits, citrus, and wine grapes) a wide genetic gap may also exist between modern cultivars and the wild progenitors. However, in crops of relatively recent origin (e.g., plantation forest species, tea, coffee, cocoa, rubber), the cultivars are similar to the wild progenitors and much time can be saved by screening existing populations, either wild or cultivated, for promising individuals. Such an approach in Ethiopian coffee has already been described (Robinson, 1976) and is now close to successful completion (van der Graaff, 1977). Some 2 to 3 million coffee trees in farmers' crops were examined to identify 650 individuals resistant to coffee berry disease incited by *Colletotrichum coffeanum*. The first harvest of each selected tree was kept for seed and, during the 3 years required for each progeny to come into bearing, each parent tree was repeatedly tested for resistance, yield, and quality. The progenies are tested for homozygosity and the horizontal nature of their resistance. This crash program will produce some 50 to 100 new cultivars with a balanced domestication, each consisting of up to 1,000 trees bearing seed, and in only 7 years.

8.18. Examples

The whole concept of breeding for horizontal resistance is so new that relatively few examples can be quoted. Perhaps understandably, the best examples come from crops derived from a continuous pathosystem and which consequently possess no vertical resistance.

The original pioneer of horizontal resistance was J. S. Niederhauser of The Rockefeller Foundation. He worked with potato blight in the Toluca Valley of Mexico which is the center of origin of *Phytophthora infestans*. At that time, all work on blight resistance involved vertical resistance which, however, broke down so quickly in Mexico that it was valueless. Niederhauser developed high levels of horizontal resistance to blight.

In the breeding of sweet potato in the United States, Jones et al. (1976) have independently devised the methodologies presented here and have obtained good levels of a well balanced domestication. Hahn (1976) has obtained equally impressive results with cassava in Nigeria. One of the best examples is that of the Hawaiian sugarcane breeders who first used a random polycross in this crop. They have produced a cane husbandry that is virtually and permanently free from parasites and that has the highest yields of sucrose in the world. Finally, the natural accumulation of horizontal resistance to *Puccinia polysora* in the maizes of Africa is the example on which all the new methodologies are based (Robinson, 1976).

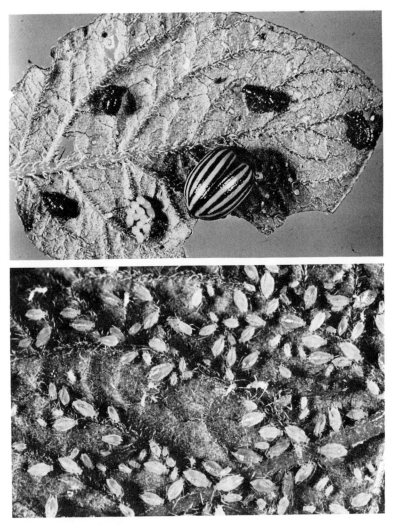

Two variable pests are the Colorado potato beetle (eggs, larvae, and adults, upper photo) and the Green peach aphid, shown on potato (lower photo) Photo courtesy of R. R. Kriner.

9

THE PROBLEM OF VARIABLE PESTS

D. R. MacKenzie

Department of Plant Pathology, The Pennsylvania State University, University Park, Pennsylvania

1. GENETIC MANIPULATION OF CROPS TO CONTROL PESTS

The domestication of crop species is thought to have occurred on many occasions throughout history. One desirable characteristic of those individual plants that may have received attention during selection was improved yield. Such improvement may have reflected differences in resistance to various pests (e.g., insects, fungi, bacteria). As agricultural sophistication increased, more willful selection of individual varieties undoubtedly occurred. In the nineteenth century, certain cultivars were found to be resistant to specific pests. Several wheat cultivars reported to be resistant to Hessian fly (*Mayetiola destructor*) were recommended for commercial use. However, it was not until this century that man's scientific skills advanced to the state of intentional manipulation of crop species for controlling pests.

1.1. Mendelian Genetics First Used to Breed Rust Resistant Wheat

Seven decades ago Sir Roland Biffen initiated the systematic approach to controlling crop pests by the application of Mendelian genetics to plant breeding. In 1907, Biffen (Johnson, 1961) reported on the progress of crossing wheat (*Triticum aestivum*) for resistance to stripe rust (caused by *Puccinia striiformis* = *P. glumorum*). Biffen had discovered that the resistance of the wheat cultivar 'American Club' was controlled by a single gene that was inherited independently of other factors. Rust resistance could therefore be combined with other desirable and genetically controlled characters. Thus began humanity's willful attempts to alter the course of the coevolution of parasitism: coevolution between plant species and their pests. Some breeding attempts have been classic success stories. Others have been colossal failures.

1.2. "Physiological Races" Attacked Resistance

If Biffen's classical work initiated the age of scientific plant breeding for resistance to plant pests, it also inaugurated the age of understanding the pest population. It took only 9 years before it was realized that this new approach to pest control was threatened by the problem of variable pests. By 1916, Stakman and co-workers (Johnson, 1961) discovered that, for stem rust of wheat, "physiological races" of the causal agent, *Puccinia graminis tritici*, could be identified on "differential host" lines of wheat.

Similar discoveries with the Hessian fly (*Mayetiola destructor*) on differential wheat cultivars (Painter, 1930) established a parallel system in insect pests. Although the terminology differs greatly between the disciplines of entomology and plant pathology, the ever-extending list of cultivars with "broken resistance" points to some basic underlying biological principles. It is the intent of this chapter to identify those general principles of pest variation, to unify the terminology for interdisciplinary usage, and to point to new directions for battling variable pests such as insects or fungi.

1.3. The "Breakdown" of Resistance

New cultivars are often released with claims for resistance to specific pests. Often that resistance is not lasting. A previously resistant cultivar may become susceptible to that pest. To the casual observer the transformation would appear to be a "breakdown" of resistance. Actually a new form of the parasite, one capable of attacking that resistance, has risen to predominate the pest population; this is the manifestation of the problem of variable pests.

2. THE POPULATION AS A CONCEPT

Throughout this chapter the term *population* denotes a community of interbreeding or potentially interbreeding individuals. This community shares two important attributes, a gene pool and gene frequencies.

The gene pool is the source of the gametes for the next generation and hence the source of variation, whether it be for parasitism or for host resistance. Gene frequencies are simply the proportions of the different alleles of the genes in that population. The concept of a population is extremely important in pest control, for without an understanding of the extent of the gene pools or the gene frequencies, control strategies can and do fail.

3. COEVOLUTION OF HOST AND PARASITE

The history of genetic manipulations of crop species points to three breeding objectives that now appear to be the root causes of the variable-pest problem: (1) crop uniformity, (2) no yield losses, and (3) simple inheritance for ease of genetic manipulation. All three objectives are understandable and the justifications are many. The problems they create can be viewed in terms of the interaction between the host population and the pest population. Man's breeding for pest resistance is a deliberate attempt to upset a well-established equilibrium. A new equilibrium would then be established between host and pest through the process of coevolution.

3.1. Coevolution as a Process

The evolution of parasitism was not a singular, chance event, but most likely has occurred time and time again on numerous occasions. We can visualize two competing saprophytic fungi on the tip-burned margin of a leaf. One of these could possess, by chance, more staling products than the other. The competitive advantage of the staling products might be in their toxic effects on the living portion of that leaf. More dead tissue in the area of the competitor would allow more reproduction relative to the other saprophyte. This competitive advantage translates directly to fitness. Fitness is relative reproductive success and, in this example, the toxic action of the staling products imparts increased fitness. Thus we see a step toward parasitism.

Similar events undoubtedly took place within and between insect populations. The competitive advantages that could be expressed on the periphery of dead tissue is one easily envisioned circumstance for the evolution of phytophagous individuals from nonphytophagous insects.

Selection for survival of the host would also be expected to operate. As selection pressure increased from the new fungus or insect population, those individuals in the host population that suffered less damage would be expected to contribute more individuals to the next generation. Their reproductive success would be greater than that of other, susceptible individuals, and their relative numbers would increase.

The ebb and flow between host population and pest population would continue over many generations. First one side would be disadvantaged and then the other. Neither would, in the end, become the victor. The result of such a series of trials is the establishment of an equilibrium. This is the final product of extended coevolution.

3.2. The Units of Coevolution; the Genes

The weapons used to wage this war of coevolution are genes, genes for resistance in the host and genes for parasitism in the pest. Two general types of genes are recognized by man the classifier. Major gene systems give qualitative differences in reaction. For example, a host individual having a major gene for resistance would be strikingly distinct from the allelic form of the gene conditioning susceptibility. Major gene reactions in the pest population would also be commonly observed mostly as all-or-nothing responses.

The second system of inheritance is of the minor gene type. The contribution of each gene is small by itself, but the effects are additive. Individuals possessing many minor genes are distinct from other individuals in the population. How such genes operate within the individual is not known. We can only observe how the effects vary with numbers of genes and by the heritability associated with the measured statistical variances.

4. HOST–PARASITE COMPATIBILITY

Wild populations of plants generally possess both major and minor gene systems for pest protection. Major genes can be thought of as the first line of defense in the host. From the viewpoint of the host, newly arriving pest individuals must possess genes for a compatible reaction with that host plant. Individuals lacking a compatible reaction are removed from the pest population. In this chapter such differences in compatibility between individuals in the parasite populations are called *race differences.*

The corresponding differences in compatibility between individuals in the host population have been termed vertical resistance (Van der Plank, 1963, 1968). This term is used to suggest the sharp differences in reactions of cultivars to specific races of a parasite. A histogram of differences in compatibility of races on different cultivars is suggestive of vertical distinctions (e.g., all or nothing).

Figure 1 depicts race differences on a vertically resistant host plant. Race 2 is incompatible. It cannot become established. Race 1 is compatible and

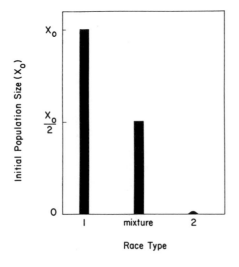

Figure 1. Histographic representation of the relative initial success of two races on a veretically resistant cultivar. Race 1 is successful in becoming established at a level of X_0. Race 2 is not compatible and does not become established. Such differences in histogram plots have led to the term vertical resistance. A mixture of races 1 and 2 on the vertically resistant cultivar, as depicted by the middle bar, would result in the establishment of only race 2. Such effects are considered genetic sanitation.

becomes established. A 1:1 mixture of race 1 and race 2 on this vertically resistant host would be expected to be only one-half as successful as the race 1 alone since all of the race 2 of the mix would be eliminated. Only the race 1 portion of the mixture would become established. This effect, which could be called genetic sanitation, is the apparent function of vertical resistance in the host population.

It is now clear that pest variants capable of overcoming the specific resistance of a cultivar are only part of the host–parasite interaction. For a pest variant to be of consequence, it must be not only compatible with that specific cultivar, but epidemiologically competent.

5. HOST-PARASITE COMPETENCE

The second line of defense operates in natural populations to restrict the increase of the pest population subsequent to its establishment. This form of host resistance has been described as horizontal resistance (Van der Plank, 1963, 1968). The intent of this term was to imply that the resistance operates against all "races" of the parasite. It functions to restrict pest population buildup. It is considered to be more stable than vertical resistance. Much work remains to be done to prove the stability of horizontal resistance, a subject addressed later in this chapter. Much of the discussion about the stability of horizontal resistance focuses on the host response. Little is known about the complement in the parasite population. More research is needed to unravel the underlying principles of parasitism so that intelligent use of our genetic resources can be more fully realized.

6. POPULATION DYNAMICS

In addition to having a gene pool and gene frequencies, a pest population is at any given moment either growing, shrinking, or static. Of importance to the study of pest populations is the rate at which the population is changing. Control procedures can be judged by their effects on the pest population size over a period of time. Often one or more control procedures may not result in a negative growth rate, that is, a population that is shrinking in size. The choice of control practices is likely to be the lesser of two evils (i.e., the lesser of two positive growth rates).

These changes in a pest population size are the concern of a mathematical branch of ecology referred to as population dynamics.

6.1.   Population Growth Model

To describe the changes in the numbers of individuals of a pest population we can apply the general population growth model:

$$X = X_0 e^{rt} \tag{1}$$

which states that a population of X_0 size will increase to size X over time t if r is positive. It will shrink if r is negative or will remain unchanged if $r = 0.0$. In this model r is the continuously compounded interest rate often referred to as the Malthusian parameter. Its importance to the model is that it operates as an exponent, and a small change in r has large (i.e., exponential) effects on the size of X. The remaining parameter of the equation is the mathematical constant e, which is equal to 2.7182818 and serves as the base for the natural log system.

This growth model has been applied to many biological growth systems. It has limitations and different forms. Precise descriptions of population growth patterns are not possible and more sophisticated models are available. However, for purposes of explaining pathogen variation, we shall apply, with modifications, the general growth model of equation 1. When

$$r = 0.0$$

then

$$rt = 0.0$$

and

$$e^{0.0} = 1.0$$

so

$$X = X_0$$

or the population is not growing. When r is greater than 0.0 the population is growing. When

$$rt = 0.693$$

then

$$e^{0.693} = 2.0$$

and

$$X = 2X_0$$

or the population doubles. Therefore, the doubling time of a population can be estimated as $t = 0.693/r$.

Demographers apply this relationship to world population statistics when they project human population doubling time. Given today's crude human population growth rate (r = 0.02 units/year or 2 percent/year), we can expect the world's population to double in 35 years.

This population growth model can also be used to characterize a shrinking population. In this case r would be negative. Since $e^{-0.693}$ = 0.5, X_0 would be halved when rt = -0.693. The half-life of a population is then estimated as

$$t = \frac{-0.693}{r}$$

Given a population of X_0 = 1,000 individuals and a rate of decrease (r) of -0.02 units/day we would project the half-life of that population to be

$$t = \frac{-0.693}{-0.02} = 34.65 \text{ or } 35 \text{ days}$$

Thus 1,000 individuals will be reduced to 500 after 35 days, 250 after 70 days, and 125 after 105 days. Figure 2 plots the function $X = X_0 e^{rt}$ for r = 0.02 and -0.02 units/day.

6.2. Linear Transformation of the Population Growth Model

The function of $X = X_0 e^{rt}$ can be straightened by natural log transformation. In its transformed state this population growth model becomes

$$\ln X = \ln X_0 + rt$$

which is the equation for a straight line. The characteristics of this line are an intercept at $\ln X_0$ (which should be read as "the natural log of X_0") and a slope of r. Because it is a straight-line model, linear regression analysis can be applied with the independent variable as time t and the linear regression coefficient describing the rate of population growth r (when positive) or decline (when negative). Figure 3 plots the transformed values of Figure 2 to demonstrate the effect of transformation on the relationships of X (number of individuals) to time t and rate of increase r.

Algebraic rearrangement of the natural log transformed equation can be used to solve for the rate of growth r, which gives

$$r = \frac{1}{t} \left(\ln X - \ln X_0 \right) \tag{2}$$

which is shown graphically in Figure 3 for the relationship given in Figure 2 under r = 0.02 units/day.

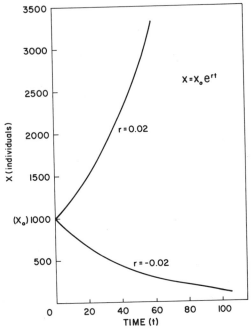

Figure 2. The curvilinear relationship of time (t) to numbers of individuals (X) for the two values of r (the Malthusian parameter) is indicated by the plotted values. Note that the curves are not complements for apparently complementary r values (i.e., 0.02 versus −0.02).

6.3. Population Dynamics and Pest Control Strategies

The mission of pest control programs is to limit the final amount of that pest, be it a phytophagous insect or a pathogenic fungus. Various approaches are used. Some are aimed at reducing the initial numbers X_0. Quarantines have in many instances been very effective in excluding pests from areas where susceptible crops are grown. Sanitation practices also reduce X_0 as do eradicant pesticides. Genetic sanitation (i.e., vertical resistance) also reduces X_0 by eliminating those races which are incompatible with that race-specific resistance. Reducing X_0 diminishes the final amount of pest (X) as seen by our growth model given as equation 1:

$$X = X_0 e^{rt}$$

The effects of reducing X_0 on the buildup of the pest population are direct and obvious.

Other control efforts are aimed at reducing the rate of increase r. Many examples (but not all) of antibiosis in insects and the horizontal resistance mentioned previously act to reduce the rate of buildup of pests. Protectant pesticides also operate to reduce r.

Some crop management systems are effective by reducing the time to harvest (reduced t) by early maturity. Still other practices result in escape or minimization of the pest problem by adjusted planting dates. Depending on the particular pest problem this could reduce X_0 or r or t or any combination of the three.

Tolerance of a pest, as a lessened impact on yield (and/or quality) in the presence of an intense pest population, would not be expressed by this population growth model. Tolerance, in this usage, is not a resistance to the pest population, but a forbearance on the part of the host to that pest. It would not be properly classified as a form of pest resistance.

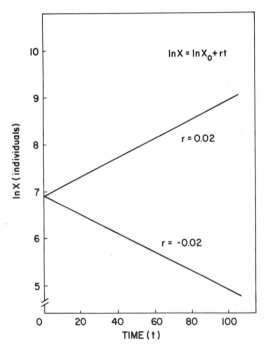

Figure 3. Natural log transformation of the plotted values of X in Figure 2 gives a straight-line relationship with time. Linear regression analyses of such transformed data simplify quantification of r, the Malthusian parameter, and allow statistical evaluation of experimental error.

6.4. Logit Transformation of Pest Progress Curves

One severe restriction on the above growth model is that the population must be allowed to grow unbounded. This assumption is, for most pest populations, unrealistic. Constraints on growth are expected. As a population grows, food supplies become depleted or waste products accumulate to restrict growth. Disease progress curves and insect pest numbers typically increase in a sigmoid curve (Figure 4) rather than the unbounded upward curve of Figure 2 where r was positive.

To correct for this sigmoid feature of most growth curves the logit transformation is often used. The logit of X is given as $\ln [X/(1 - X)]$ when X is measured as a proportion and, for a sigmoid-shaped pest progress curve

$$\ln \frac{X}{1 - X} = \ln \frac{X_0}{1 - X_0} + rt$$

which describes a straight line.

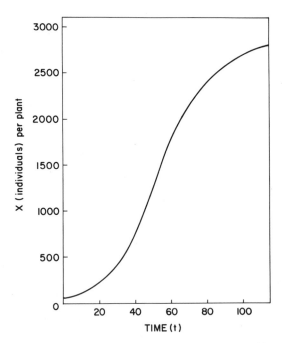

Figure 4. A pest population progress curve demonstrating the sigmoid shape of population growth. Rapid early population growth is followed by constraints on population size, limiting total growth to some upper limit.

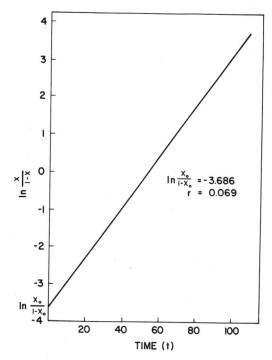

Figure 5. Logit transformation of pest progress curves can be used to straighten sigmoid curves for linear regression analyses. Logit transformation of X is given as $\ln [X/(1 - X)]$, where the total population size is expressed as 1.0 and X is expressed as a proportion (i.e., between 0.0 and 1.0).

Algebraic manipulation gives:

$$r = \frac{1}{t} \left(\ln \frac{X}{1 - X} - \ln \frac{X_0}{1 - X_0} \right) \tag{3}$$

where r is the apparent infection rate for plant pathogens and the apparent infestation rate for phytophagous insects. The qualifier "apparent" must be used to indicate that this is the infection rate when accounting for multiple infections (i.e., repeat infections on already infected tissue) and multiple infestations (i.e., intra-population competition for resources limiting real growth of the population). Analysis of epidemics by this method has come to be known as van der Plankian, in honor of the individual who first comprehensively described the applications of logit analysis to plant disease progress curves (van der Plank, 1963). Figure 5 plots the logit transformed values of Figure 4.

7. TYPES OF RESISTANCE

Host resistance can then be defined in terms of its effect on the buildup of the pest population. Those types of resistance that reduce X_0 are referred to as vertical resistance. Those that reduce r are referred to as horizontal resistance.

This system of classification is not intended to replace other terms. One should not conclude that the multiplicity of terms reflects confusion or ignorance about resistance. Convenience dictates that we employ many seemingly contradictory terms for resistance and hence pest variation. Within this chapter resistance is defined in terms of the population dynamics (i.e., pest buildup). Other terms have been used, such as slow-rusting and rate-limiting types of resistance. Other applications for terminology require alternative approaches. Some forms of resistance are defined in terms of their inheritance (e.g., polygenic, oligogenic, minor gene resistance). Resistance may be defined in terms of the mechanism (active or passive, hypersensitive, antibiosis, nonpreference, etc.), or in terms of the host–pest interaction (e.g., race-specific vs. race-nonspecific, generalized). Those authors who point out the extent of contradictive terminology fail to recognize the need for different classification schemes to better communicate the concepts of pest management through breeding for pest resistance. Here resistance is defined in terms of population dynamics to allow for a better description of pest variation.

8. POPULATION EQUILIBRIUM

Armed with these concepts of population dynamics we can now turn our attention to population stability and the consequences of breeding for pest resistance and the problem of variable pests. Populations whose gene frequencies do not change with time are said to be in equilibrium. There are four general reasons why a population would maintain equilibrium. [For a general treatment of these topics see Strickberger (1968).]

8.1. Hardy–Weinburg Equilibrium

The first reason is known as the Hardy–Weinburg law, which states that unless acted upon by some outside force, gene frequencies will remain unchanged over generations. Outside forces that might upset this equilibrium are nonrandom mating, differential mutation rates between alleles, migration, and/or selection.

One can see immediately the restrictions placed on the application of this law to most pest species. These restrictions are very important considerations for all scientists contemplating the application of population genetics theory to pest population analysis. Almost all plant-pathogenic fungi and many insect pests would not qualify as random mating systems. It is for this reason that, for purposes of developing the concepts of population dynamics, the individual rather than a specific genetic locus of a single gene is taken as the unit of frequency. Nonrandom mating, and asexual propagation in particular, causes the unit of reproductive success to become the individual (i.e., the whole genome) or the isolate rather than a single gene. Hardy–Weinburg equilibrium would not be expected for specific alleles, and hence between individuals, unless all else in the genome was equal.

8.2. Stabilizing Selection Equilibrium

A second cause of genetic equilibrium could be from stabilizing selection. Figure 6 demonstrates the action of stabilizing selection which operates against the extreme classes. The effect of stabilizing selection is to maintain the mean (μ) and reduce the standard deviation (σ) of the population distribution. Maintenance of the population mean is maintenance of an equilibrium.

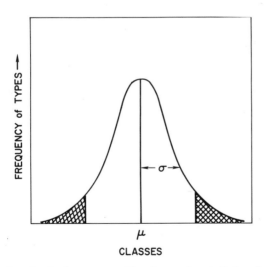

Figure 6.   Population distributions are considered normal or bell-shaped. Normal distributions are characterized by their mean (μ) and standard deviation (σ). Stabilizing selection operates against the extreme types (crosshatched areas). The effect is to maintain the mean or the population equilibrium.

8.3. Balanced Polymorphism

A third cause of genetic equilibrium is balanced polymorphism. Heterozygote advantage in diploids is one way in which balanced polymorphism is maintained. The heterozygote (*Aa*) of one generation produces less fit homozygotes (*AA* and or *aa*) of the next generation. The population is polymorphic (i.e., has many forms) but balanced or in equilibrium. Other mechanisms to promote balanced polymorphism include changes in selective advantage (e.g., changes during the rust season in environment along the Mississippi Valley *Puccinia* pathway from south to north could cause changes in rankings for fitness). Another mechanism that acts to give balanced polymorphism is that of gene interaction and linkage in diploids, as modeled by Mode (1958).

8.4. Mutation–Selection Equilibrium

The fourth cause of genetic equilibrium mentioned here is the result of a balance between mutation rate and selection. Maintenance of this equilibrium depends on mutation supplying an allele at the same rate that it is being lost from the population because of selection.

9. PEST POPULATION RESPONSE TO HOST RESISTANCE

Whatever the cause of genetic equilibrium within a pest population, introduction of a resistant variety tends to upset that equilibrium. Selection proceeds, when possible, in the direction of increased fitness in the pest population. This effect has been termed directional selection and is depicted in Figure 7. The "old population" of the pest on the old variety experiences directional selection on the new variety producing the "new population." This is selection for parasitic competence; this is the dynamics of variable pest populations.

Directional selection for increased parasitic fitness would operate only in those populations with some genetically different types. Genetic variance is necessary for selection to proceed. This variance could come from several sources, with mutation and sexual recombination being two very important and obvious sources.

10. THE HOST-PARASITE HIERARCHY

The differences in terminology among the different pest control disciplines are a barrier to interdisciplinary communication. For this reason, I pro-

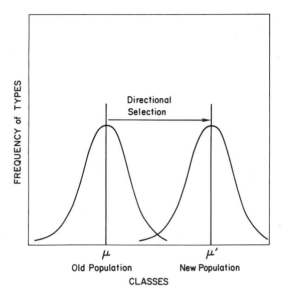

FREQUENCY of TYPES

Directional
Selection

μ
Old Population

μ'
New Population

CLASSES

Figure 7. If a pest population is in equilibrium on an old cultivar and a new cultivar is introduced, directional selection may operate to create a new pest population distribution of mean μ. Directional selection in the parasite population is observed as a "breakdown" of the host resistance.

pose a new order of terminology so that entomologists, pathologists, nematologists, and others can better communicate the concepts of pest variation.

10.1. Parasitism Is An Attribute of a Species

Puccinia graminis tritici is a parasite of wheat. *Mayetiola destructor* is a parasite of wheat. However, not all individuals of a population are necessarily compatible with a specific host cultivar.

10.2. Parasitic Compatibility Is An Attribute of a Race

Race '15B-3' of *Puccinia graminis tritici* is compatible with the wheat cultivar 'Selkirk.' Race 'Great Plains' (biotype in some usage) of *M. destructor* is incompatible with the wheat cultivar 'Benhur.' Growing evidence now indicates that this level of compatibility is characterized genetically as a gene-for-gene relationship. Flor first noted that in the flax (*Linum usitatissimum*) rust (caused by *Melampsora lini*) interaction was on a one-for-one matching in the host (for resistance) and the pathogen (for virulence). This gene-for-gene relationship is now accepted by many as a basic principle of parasitism (Flor, 1971). Burnett tabulated the known and suggested examples of gene-for-gene relationships for plant pathogens (Burnett, 1975). Hatchett and Gallun have described an equivalent gene-for-gene relationship for the Hessian fly interaction on wheat (Hatchett and Gallun, 1970). As this body of evidence accumulates to support more gene-for-gene relationships with other host–parasite interactions, it suggests a broad genetic relationship of parasitism.

One practical application of the understanding of the gene-for-gene relationship in the host–parasite relationships is to help us better comprehend the speed with which some pest populations manage to "break" resistance. Single gene resistance in the host need only be matched by a single gene for compatibility in the pest. Random mutation is one very likely source of newly compatible individuals.

10.3.   Parasitic Aggressiveness Is An Attribute of a Biotype

Not all individuals within a race classification are equal in aggressiveness. Individuals within an insect pest population can be thought to differ in aggressiveness. More aggressive individuals might take less time to reach adulthood, produce more eggs/adult, or have a greater proportion of eggs hatch. Corresponding differences in aggressiveness of plant-pathogenic fungi are also viewed as increased infection efficiency, increased sporulation, or shortened time for each step in the progression of disease development. Two isolates of race '15B-3' of *Puccinia graminis tritici* may differ in the rate at which they cause disease. The overall effect is to increase the aggressiveness of that individual. Such differences in aggressiveness are attributes of a biotype.

Horizontal resistance in the host is matched by parasitic aggressiveness by the pest. Although attempts have been made to define horizontal resistance and parasite aggressiveness in terms of the number of genes conditioning each character, sufficient exceptions exist to limit such efforts. Let it suffice here to note that horizontal resistance and aggressiveness are commonly observed to be polygenic, which is said to account for the stability of this type of resistance.

10.4.   Parasitic Fitness Is An Attribute of an Individual

Compatibility and aggressiveness together determine an individual's parasitic fitness. Parasitic fitness is an attribute of an individual genome or isolate. For an isolate to be parasitically fit it must be compatible with that host genome and be aggressive. It is from this vantage that we can view the ebb and flow of the problem of variable pests. But to do this we must be able to measure parasitic fitness.

11. ANALYSIS OF PARASITIC FITNESS

If r (the apparent infection rate) measures differences in horizontal resistance between two varieties for the same pest isolate in the same environment, then it must be equally true that differences in r between two isolates for the same variety in the same environment measure isolate aggressiveness. In other words, we can apply the epidemic analysis of van der Plank to the quantification of parasitic fitness. To do this we return to equation 1:

$$X = X_0 e^{rt}$$

where X_0 is now a measure of compatibility, taking on a value of 0.0 for incompatible reactions (i.e., genetic sanitation).

11.1. Absolute Parasitic Fitness

The parasitic fitness (F) of an isolate from one time period to the next is given as

$$X = X_0 F \qquad (4)$$

for that one time period. The fitness (F) of that isolate (population) could be greater than 1.0 (growth), less than 1.0 (decline), or equal to 1.0 (static).

The second time period would be given as

$$X = X_0 FF$$

or

$$X = X_0 F^2$$

Generalizing the equation for t time periods gives

$$X = X_0 F^t \qquad (5)$$

Inspection of this equation and the one used to express population growth shows that fitness (F) is related to the instantaneous population growth rate (r) as

$$F = e^r$$

or

$$\ln F = r$$

which states that if r is positive, F is greater than 1.0 and the population is experiencing growth. If r is negative, F is less than 1.0 and the population is declining. And if r is 0.0, $F = 1.0$ and the population size is static.

The values for F and r would be determined for a pest isolate by the compatibility of that race on that host and by the aggressiveness of that biotype on that host. By observing the changes in absolute numbers of individuals on a host we can quantify that isolate's absolute parasitic fitness (F).

11.2. Relative Parasitic Fitness

More commonly, researchers study mixtures of biotypes on susceptible hosts to investigate one isolate's abilities relative to that of another. Such studies should determine the parasitic fitness of an isolate relative to another. Data are usually expressed as a relative frequency rather than absolute numbers (as is done for absolute parasitic fitness). It is for this reason that the term relative parasitic fitness (W)* is used to describe such relationships.

To compute relative parasitic fitness we choose a standard value for the frequency of the more fit isolate to simplify some later steps. Data for an insect (rice brown planthopper) study (International Rice Research Institute, 1975) and a plant pathogen (stem rust of wheat) study (Loegering, 1951) are used to demonstrate how such calculations are made. The data for the rice brown planthopper (*Nipaparvata lugens*) (Table 1) indicate that isolate B4 (biotype 4 was the term used by the original text but it is inconsistent with the present usage) was more fit than B1 on all four rice cultivars (i.e. greater survival). Therefore, B4's relative parasitic fitness (W) is set equal to 1.0 (i.e., no relative change in frequency over the period studied). If p is the frequency of B4, then p/p equals 1.0. If q is the frequency of B1, then q/p is the frequency of q relative to the frequency of p. The change in q (relative to p) can now be given as

$$\frac{q}{p} = \frac{q_0}{p_0} W^t$$

where W is the relative parasitic fitness and, since we arranged it this way,

---

* Standard notation for relative biological fitness would be lowercase w. However, it is recognized that far more than parasitic abilities determine biological fitness (e.g., overwintering ability) and the use of capital W recognizes this fact.

Table 1.   The Percent Survival of Brown Planthopper (*Nilaparvata lugens*) Isolates[a] when Caged on the Rice Cultivars 'ASD-7', 'IR26', 'Mudgo' (Reported as Resistant to Isolate B1) and the Susceptible Cultivar 'Taichung Native 1' ('TN1'). Isolate B4 was Reported to have a High Rate of Survival on all Four Rice Cultivars. (Data from Figure 21 of The International Rice Research Institute, 1975)

Days After Caging	TN1		ASD-7		IR26		Mudgo	
	B4	B1	B4	B1	B4	B1	B4	B1
1	100	100	95	88	95	80	98	68
2	98	100	92	88	95	65	97	62
5	98	100	92	68	92	58	92	22
10	96	92	82	68	89	38	90	10
15	96	82	82	41	80	24	87	10

[a] The authors referred to the collections as biotypes 1 and 4. This use of the term biotype is inconsistent with its use in this text and are therefore noted here as isolates B1 and B4.

W is less than 1.0 because it measures the parasitic fitness of the less fit isolate.*

When p and q are expressed as proportions, $p + q = 1.0$ and then $p = 1.0 - q$. We can, therefore, rewrite the relative fitness expression above to

$$\frac{q}{1-q} = \frac{q_0}{1-q_0} W^t$$

and since $W = e^r$,

$$\frac{q}{1-q} = \frac{q_0}{1-q_0} e^{rt}$$

and

$$\ln \frac{q}{1-q} = \ln \frac{q_0}{1-q_0} + rt$$

* Another way of viewing these interrelated concepts is, when $r' > r$ and $r - r' = -r''$ and S is the coefficient of selection,

$$\frac{e^r}{e^{r'}} = e^{r-r'} = e^{-r''} = W = 1 - S$$

or

$$r = \frac{1}{t} \left(\ln \frac{q}{1-q} - \ln \frac{q}{1-q_0} \right) \tag{6}$$

This mathematical relationship is equivalent to the logit transformation analysis of epidemics given by van der Plank (1963). Thus the relationship between the host's horizontal resistance and the relative parasitic fitness of the pest becomes obvious through the connecting link of the logit analysis. The logit transformed plot of the data for rice brown planthopper isolates B1 and B4 is given in Figure 8. A tabular summary of the statistical analyses of these relationships is given in Table 2. Of specific interest to this discussion is that the parasitic fitness of isolate B1 on all four rice cultivars is not very different from that of B4. (Analyses of the regression coefficients indicate no statistically significant differences.) This suggests that the two isolates are nearly equal in parasitic fitness. Moreover, the ranking pattern of relative parasitic fitness remained constant, with B4 in all cases more fit than B1. Had there been any significant differences in relative parasitic fit-

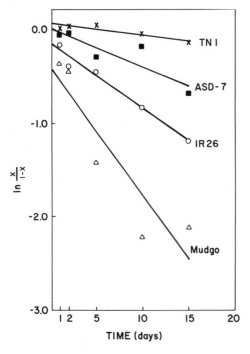

Figure 8. Linear regression analyses of the logit transformed survival values of rice brown planthopper isolate B1 (relative to B4) on the rice cultivars 'Taichung Native 1' ('TN1'), 'ASD-7,' 'IR26,' and 'Mudgo.' Statistical parameters for the regression fit and the values for the relative parasitic fitness of isolate B1 are given in Table 2 (data from Table 1).

Table 2. Statistical Analysis Summary of the Relative Parasitic Fitness of Isolate B1 of the Rice Brown Planthopper (*Nilaparvata lugens*) Relative to Isolate B4 on Four Rice Cultivars. Linear Regression Analyses of the Logit-Transformed[a] Proportion of Surviving Individuals over a 15-Day Study were used to Determine the Relative Parasitic Fitness and the Coefficient of Selection for B1 (Data from Table 1)

Rice Cultivar	Intercept	Coefficient of Regression, r	Coefficient of Determination, R^2 (Percent)	Relative Parasitic Fitness[b] ($W = e^r$)	Coefficient of Selection ($S = 1 - W$)
Taichung Native 1	0.045	−0.012	82.0	0.988	0.012
ASD-7	0.001	−0.039	76.6	0.962	0.038
IR26	−0.157	−0.069	97.9	0.933	0.067
Mudgo	−0.407	−0.138	83.8	0.871	0.129

[a] To derive the logit values from Table 1 the percent survival of *B4* and *B1* can be expressed as X and Y, respectively. The logit of X as $\ln [X/(1 - X)]$ would then be $\ln \{X/[(X + Y)]\} = \ln X/Y$, saving several unnecessary steps of calculating.

[b] Expressed in units per day.

205

ness they could be thought of as a parasitic aggressiveness–horizontal resistance interaction.

A more complete expression of the differences in parasitic fitness of isolates B1 and B4 would have been obtained from a more extended study across generations. Much more research emphasis on differential survival over pest generations is needed. Such effort would lead to a better quantification of parasitic fitness and hence a better description of pest variation.

Table 3 gives an analysis of the data of Loegering (1951) for mixtures of *Puccinia graminis tritici* "races" 'R17' and 'R19,' with the caveat that the comparison is between isolates, not races. Figure 9 plots the logit-transformed data over asexual generations. Nearly straight lines were fitted for the six generations of stem rust reported for cultivars 'Little Club' and 'Fulcaster.' However, the last two generations for 'Mindum' appear inconsistent in pattern. Regression analysis was, therefore, applied to the

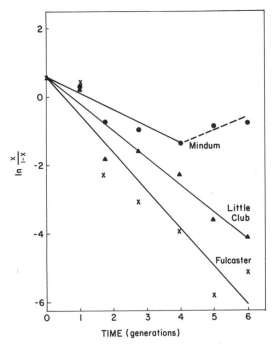

Figure 9. Linear regression analysis of the logit transformed relative proportions of *Puccinia graminis tritici* (stem rust or wheat) isolate R19 (relative to R17) on the three wheat cultivars 'Mindum,' 'Little Club,' and 'Fulcaster.' Statistical parameters for the regression fit and the values for the relative parasitic fitness of isolate R19 are given in Table 4 (data from Table 3).

Table 3.   The Percent of Isolates R17 and R19^a of *Puccinia graminis tritici*
(Causal Agent of Stem Rust of Wheat) in a Mixture Studied for Six Asexual
Generations on the Wheat Cultivars 'Mindum', 'Little Club', and
'Fulcaster.' (Data Expressed as an Average of Three Replications from
Figure 2 of Loegering, 1951)

Asexual Generation	Mindum		Little Club		Fulcaster	
	R17	R19	R17	R19	R17	R19
1	44	56	44	56	44	56
2	67	33	86	14	90	10
3	72	28	83	17	96	4
4	79	21	90	10	98	2
5	70	30	97	3	99	1
6	68	32	98	2	99	1

^{a} The author referred to the collection as races 17 and 19. The use of the term race is inconsistent with its use in the text and are therefore noted here as isolates R17 and R19.

first four generations only. This inconsistency in the data may well reflect a change in the environmental conditions that reversed the relative parasitic fitness of the two isolates studied. A statistical summary of these analyses is given in Table 4. Of interest are the large differences in relative parasitic fitness of isolates R19 on the three wheat varieties. It is these host effects on isolate fitness that can be employed to control variable pests through plant breeding. Of concern to these discussions is the change of pattern on the cultivar 'Mindum.' Some authors (van der Plank, 1963, 1968, 1975; Robinson, 1976) have argued for the stability of horizontal resistance and its nondifferential interaction with isolates. In this particular study some differential interaction is suggested. Much more research is needed before definite conclusions can be advanced as to the expected pattern of interactions for parasitic aggressiveness and horizontal resistance.

12.   ENVIRONMENTAL BIOTYPES

We know that the environment contributes a large component to the growth dynamics of a population. Under some environmental conditions some pest populations cease to increase. Such conditions are easily identified for the

Table 4. Statistical Analysis Summary of the Relative Parasitic Fitness of Isolate R19 of *Puccinia graminis tritici* Relative to Isolate R17 on Three Wheat Cultivars. Linear Regression Analyses of the Logit-Transformed Proportion of Stem Rust Lesions over Six Generations was Used to Determine Relative Parasitic Fitness (W) and the Coefficient of Selection (S) for R19. (Data from Table 3)

Wheat Cultivar	Intercept	Coefficient of Regression, r	Coefficient of Determination, R^2 (Percent)	Relative Parasitic Fitness[b] ($W = e^r$)	Coefficient of Selection, ($S = 1 - W$)
Mindum[a]	0.540	−0.496	90.6	0.611	0.389
Little Club	0.581	−0.786	90.9	0.456	0.544
Fulcaster	0.521	−1.097	88.8	0.334	0.666

[a] Generations 5 and 6 removed from data set (see text for explanation).
[b] Expressed in units per generation.

late blight disease of potato, *Solanum tuberosum*, caused by *Phytophthora infestans*.*

It follows that under a given set of environmental conditions some individuals in a pest population will be more capable than others of growth. If relative reproductive success under those conditions is greater, then it can be said that fitness is greater. Therefore, groups of isolates with different sensitivity to environmental factors should be termed *environmental biotypes*. This would distinguish such variants from host-mediated biotype differences. To avoid lengthy phraseology, the implied meaning for the term "biotype" should be reserved for the host-mediated biotype differences, and "environmental biotype" is best limited to environment-mediated biotype differences. All factors mediating parasitic differences must, however, be considered when the problem of variable pests is discussed.

13. APPLICATION OF CONCEPTS TO PEST CONTROL

How do we use this information to protect crops from variable pests? Several very exciting approaches have been suggested and/or explored to utilize parasite variation to manipulate pest populations (or at least stay ahead of their destructive impact).

13.1. Stabilizing Selection

Pest races with many genes for compatibility sometimes appear to be less fit to survive on cultivars for which those compatibility genes are unnecessary. Van der Plank (1968) has inappropriately applied the term "stabilizing selection" to this phenomenon. This is unfortunate since much confusion has resulted from this use of the terminology. Van der Plank's argument could be appropriate if the population had been originally in an equilibrium maintained by stabilizing selection (see Figure 6). Then, following the introduction of a new cultivar, directional selection could operate to produce a new pest population distribution (Figure 7). In this event, the cultivar's resistance would appear to have "broken." The argument continues that removal of the now-susceptible cultivar would allow the reestablishment of the old pest population through directional selection (not stabilizing selection, as van der

---

* A disease forecasting system based on these parameters is now offered as a service by The Pennsylvania State University, Department of Plant Pathology to commercial growers in many areas of the eastern United States. Growers need not spray fungicide on potatoes when environmental conditions limit r to zero.

Plank called it). Once reestablished by directional selection, that old equilibrium would then be maintained by stabilizing selection.

The applications of this concept to disease control are many. Removal of a "broken" cultivar from commercial production could, after directional selection has returned the population to the previous equilibrium, lead to its reuse.

The resistance of pests to pesticides is an often-cited example of how unnecessary genes "drop out of a population." Generalization from the available records is avoided here because the evidence in the literature is contradictory, but the explanation for these contradictions is simple. As previously explained, a population can be expected to maintain equilibrium for one of several reasons. These factors may act singly or in combination. Given a pest population in Hardy-Weinburg equilibrium that is subsequently exposed to a newly released resistant cultivar, directional selection would be expected to function. Resistance would appear to "break down." However, one would not necessarily expect a return to the old equilibrium levels of the population by removal of that new cultivar. The gene(s) for compatibility would remain in relatively high frequency.

To predict the response of a population to cultivar management, one must know the dynamic forces controlling that population's gene frequencies. Generalization of the predictions can only lead to contradictions. Much more must be learned before adequate population kinetics predictions can be made. However, one hypothesis can be advanced regarding those population equilibria now maintained by stabilizing selection. If possession of a gene for compatibility by the parasite leads to reduced relative parasitic fitness (i.e., W and hence a larger negative r), then it seems reasonable to suggest that one use of vertical resistance genes is to let them "break down." In those instances where specific races overcome specific genes for resistance (i.e., vertical resistance) the compatibility race would be relatively less fit and the epidemic would proceed at a lessened rate. This effect has in fact been reported (Martin and Ellingboe, 1976) for powdery mildew of wheat (caused by *Erisiphe graminis tritici*). Using isogenic lines of wheat, it was discovered that the "race" p4, which is compatible with the vertical resistance gene *Pm4*, had only half the infection efficiency that it had on the recessive *pm4* allele. Reduction of infection efficiency would have direct effects on the rate of buildup of an epidemic. This would be measured as a reduction of the apparent infection rate and hence, horizontal resistance. Perhaps, then, once overcome by race p4 it would be best to continue with the use of vertical resistance gene *Pm4* in the "broken resistance" state rather than reintroduce old cultivars without gene *Pm4*. In any event, the report by Martin and Ellingboe points out the difficulty of defining types of resistance by their inheritance and then applying that definition to epidemic progress analysis. Horizontal resistance, in this case, would be conditioned by the single host gene *Pm4*.

13.2. Conditional Lethality

Another approach suggested for the manipulation of pest populations employs the "conditional lethality" of some pest compatibility systems. Foster (1976) recently discussed in detail an approach to the manipulation of the native Hessian fly population. Six races of Hessian fly have been reported. In studies of the inheritance of racial differences (Hatchett and Gallun, 1970; Gallun and Hatchett, 1969) it was found that the incompatibility of the 'Great Plains' race to survive on resistant wheat cultivars was dominant to compatibility. The expression of this trait is a "conditional lethal" since larvae of the 'Great Plains' race survive on the susceptible cultivar but not on the resistant ones.

It is known that the pest population has only two generations per year, with emergence restricted to two 1-month periods throughout most of the winter wheat region. Moreover, in many regions cultivars resistant to the 'Great Plains' race are now grown. Given these circumstances it has been suggested (Gallun and Hatchett, 1968) that the massive introduction of the 'Great Plains' race into these areas would result in progeny incompatible with the wheat cultivars of that region. Preliminary research of this scheme looks exciting (Foster, 1976).

13.3. Multilines

Jensen (1952) and Borlaug (1959) suggested using multilines for the control of highly variable plant pathogens. Multilines are mechanical mixtures of phenotypically similar component lines, each differing for specific genes for resistance to the pest population. The power of the multiline is in the blocking of infection between plants [i.e., slowing the exodemic in the terminology of Robinson (1976)]. Some spectacular successes controlling oat rusts with multilines in Iowa (Browning, 1974) suggest that for some plant pests this approach may block rapid population buildup. Moreover, as the pest population shifts in response to selection, other component lines can be substituted. The effect, if done with finesse, would be to manage the pest population for the purpose of avoiding buildup of specific races. The application of multilines to insect pest problems awaits further research.

13.4.   Selection for Horizontal Resistance to Pests

Other applications for an understanding of the problem of variable pests to breeding crops for pest resistance are to be found in limiting the rate of pest buildup in plant populations. By partitioning horizontal resistance into its components (e.g., latent period, sporulation, and disease efficiency for a plant pathogen or, for insect pests, the number of eggs laid, number of eggs hatched), selection techniques for segregating host populations can be developed. Extension of the time required for a fungus pest to sporulate would be expected to have a negative exponential relationship with disease increase. Slight modification of a few hours to a day or two could have sizable effects on some host–parasite systems. Uniform and controlled inoculation of F_2 wheat populations with selected isolates of *Puccinia graminis tritici* could allow for the identification of those differences. Similar host-caused delays in life cycle completion of an insect pest population could have equivalent benefits. Such techniques are needed by plant breeders for all of the components of horizontal resistance. Once they have been identified, they must be studied for differential interaction with specific isolates of the parasite. Such isolates should allow for an appropriate evaluation of the stability of that resistance when faced with the variable pest population.

14.   MONITORING PARASITIC FITNESS

An additional area that needs immediate attention is that of monitoring pest potential. Currently, pest variation is described in terms of race differences. Each year hundreds of isolates of many pest species are "identified" on differential varieties. Little or no attention is paid to the differences attributable to aggressiveness and/or fitness. The existence of compatibility on the wheat stem rust resistance gene *Sr6* in the Mississippi Valley says little about the parasitic fitness of those *Puccinia graminis tritici* isolates. Large differences could exist in their ability to incite an epidemic.

15.   FINAL THOUGHTS

More than 38 years ago H. S. Smith (1941) in his presidential address to the annual meeting of the American Association of Economic Entomologists, pointed out the inevitable consequences of parasitic variation when he stated that "A Shift of the characteristics of the population in the direction of greater fitness *must* (Smith's italics) occur if variants are present which are superior to the general population in their ability to persist and to leave progeny in the presence of the changed conditions; if the superiority of the variants is hereditary; and if the frequency of interbreeding with other races is retarded." This statement remains absolutely true today!

Plant and insect models complement mass rearing techniques used in resistance studies. Photo courtesy of CIMMYT.

10

THE USE OF PLANT AND INSECT MODELS

Johnie N. Jenkins

Boll Weevil Research Laboratory, U.S. Department of Agriculture, and Mississippi State University, Mississippi State, Mississippi

1. INTRODUCTION

When one or more sources of resistance to an insect pest have been discovered, the task of breeding the source or sources into agronomically acceptable cultivars begins. Cultivars of plants grown in commercial agriculture are bred and selected to produce high yields under given production practices. Thus these cultivars have been "finely tuned" for quantity and quality. Incorporating a source of resistance to a particular insect may have a positive, negative, or neutral effect on resistance to other insects as well as on quantity and quality of the harvested product. Plant growth is a complex process affected by a number of environmental factors including insect damage.

The way in which an insect develops and successfully utilizes a crop plant for food, shelter, or oviposition is also complex. Whenever sources of resistance are discovered and one begins to breed these into commercial cultivars of crop plants, one expects to change the manner in which the pest insect utilizes the host plant. The growth of the insect is a complex process of physiological as well as behavioral functions. An interaction exists between plant growth and insect growth; each can influence the other. The host-plant resistance team must deal with these complex plant–insect interactions and turn them to their advantage. Except in cases of tolerance, the objective is to favor plant growth and development at the expense of insect growth and development.

For many crops alternative methods of control, such as insecticides, are available and host plant resistance must compete with these. For many crops, there is also more than one pest insect. The additional complicating effects of parasites and predators of the pest insects deepen the complexity of the agro-ecosystem which we are trying to manipulate to our advantage. The host-plant resistance team must attempt to consider the total plant–insect growth system and to optimize it to the advantage of the producer; otherwise, the producer will not choose the resistant variety.

Limited situations exist in which only one or two pest insects affect a crop and host-plant resistance is the only means of control available, but these are exceptions rather than the rule in agricultural production. In such situations, the grower may not actually have a choice of methods of control or of varieties higher yielding than the resistant ones. This chapter is primarily devoted to situations more complex than these.

Simulation models, widely used in the physical sciences, are emerging as tools in the biological sciences. Plant and insect simulation models offer the host-plant resistance team an excellent tool to advance their breeding work and to save them time and money. This chapter defines models, discusses their physiological basis, considers the presently available plant and insect models, and presents selected examples of their use. In the systems approach to modeling one considers the complete picture and includes all possible relevant elements that influence a decision or response in a particular situation (Witz, 1973). Using modeling as a tool in host-plant resistance does not mean that we can do away with the need for good management, high-quality research, clear thinking, and hard work. It gives the host-plant resistance team an added tool to understand better the complex plant–insect growth situations and to enhance their ability to make correct decisions.

2. DEFINITIONS

The following definitions have been gleaned from several sources (ARS Modeling Committee, 1977; Jenkins, 1975; Edminster, 1978; and *Technical Manual* 520).

Model.   A representation of something that exists.

Mental model.   A mental image of how something that exists is *imagined* to be. Thus each person may have a mental image or mental model of the same thing, and yet they may differ. For example, each person can develop a mental model of a particular city or a particular insect, but none of these mental models would be identical.

Tangible model.   A mental model transformed into an explicit form such as a word description, physical scale model, or a set of equations. Tangible models may be either abstract or physical scale models.

Physical scale model.   A tangible model that reproduces the form of the original in some scaled way. For example, physical models of biological organs are used to teach anatomy of animals and plants. The U.S. Army Corps of Engineers has a physical scale model of the Mississippi River at Vicksburg, Mississippi, which they use to develop and test theories dealing with water management in the Mississippi River and its tributaries.

Abstract model.   A tangible model in which symbols are used to represent the form or function of the original in a nonphysical manner.

Qualitative model.   An abstract tangible model that represents, as a minimum, the qualitative aspects of what is being modeled.

Mathematical model.   An abstract tangible model (nonphysical) that uses mathematical notation to represent quantitative as well as qualitative aspects of what is being modeled. Examples are (1) a single-gene model of Mendelian inheritance that gives a 3:1 ratio in the F_2 generation, (2) a set of equations that describes the sigmoid growth curve of plants, and (3) a components-of-variance model in the statistical analysis of variance. All scientists use mathematical models in one form or another.

Computer model.   A mathematical model that has been programmed to operate on a computer. The computer may implement a mathematical model in a dynamic form, with iteration or time steps which are a small part of the total time period of interest. For example, hourly or daily iteration may be used in a simulation of a season's crop growth. Such a model typically involves much logic as well as mathematical functions in its simulation. This is the type of plant and insect models that are discussed in this chapter as examples of useful resistance breeding tools which are emerging in the biological sciences.

Verification. Comparison of a tangible model with the mental model used in its development to be certain the mental model is accurately represented in the tangible model. When a multidisciplinary team develops a model as complex as plant or insect growth, verification should be considered a separate and important step in model building. Each researcher has his own mental model based on his research knowledge of the system being modeled. Verification assures that each one's mental model is represented in the abstract model.

Validation. Comparison of a verified model to "real world" processes, to determine if the model suits its intended purposes. Validation involves experimentation under practical conditions.

3. TRANSFORMING MENTAL MODELS INTO TANGIBLE MODELS

All plant and insect models represent attempts to describe and connect the physiological life processes in terms of mathematical equations. This is necessarily complex and may not be feasible in some situations. Thus a logical question is, "Do we know enough about a particular plant or insect to develop a model?" Rather than answer this question directly, we consider three points:

1. Plants, insects, and populations of each *grow.* We conduct much of our research on growing plants or insects and we attempt to apply our results to crops and insect populations. Are we applying research results to a system that we do not understand well enough to describe in tangible terms? Is the experimental information really applicable to the development of a system model? If so, are all the major components covered in some way?
2. We should build the best model we can, but we do not need a perfect model to benefit from its use. Initially we need only a model that will further our understanding of plant and insect growth processes and interactions, that is, a model that will help us to identify information gaps.
3. Converting a mental model to a tangible model takes it out of the mind and puts it in a form in which all who wish can see it, conduct research with it, improve on it, and advance our understanding of the processes described.

4. USES OF MODELS

The following uses of tangible computerized models were cited by the U.S. Department of Agriculture ARS Committee on Modeling (1977):

1. Models help to sharpen the definition of hypotheses. For example, several alternate hypotheses can be developed and simulated. One can then conduct research under practical conditions on the most likely hypothesis.
2. Models enhance communication, especially among interdisciplinary team members. Consider as an example the term "insect control." If a biological engineer, agronomist, entomologist, toxicologist, insect pathologist, and host-plant resistance team are asked to control the insects on a particular crop, each would approach the job differently because they do not have the same mental image of insect control. However, if the mental model of the insect control requested is converted to a simple tangible model of 123g/hectare of azinphosmethyl, applied every 3 days with ground equipment, communication has been facilitated. Obviously, mental models of plant or insect growth are much more complicated, which is all the more reason to convert them to tangible models.
3. Models help to define and to categorize the state of our knowledge. For example, when one attempts to build a model, vague impressions of how systems operate are difficult to use and one must attempt to quantify the knowledge or admit to a gap in knowledge.
4. Models provide an analytical mechanism for studying the system of interest, for example, determinations of (1) the process or rate at which a plant converts photosynthate into both vegetative and fruit growth, or (2) how an insect converts plant tissue into body weight and reproductive growth.
5. Models can be used to simulate experiments in situations in which actual experimentation would be very costly or dangerous to conduct. Simulation models can be used to gain a better understanding of the situation, in place of, or before the "real world" experiments.
6. Models can be used to plan efficient experiments.
7. Models can provide a key to determining the progress of research.
8. Models can provide a mechanism for disseminating information.
9. Models can be used for prediction.

In this chapter, examples of the use of plant and insect models to accomplish several of these uses are given. Witz (1973) succinctly stated that models can provide an extension of a researcher's power for handling complexity, a framework for collecting bits and pieces of information, and a tool for evaluating and experimenting with a system to demonstrate new ideas or hypotheses. Moreover, a proven or validated model can be further applied in a predictive mode to improve or to optimize the control or use of the modeled system.

A word of warning concerning the use of models was given by Krigman (cited by Witz, 1973): "Risks are high of confusing mathematical precision with strenth of empirical verification by forgetting that the results are not explanations of phenomena but only consequences of assumptions." This warning is also applicable to traditional research. He also mentioned Maxwell's warning that "observing a phenomenon through an analogous medium encourages blindness to facts and rashness of assumptions." One should keep these admonitions in mind as models are used to enhance our understanding of the biological systems of plant–insect interaction we are studying in host plant resistance.

5.   AVAILABILITY OF PLANT AND INSECT MODELS

The availability of plant and insect models has taken on a new perspective with the formation of FARM (File of Agricultural Research Models) by the U.S. Department of Agriculture, Agricultural Research Service. This group plans to keep an updated listing of agricultural models and to make it available to interested researchers.* FARM is an information retrieval system established by the Data Systems Application Division (DSAD) in cooperation with the ARS Committee on Modeling. The purpose of the retrieval system is to provide a complete and convenient source of information on agricultural simulation models.

The original intent of FARM was to make an inventory of ARS models. However, other organizations, particularly universities and state agricultural experiment stations, should feel free to participate. Simulation models generally involve a set of state variables whose values change with time as a result of explicit cause–effect relationships. FARM lists only those models that have been implemented on a digital computer and that simulate some biological system related to agriculture. Its inventory consists of descriptions rather than complete models themselves. The DSAD does not act as a distribution center for models; interested individuals should contact the authors of the models directly.

FARM information on each model forms a single record with the nine fields shown in Figure 1 (FARM 1977).

Figures 2 and 3 show examples of a plant growth model and of an insect growth model as they are listed in the FARM printout. They indicate the type of information one can obtain from FARM and show some of the physiological bases of models.

---

* The information can be obtained by writing to Mr. Bruce Crane, Data Systems Applications Division (DSAD), National Agricultural Library, Room 013, USDA, Beltsville, MD 20705.

Field	Name	Contents
1	AUTH	Authors' names.
2	SUBJ	Title, objectives, main variables, inputs and outputs.
3	METH	Mathematical methods, computer type, computer language, storage requirements, step size, time span of valid operation, and solution timing.
4	KEY	Model name, important key words, and phrases.
5	ADDR	Address, telephone of contact, and possibly the computer center where the program is operational.
6	CODES	Organizational code: ARS-Code, WRU-Code, CRIS-Code (if ARS); ERS, CSRS, FS, etc. (if USDA); Univ.-Name (if a University); Other (if none of the above).
7	DOCU	Description of program documentation and a description of publications that have resulted from use of the model.
8	STUS	Description of the completion status of the model and how well it has been validated. Current level of effort in scientist years.
9	UPDT	Month and year of last update of this record (not of the model).

Figure 1. Format of record of agricultural research model in FARM information retrieval system.

The April 1977 listing of FARM contained 109 models, at least 27 of which should be useful in host-plant resistance research. Plant models vary from one of the growth of a complete crop to the model of a root growing in the soil. Insect models vary from one of complete insect growth to a model to predict times of peak emergence and flight of a particular moth.

6. A SELECTED EXAMPLE OF THE USE OF MODELS IN HOST-PLANT RESISTANCE

The examples chosen to illustrate the use of models in host-plant resistance research come from cotton and cotton insect pests. Several attributes of this system make these examples especially appropriate:

1. Cotton has a complex growth pattern, producing fruiting buds over a 90-day period. A crop will contain squares (buds), flowers, young bolls, and open bolls at the same time during some stages of the growth process.
2. Cotton has the ability to compensate for insect damage at certain stages of growth and when certain fruit is lost to insects.
3. Most pest insects of importance feed on the reproduction structures (buds or bolls). Thus they do not kill the plant and the damage is difficult to measure since it may show up either as a delay in maturity or a reduction in yield.

4. Several morphological characters present in cotton affect different pest insects in different and sometimes opposing ways. For example, the smooth leaf character increases resistance to *Heliothis* spp., but increases susceptibility to *Lygus lineolaris*; pubescence increases resistance to *L. lineolaris*, but increases susceptibility to *Heliothis* (Table 1).

5. Two of the primary pest insects of cotton are boll weevil, *Anthonomus grandis*, and the *Heliothis* complex (*H. virescens* and *H. zea*). These insects have strong flying capabilities and the capacity to build a population rapidly. Because of these factors, either large plots or isolation are necessary to conduct many field experiments. Field experiments are thus costly and difficult to interpret and relate to practical conditions.

6. Good cotton plant simulation models are available.

7. Fair-to-good insect simulation models are available for two major pest insects, *Heliothis* and boll weevil.

AUTH	Baker, D. N.; Lambert, J. R.; McKinion, J. M.; Alexander, G. B.
SUBJ	Gossym. Purpose: The simulation of physiological response in cotton to climate and pest influence. Simulates the processes of canopy light interception, photosynthesis, respiration, evapotranspiration, morphogenesis, and growth. Addresses the subject of stress physiology and simulates the abscission of squares and bolls. Simulates in two dimensions the movement of roots, water, and nitrogen in the soil. Inputs required: soil desorption data, hydraulic conductivity data, dates and rates of planting, fertilizer application, irrigation, daily standard weather station solar radiation, maximum and minimum temperature, pan evaporation, and rainfall. Outputs: daily weights of total plant, leaves, stem, roots, squares, and bolls. Plant height, leaf area index, average nitrogen content of leaves, stems, roots, squares, and bolls. Plant map diagrams and spatial representations of root, water, and nitrogen distribution in the soil. Gossym provides a basis for the dynamics of the effects of insect damage on physiology and yield in cotton.
KEY	Gossym, Photosynthesis, Cotton, Root Growth, Rhizos.
ADDR	D. N. Baker, Cotton Production Research, P.O. Box 5367, Mississippi State, Ms 39762, Phone 601-323-2230 / J. R. Lambert, Dept. of Agr. Eng., McAdams Hall, Clemson University, Clemson, SC 29631, Phone 803-656-3251.
CODE	ARS-7502/07, WRU-12330, CRIS-7502-12330-001, Univ.-Clemson.
DOCU	Documentation consists of program listing, computer-drawn flow charts, and a dictionary of terms.
STUS	The source listing is well documented with comment cards. The program is verified and partially validated. 3 SMY.
UPDT	12 1975

Figure 2.   Example of FARM printout on a plant growth model.

AUTH	Jones, James W.
SUBJ	Boll weevil population dynamics as influenced by the dynamic cotton crop status. Field level model. Components of weevil behavior related to crop status (availability of squares and bolls). Reproduction of boll weevils determined by age of females, temperature, and diet. A bioenergetic model of protein, fat, and carbohydrate ingestion and utilization is included. Development of immature stages, adult longevity, and mortality are primarily temperature functions. Mortality of immature stages is increased when multiple eggs are in a site. Insecticide application dates, crop status data, and weather data are inputs. Variations in developmental times are considered.
METH	Computer was Univac 1106 Exec 8 using Fortran V. 50K words storage required insect development and mortality calculated using 3-hour temperature data, but model is updated daily. Model designed to simulate one cotton growing season (150 days) and requires about 9 minutes for solution.
KEY	Crop-Insect Interface, Boll Weevil, Insect Nutrition, Population Dynamics, Cotton Crop Damage, BWSIM, Insect.
ADDR	Agricultural and Biological Engineering Dept., P.O. Box 5465, Mississippi State, Ms 39762, Phone 601-325-5246.
CODE	ARS-7502/05, WRU-15480.
DOCU	Documentation consists of a Ph.D. dissertation containing complete model development, assumptions, validation data, and program listing; a simulation model of boll weevil population dynamics as influenced by the cotton crop status. Unpublished Ph.D. Dissertation, Biol. and Agr. Engr. Dept., North Carolina State Univ., Raleigh, N.C., 1975.
STUS	Program is complete, debugged, and validated. Model development took 1.8 MY. with 1.5 SMY. model usage took 0.4 MY. with 0.3 SMY.
UPDT	05 1975.

Figure 3. Example of FARM printout on an insect growth model.

Table 1. The Effects of Three Morphological Characteristics in Cotton on Pest Insects and Insecticide Coverage (Lukefahr et al., 1965; Jenkins et al., 1973; Parrott et al., 1973)

Characteristics	*Heliothis* spp.	*Lygus lineolaris*	Boll Weevil	Insecticide Coverage
Smooth leaf	Reduce	Increase	None	None
Nectariless	Reduce	Reduce	None	None
Frego bract	None	Increase	Reduce	Increase

223

Table 2. Simulated Boll Weevil Population Dynamics on Normal-Bract and Frego-Bract (Resistant) Cotton from the Interacting Boll Weevil and Cotton Crop Model (See Table 3)

Cotton Type	Days from Emergence	Eggs	Larvae	Pupae	Number/Hectare Pre-ovipositing	Adults Ovipositing	Diapausing	Over-wintering
Frego	35	0	0	0	0	52	0	52
Normal	35	0	0	0	0	52	0	52
Frego	49	760	8	0	0	636	0	640
Normal	49	1293	13	0	0	636	0	640
Frego	63	3779	5169	433	0	884	0	899
Normal	63	7555	10200	745	0	884	0	899
Frego	70	4231	6500	2719	265	955	0	995
Normal	70	8461	13000	5307	452	955	0	955
Frego	77	5152	6571	3540	1578	1700	0	1055
Normal	77	13450	13300	7077	3139	2490	0	1055
Frego	84	20740	13850	3848	1493	4133	200	1081
Normal	84	76140	42810	7711	2985	7704	380	1081
Frego	91	28950	33680	6246	1397	5281	448	0

Normal	91	112200	125600	17210	2794	10228	874	0
Frego	98	27730	46090	15230	2148	5994	1080	0
Normal	98	109100	179100	54840	5908	11886	2361	0
Frego	105	17320	38950	23380	5385	3917	2756	0
Normal	105	65320	152800	89890	19141	9074	8360	0
Frego	112	22210	29100	23780	8281	6640	4712	0
Normal	112	133200	118100	93370	31943	22663	17244	0
Frego	119	24770	39530	17490	7606	7051	6028	0
Normal	119	165000	255400	67800	29851	26385	23129	0
Frego	126	21860	35970	18190	5174	3434	4158	0
Normal	126	162500	240800	98180	197111	13228	15911	0
Frego	133	5873	29090	17770	4408	2159	3928	0
Normal	133	45990	199800	121500	20596	8092	17161	0
Frego	140	1961	10050	18400	3730	969	3443	0
Normal	140	16890	73580	126700	25575	4374	21943	0
Frego	147	1267	4222	13950	2469	503	1756	0
Normal	147	1546	33570	96370	16663	2810	11507	0
Frego	161	341	772	2068	1464	402	878	0
Normal	161	0	1701	16060	10335	2620	6097	0
Frego	175	778	590	320	403	235	300	0
Normal	175	9624	3906	852	3137	1715	2283	0

Table 3. Simulated Plant Growth on Normal-Bract and Frego-Bract (Resistant) Cotton from the Interacting Boll Weevil and Cotton Crop Model (See Table 2)

| Cotton Type | Days from Emergence | Fruit | | | Number/Hectare | | | |
		10 to 26 Days (Squares)	27 to 42 Days (Bolls)	Open Bolls	Total Fruit	5-Day Cumulative Oviposition Damaged Squares	Damaged Bolls	Oviposition Damaged Square (Percent)
Frego	35	0	0	0	22060	0	0	0
Normal	35	0	0	0	22060	0	0	0
Frego	49	35860	0	0	232920	540	0	1.5
Normal	49	35840	0	0	232890	1080	0	3.0
Frego	63	322910	20	0	1234310	4640	0	1.4
Normal	63	313500	20	0	1224900	9090	0	2.9
Frego	70	904320	54890	0	2180390	6150	0	0.7
Normal	70	892060	49270	0	2164440	12190	0	1.4
Frego	77	1378110	209600	0	2451900	6870	0	0.5
Normal	77	1368540	196340	0	2432870	17000	0	1.2
Frego	84	1238870	562290	0	2509300	27260	0	2.2
Normal	84	1205770	536390	0	2453040	95600	0	7.9
Frego	91	781000	696710	0	1724350	38520	0	4.9

Normal	91	705240	628380	0	1574230	135310	0	19.2
Frego	98	358010	578140	0	1227720	33500	60	9.4
Normal	98	257440	466780	0	1007270	98930	1130	38.4
Frego	105	132960	525080	0	1222170	7110	2860	12.9
Normal	105	68120	338650	0	944250	32260	32000	47.4
Frego	112	54500	194120	0	1015140	16060	22710	29.5
Normal	112	37490	51350	0	733490	35710	51030	95.3
Frego	119	65260	48490	10370	964470	19130	17290	29.3
Normal	119	40400	5390	9260	772590	35890	10680	88.8
Frego	126	36900	9490	38770	873710	10610	5930	28.8
Normal	126	18210	110	34930	718450	42830	120	100.0
Frego	133	20460	2670	75690	830590	3080	2580	15.1
Normal	133	19630	0	69510	610410	19280	0	98.2
Frego	140	11820	2330	163380	802790	550	1730	4.7
Normal	140	8050	0	153980	568930	4060	0	50.4
Frego	147	5250	840	271870	815090	60	670	1.2
Normal	147	200	0	258730	577230	0	0	0
Frego	161	2420	420	643040	933010	0	370	0
Normal	161	190	0	532190	692940	0	0	0
Frego	175	58590	1570	788440	1129490	630	410	1.1
Normal	175	40960	0	560870	922030	6210	90	15.2

8. A cotton crop model and a boll weevil model have been interfaced to run interactively on the computer; a cotton crop model and a *Heliothis* model have also been run together. A cotton plant model and models of both these insects are in the process of being interfaced on the computer so that the three models will run interactively.

The example of a cotton insect model and plant model interactions given here is intended to show the types of data obtainable from their use. Each of the models, COTCROP (Brown et al., 1977), and BWSIM (Jones, 1975), are in early states of use and validation is incomplete. This means that the actual output in terms of quantitative data may fit actual field situations to varying degrees. One should not expect close correspondence at this stage of model development. However, with one or two seasons of validation data, these models should be expected either to fit in the proper quantitative ranges of field situations or to point out gaps in our scientific knowledge.

The use of plant and insect models is just now beginning in host-plant resistance research. As with any new tool or technique used to understand better the plant–insect relationship, its precision will improve. At present, it should be used as widely as possible to help define its applications and to encourage the further development and use of the systems approach.

Frego-bract strains of cotton are resistant to boll weevil damage because they cause a change in boll weevil behavior which results in a 50 percent reduction in oviposition. The bracts of frego-bract cotton (one recessive gene) are rolled and twisted, and expose the cotton bud (square). The output data in Tables 2 and 3 represent the types of information obtainable from the simulation. Both insect and plant data are available in a quantitative fashion on a daily basis throughout the growing season. A comparison of the data in Tables 2 and 3 shows the effects of the resistance to boll weevil in frego-bract cotton. The 50 percent reduction in oviposition is translated into a large difference in population dynamics of the boll weevil and in yield of cotton. These plant data are based on a properly irrigated and fertilized cotton crop simulated under Mississippi weather conditions. They would represent yields similar to ideal conditions for an excellent grower in Mississippi. The simulated yield was 6.27 bales/hectare without any insect damage. The amount of information on the plant growth and the daily insect population structure indicates that the use of plant–insect simulation models, along with traditional research approaches, should enhance the ability to develop useful hypotheses, to advance the development of resistant plant types, and to determine the best ways of growing these strains.

Table 1 shows that frego-bract cotton confers boll weevil resistance but increases the susceptibility of cotton to damage from *Lygus lineolaris*, tarnished plant bug. Another morphological trait in cotton is nectariless.

This character, controlled by two recessive genes, results in the absence of the nectary in the leaf midrib as well as the three nectaries at the base of the calyx and at the base of the square (Meyer and Meyer, 1961). Nectar is high in carbohydrates and amino acids (Hanney and Elmore, 1974). Thus the removal of the nectaries removes a food source for insects. Oviposition by tarnished plant bugs is reduced 75 percent on nectariless cotton (Calderon, 1977), whereas oviposition by *Heliothis* is reduced 45 percent (Lukefahr et al., 1965). Thus nectariless, frego-bract cotton should alleviate the increased plant bug susceptibility of frego strains. Unfortunately, we do not have a plant bug model to obtain simulation data on the degree of resistance needed in frego bract cotton. We do have sources of resistance other than nectariless cotton. Thus in the absence of information from simulation from cotton–*Lygus* models, the breeder should attempt to combine both types of plant bug resistance with the frego-bract character. Fortunately, we do have *Heliothis* spp. models and can thus obtain simulation of the nectariless effect on *Heliothis*. The breeder can be guided by the simulation of nectariless effects on *Heliothis*. He could then combine the boll weevil, *Heliothis*, and cotton crop models and simulate the effects of frego-bract, nectariless, and smooth-leaf characteristics on boll weevil and *Heliothis*.

APPENDIX: TITLES AND OBJECTIVES OF SELECTED AVAILABLE MODELS APPLICABLE TO HOST-PLANT RESISTANCE

1. *Root gro.* Growth and function of root system in a heterogeneous soil matrix.

2. *Soil–plant–atmosphere simulation model.* Prediction of microclimate at atmosphere interfaces in a plant community and plant community activities.

3. *Simulation of alfalfa crop growth.* Field level model developed for use in pest management modeling can be used to predict dry matter accumulation of alfalfa.

4. ALSIM 1 (Level 1). Simulation of alfalfa growth and cutting management. Developed for use in studying alfalfa insect management.

5. *Sorghum.* Simulation of daily growth and development of a typical grain sorghum plant in a field stand.

6. *Crops.* A *gasp* IV based simulation language which emphasizes the development of equations that model crop physiology.

7. *Gossym.* Simulation of physiological responses in cotton to climate and pest influences.

8. *Rhizos.* Simulation of the movement in two-dimensional space of roots, water, and nutrients in the rhizosphere.

9. *Cotton germination and emergence.* Simulation of germination which extends from planting time until the average radicle length reaches 3 mm.

10. *Boll weevil population dynamics as influenced by the dynamic cotton crop status.* Field level model.

11. *Pheromone trapping model.* Model of the cumulative effects of pheromone sources and calculates the probability of a boll weevil being caught.

12. *Boll weevil eradication model.* Field level model which simulates boll weevil behavior in relation to pheromone sources.

13. *Temperature-driven insect population simulator.* For studies of insect population dynamics using a temperature-dependent model for insect development.

14. *Multifield pest control simulation.* For studies of the effects of cropping density in an area as it affects the tendency of an insect to become a pest.

15. *Pest detection and insecticide control simulation.* For studies of population dynamics with insect detection and control measures (given statistics) using a temperature-dependent model.

16. *Genotype interbreeding simulation.* To study population relationships and dynamics using a model that simulates breeding among dominant, hybrid (mixed), and recessive genotypes of an insect.

17. *Linked populations (parasite/host, predator/prey) dynamics.* To determine if generalizations (patterns, rules, etc.) could be recognized from the interplay of two linked populations.

18. *Insect development and control.* A general program for the population dynamics of an insect.

19. *Simulation of the effects of weather on european corn borer populations.*

20. *Model of Heliothis zea.* Simulation of the population dynamics of *H. zea* from the time of emergence of the diapausing moths in spring to the end of the growing season.

21. *Sex pheromone emission response model.* A mathematical model describing the response of male Lepidoptera to female or synthetic release of pheromone as a function of environmental factors such as temperature and wind velocity.

22. *A dynamic model of Heliothis zea and H. virescens.* Simulation of the population dynamics of the two species.

23. *Population reduction by release of sterile insects.* Population simulation model with which various pest control strategies and outcomes can be analyzed in advance of field application.

24. *Population simulation model of Diatraea saccharalis and its larval parasite Lixophaga diatraea.* Testing by simulation of the hypothesis that properly timed releases of a selective parasite species can offer a practical method of managing pest insect populations.

25. *Population dynamics of scale insects (Diaspididae) and a parasite (Aphytis lingnanensis).* Study of changes in sex ratio and population density.

26. *"Epidemic."* A general plant disease simulator to model epidemics rather than specific diseases.

27. COTCROP. Simulation model of the growth and yield of a cotton crop designed for interfacing with cotton insect pest models for studying integrated pest management.

Pubescent cotton varieties have a dense covering of short trichomes that provides resistance to the boll weevil. Photo courtesy of G. A. Niles.

11

RESISTANT VARIETIES IN PEST MANAGEMENT SYSTEMS

Perry L. Adkisson

Department of Entomology, Texas A&M University, College Station, Texas

V. A. Dyck

The International Rice Research Institute, Manila, Philippines

In the evolution of agriculture toward intensively farmed monocultures, there has been a trend away from traditional methods of pest control (e.g., crop rotation, stalk destruction, elimination of pest habitats, and use of tolerant crop varieties) to an almost complete dependence on pesticides. With some crops, the older methods of control have been replaced entirely by chemicals. In many instances, chemical control has been applied in an attempt to achieve total elimination of the pest of concern from a crop. Pests have adapted to these tactics in various ways. Many of the world's most important crop pests have developed insecticide resistant strains. Secondary pests have been released from natural biological control, and some now are more important pests in some crops (e.g., *Heliothis* spp. in cotton) than the primary pests. Because of these problems, farmers have been forced to use increasing quantities of toxic chemicals. The end result has been increasing insecticidal poisoning of agricultural workers, contamination of the environment, destruction of nontarget species, and major disruption to the ecosystem.

In retrospect, we know that complete reliance for crop protection on a single-component control system has seldom succeeded over the long term. For this reason, knowledgeable plant protection specialists are developing systems of integrated pest management which utilize a combination of cultural, chemical, and biological control methods. The integrated system is designed to suppress pest numbers below crop damaging levels. It is not intended to replace chemical pesticides. In the integrated system, chemicals are used only as needed and in a manner that causes the least disruption to natural control agents. The major feature of the integrated system is purposeful manipulation of the environment (including host plants) to make it as unfavorable as possible to the pest species or more favorable to their natural enemies. The objectives are to reduce the rate of pest increase and the amount of damage to the crop (Isely, 1948; National Academy of Science, 1969; Stern et al., 1977).

In applying integrated control, plant protection specialists should use a variety of measures to take advantage of "weak links" in the seasonal cycles of the pests of crops to reduce pest numbers at a time or in a place where natural enemies of insects are little affected. To accomplish this, primary hosts, alternate hosts, and habitats may be manipulated to suppress the major pests while preserving natural control agents. A resistant variety can provide a foundation on which to build an integrated control system and, in fact, may be most productive when used in adjunct with cultural, chemical, and biological control methods. With some crops, particularly those having low cash value per hectare, the use of resistant varieties may offer the best (and perhaps only) economical method of control of certain pests.

In this chapter conceptual models and case histories in a few selected crops are discussed to illustrate the use of resistant varieties in integrated pest management systems.

1. USE OF RESISTANT VARIETIES FOR PEST MANAGEMENT

Plant resistance to insects has been discovered in many crops in various parts of the world. The resistance is usually moderate, but is sometimes high. Many examples of insect resistance are found in the older traditional varieties, indicating that plant resistance has been a factor in pest control for a long time; but farmers usually have not fully realized or taken advantage of the pest suppression that resistant varieties afford.

Plant resistance as a method of insect control offers many advantages. In some cases it is the only method that is effective, practical, or economical (Horber, 1972; Pathak and Saxena, 1976). Resistance developed in plants for one pest species may provide resistance to several others (Way and Murdie, 1965; Gahukar and Chiang, 1976). For crops grown in developing countries perhaps the most attractive feature of using pest resistant plants is that virtually no skill in pest control or cash investment is required of the grower. Effective pest control may be easily achieved by planting a variety having a high level of resistance, such as the modern rice varieties resistant to the planthopper, *Sogatodes orizicola.*

There are some problems, however, in relying exclusively on plant resistance for pest control. High levels of resistance may lead to the development of new insect biotypes, as has happened with the brown plant-hopper on rice in the Philippines and with the chestnut gall wasp, *Dryo-cosmus kuriphilus,* in Japan (Shimura, 1972). Also, resistance may not be expressed in every environment in which the variety is grown (Horber, 1972; Kogan, 1975; Coppel and Mertins, 1977).

Painter (1951) stressed that resistant varieties are not a panacea for all pest problems. To be most effective they must be carefully fitted into control systems designed for specific pests and into the plant improvement programs of particular crops. Other means of pest suppression also must be considered, and these often are the first line of defense against injurious insects. Painter categorized the use of resistant varieties in pest management as (1) the principal control method, (2) an adjunct to other measures, and (3) a safeguard against the release of more susceptible varieties than exist at the present time. Resistant varieties usually have to be integrated with other methods of pest control to achieve stable pest suppression.

1.1. Resistant Varieties as a Principal Control Method

Plant resistance to insects was used as a primary method of pest insect control long before the advent of synthetic organic insecticides. A few pests have been controlled for many years by use of resistant varieties alone. Insects for which this has been true most often have been those with a high host specificity, such as aphids and scales (Painter, 1951).

Among the best examples of useful resistance are the control of the grape phylloxera, *Phylloxera vitifoliae,* in France by resistant rootstocks imported from the United States; control of the cotton jassid, *Empoasca facialis,* in

Africa by resistant cotton varieties; and use of the resistant apple variety 'Northern Spy' to control the woolly apple aphid, *Eriosoma lanigerum* (Painter, 1951; Martin, 1973).

Resistant varieties of wheat in the United States provide the principal method of control of the Hessian fly, *Mayetiola destructor.* 'Pawnee,' 'Ponca,' 'Poso 42,' and 'Big Club 43,' developed in Kansas and California, were the first Hessian fly resistant varieties released (Painter, 1958; Maxwell, 1972). These were followed by the release in Indiana of the resistant varieties 'Dual' and 'Benhur.' In recent years several other varieties of wheat resistant to Hessian fly have been released in the major wheat growing regions of the United States (Maxwell et al., 1972).

As a consequence of the use of resistant varieties, the Hessian fly is now considered a minor pest of wheat. Currently, more than 4 million hectares of wheat in the United States are planted each year with more than 20 different Hessian fly resistant varieties. Losses inflicted by Hessian fly to wheat have been reduced to less than 1 percent (Maxwell et al., 1972). Luginbill (1969) reported the value of increased wheat yield produced by the growing of resistant varieties in the United States to be about $238 million annually.

The spotted alfalfa aphid, *Therioaphis maculata* is another pest that has been controlled by resistant varieties. First found in North America in 1954, this aphid rapidly spread across the United States, causing millions of dollars in damage to alfalfa and becoming the most serious pest of the crop in the country. Insecticides were used first as the principal method of control but were unsatisfactory because of costs, pesticide residues left on the crop, and the rapid development of insecticide resistant strains of the aphid.

Howe and Smith (1957) found that the variety 'Lahontan' and its parental clones have a high degree of resistance to the spotted alfalfa aphid. Shortly thereafter, a number of resistant varieties including 'Moapa,' 'Zia,' 'Bam,' 'Sirsa 9,' 'Sonora,' 'Caliverde,' 'Mesa-Sirsa,' 'Washoe,' 'Nemastan,' 'C104,' and 'Cody,' were released in the major alfalfa producing areas of the United States (Howe and Pesho, 1960a; Hunt et al., 1966; Hackerott et al., 1958).

Resistant varieties are now the principal method for controlling the spotted alfalfa aphid. They outyield susceptible varieties by as much as 50 percent. The value of the resistance, in terms of averting insect damage and reducing costs of control, has been estimated to be over $100 million per year in the major alfalfa producing states alone (Maxwell, 1972).

Control of the European corn borer, *Ostrinia nubilalis,* a major pest of corn in the United States, also depends on the use of resistant varieties. Control of the borer by insecticides and cultural methods generally has not been satisfactory. Thus a major research effort was directed towards the development of borer resistant varieties (Painter, 1951; Brindley and Dicke,

1963). Several hybrid corn varieties resistant to the first-brood borers have been released and grown successfully over more than 12.1 million hectares. Luginbill (1969) estimated the value of the resistance during the period 1962–1969 to have exceeded $150 million annually. Major efforts are now directed toward the development of corn varieties resistant to the second-brood borers. Combined resistance to first- and second-brood borers should greatly increase the value of resistance as a means for controlling the European corn borer.

Other pests for which resistant varieties are used as the principal control measures include the wheat stem sawfly, *Cephus cinctus* (seven resistant varieties of wheat); pea aphid, *Acyrthosiphon pisum* (five resistant pea varieties); and corn earworm, *Heliothis zea* (several resistant corn hybrids) (Maxwell, 1972). Others are mentioned in various chapters of this volume.

1.2. Resistant Varieties in Integrated Control Systems

General Considerations

Ideally, resistant varieties should provide complete and permanent control of the major crop pests of the world. However, such high levels of resistance are present in only a few crop varieties.

Fortunately, high levels of resistance are not necessary for a crop variety to have value in an integrated control system. Varieties with low or moderate levels of resistance, or those that may be grown to evade pest attack, can be used to good advantage for pest suppression. The key to success lies in their incorporation into management systems involving other control measures such as regulated planting dates, early harvesting and crop residue disposal, manipulation of alternate hosts, host-free periods, and destruction of overwintering pest insects. The system should suppress pest numbers and conserve their natural enemies. If this is achieved, insecticides then might be used more selectively and less frequently for crop protection.

Resistant varieties, even those with low and moderate levels of resistance, offer a number of advantages to an integrated control system. The reduction in pest numbers achieved through resistance is constant, cumulative, and practically without cost to the farmer. The reduction in pest numbers makes control by chemical and cultural methods easier, and the level of natural biological control required to hold pest numbers below crop-damaging levels need not be so great (Pimentel, 1969; Maxwell, 1972; Horber, 1972; Dahms, 1972).

Varietal Resistance with Insecticidal Control

The most common form of integrated control involving resistant varieties is the use of carefully supervised treatments of insecticides to control outbreaks of pests on varieties having low or moderate levels of resistance or tolerance. The major advantage of using the resistant variety is to induce a constant level of suppression on each pest generation. This reduces the numbers of pests in each generation and slows population growth. If all the crop area is planted to a resistant variety, the reduction in pest numbers will be cumulative over time, and numbers within the area should become smaller each succeeding year (Painter, 1951; Coppel and Mertins, 1977).

The impact of even a moderately resistant variety on the population dynamics of a pest may be demonstrated conceptually by using simple insect models of the type devised by Knipling (1964). As shown in Table 1, if an area is infested with an overwintering pest population averaging 100 insects/hectare, and they increase at a fivefold rate per generation without control, the buildup will be very rapid on the susceptible variety. By the end

Table 1. Theoretical Rate of Increase by a Hypothetical Insect Population on a Susceptible and a Resistant Variety that Reduces Population Size by 50 Percent in Each Generation. Assume Fivefold Rate of Increase per Generation (Adapted from Knipling, 1964)

	Number of Insects/Hectare	
Generation	Susceptible Variety	Resistant Variety
First year		
Parent	100	100
F_1	500	250
F_2	2,500	625
F_3	12,500	1,563
F_4	62,500	3,906
Second year[a]		
Parent	1,625	105
F_1	8,125	263
F_2	40,625	656
F_3	203,125	1,641
F_4	1,015,625	4,102

[a] Assume 10 percent of F_3 and 50 percent of F_4 are diapausing individuals of which only 5 percent survive the winter.

of the season, the pest density in the area growing the susceptible variety would be 16-fold greater than in the one growing the resistant variety. The most dramatic advantage of growing resistant varieties occurs after the first year of planting. Assuming that winter mortality is the same for the pest populations in both areas, the susceptible variety would start the second growing season with 15 times as many pests, and would end with 248 times as many pests as the resistant variety. There is, of course, a limit to the number of pests that can infest a unit area of crop. This limit probably would be achieved during the F_3 and F_4 generations of the second year in the area growing the susceptible crop. This would occur because of competition, lack of food, and other density-dependent factors. Insecticides would have to be applied to the susceptible variety very early in the season, especially in the second year, to prevent yield losses.

The numbers of pests produced on a resistant variety usually decline over time, making control with insecticides much easier. Also, once the pests are controlled future increase will be slower. Moreover, if the pest population should be subjected to a natural calamity, such as a disease epidemic, heavy attack by parasites and predators, or adverse weather, insecticides may not be needed because resurgence to original levels will occur slowly, if at all.

The comparative effectiveness of insecticidal control of a pest on a susceptible and resistant variety is illustrated in Table 2. The model illustrates a hypothetical situation in cotton in which adults of the overwintering generation of the boll weevil, *Anthonomus grandis* are treated with an insecticide. The model assumes that the insecticide kills 90 percent of the females before reproduction and the survivors reproduce at a fivefold rate. The resistant variety is a frego-bract cotton which reduced boll weevil numbers by more than 50 percent in a number of experiments (Jenkins and Parrott, 1971; Maxwell et al., 1972).

The model in Table 2 shows that the boll weevil on the susceptible variety recovers from the early-season insecticide treatment by the F_3 generation. Additional treatments in late season would be needed to protect immature bolls. On the resistant variety, treatment of overwintered adults in the early growing season is sufficient to hold numbers below crop-damaging levels for the entire season. This is important because insecticide treatments may be avoided during the period when destruction of natural enemies may induce attacks by the bollworm, *Heliothis zea,* and tobacco budworm, *H. virescens,* both of which are very difficult to control economically with any insecticide (Adkisson, 1972).

The most dramatic advantage occurs in the second year, when overwintering weevil numbers on the resistant variety will be a twelfth of those on the susceptible variety such that an insecticide may not be needed at all. However, if the overwintering population is subjected to early-season

Table 2.   Effectiveness of Early-Season Insecticide
Control Against a Population of Boll Weevils on a
Susceptible and on a Resistant Frego-Bract Variety of
Cotton which Reduces Population Size 50 Percent in
Each Generation. Assume Fivefold Rate of Increase
Per Generation (Adapted from Knipling, 1964)

	Number of Weevils/Hectare	
Generation	Susceptible Variety	Resistant Variety
First year		
Parent[a]	100	100
F_1	50	25
F_2	250	63
F_3	1,250	156
F_4	6,250	390
Second year[b]		
Parent[a]	38	3
F_1	19	1
F_2	95	3
F_3	475	8
F_4	2,375	20

[a] Assume 90 percent of the parents are killed by an insecticide treatment before they can reproduce.
[b] Assume 10 percent of the F_3 and 50 percent of the F_4 are diapausing adults of which only 5 percent survive the winter.

insecticide treatment, numbers in the area growing only the resistant variety would by the end of the second year, be less than 1 percent of the numbers in the area producing the susceptible variety. The model clearly demonstrates that insecticides may be used with much greater efficiency on resistant than susceptible varieties. Fewer pest insects develop on the resistant variety than on the susceptible one, and costs for insecticide control may be greatly reduced.

Another advantage of growing a resistant variety is that the reduced rate of pest increase may greatly prolong the time required by the pests to reach the economic threshold for crop damage. This is especially true if the resistance characters in the variety produce mortality to the immature stages and prolong the developmental period of the survivors. This is demonstrated by the hypothetical insect population depicted in Figure 1. If the economic injury threshold is assumed to be 2,500 adults/hectare then insecticides would have to be applied to the susceptible variety in July against the F_2 generation. In some crops, such as cotton, insecticides applied against a key pest during this period may induce severe secondary outbreaks of *Heliothis* spp. and other lepidopterans because of destruction of parasites and predators. This, in turn, may require additional treatments for the remainder of the season. Thus there is great economic advantage to any control system that averts secondary pest outbreaks. The pest population on the resistant variety undergoing 50 percent reduction in population size in each generation would not achieve the economic injury level until the F_4 generation in early September. When the resistant variety results in a 50 percent reduction in numbers in each generation and a 50 percent increase in duration of the developmental period of survivors, the economic threshold is not attained until November.

With many pests and crops the situation shown by the model for the two

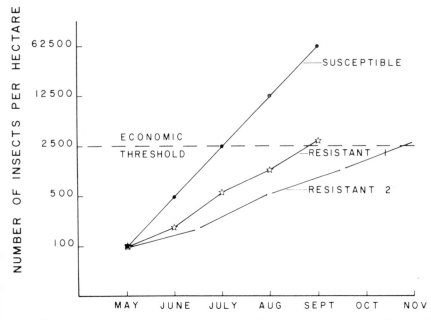

Figure 1. Theoretical population trends of a hypothetical insect population on a (1) susceptible variety, (2) resistant variety 1 which reduces the pest insect population size by 50 percent in each generation and (3) resistant variety 2 which reduces pest population size by 50 percent in each generation and increases the duration of the developmental periods of the survivors by 50 percent. Assume five-fold increase in pest numbers per generation.

resistant varieties probably would never occur. The pest would not be able to attain the economic injury threshold so late in the season because the maturing plants would no longer be attractive to the insects for oviposition of feeding. In this case, no insecticides would be needed to protect the crop.

Other ways in which plant resistance enhances the effectiveness of insecticide treatments follow: (1) morphological changes in the plant that provide pest resistance (e.g., frego-bract cotton) also may allow better insecticide coverage of the fruiting structures; (2) changes in plant morphology may cause the insects' predators to be more active on the plant, thereby increasing their contact with pest species; and (3) toxic substances or nutritional inadequacies in the resistant plant may make the pest more susceptible to certain chemicals or naturally occurring pathogens (Maxwell, 1972).

· All the above-mentioned characteristics of resistant plants may be used to reduce the number of insecticide treatments, as well as the amounts applied. The incorporation of a resistant variety into an integrated pest control system lessens the expense of producing the crop, conserves insect natural enemies, preserves environmental quality, and slows the rate of development of insecticide resistant pest strains. This is a much better system of crop protection than complete reliance on insecticides to protect a susceptible variety (Pimentel, 1969; Dahms, 1972; Maxwell et al., 1972; Pathak, 1975; International Institute of Tropical Agriculture, 1977).

Rice well illustrates the use of resistant varieties in combination with insecticide treatments to produce economical yields. Plant resistance to many species of rice insect pests has been discovered, mostly within the past 15 years. Considerable success has been achieved in breeding high-yielding rice varieties with resistance to the stem borer, *Chilo suppressalis*; the gall midge, *Orseolia oryzae*; the brown planthopper, *Nilaparvata lugens*; the planthopper *Sogatodes oryzicola*; and the green leafhopper, *Nephotettix virescens*. These varieties have been integrated into the pest and crop management systems of rice.

The rice variety 'IR20,' named in 1969, was the first high-yielding variety bred for insect resistance. It is moderately resistant to the stem borer and highly resistant to the green leafhopper. The variety was widely grown in several Asian countries during the early 1970s, especially in the Philippines, Bangladesh, and Vietnam (Pathak et al., 1973). Fewer insecticide treatments are needed on 'IR20' than on borer-susceptible varieties (Pathak and Dyck, 1973). Gallun et al. (1975) noted that 'IR20' was grown on a total area of more than 2 million hectares with an estimated annual savings for pest control of about U.S. $5 million.

Since the introduction of 'IR20,' a number of rice varieties resistant to stem borers, gall midges, the rice whorl maggot (*Hydrellia phillippina*), and planthoppers have been widely grown throughout Asia and South America

(Mishra et al., 1976; Khush, 1977a, 1977b; Pathak and Khush, 1977; Mohanty and Roy, 1977). Pest suppression in the field produced by the resistant varieties has been dramatic, and the number of hectares planted to them has rapidly increased (Oka, 1976; Pathak and Saxena, 1976). For example, approximately 6 million hectares of brown planthopper-resistant varieties are being grown in the Philippines and parts of Asia (Khush, 1977b).

The pest management system developed in the Philippines illustrates the key role of varietal resistance in rice pest control (Sanchez et al., 1976). All varieties recommended for cultivation are classified according to the kind of resistance they possess. Currently, resistance to tungro disease, the brown planthopper, and grassy stunt disease are the key factors governing pesticide recommendations. Economic thresholds are given for sporadic pests or when the brown planthopper biotype is unknown. It is expected that stem borer resistance in many varieties will reduce the need for insecticides, but since the resistance level is only moderate, treatment may be needed during the crop's vegetative stage. Prophylactic insecticide treatments are required for regular pest problems. Narrow spectrum chemicals are preferred over those with wide spectrum.

Several rice varieties are available in Asia with multiple resistance to several pests and insect vectored viruses, greatly reducing the need for insecticides. Insecticides are applied only where the resistance level is low or moderate. There is no variety immune to all pests, so some insecticidal protection may be needed even for varieties with multiple resistance. A series of experiments has shown that inexpensive insecticide treatments on a variety with resistance to several pests were almost always as economical, if not more so, than costly treatments on susceptible varieties (Dyck et al., 1976). Varietal resistance to the major pest problems, such as vectored diseases and brown planthoppers, could greatly reduce the cost of insecticides needed to produce a crop. Future control of rice insects will depend on appropriate combinations of varietal resistance, cultural control, conservation of important natural enemies, and judicious use of insecticides (Way, 1976).

Varietal Resistance with Biological Control

Resistant varieties are highly compatible with biological control since they usually do not greatly affect the natural enemies of the pest species. Varieties with only moderate levels of resistance or tolerance allow a few pests to remain on the crop at subeconomic levels and serve as food for the natural enemies. The natural enemies help to control the target pest, as well as other pests to which the variety is not resistant, and even pests of nearby crops (Horber, 1972; Maxwell, 1972).

By reducing the pest population somewhat, varietal resistance enables the natural enemies to be more effective because of an improved pest/natural enemy ratio. This may be very useful, especially if the crop is able to tolerate some damage (Kogan, 1976). In this way plant resistance is selective (Coppel and Mertins, 1977). Van Emden and Wearing (1965) showed theoretically how a low level of resistance combined with the action of natural enemies may control a pest when either method alone is ineffective.

Pathak (1970, 1975) suggested that the restless behavior of pests on resistant varieties may expose them to predators which may devour more small insects on resistant plants than larger insects on susceptible hosts. Resistant plants also may reduce the pests' vigor, improving natural enemy efficiency (Maxwell, 1972), and predators that are relatively inefficient on a susceptible variety may be more effective on a resistant one (Kogan, 1975). Dahms (1972) pointed out that when the rate of pest nymphal development is reduced, immature stages are exposed longer to natural enemies. Resistant varieties may better synchronize parasite activity and pest development. Resistant varieties may directly influence the density of natural enemies by being more attractive to them (Way and Murdie, 1965). The resistant plants' morphology also may make it easier for predators and parasites to find the host and may therefore favor the spread of insect pathogens (Johnson, 1953; Maxwell, 1972).

A major advantage of using a resistant variety in an integrated control system is the preservation of the insect natural enemies of key and secondary pests. Unlike insecticides, the resistant variety may be managed to work in harmony with nature so that key pest infestations are suppressed without unleashing secondary pest outbreaks.

Varietal Resistance with Cultural Control

Resistant varieties, including those that can be manipulated to evade pest attack, are highly useful in cultural control systems designed to maintain key pest numbers below the economic threshold while preserving insect natural enemies. Insecticides may be required in this type of management system but are applied only at minimal effective dosages when absolutely needed. Treatments are carefully timed to avoid unleashing secondary pest outbreaks.

The production of cotton in the irrigated deserts of the western United States offers an excellent example of how a resistant variety might be used to manage the pest complex in a crop with minimal use of insecticides. In a large part of this region, the pink bollworm, *Pectinophora gossypiella,* is the key pest and insecticides must be applied each year for its control. The insecticides kill insect natural enemies, unleashing the bollworm and tobacco budworm. These insects are difficult to control and the costs of the insecticides may represent a substantial percentage of the gross return produced by the crop.

The pink bollworm can be controlled by cultural practices designed to reduce overwintering numbers. This involves early uniform planting, early maturity, defoliation, and stalk destruction of the cotton in late August and September before the larvae are forced into diapause by short days and cool nights (Adkisson and Gaines, 1960). Farmers in the irrigated areas neglect these practices as they strive for high yields by growing highly fertilized and irrigated, indeterminate varieties over a long growing season.

Wilson and Wilson (1976) found that certain nectariless varieties may reduce the numbers of pink bollworms to develop in a cotton field by more than 50 percent when compared with susceptible varieties. In Table 3 a theoretical model has been developed to show how these varieties might be used to suppress the pink bollworm. The model assumes that the nectariless variety will reduce population size by 50 percent in each generation and the short-season nectariless cotton will be ready to defoliate and harvest in August or early September, before the pink bollworm enters diapause.

The model indicates that even with stalk destruction, plowing, and a host-free period, the pink bollworm population will increase to high levels each year on the susceptible variety. This occurs because a large F_4 generation of diapausing larvae are allowed to develop. The use of the nectariless variety alone should greatly reduce the size of the pink bollworm population and, when combined with an effective stalk destruction program, should dramatically decrease the numbers of pink bollworms in the area. The model indicates that a 16-fold reduction might be possible within the second year. It also suggests that the use of the resistant variety, combined with appropriate phytosanitation practices and host-free periods, should result in a lowering of the population density in each succeeding year.

The greatest impact on the pink bollworm in these areas could be realized by combining the nectariless character with varietal earliness as described by Walker and Niles (1971). In this case, the variety would be mature approximately 2 to 3 weeks earlier than the long-season cotton. For the pink bollworm, this would mean that only three generations would develop in a field because the early cotton could be defoliated and harvested, and the stalks shredded and plowed under before daylengths and temperatures are appropriate to force the larvae into diapause. The combined reduction in numbers of the F_3 generation produced by defoliation (90 percent), stalk shredding and plowing (90 percent), winter kill, and "suicidal" emergence of moths in the spring (90 percent) amounts to 99.9 percent. Under this system, at the beginning of the second year there would be 24-fold fewer pink bollworms in the area growing the short-season nectariless variety as compared with an area producing the nectariless long-season variety and

Table 3.   Theoretical Rate of Increase by a Pink Bollworm Population on A Suscepitble Variety of Cotton, A Nectariless Variety that Reduces Population Size by 50 Percent in Each Generation, and A Nectariless Short-Season Variety. Assume Fivefold Rate of Increase Per Generation (Adapted from Knipling, 1964)

	Number of Insects/Hectare		
Generation	Susceptible Variety	Nectariless Variety	Short-Season Nectariless Variety
First year			
Parent	100	100	100
F_1	500	250	250
F_2	2,500	625	625
F_3	12,500	1,562	$1,562^a$
F_4	62,500	3,905	Stalks shredded
Overwintering	Assume early defoliation will reduce numbers of diapausing larvae on short-season variety by 90 percent. Stalk destruction and plowing will kill 90 percent of the potential overwintering larvae on all varieties. This will be followed by 90 percent mortality of survivors before reproduction in the second year.		
Second year			
Parent	625	39	1.6
F_1	3,125	98	4.0
F_2	15,625	245	10.0
F_3	78,125	613	25.0^a
F_4	390,625	1,533	Stalks shredded

^{a} There would be no F_4 with the short-season cotton because it may be defoliated and harvested before this generation can develop. Assume only 10 percent enters diapause because of shortened season.

390-fold fewer than in the area producing a susceptible long-season variety. At the end of the second season, the magnitude of these differences would be considerably greater.

The important feature of the model is that insecticide treatment would not be needed on the short-season nectariless variety in the second year. It might be needed late in the season on the long-season nectariless cotton and most surely would be needed on the susceptible variety. This pattern of need for insecticide treatment on each of the varieties might be expected to repeat each succeeding year.

The differences in the magnitude of infestations predicted by the model show the great value of incorporating an additional suppressive measure against a specific generation of a pest. The value of adding earliness to the nectariless variety so that early defoliation may be practiced can be shown by a simple model. Assume that both the nectariless and the short-season nectariless varieties have the potential of producing 1,000 overwintering pink bollworms per hectare. But early defoliation reduces this number in the short-season nectariless variety by 90 percent, leaving only 100 pink bollworms per hectare. If this is followed in both varieties by a 90 percent reduction owing to stalk shredding and plowing, and this is followed by another 90 percent reduction owing to winter kill and "suicidal" emergence of moths, there would be 10 moths per hectare to infest the nectariless cotton but only one moth per hectare to infest the short-season nectariless cotton. The combined reduction in the short-season nectariless would be 99.9 percent compared with 99 percent in the nectariless cotton, a difference of only 0.9 percent. This becomes significant when one realizes that the size of the population in the nectariless cotton would be nine times larger than in the short-season nectariless cotton. The value of utilizing more than one control measure against a specific generation of a pest then becomes obvious.

According to Painter (1951) the early maturity of the short-season nectariless cotton variety discussed above may not be classed as true resistance. He designated characteristics of this type as a form of pseudoresistance termed "host evasion." Early crop maturity is such a characteristic, and one that can be effectively used to evade or alleviate insect problems on a crop. An outstanding example is the use in Texas of short-season cotton varieties in an integrated system designed to control the boll weevil and cotton fleahopper, *Psuedatomoscelis seriatus,* without inducing outbreaks of the bollworm and tobacco budworm.

In Texas the boll weevil and cotton fleahopper are key cotton pests that require insecticide treatment each year to prevent crop loss. These treatments kill insect natural enemies, unleashing outbreaks of the bollworm and tobacco budworm. The latter pests can cause tremendous damage to the crop. Because they are highly resistant to most insecticides, treatment must be made often and with expensive chemicals. As many as 20 insecticide treatments may be required in irrigated cotton. The insect problem is compounded by the growing of indeterminate cotton varieties that fruit over a long period and are slow in maturing. In addition, these varieties usually are heavily fertilized and may be irrigated, producing a lush plant with late-set fruit that is highly vulnerable to damage by late-season infestations of boll weevils, bollworms, and budworms. If maximum yields are to be produced by these varieties, they must set fruit during the period when they are subject to the greatest numbers of pest insects (Adkisson, 1972).

The use of fast-fruiting, short-season cotton varieties is one of the best methods for dealing with the above problems, especially if these varieties may be placed into an integrated control system that utilizes several other pest suppression tactics. Accordingly, a plan of research was developed in Texas with three objectives: (1) develop a variety of early-maturing, fast-fruiting cotton that would make a good yield from the bolls set during the first 30 to 40 days of flowering; (2) develop a method for suppressing the boll weevil and fleahopper during this period, without inducing an outbreak of the bollworm-budworm complex, so that maximum fruit set could be obtained; and (3) defoliate, harvest, and shred stalks in August and early September before environmental conditions force the boll weevil (and pink bollworm) into diapause. The early maturity also provides considerable escape from late-season infestations of the most damaging generation of the bollworm and tobacco budworm. The plan of attack was simple, that is, to develop a variety of cotton that would fruit so early and so fast as to "outrun" early pest infestations, avoid mid- and late-season insecticide treatments that induce bollworm-budworm outbreaks, harvest before late-season diapausing pest generation occur, shred stalks and plow under residue to reduce overwintering pest numbers, and create a long host-free period for boll weevils and pink bollworms.

Figure 2 compares the flowering rate of a short-season variety with a conventional indeterminate cotton (Walker and Niles, 1971) and shows population trends of reproductive and diapausing boll weevils. The short-season cotton can set 50 percent of the harvested crop within the first 30 days of flowering. Full yields can be made within the first 45 days provided the boll weevil and fleahopper are held below damaging levels. The long-season cotton, however, will set only about 30 percent of the bolls during this period, with the remainder highly vulnerable to attack by F_2 and later generations of the boll weevil as well as the bollworm and tobacco budworm.

Walker and Niles (1971) have developed a method in which one to three insecticide treatments are made to reduce numbers of fleahoppers and overwintering boll weevils before the cotton has produced squares of suitable size for boll weevil oviposition. This reduces the F_1 weevils below damaging levels for 30 to 45 days, sufficient time for the short-season variety to set a full crop of bolls and mature them to the point that they are not vulnerable to damage by the F_2 and later weevil generations or by other pests. The early-season treatments do not induce bollworm and budworm attack, so no mid- or late-season treatments are needed. By contrast, the

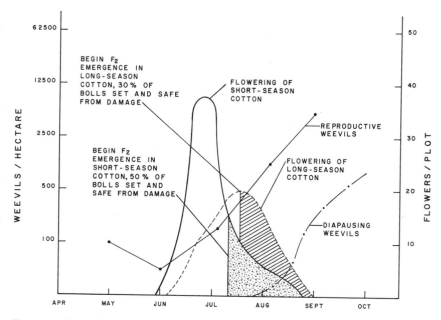

Figure 2. Comparison of flowering rates of short- and long-season cotton varieties showing theoretical population trends of reproductive and diapausing boll weevils (adapted from Walker and Niles, 1971).

indeterminate variety must be treated throughout the fruiting period or a major portion of the yield will be lost.

The full impact of the short-season cotton on the population dynamics of the boll weevil may best be illustrated by the model presented in Table 4. Early-season insecticidal control of the boll weevil on the long-season cotton will not suffice to prevent damage by the F_3 and F_4 generations, even when the weevil population of the season starts from a relatively low level. If it starts at a high level, as in the second year of the model, then insecticide treatments must begin during the F_2 generation. This too is a common occurrence on long-season varieties. On the short-season variety, however, early-season control is sufficient to provide full protection for the crop and, because of the impact of the short-season cotton on the development of diapausing forms (Figure 2), there will be few weevils left to infest the next year's crop.

The above describes how the first of these cottons, 'Tamcot SP-37,' developed by Dr. L. S. Bird of Texas A&M University, is being managed in Texas. The crop now may be produced with one to four insecticide treat-

Table 4.   Theoretical Population Trends of the Boll Weevil on a Conventional Long-Season Cotton and a Short-Season Variety Under an Early-Season Insecticide Control Program and a Combined Program of Control of Overwintering Survivors in Spring and Diapausing Adults in the Fall. Assume Fivefold Increase Per Generation (Adapted from Knipling, 1964)

	Weevils/Hectare			
	Early-Season Control		Early-Season and Fall-Control	
Generation	Long-Season Variety	Short-Season Variety	Long-Season Variety	Short-Season Variety
First year				
Parent[a]	100	100	100	100
F_1	50	50	50	50
F_2	250	250	250	250
F_3	1,250	1,250	1,250	1,250
F_4[b]	6,250	—	6,250	—
Fall control	No	No	Yes	Yes
Second year				
Parent	325	13	16	.63
F_1	163	7	8	.32
F_2	815	35	40	1.60
F_3	4,075	175	160	8.00
F_4	20,375	—	460	—

[a] Early season insecticide treatment kills 90 percent of overwintering parents before they can reproduce.

[b] Assume 10 percent of F_3 and 50 percent of F_4 are diapausing adults. Insecticides and defoliants applied during harvest period kills 90 percent of these. This is followed by 95 percent winter mortality of survivors.

ments in many areas, whereas 10 to 20 were used on conventional varieties. Another important feature is that, since the plants are not fruiting over such a long period, they do not require as much fertilizer and irrigation. In fact, irrigation has been reduced as much as 50 percent and amounts of fertilizer by 80 percent. The combined reduction in costs of insecticides, fertilizer, and irrigation has increased producer profits by more than $500/hectare (Sprott et al, 1976).

Although an individual farmer can benefit by growing the short-season variety, greater benefits will be obtained when all producers in an area grow the resistant variety and join together in an areawide integrated program for suppression of the total pest population. This will require uniform planting of the short-season variety, early-season control of overwintering boll weevils, and early defoliation and harvest of the crop followed by immediate stalk destruction. The defoliant should be mixed with an effective insecticide to kill diapausing weevils before they can leave the fields for hibernation sites. As shown in Table 4, a program of this type, combining early-season and fall control, would be devastating to the boll weevil. After the second year of this type of program, the early-season insecticide treatments might be discontinued except for localized areas of heavy infestation nearest to the most favorable hibernation sites. Control then would be directed solely towards the prehibernating adults in the harvest period when there is no danger of inducing outbreaks of other pests.

2. CONCLUSIONS

The above examples were selected to demonstrate how resistant varieties may affect the population dynamics of an insect pest and to show how they may be used in integrated pest management systems. The use of a resistant variety alone should not be expected to control pests under all conditions or in all locations where the crop may be grown. Instead, resistant varieties should be used in concert with other pest suppression measures. With a proper understanding of the pest, the host plant, and the environmental factors that affect them, a number of suppression measures may be used to suppress a pest without creating serious disruptions to other parts of the ecosystem. This provides the basis for integrated pest control.

ACKNOWLEDGMENTS

A portion of the research reported herein on cotton insects was supported in part by grants to the Texas Agricultural Experiment Station, Texas A&M University, from The Rockefeller Foundation, the National Science Foundation, and the Environmental Protection Agency (NSF GB-34718).

CIMMYT's maize germplasm bank contains about 12,000 accessions kept at ± 3°, and 50% relative humidity. Photo courtesy of CIMMYT.

12

GERMPLASM RESOURCES AND NEEDS

Jack R. **Harlan**

Department of Agronomy, University of Illinois, Urbana, Illinois

Kenneth J. **Starks**

Department of Entomology, Oklahoma State University, Stillwater, Oklahoma

1. INTRODUCTION

Genetic diversity supplies the raw materials for plant breeding programs. Most of the major crops of the world have been grown over extensive geographical regions and in numerous ecological environments for millennia and have generated enormous reserves of genetic diversity. When systematically sampled and stored in world collections, this diversity becomes accessible to plant breeders and can be exploited in plant improvement programs. Unfortunately, world collections have seldom been systematically assembled and none is as complete as it should be. There are great differences among the major crops in the quality and quantity of the genetic sampling as found in current collections. We shall attempt to assess a select list of crops in these respects, but we wish to make it clear that this assessment could be outdated in a year or two. Collecting activities are going on, and filling some conspicuous gaps could quickly upgrade a collection substantially.

That the genetic diversity deployed in many crops is eroding rapidly is widely known. Much valuable material has already been lost as modern, high-yielding cultivars have replaced old-fashioned landrace populations and primitive cultivars. Urgent appeals have repeatedly been made for action that would salvage threatened germplasm, fill recognizable gaps, and refine the genetic sampling of world collections (Harlan, 1972a). Response has been slow, but during the last decade a number of developments have taken place that should improve genetic resource management on a global scale (Harlan, 1975).

A Panel of Experts on Plant Introduction and Genetic Conservation was appointed to advise units of the Food and Agriculture Organization (FAO) of the United Nations on matters of plant genetic resources. The panel and FAO, together with the International Biological Program (IBP), were responsible for two international meetings held in Rome, each resulting in a book on the subject (Frankel and Bennett, 1970; Frankel and Hawkes, 1975). The panel, by various means, repeatedly called attention to the hazards of genetic erosion. It also drew up a plan for the development of a global network of genetic resources centers which was presented to the Consultative Group for International Agricultural Research (CGIAR). CGIAR accepted the plan, with some modifications, and as one result an International Board for Plant Genetic Resources (IBPGR) was established in 1973 with a secretariat in FAO.

The International Board has established priorities by regions and crops, appointed crop advisory committees for several of the major crops, supported the development of information retrieval and documentation systems, developed guidelines for long-term cold storage of base collections, and activated portions of the global network plan. It is currently supporting some collecting activities and will, no doubt, support more in future (Secretariat IBPGR, 1977). The next few years should see a marked improvement in the holdings of selected high priority crops. Material already lost, of course, cannot be recovered.

Meanwhile, a number of countries have tried to improve holdings of selected crops, and the international institutes have attempted to improve the genetic samplings of the crops for which they have international responsibility.

As material is collected from various parts of the world and brought to either national or international genetic resources centers, problems of protection from introduced pests and diseases become critical. No one wishes to introduce insects or diseases into countries where they are not present, but the germplasm is assembled with the intention of its being used. As plant improvement programs become more and more international, seed lots or vegetative clones are increasingly shipped around the world. This is a trend that is likely to accelerate greatly in the future with a consequent increase in the likelihood of introduction of unwanted diseases and pests. The problem is not easily resolved. A special FAO/IBPGR Task Force was established to consider the hazards and make recommendations. Results are soon to be published (FAO/IBPGR, in press), but the problem will remain indefinitely.

Collections of seed crops are most easily preserved as seed. Vegetatively propagated crops such as potato, sweet potato, yam, and sugarcane may produce seed that can be stored, but the seeds do not breed true, and if a particular clone is desired it must be maintained vegetatively. Some highly selected clones of tuberous crops have never been known to produce seed and must be preserved in the vegetative state. This is always more difficult and expensive than seed storage.

For the more important crops, base collections are usually set aside for long-term storage. These constitute world collections and, in the United States, seeds are stored at low temperature and low humidity at the National Seed Storage Laboratory at Fort Collins, Colorado. The potato collection is maintained at Sturgeon Bay, Wisconsin. The base collections are not generally available since the samples would soon be exhausted if distributions were frequently made. Materials generally available to plant breeders are maintained in working collections of the individual plant breeders or in short- to medium-term central storage laboratories. The small grains collections, for example, are maintained at Beltsville, Maryland, for distribution to breeders and at Fort Collins, Colorado, for long-term preservation.

Some of our collections are already very large (Reitz, 1976) and screening them on a global basis presents serious problems. Most reports published so far are extremely limited and unsystematic. Fragments of collections are challenged by natural or artificial infestations under all kinds of conditions so that one report can seldom be compared with another. Screening whole world collections under controlled conditions has seldom been done with diseases and almost never with insects. Until such work is conducted more uniformly and systematically little information will accumulate that would be helpful in generalizing on genetic sources of resistance.

Efficient, rapid, and reliable screening techniques are needed to make effective uses of large collections. In some cases it is relatively easy to rear large numbers of insects to challenge vast numbers of accessions, but in many cases this is a difficult, slow, and expensive process. We do know that higher plants contain enormous numbers of secondary metabolites. Several thousand compounds have been identified so far. Many, if not most of them, function as plant protectants. There are vast arrays of toxins, attractants, repellents, and so on. If the basis of insect resistance could be adequately understood at the chemical level, modern chemical techniques might permit rapid, inexpensive and accurate screening. Too often the fundamental causes of resistance are not known, and investigations into the biochemistry of resistance would appear to be an area in urgent need of expansion.

Some of the naturally occurring plant defense substances can be toxic to human beings as well as to insects. Selection for high levels of the protective agent could conceivably develop well-protected plants that are poisonous to man or his domestic animals. In other cases, the plant protectant may not actually be toxic but may lower digestibility and reduce food value. Bird-proof sorghum, for example, accumulates polyphenols to the point that birds avoid it, but at the same time digestibility in cattle is reduced to 40 percent. If this is the price of protection, one might as well let the birds have it. Similar examples may occur in plant–insect interaction and coevolution.

The example of L-canavanine in leguminous seeds may be an extreme case, but it illustrates the point. The compound is a nonprotein amino acid analogue of -arginine and is extremely toxic to insects. *Dioclea megacarpa* is a plant of the American tropics that evolved protective measures to the point that more than 60 percent of the seed nitrogen was tied up in canavanine, which constituted more than 8 percent of the dry weight of the seed. Yet one species of bruchid beetle, *Caryedes brasiliensis*, evolved whose larvae can discriminate between L-arginine and the deadly L-canavanine (Rosenthal et al., 1976). One might predict from this and other examples that high levels of resistance to some insects may produce high levels of susceptibility to others. All these areas must be explored before efficient use of large collections can be expected. The research will require more than a team of plant breeders and entomologists. Chemists, biochemists, nutritionists, physiologists, and perhaps pharmacologists must be involved as well.

Information on host range and insect resistance with respect to the wild relatives of crop plants is also often fragmentary. The wild races and species related to domesticated races can be expected to provide sources of insect resistance as they have provided sources of disease resistance (Harlan, 1976). The wild populations have been challenged over an evolutionary time scale much longer than the domestic ones, and they have adapted to the local insect populations. The use of wide crosses to incorporate resistance has been emphasized less in entomology than in plant pathology. One reason information on host range in the wild relatives is fragmentary is that the wild forms are very poorly represented in most world crop collections. Search for insect resistance may demand more complete and better balanced collections than we now have.

Experience would suggest that breeders interested in insect resistance are not likely to be tempted very often by vertical resistance as Robinson (1977) defines it. According to Robinson, vertical resistance is a defense *within* the capacity of microevolutionary change of the parasite, whereas horizontal resistance is a defense *beyond* the capacity for change of the parasite. So far, the Hessian fly (Gallun, 1977), the brown rice planthopper (Pathak, 1977), and several species of aphids seem to be the only examples of vertical resistances that break down because of changes in the insect genotype. More examples may be discovered in the future, but it seems at present that plant breeders will be dealing primarily with horizontal resistances that may persist, but may depend on quantitatively inherited traits. Resistance is likely to be relative, and depend to a considerable degree on population density of the insects, the environment, cultural practices, and other factors.

In this chapter we present a few examples of insect resistance and their geographic and taxonomic sources. These examples may provide clues for future search. We then select a sample of the most important food plants of the world and attempt to assess the current status of the available germplasm resources. Some of the prominent deficiencies are noted.

2. GEOGRAPHIC SOURCES OF RESISTANCE

Much has been written about the importance of centers of origin and/or centers of diversity as sources of useful germplasm. Such centers do exist for some crops and have, indeed, been rewarding as sources of germplasm for breeding programs. Biological theory would suggest that the most likely region in which to find resistance to a given insect would be where the insect is endemic (always present). In actual practice, there may or may not be a correlation.

The European corn borer (*Ostrinia nubilalis*) was introduced into North America early in this century. Selection for resistance resulted in inbreds with good levels of resistance (Scott and Guthrie, 1967). It was found, eventually, that some of the resistance is conferred by the chemical DIMBOA,* and that increasing the level of production of this material improved the protection of the plant (Tingey et al., 1975). More recently, tropical corn genotypes have been found that also confer resistance that is due to some agent other than DIMBOA (Tingey et al., 1975).

Similarly, Pathak (1977) reported that some 300 accessions of African rice (*Oryza glaberrima*) were highly resistant to the green leafhopper (*Nephotettix virescens*), which is not known to occur in Africa, and many cultivars of *O. sativa* from Asia were highly resistant to the delphacid *Sogatodes oryzicola* in South America (see also Jennings and Pineda, 1970). On the other hand, historical associations can be important in the evolution of resistance. A search among the rice cultivars of Southeast Asia has yielded sources of resistance to several of the most serious insect pests of the region (Pathak, 1977).

* 2,4-Dihydroxy-7-methoxy-2H-1,4-benzoaxazin-3(4H)-one.

In alfalfa there appears to be some geographic concentration of resistances. Accessions from the Turkestan region are likely to contain genotypes resistant to the spotted alfalfa aphid (*Therioaphis maculata*), the alfalfa seed chalcid (*Bruchophagus roddi*), nematodes, and some diseases (Sorensen et al., 1972). Resistance to the woolly pear aphid (*Eriosoma pyricola*) was found in wild pears of Asia Minor (Westwood and Westigard, 1969) and greenbug *Schizaphis graminum* resistance was found in wild races of sorghum in Africa (Weibel et al., 1972).

On the whole, however, geographic sources are not readily correlated with centers of origin, centers of diversity, or the geographical range of the insect pest. Resistance is where you find it, and good levels of resistance may evolve in the total absence of the insect. This may be related to the nature of insect resistances. A chemical that is toxic to one insect may be toxic to another. The protective agents may not necessarily be highly specific. But, once again, we emphasize that we know too little about the biology and biochemistry of resistance to screen large collections efficiently. Considering our present levels of understanding, the geographic origins of accessions assist relatively little in our search. We are reduced to screening massive collections to locate the occasional useful source.

3.   TAXONOMIC SOURCES OF RESISTANCE

3.1.   Genotypes within Breeding Populations

It is not uncommon to find genetic resistance within populations that are largely susceptible. Several examples from alfalfa could be cited. The cultivar 'Moapa' is founded on nine clones resistant to the spotted alfalfa aphid that were selected from populations of 'African.' In the same way, 'Zia' was selected from 'Lahontan,' and 'Cody' from 'Buffalo.' Resistant plants may be found in most alfalfas, but the highest levels seem to occur in alfalfas of Turkestan origin (Sorensen et al., 1972). Good levels of resistance to pea aphid (*Acyrthosiphon pisum*) can be found in individual plants in many cultivars, but the best sources seem to trace to the Flemish types or to Turkestan derivatives. For spittlebug (*Philaenus spumarius*), resistant plants may be found in most alfalfas but the best sources for antibiosis are *falcata* types. Reasonably good tolerance to potato leafhopper (*Empoasca fabae*) may also be found in *falcata* derivatives such as 'Culver,' 'Rambler,' 'Rhizoma,' and 'Teton.' However, tolerant clones may be simply selected out of adapted populations, as in 'Cherokee' (Sorensen et al., 1972).

Similar experiences have been obtained in maize breeding where inbreds have been developed with resistance to European corn borer (Zuber et al.,

1971a), western corn rootworm (*Diabrotica virgifera*), and corn earworm (*Heliothis zea*) (Widstrom and Wiseman, 1973). Resistances are enhanced by selection within breeding populations. Such sources are, in fact, rather common in breeding for insect resistance.

3.2.   Resistance among Cultivars

It is perhaps even more common to find some cultivars resistant and others not. We have already mentioned the Turkestan, Flemish, and *falcata* alfalfas. Other examples include greenbug resistance in 'Omugi' barley (Gardenhire, 1965), resistance to boll weevil (*Anthonomus grandis*) in frego-bract cottons (Jenkins and Parrott, 1971); resistances to the green leafhopper, the brown planthopper (*Nilaparvata lugens*), and the stem borers, *Chilo suppressalis* and *Tryporyza incertulas*, found among rice cultivars (Pathak, 1977), and chinch bug (*Blissus leucopterus leucopterus*) resistance among kafirs and fetrita sorghums (Dahms and Martin, 1940). In maíze, resistance to grasshopper feeding is reported for 'maíz amargo' (Horowitz and Marchioni, 1940), and inbreds differ strikingly in resistance or tolerance to several insects as indicated above.

In a few cases special morphological (taxonomic) features are associated with resistance. The solid-stemmed wheats are more resistant to wheat stem sawfly (*Cephus cinctus*) than other wheats, and resistance to the cereal leaf beetle (*Oulema melanopus*) is associated with dense trichome development on the leaves (Hoxie et al., 1975; Webster, 1977).

3.3.   Resistance in Wild Races

The use of wild races within the primary gene pool (Harlan and deWet, 1971) has apparently not been explored much at present. This is partly because of the very poor representation of wild relatives in our collections. The case of greenbug resistance in sorghum (Weibel et al., 1972) may be cited as an example. A more systematic search among the wild races of our crop plants would probably yield a number of sources of resistance.

3.4.   Resistance in Related Species

We have already mentioned resistance in green leafhopper in African rice (Pathak, 1977). Other examples include resistance to the woolly pear aphid derived from wild pear species (Westwood and Westigard, 1969), to the pine weevil (*Cylindrocopturus furnissi*) in a complex three-way species hybrid in pines (Smith, 1960), to the rubus aphid (*Amphorophora rubi*) in raspberry derived from other species of *Rubus* (Keep and Knight, 1967), and to the green peach aphid (*Myzus persicae*) in tobacco derived from wild species of *Nicotiana* (Thurston, 1966b).

3.5.   Resistance in Related Genera

Wide crosses have not been exploited as much for insect resistance as for disease resistance, but more attention will probably be given to these sources in the future. One outstanding example is the transfer of greenbug resistance from *Secale* to *Triticum* by way of X-radiation of triticale (Sebesta, in press). The Crop Evolution Laboratory of the University of Illinois has developed some maize material derived from *Zea* × *Tripsacum* hybrids and sorghum lines derived from *Saccharum* × *Sorghum* crosses. Evaluation for insect resistance has not yet been completed, although disease resistance has already been found in these derivatives.

Although the list of examples is far from complete it does suggest that resistances can be found at all taxonomic levels within reach of the plant breeder. The preference, naturally, is for resistance within populations of adapted material. The farther one must reach, the more difficult the breeding task. Wide crosses may require decades for exploitation and will be used as a last resort by the applied plant breeder. Such studies are more likely to be undertaken by geneticists interested in evolutionary relationships and theoretical biology. Nevertheless, some serious disease problems have been solved by wide crosses (Harlan, 1977) and it is reasonable to expect that similar solutions may be developed for insect problems.

4.   CROPS THAT FEED THE WORLD

It is surprisingly difficult to obtain accurate data on even the most important food plants. Some figures on gross production extracted from the 1974 Production Yearbook of FAO are shown in Table 1. These are revealing in showing how the human population depends heavily on a few major crops. The first four crops produced more tonnage than the next 50 crops combined, a phenomenon not characteristic of subsistence agriculture of a few centuries ago. The entire human population is being nourished by fewer and fewer crops. A serious problem with any one of the major food crops now means automatic starvation for millions. Not only are we being

Table 1.   Twenty-two Major Crops: Gross Production (Million Metric Tons)

1.	Wheat	360	12.	Millet	45
2.	Rice	320	13.	Banana	35
3.	Maize	300	14.	Tomato	35
4.	Potato	300	15.	Sugar beet	30
5.	Barley	170	16.	Rye	30
6.	Sweet potato	130	17.	Orange	30
7.	Cassava	100	18.	Coconut (nut)	30
8.	Soybean	60	19.	Cottonseed	25
9.	Oat	50	20.	Apple	20
10.	Sorghum	50	21.	Yams	20
11.	Sugarcane	50	22.	Peanut	20

fed by fewer and fewer species, but the crops we do grow are becoming more and more vulnerable to pests (National Academy of Sciences, 1972).

However, the table on gross production is misleading in many ways. Rice, for example, is reported as paddy in the husk; potato has very high moisture content; only 80 percent of the cassava tuber is usable; and so on. In Table 2, we have attempted to correct the figures for both waste and moisture content. The corrections may not be altogether accurate, but the figures are more meaningful than those in Table 1, in indicating food production in terms of edible dry matter.

The first 12 crops are the same in both lists although relative positions are changed. However, tomatoes, oranges, and apples have disappeared and are replaced by beans, peas, and sunflowers. We must recognize in addition

Table 2.   Twenty-two Major Crops: Edible Dry Matter (Million Metric Tons)

1.	Wheat	325	12.	Cassava	30
2.	Maize	270	13.	Sugar beet	30
3.	Rice	230	14.	Rye	25
4.	Barley	110	15.	Cottonseed	15
5.	Potato	60	16.	Peanut	14
6.	Soybean	55	17.	Coconut	9
7.	Sugar cane	50	18.	Bean	9
8.	Sorghum	45	19.	Pea	9
9.	Sweet potato	40	20.	Banana	8
10.	Oat	35	21.	Sunflower	7
11.	Millet	30	22.	Yam	6

Table 3.   Twenty-two Major Crops: Protein Production (Million Metric Tons)

1.	Wheat	39	12.	Sweet potato	2.7
2.	Maize	27	13.	Rye	2.7
3.	Soybean	22	14.	Bean	2.0
4.	Rice	20	15.	Pea	2.0
5.	Barley	11	16.	Sunflower	1.0
6.	Potato	6.0	17.	Cassava	0.8
7.	Oat	5.5	18.	Coconut	0.5
8.	Sorghum	4.5	19.	Yam	0.4
9.	Peanut	4.2	20.	Banana	0.3
10.	Millet	3.0	21.	Sugar cane	0.0
11.	Cottonseed	2.7	22.	Sugar beet	0.0

that there are vast differences in the amounts actually consumed directly by humans. For example, about half the maize is fed to livestock. Yet production is so enormous that reduction by 50 percent would not alter the relative importance of maize appreciably. The same may be said of barley, soybeans, oats, and sorghum. Most of the cottonseed is not processed for human consumption and a considerable amount of the coconut production goes into soaps and other commercial products.

Finally, there are huge differences in food value among the major crops. In Table 3, we attempt to give approximate protein production for the crops shown in Table 2. Again, the data may not be very accurate, but even sizable errors would not change the overall picture appreciably. The list fairly represents the most important food sources in the world, and we shall attempt to appraise the germplasm resources for each.

5.   APPRAISAL OF MAJOR GERMPLASM COLLECTIONS

5.1.   Wheat

According to Reitz (1976), the USDA has about 35,000 accessions of wheat. Recent collections in the Near East have greatly improved the situation with respect to wild and weedy forms poorly represented in the past. We are still deficient, however, in wild materials from Jordan, Iraq, and Transcaucasia in the USSR. The main collection is located at Beltsville, Maryland, with additional, usually duplicate, holdings at many experiment stations around the United States. Duplicates of most of the material have been placed in the National Seed Storage Laboratory at Fort Collins, Colorado, for long-term conservation.

The Soviet Union probably has holdings of about the same magnitude with the main repository at the Vavilov Institute of Plant Industry (VIR), Leningrad. Some other major collections include the Laboratorio del Germoplasmo at Bari, Italy; the National Seed Storage Facility, Hiratsuka, Japan; the World Collection of Wheat, Tamworth, Australia; the Central Experimental Farm, Ottawa, Canada; the Crop Research and Introduction Center, Izmir, Turkey; the Plant Breeding Institute, Cambridge, England; the Foundation of Agriculture and Plant Breeding, Wageningen, the Netherlands; and the Swedish Seed Association, Svalov, Sweden. Other collections are maintained in the Federal Democratic Republic of Germany and in Poland, Bulgaria, Romania, and elsewhere. The Centro Internacional de Mejoramento de Maíz y Trigo (CIMMYT) has international responsibility for wheat improvement but maintains only working collections. The center uses the U.S. Department of Agriculture (USDA) collection as a base resource.

Wheat is a cereal of the developed countries, and many nations with considerable resources are interested in it. There is no reason why collections and facilities for preservation should not be adequate. Yet when new high-yielding culivars started to replace the genetic diversity of traditional germplasm sources some years ago, there were few and sporadic attempts to salvage the discarded germplasm. A good deal of genetic diversity was lost that should not have been lost. There is still uncollected material in remote regions of Asia and Africa; Andean highland wheats are poorly represented, and although the Soviet Union may have a good sample of the strange and interesting wheats of the Caucasus, the rest of the world does not.

One resource frequently overlooked when we assess our genetic holdings is the enormous amount of wheat under cultivation (Reitz, 1976). The crop is grown on more hectares than any other in the world. Mutations are being generated continuously and genetic diversity deployed in the field is enormous. Even when new cultivars replace old landraces, the populations continue to change and evolve. Wheat cultivars are usually grown in a region for several years before they are replaced by newer ones. This permits some local adaptation and short-term evolution. By contrast, in maize, sorghum, and other crops in which hybrids are used, the farmer must get his seed each year from the hybrid seed producer and the evolution takes place in the breeding nurseries and not in the field.

Although some of the collections mentioned are rather modest and there is a great deal of duplication among them, the total supply of diversity in wheat is very large. Some accessions have been lost over the years for failure to monitor viability or because of poor storage conditions; some have lost identity through mixing and mislabeling; yet the situation with respect to wheat is better than for most of our crops.

5.2. Maize

The largest holdings are maintained at CIMMYT, El Batán, Mexico. Cold-storage facilities are available, although not at temperatures as low as now recommended. The CIMMYT collection contains more than 12,000 accessions and includes the bulk of the collections made during the systematic studies on the races of maize in Latin America during the 1940s and 1950s. In many cases a number of original collections were bulked to represent a race and reduce cost of maintenance. In the case of Mexican and Guatemalan races, however, the original collections were maintained by the Instituto Nacional de Investigaciones Agricolas (INIA) at Chapingo, Mexico, with holdings of more than 7,000 accessions (Brown, 1975).

Most of the Andean races and collections are maintained by the Instituto Colombiano Agropecuario (ICA), Medellin, Colombia, with additional accessions in Peru. Argentina has a small national collection. Most of the Brazilian and lowland South American material has been transferred to CIMMYT (Brown, 1975). The USDA has a modest collection of some 2,500 accessions maintained at Ames, Iowa, and there is a maize genetic-stocks collection at the University of Illinois. Small collections are found scattered across Europe and wherever maize is a significant crop.

Maize has probably been the most systematically collected crop of all. This is due largely to the foresight of the early programs sponsored by the Rockefeller Foundation in Mexico, Colombia, and elsewhere, and to the interest taken by the National Academy of Sciences and National Research Council with a strong input by Professor Paul C. Mangelsdorf. Maize was collected country by country, analyzed, and classified into races. It was a monumental work not duplicated in any other crop. There has been some loss and attrition. As a cross-pollinated crop, maize readily loses authenticity. The populations shift easily with changes in environment and even techniques of maintenance. There are regions near the Amazon and the Orinoco that have not been sampled, and some interesting maize races of Assam and, perhaps, Burma that have been collected in India but are not available elsewhere.

5.3. Rice

The primary world collection has been assembled and is being maintained by the International Rice Research Institute (IRRI), Los Baños, Philippines. This collection now includes some 35,000 items and the situation is much better now than it was a few years ago (Chung, 1976). Duplicates are being placed in the National Seed Storage Laboratory, Fort Collins, Colorado, for long-term holding at low humidity and temperature. Other rice collections are maintained at the Central Agricultural Station, Non Repos, Guyana; Hiratsuka; Beltsville; VIR; Cuttack, India; and the People's Republic of China. There are modest national collections in Sri Lanka, Burma, Thailand, Cambodia, and elsewhere. Efforts are being made to maintain a collection of African rice in West Africa since it is so difficult to culture at IRRI. There are still unsampled regions but the recent collections have improved the situation enormously. The USDA collection includes about 10,000 accessions. The most important materials not generally available are in the People's Republic of China.

5.4. Barley

The centers that maintain wheat germplasm usually have a barley collection as well. One might note the Research Institute of Cereals, Kramery, Havlivkovo, Czechoslovakia; the Waite Agricultural Research Institute, Adelaide, Australia; and the Ohara Institute for Agricultural Biology, Okayama University, Kurashiki, Japan as additional centers for the crop. The USDA collection includes some 18,000 items held at Beltsville (Reitz, 1976) and there is a barley genetic stock collection of about 2,000 items at Fort Collins. Recent collections in Ethiopia have improved the coverage considerably, but we are still weak in wild and weedy races. Recent reports of wild and weed barley in southwestern China have been published but no accession is recorded for western collections (Hsu, 1975; Shao et al., 1975). Among cultivated races, the largest gaps in our sampling include the Atlas mountains and oases in northern Africa, the Himalayas, and the People's Republic of China.

5.5. Potato

This is an Andean crop and some of the materials are difficult to maintain outside of the region. The Centro Internacional de Papas (CIP) has recently been established to assemble and maintain materials for use in international improvement programs. The work is just beginning, but there is an enormous wealth of genetic diversity in the region. In the United States, at the special potato station at Sturgeon Bay, Wisconsin, some 3,000 accessions are maintained. Europeans are particularly interested in the crop and collections are maintained in a number of countries including the United Kingdom, the Netherlands, France, Germany, Poland, the Soviet Union, and others. These collections, in general, are rather modest in size and represent a minute fraction of the germplasm of the species. The bulk of the potato genetic resources is still in the Andes. It is threatened locally by changes in cultural practices and farming systems and should be carefully monitored. These resources are too important to lose.

5.6. Soybean

The largest collection in the world is maintained by the U.S. Regional Soybean Laboratory at the University of Illinois. There are only about 4,000 items in the collection. Some of the late-maturing cultivars better suited to southern conditions are grown at Stoneville, Mississippi. The crop is of Chinese origin, but the Chinese apparently have working collections only and not many of these. Until recently, the crop was largely confined to regions of Chinese influence. More collecting could be profitably done in Indonesia, Indochina, Thailand, Burma, India, Nepal, Taiwan, Japan, Korea, and Siberia, but the bulk of the genetic diversity is in the People's Republic of China. Related wild species are found across the Pacific Islands and in Australia. These have been poorly sampled to date.

5.7. Sugarcane

Collections tend to be relatively small because of the difficulty and expense of maintenance. Few are likely to carry more than 1,000 clones, but collections of some sort are maintained wherever sugarcane breeding programs are carried out. In the United States, Beltsville, Canal Point, Florida, and Hawaii are the main stations. Other collections are found at Coimbatore, India; Queensland, Australia; Java, Indonesia; Fiji; Barbados; Mauritius; and Taiwan. Wide crosses have been used a great deal in protecting the crop against disease so that many of the collections include clones of related species and genera. These would be available for breeding for insect resistance as well.

5.8. Sorghum

The International Crops Research Institute for the Semi-arid Tropics (ICRISAT), Hyderabad, India, has international responsibility for the world sorghum collection. The Institute is rather new, however, and has yet to conduct systematic collection work. The world collection, such as it is, is also being maintained cooperatively by the USDA, Texas Agricultural Experiment Station and the Federal Experiment Station of Mayagüez,

Puerto Rico. It includes some 17,000 items, but large sections of Africa and Asia are not represented and many of the known races are missing. Some of the most conspicuous gaps include Mali, Chad, southern Sudan, Zaïre, Tanzania, Somalia, Yemen, Saudi Arabia, Indonesia, and the People's Republic of China (Harlan, 1972b). Improvement in the situation is urgently needed, and initial steps are being taken through assistance of the International Board for Plant Genetic Resources.

5.9. Sweet Potato

As indicated in Tables 1 and 2, the sweet potato is one of the major crops nourishing mankind, yet it is one of the most poorly collected. Plant breeders, of course, have their working collections and Yen (1974) has assembled material from the Pacific Region, as well as from the Americas. There is no genuine world collection for the crop, and working collections number accessions in the hundreds rather than in thousands. Though sweet potatoes make seed, the crop is usually propagated vegetatively so that a cherished clone may be preserved more or less indefinitely. As noted earlier, maintenance of large collections of vegetatively propagated plants is much more difficult and expensive than maintenance of seed crops.

5.10. Oat

Institutes and stations that carry collections of wheat and barley usually have oat collections as well. The Central Experiment Station at Ottawa has been particularly active in exploration of wild and weedy races and species. For those who believe some parts of the world are overcollected and have nothing to offer, the experience of collectors of oats should provide a warning. No less than three undescribed species of *Avena* have recently been discovered in the western Mediterranean and the Canary Islands, where the flora have been intensively studied for many years. The weed *A. sterilis* apparently has a good deal to offer in the improvement of cultivated oats and a wider sampling of wild oats may be desirable. The USDA holdings number about 14,000.

11. Millet

On the whole, millet germplasm collections are in remarkably bad shape. The most extensive collection in the United States is a pearl millet collection assembled and maintained by G. W. Burton at Tifton, Georgia. From the point of view of human nutrition, pearl millet is the most important of the millets. Millions of people in both Africa and Asia depend on it as the staff of life. Collections have been assembled, but much of the material has been lost through improper maintenance. The crop is allogamous and maintenance of large numbers of accessions is difficult and expensive. Recent collections in Africa by French teams should improve the holdings considerably. It is expected that ICRISAT will maintain world collections of millets for international programs.

Italian or foxtail millet is still a major crop in the People's Republic of China where the bulk of production is centered. Few Chinese accessions are available to the West. A modest collection in the United States is maintained in Colorado. The crop was once common as a short-season hay crop, but is little used for that purpose now; its chief use in the United States is as birdseed.

Ragi or finger millet is more than a trivial crop in India and the East African highlands. Working collections are maintained at several Indian research stations, but the combined holdings are relatively small.

Proso, kodo, and other millets have received but little attention and collections are small.

5.12. Cassava

This is another example of a major crop on the world scene that is poorly collected and not even very well known genetically, taxonomically, or agronomically. A collection has been assembled at the Centro Internacional de Agricultura Tropical (CIAT), Cali, Colombia and another at the International Institute of Tropical Agriculture (IITA), Ibadan, Nigeria. Smaller collections are maintained by the USDA in Puerto Rico and Florida and by the Brazilian Ministry of Agriculture. None is really large, and cassava germplasm is not well sampled at present. A major difficulty in germplasm collection is the serious infestation of virus diseases, especially in Africa.

5.13. Sugar Beet

The bulk of the sugar beet germplasm is in Europe where the crop originated. Early in the development of the sugar beet industry of the United States it was impossible to produce seed, and tons of sugar beet seed were shipped from Europe every year. This has provided the base for sugar beet breeding in the United States, which is primarily in the hands of sugar companies. European sugar corporations and cooperatives also maintain their working collections. Several species of wild *Beta* have proved useful in breeding programs.

5.14. Rye

Rye collections are usually found at the same institutions as wheat, barley, and oat collections. Rye has received most attention in northern Europe where it probably originated as a crop. Wild forms from the Near East have been used in breeding programs but are not well sampled.

5.15. Cotton

Cotton is primarily a fiber crop but, as shown in the tables, is also an important source of oil and protein. There are substantial working collections around the world wherever active breeding programs are established. The genus is rather poorly sampled and species collections have been decimated for lack of a tropical facility for maintenance. The United States collection of wild diploids is held at College Station, Texas, and Tucson, Arizona. Primitive tetraploids and obsolete cultivars are maintained at Stoneville, Mississippi, and College Station, Texas, and the barbadense materials are held by the University of Arizona. Other collections are in the hands of private cotton breeding companies. Sizable collections are found at Shambat, Sudan; Tashkent, USSR; the Chaco Station, Argentina; and in India (Harland, 1970).

5.16. Peanut

The USDA holdings amount to about 4,000 accessions, most of which are maintained at the U.S. Southern Regional Plant Introduction Station, Experiment, Georgia. Duplicates are placed with the National Seed Storage Laboratory. There are numerous collections around the world, for example, in Argentina, Venezuela, Senegal, Nigeria, Malawi, South Africa, Rhodesia, Israel, India, and China. The genus is still poorly understood. The largest species collections are maintained by Gregory in North Carolina and Krapovickas in Córdova, Argentina (Krapovickas, 1969). Resistances to several insects have been identified among the wild species (Banks, 1976), but it is not yet known if they can be transferred to cultivated races. The crop is widely grown in South America and many of the less accessible areas have not been sampled at all. Peanuts have been grown long enough in West Africa to develop a striking diversity that is also inadequately collected. ICRISAT has recently accepted international responsibility for the crop and more collections are being sponsored by the International Board for Plant Genetic Resources.

5.17. Coconut

The crop has been neglected from the point of view of breeding and genetics. Coconut research stations have been established in India, Sri Lanka, Malaysia, and Philippines. Each of these has rather small collections, and Indonesia, Melanesia, and Polynesia have been sampled on a limited scale. It is safe to say that most of the variation is still in the field and not assembled in conservation collections (Harries and de Poerck, 1971; Williams, 1975)

5.18. Bean

Most of the breeding work with beans has been done in the more developed countries, and these carry the largest collections, as would be expected. Total holdings in the United States, Europe, and the Soviet Union are considerable. Collections among the tropical races are more fragmentary. CIAT has recently assumed international responsibility for the crop, and it is hoped that a truly adequate sampling of tropical beans will emerge. In addition, there are some national collections of importance in Mexico, Guatemala, Costa Rica, and elsewhere in Latin America. The diversity of beans in Turkey is rather remarkable and approaches that of a secondary center. Beans in Turkey have been sampled, but perhaps not adequately. Wild beans have been fairly well collected in Mexico but are not well sampled in South America (Gentry, 1969; Bergland-Brücher and Brücher, 1976).

5.19. Pea

Breeding work with peas is even more than beans confined to the developed nations, since the crop has much less potential in the tropics. The largest collections are in the United States, Canada, and Europe. VIR has extensive holdings. The wild races have been sampled in only a fragmentary fashion.

5.20. Banana

As with other vegetatively propagated crops, banana collections tend to be small, but there are many of them scattered through the tropics, for example, in India, Malaysia, Thailand, Philippines, Indonesia, Hawaii, Honduras, Jamaica, Trinidad, Cuba, and elsewhere. There has been no large, extensive, or systematic effort to collect and maintain banana germplasm (Williams et al., 1975).

5.21. Sunflower

Although the sunflower was probably domesticated in the American Midwest, it has not been a major commercial success in the United States. Most of the breeding work has been done in eastern Europe and the Soviet Union, and that is where the bulk of the cultivated germplasm is located. We do have vast amounts of wild and weedy races should they turn out to be useful. Weed sunflowers have been shown to carry genetic resistance to some diseases (Harlan, 1972), and it may well be that they have resistances to insects as well.

5.22. Yam

This is another of the neglected tropical crops. There are several species grown for food, but by far the most important are *Dioscorea alata* and *D. esculenta* of Southeast Asia and *D. rotundata* (= *D. cayenensis*) of West Africa. Some collections have been assembled in Nigeria, Ghana, Puerto Rico, Tahiti, China, and elsewhere, but the sampling has been fragmentary. IITA has assumed international responsibility for germplasm assembly and more systematic collection is underway.

Additional information on germplasm collections will be found in the chapters on individual crops in Part 2, especially Chapter 13 on alfalfa and Chapter 15 on cotton.

The International Board for Plant Genetic Resources (IBPGR)* is the world coordinating body for germplasm collection and maintenance activities.

* IBPGR may be addressed care of the Crop Ecology and Resources Unit, Food and Agriculture Organization, Via Delle Terme di Caracalla, 00100 Rome, Italy.

6. CONCLUSIONS

Although some of the world collections are very large, none is as complete as it could be. There is usually a great deal of duplication and there are arrays of closely related materials, whereas some taxa and geographic regions are essentially unsampled. Collections tend to be weakest in wild races and related species, yet these sources have repeatedly yielded resistances to diseases, nematodes, and insects. Remote and inaccessible places tend to be the least sampled, as expected, yet conspicuous geographic gaps in coverage may be close at hand and easily reached. Only a few collections have been systematically assembled according to a plan. Most of them have simply accumulated over the years. Some very important tropical crops have been badly neglected and collections are grossly inadequate.

We have not had enough experience with insect resistance to be guided to the most likely sources. Useful resistance may be found within breeding populations, among cultivars, in wild races of the primary gene pool, or in related species and genera. Resistances may be found in regions where the insect is endemic, occasional, or absent. There are indications that more complete collections may be required for breeding for insect resistance than for breeding for disease resistance.

Genetic erosion has been critically serious in some crops and useful germplasm is lost every year as new cultivars replace old landraces and primitive sorts. This is a threat of global importance and should be taken seriously.

Part Two

BREEDING SYSTEMS FOR RESISTANCE

Breeding for Resistance
in Specific Crops

Alfalfa susceptible to the alfalfa weevil (bottom) is shown in contrast with the resistant variety (top). Photo by Murray Lemmon, courtesy of U.S. Department of Agriculture.

13

BREEDING APPROACHES IN ALFALFA

M. W. Nielson

Forage Insects Research Laboratory, U.S. Department of Agriculture, Tucson, Arizona

W. F. Lehman

Department of Agronomy and Range Science, University of California, Davis, California

1. ALFALFA

Alfalfa, *Medicago sativa,* the most important single forage crop in the world, was cultivated long before recorded history. According to Bolton et al. (1972), the plant originated in the "Near Eastern Center" which includes Asia Minor, Transcaucasia, Iran, and the highlands of Turkmenistan. Iran is often mentioned as the home of alfalfa.

Alfalfa occurs in all major geographical areas of the world, particularly in the temperate zones, where it was largely introduced by the migration of man and his domesticated animals. After establishment in Europe, Asia, and North Africa, alfalfa was introduced to the New World in the sixteenth century by the Spaniards. Later, the plant was brought to South Africa and Oceania around 1800 (Bolton et al., 1972).

More than 33 million hectares of alfalfa are under cultivation in the world. Approximately 21 million hectares are grown in the Northern

Hemisphere, mainly in United States and Europe; 12 million hectares are cultivated in the Southern Hemisphere, primarily in Argentina. The United States, USSR, and Argentina contribute about 70 percent of the world's total production. France, Italy, Canada, and Australia produce about 20 percent; and South Africa, Oceania, Iran, and a few other countries contribute the remaining 10 percent (Bolton et al., 1972).

Alfalfa is used primarily as dried hay, pasture, or greenchop for a wide variety of livestock, including dairy cows, beef cattle, horses, sheep, swine, poultry, and rabbits. Recently it has become popular for human consumption in the form of "alfalfa sprouts." The high nitrogen-fixation capabilities of alfalfa have labeled it as a good source of soil fertility. It is grown in many areas of the United States as a green manure crop. Alfalfa also produces excellent honey from the nectar that is gathered by honeybees.

2. INSECT PESTS

Numerous insects are attracted to alfalfa. They utilize the plant directly for food, for oviposition, for shelter, or to prey on other insects that inhabit the crop. App and Manglitz (1972) stated that more than 100 species of insects are injurious to alfalfa in the United States but that only about 30 species periodically cause serious economic injury. Annual loss of the crop has been estimated at $260 million. In northeast Africa and Southeast Asia, 108 species of insects were listed by Gentry (1965) as pests of alfalfa and two other legume species of plants. Table 1 lists some of the common insect pests of alfalfa, along with information on geographical distribution, pest status, and reported resistance either as resistant sources, germplasm available, or as released varieties.

3. ALFALFA GERMPLASM

3.1. Classification

Cultivated alfalfa (*Medicago sativa*) is a highly heterogeneous legume that is adapted to a wide range of climatic conditions. Because of this, it has been classified in many ways, for example, by flower color, winter hardiness, area of origin, spring growth, dormancy, recovery after cutting, flowering, growth type, rainfall conditions, and soil type (Bolton, 1962; Busbice et al., 1972; Lowe et al., 1972). A practical classification used by most alfalfa workers and applicable over a wide variety of plant types and growing conditions is one based on a combination of winter hardiness and dormancy (Lowe et al., 1972; Marble, 1977). These two characteristics are related, and the classifications based on them are usually called hardiness classifications. Other characteristics such as area of origin, flower color, and growth type have also been used for further subdivision within the hardiness classes.

Marble (1977) presented a detailed classification that included three major hardiness or dormancy groups: (1) winter hardy or dormant; (2) medium winter hardy or semidormant; and (3) nonwinter hardy or nondormant. Examples of varieties in Group 1 are 'Grimm,' 'Ladak,' 'Ranger,' 'Vernal,' and 'Dawson;' in Group 2 are 'Buffalo,' 'Lahontan,' 'Mesilla,' and 'Cody;' and in Group 3 are 'Caliverde 65,' 'Moapa 69,' 'Mesa Sirsa,' 'Hairy Peruvian,' and 'CUF 101.' These groups have also been further subdivided to reflect the differences and requirements of cultivars in specific areas of adaptation. For example, the winter hardy group was subdivided into very winter hardy ('Ladak'), winter hardy ('Ranger'), and moderately winter hardy ('Saranac') on the basis of fall growth after cutting the first week in September (anonymous, 1978). Similarily, Marble (1977) subdivided the nonwinter dormant class into intermediate dormancy, moderately nondormant, and very nonwinter dormant represented by 'Caliverde 65,' 'Moapa 69,' and 'CUF 101,' respectively. Cultivars within these groups have also been classed for other commonly used characteristics, such as cultivar origin, parental germplasm, developing agency, and year the first seed was available (Barnes et al., 1977).

All germplasm identified as *Medicago sativa* and *M. falcata* will intercross (Barnes et al., 1977). *M. varia* (*M. media*), originated from material crossed between *M. sativa* and *M. falcata,* will also cross with *M. sativa.*

Certain species of *Medicago,* or germplasm often classed as species of *Medicago,* have been used as sources of genes for cultivated alfalfa. The most successful species has been *Medicago falcata.* The cultivars 'Vernal' and 'Rambler' are thought to have derived improved winter hardiness and broad crowns from *M. falcata. M. glutinosa* will also cross readily with *M. sativa* but has been used with limited success. Studies have been conducted on the transfer of genes from more distantly related species of *Medicago* to *M. sativa,* but this work is in the experimental stage and has not yet contributed important characteristics to alfalfa cultivars.

Successful interspecific crosses are most likely with the perennial species of *Medicago.* Of these *M. glomerata, M. prostrata, M. rhodopea, M. pironae,* and *M. daghestanica* appear most promising (Lesins and Gillies, 1972).

Table 1. Important Insect Pests of Alfalfa of the World, Their Distribution, Pest Status, and Reported Resistance in Plants

Common Name	Scientific Name	Geographical Distribution	Pest Status[a]	Resistance Reported
Insects that suck sap				
Spotted alfalfa aphid	Therioaphis maculata	SW Asia, North America, Australia	Key	Yes
Pea aphid	Acyrthosiphon pisum	Cosmopolitan	Key	Yes
Blue alfalfa aphid	Acyrthosiphon kondoi	Eastern Asia, United States, Argentina, Australia	Key	Yes
Cowpea aphid	Aphis craccivara	SW Asia	Occ.	No
Potato leafhopper	Empoasca jabae	North America	Key	Yes
Alfalfa leafhopper	Empoasca mexara	SW United States	Key	No
Leafhopper	Aceratagallia curvata	SW United States	Occ.	No
Meadow spittlebug	Philaenus spumarius	Midwestern and eastern United States	Key	Yes
Three-cornered alfalfa hopper	Spissistilus festinus	Southern United States	Key	No
Alfalfa plant bug	Adelphocoris lineolatus	North America, Europe, SW Asia	Occ.	No
Potato capsid	Calocoris norvegicus	SW Asia, Europe	Occ.	Yes
Lygus bug	Lygus hesperus	Western United States	Key	Yes
Tarnished plant bug	Lygus lineolaris	North America	Key	Yes
Say stinkbug	Chlorochroa sayi	North America	Key	No
Pyrrhocorid bug	Pyrrhocoris apterus	SW Asia	Occ.	No
Western flower thrips	Frankliniella occidentalis	United States	Key	No

Insects that chew leaves and stems

Common name	Scientific name	Distribution	Status	Economic
Alfalfa weevil	*Hypera postica*	United States, Europe, SW Asia	Key	Yes
Egyptian alfalfa weevil	*Hypera brunneipennis*	SW United States, SW Asia	Key	Yes
Cloverleaf weevil	*Hypera punctata*	United States	Occ.	No
Alfalfa caterpillar	*Colias eurytheme*	North America	Key	No
Clouded yellow butterfly	*Colias croceus*	SW Asia, North Africa	Key	No
Alfalfa looper	*Autographa californica*	United States	Occ.	No
Green clover worm	*Plathypena scabra*	United States	Occ.	No
Alfalfa webworm	*Loxostege commixtalis*	United States	Occ.	No
Yellowstriped armyworm	*Spodoptera ornithogalli*	United States	Occ.	No
Army cutworm	*Euxoa auxiliaris*	United States	Occ.	No
Grasshopper	Various spp.	Cosmopolitan	Occ.	Yes
Insects that feed on roots				
Clover root curculio	*Sitona hispidulus*	North America	Occ.	No
White-fringed beetle	*Graphognathus* spp.	North America	Occ.	No
Alfalfa snout beetle	*Otiorhynchus ligustici*	North America	Occ.	No
Root curculio	*Sitona* spp.	North America, Europe, SW Asia	Occ.	No
Seed insects				
Alfalfa seed chalcid	*Bruchophagus roddi*	Cosmopolitan	Key	Yes
Seed midge	*Dasineura* sp.	Western United States	Occ.	No

a Occ. = occasional.

283

3.2. Germplasm Collections

Alfalfa germplasm collections are maintained in three primary locations:
(1) The National Seed Storage Laboratory at Fort Collins, Colorado, maintains a collection of most released varieties, breeding lines, germplasm
sources, and reserve seed of introductions. This is usually considered long-
term storage, and material from this collection is used only when other
sources are unavailable. (2) Regional Plant Introduction Stations are
located in the United States at Ames, Iowa; Geneva, New York; Pullman,
Washington; and Experiment, Georgia. They maintain seed of about 1,500
to 1,600 *Medicago* plant introductions. These are working collections from
which requests for research purposes from qualified individuals or groups
from within the United States will be filled. Requests from other countries
for seed should be made through the Germplasm Resources Laboratory at
the Agricultural Research Center in Beltsville, Maryland. Seed of the
perennial *Medicago* lines is primarily held at the Ames, Iowa location, and
that of the annual species is held at Experiment, Georgia. However, each
location may hold some *Medicago* introductions. (3) Many individuals,
especially plant breeders, working with *Medicago* maintain limited germplasm collections and may be willing to provide seed upon request. The
Food and Agriculture Organization of the United Nations at Rome, Italy,
publishes the *Plant Genetic Resources Newsletter,* which contains information on genetic resources, as the name indicates. However, *Medicago* has
received little attention in past newsletters.

A list of *Registered Field Crop Varieties: 1926-1974* has been published
by the American Society of Agronomy.* Included in this list are the variety
name, registration number, year registered, originating institution, agency,
or organization, and reference where a description of the variety can be
found. A similar list for alfalfa germplasm releases has also been compiled
(Hunt et al., 1978). Small samples of seed of variety and germplasm
releases are usually available from the originators, if the seed supply has not
been exhausted. Listings and descriptions of the *Medicago* species can be
found in Sinskaya (1950), Bolton (1962), Heyn (1963), and Lesins and
Gillies (1972).

4. BREEDING CONCEPTS, PROCEDURES, AND TECHNIQUES

4.1. Genetic Basis

Alfalfa is a cross-pollinated crop with rapid inbreeding depression and
breeding behavior of an autotetraploid. Hayes et al. (1955) classified alfalfa

* 677 South Segoe Road, Madison, WI 53711.

as a cross-pollinated species with an average of 90 percent cross-pollination. Busbice et al. (1972) reported that seed production of self-pollinated alfalfa plants ranged from a high of 35 percent of the noninbred for S_1 plants (first generation of self-pollination) to a low of 2 percent in the S_4. Forage yield ranged from a high of 74 percent of the noninbred parent in the S_1 to a low of 38 percent of the noninbred in the S_2. Forage yield of an S_3 was 75 percent of the noninbred parent. In general, each successive generation of inbreeding caused a rapid decrease in vigor and an increase in mortality.

Busbice et al. (1972) also analyzed data on inheritance of alfalfa characteristics. They concluded that it was complex, owing to the autotetraploid nature of meiosis, and that the inheritance of qualitative characteristic was consistent with autotetraploid behavior. In addition, they concluded that homozygotes have not been produced in alfalfa because of severe inbreeding depression or because lethality occurred with severe inbreeding. This indicates that it would be difficult to obtain homozygosity for complex characters. The perennial nature of alfalfa, its cross-pollinated behavior, and autotetraploid type of inheritance should be important considerations in developing a breeding program and determining breeding methods.

4.2. Breeding Methods

New alfalfa cultivars can originate from three major sources, that is, introductions from foreign countries, selection within introductions or adapted cultivars, and hybridization. Introductions can be tested and, if superior to other cultivars, can be used without further work. Introductions and adapted cultivars may be a source of valuable parental material in a selection program and/or as parents in a hybridization program.

Germplasm may be handled by one or a combination of three primary breeding methods: plant or line selection, backcross, and recurrent selections. Either individual plants or seed lines can be used. Methods used for plants or seed lines would be similar in many situations.

Plant or Line Selections

With the plant or line method, the initial number of plants or lines used is very large. This number is gradually reduced until only the desired number of superior plants or lines remain. A variety is then synthesized from the surviving plants or lines. For example, the starting number of plants from a cross variety or plant introduction might be 1,000 to 10,000. If lines are used, the number may vary between 100 and 1,000. A larger number of plants can be handled if seedling methods of selection are employed. A much lower number of plants may have to be used if selections are made among space-planted material growing in the field or in some other selection situation. In these procedures, the population is successively narrowed as plants are screened for the primary insect, retested for the primary insect, progeny tested, and observed or screened for other problems such as seed production and resistance to other plant pests.

Maternal lines may originate from seed obtained from individual plants growing in a severely infested field. This may be a seed field, test plot, roadside, or abandoned area. Seed may be harvested from 100 to 1,000 plants, and these seed lines are then tested for resistance to the pest. About 10 or more lines with some resistance might be obtained. If necessary these lines could be subjected to other tests. One or all of these lines may be combined into a variety that is resistant to the pest. Resistance to the pest would be lower when the described line method is used than with the individual plant method because resistance in the seed lines would have been diluted by pollen coming from susceptible plants. Plant selection is preferred in the development of insect resistance because sources of dilution can be controlled. However, resistance in seed lines could be increased with one or more cycles of recurrent selection.

Backcross

The backcross method of breeding, first used on alfalfa by Stanford (1952), is a way of transferring one desirable or needed characteristic found in an undesirable or unadapted genetic background into a cultivar that is well adapted to a particular area. In this method an original cross is made between the unadapted (nonrecurrent) parent and the adapted (recurrent) parent. If the characteristic being transferred is dominant and easily identified, the plants with the desirable characteristic are backcrossed to the recurrent parent. Progeny with the desired characteristic from this cross are again backcrossed to the recurrent parent. Crossing of progeny with the recurrent parent is continued until a sufficient amount of germplasm of the recurrent parent has been recovered. One or two cycles of self-pollination of phenotypic recurrent selection can be done after crossing in order to intensify or increase the gene frequency of the desired character.

Theoretically, germplasm of the recurrent parent in the cross and the successive backcrosses are 50, 75, 88, 94, 97 in the F_1, BC^1, BC^2, BC^3, and BC^4, respectively.

If the character being transferred is recessive, one generation of selfing may be required after each cross in order to recognize the phenotype of the desired character.

The number of plants used in each cross should be as large as possible since alfalfa is a heterogeneous cross-pollinated crop. The actual number of parent plants used, however, will be largely determined by the number of plants that can be handled. If a large number of plants is used, more of the genetic background of the recurrent parent and desirable genes of the nonrecurrent parent will be recovered. Stanford (1952) used an average of 125 plants in each generation for the recurrent parent (female) and 275 plants for the nonrecurrent parent (male). He also used four backcrosses and selected 50 plants for synthesis of the variety 'Caliverde' in the $BC^4 F_2$.

Recurrent Selection

Recurrent selection appears to be well suited for use in a heterogenous cross-pollinated crop such as alfalfa because highly heritable, desirable genes can be easily concentrated by using large populations. This method is being widely used today to develop germplasm with multiple pest resistance. It appears to lend itself well to breeding for pest resistance because these characteristics are generally highly heritable (Hanson et al., 1972). The method also has the advantages of maintaining genetic variability; being relatively low in cost; and offering the possibility of incorporating other breeding methods, such as plant selection or backcrossing, into the procedure (Busbice et al., 1972).

In a recurrent selection program plants or lines with the desired characteristic are selected from a population such as a cross variety or broad germplasm pool. When making the selections, one must make every effort to eliminate escapes because the gene frequency for the desired character is reduced as escapes are increased (Busbice et al., 1972). Escapes will have a much greater effect on recessive than on dominant characters. The number of selected plants or lines should be kept as high as possible (25 to 100, and preferably more) in each cycle to keep the germplasm as broad as possible and reduce the possibility of the resulting population having a higher frequency of detrimental characteristics than found in the original populations.

Selected plants or lines from one cycle of selection are interpollinated, seed is produced on them, and another cycle of selection is made in the resulting progeny for the same characteristic(s). Hanson et al. (1972) have shown rapid progress in one to three cycles of plant selection for characteristics for resistance to spotted alfalfa aphid and common leaf spot. Progress for other characteristics, such as leafhopper yellowing, however, is much slower and is still possible after 10 cycles of selection.

Other Methods

Strain building is often referred to as a method of breeding. However, as indicated by Hayes et al. (1955), this involves using a combination of breeding and testing methods over a period of time.

4.3. Use of Selections

Selection of plants or lines for individual traits can be made in the field or under controlled conditions in the greenhouse or growth chamber. The latter is used primarily when large numbers of plants must be screened in a short period. In comparisons made under field and nonfield conditions, Hanson (1972) found that yield increased when selections were made in the field for vigorous healthy plants with pest resistance but that yield decreased when selections were made entirely in nonfield conditions. This "positive field effect" indicates that breeding programs should include selections made under nonfield and field conditions.

Elgin et al. (1970) described three selection procedures to use on alfalfa: (1) independent culling and successive elimination, (2) tandem selection, and (3) index selection. In independent culling and successive elimination, selections were made for one or more factors; then the remaining plants or survivors were selected for additional factors. Selected plants were mated to produce a new cycle for additional reselections, if desirable. In tandem selection one characteristic was selected through as many generations as needed to fix it in the populations. After this, reselection was started on a second characteristic that was then worked until it was fixed in the population. This process was continued with successive characteristics. In index selection plants were scored for several characteristics, which were given similar or weighted scores. Plants with the highest scores were retained and recombined; then the selection process was repeated. The index method was the best of the three, followed by independent culling, and then tandem selection.

After the desired selections have been made, one or more selected plants or lines can be combined into a cultivar. However, cultivars are seldom produced from a single plant because of possible inbreeding and yield reduction in the successive synthetic generations. Busbice et al. (1972) concluded that a minimum of four parents is required to prevent excessive inbreeding in advanced generations, and that more than four would be needed if parents are partly inbred. They further concluded that there would be little advantage to including more than 16 parents. However, if released cultivars are also intended to be used as part of a germplasm improvement program, as suggested by Hanson et al. (1972), it might be desirable to keep the number of parents as large as possible.

4.4. Evaluation

Selected plants or populations must be evaluated to determine the effectiveness of the selection procedure or breeding techniques and the value of the plant, resulting germplasm pool, or experimental cultivar. Plants resulting from specific breeding methods may be evaluated by using (1) the reaction of the selected plant or clone, (2) S_1 progeny, (3) polycross progeny, (4) open-pollinated or topcross progeny, and (5) diallel cross. Populations or lines are tested by evaluating as many characteristics as necessary, using essentially the same techniques as in progeny testing.

Clone

Evaluation of the selected plant or clone is effective and inexpensive when the selected characteristic is highly heritable. Rapid progress has been made in aphid resistance by imposing stringent criteria when making the initial selections and then, if possible, retesting the clones to screen out escapes. The superior clones can be intercrossed and the resulting population tested and possibly used as a new cultivar, as in 'Moapa' or 'CUF 101.' If more progress is desired, a cycle of recurrent selections can be made. The clonal evaluation method is rapid, relatively inexpensive, and effective.

S_1 Progeny

Selfed seed can be produced on most selected (S_0) alfalfa plants in amounts sufficient for limited progeny tests in the greenhouse or field. Self-pollinated seed, generally from S_0 plants, is usually produced in the greenhouse by hand pollination. Tests are conducted on individual plants produced from this seed. Such tests provide the best estimates of breeding value for the parent plant. Self-pollinated seed is also the best way to store the characteristics of a desirable plant if preservation of the parent clone is impossible. Disadvantages of this testing method include the time and expense required to produce and test the seed, limited amounts of available seed, and unreliable yield information.

Polycross Progeny

The polycross progeny test for general combining ability is one of the most widely used tests. A limited number of parent plants or genotypes (usually 3 to 50) are planted in replicated plots with about two to five cuttings or clones per plot and, if possible, 8 to 12 or more replications. A large number of replications provides better mixing of pollen and more seed. A fairly large amount of seed is produced on a restricted group of possible parent plants which can be progeny-tested for resistance factors, yield, general combining ability, or other characteristics. Disadvantages of this method are the larger amounts of time required to produce the seed and test the progeny.

Open-Pollinated or Topcross Progeny

This method is generally a quick way to obtain seed for progeny tests. Open-pollinated seed is obtained from female or mother plants which usually have been pollinated by using a small number of plants. Topcross seed is made in a special test in which the male is a known line or cultivar capable of providing a uniform pollen mixture. These tests measure either general or specific combining ability, depending on the pollen source. Large amounts of open-pollinated seed can be obtained within a relatively short time. However, test results from open-pollinated seed should be used with caution because the source of pollen is unknown and may not be uniform. In addition, the male plants used to produce the open-pollinated seed may be entirely different from those used in a cultivar that might be produced from the female plants.

Diallel Cross

Except for work on hybrids, this is usually the last progeny test used in a commercial breeding program. Its main use is for special studies. This test is conducted by crossing each clone in all possible combinations with all other clones in a restricted group of plants. Advantages are that both parents are known and that mixing of male and female plants in a restricted group of parents is better than for any other test. This is the best test for general combining ability, and yield tests are reliable if sufficient seed is used. However, the method is slow and expensive. The seed is generally produced by hand pollination, which is time consuming. Some crosses are hard to obtain or produce low amounts of seed.

4.5. Pollination Techniques

Selfed seed is produced by tripping alfalfa flowers with a toothpick or with a folded, pointed piece of coarse paper. Blotter paper is good and easily available. When changing from one genotype or clone to another, the tripping device (toothpick or blotter paper) should be discarded and the workers' hands or other objects that might transfer pollen should be washed in 70 percent alcohol. The object being sterilized should be in contact with the alcohol for a minimum of about 1 minute to be certain the pollen has been killed. Self-pollination may also be done by gently rolling the raceme between the thumb and forefinger. Fairly large amounts of self-pollinated seed can be produced in the field by staking the plant and covering it with a net or cheesecloth bag. The bag is opened periodically to trip the flowers, usually by rolling the racemes between the thumb and forefinger.

Cross-pollinated seed is produced with or without emasculation, a difficult and time-consuming procedure. It is used only under special circumstances, for example, where no self-pollinated seed can be tolerated. Emasculation is performed by three methods: (1) removing the anthers with a fine pointed forceps, (2) killing the pollen by immersing the flower in 57 percent alcohol for 10 seconds, or (3) removal of the anthers by using an aspirator with an opening slightly larger than the anthers. For research projects requiring large amounts of cross-pollinated seed, male sterile plants can be used if the female parent is available with the correct genotype.

Cross-pollinated seed can be produced without emasculation by using honeybees or hand pollination. In both cases isolation, such as a screened cage or pollinator-free greenhouse, is required to prevent adulteration. If bees are used as crossing agents in an enclosed structure, they must be carefully watched and generally provided with a sugar solution and water. Hand pollination is done by alternately tripping flowers on the plants being crossed, using a toothpick or folded, pointed piece of paper.

Controlled pollination without emasculation is acceptable for most breeding and research. Theoretically, the amount of self-pollination is low (about 10 percent) and many plants produced from the self-pollination will have reduced vigor. If this is unsatisfactory, marker genes such as white flower color should be used to identify the plants arising from self-pollinations.

5. SCREENING AND TESTING TECHNIQUES FOR INSECT RESISTANCE

The development of insect resistant alfalfa cultivars through selection and breeding methods has had unprecedented success in the United States during the past quarter century. Ironically, much of this success can be attributed to the spotted alfalfa aphid, *Therioaphis maculata*. This insect, by its introduction into the United States in 1954, caused the research that ultimately led to the development and release of more than 30 alfalfa cultivars, many of which have multiple pest resistance and high forage yields.

Methods of developing insect resistant alfalfas or germplasm typically require an evaluation of available sources prior to selection and breeding for resistance. These sources are planted in the field or in greenhouses when insect populations are available, either under natural conditions or reared in captivity.

The origin and distribution of introduced pest species are of paramount importance in the search for resistant germplasm. Where pest species and hosts have coexisted for hundreds of years, gene pools for resistance have accumulated through natural selection. Often these sources are found only in the pest's country of origin.

The entomological techniques for development of insect resistant alfalfa cultivars include mass screening of plant populations followed by testing of individual selections for antibiosis, nonpreference, or tolerance. The final selection of individual plants is based on performance with the test insect and on the level of resistance desired by the entomologist and breeder. The plants are then used in one of the several ways described above to produce seed for testing. Finally, progeny tests are run to determine if the level of resistance is adequate. If not, then the germplasm pools (e.g., seeds, plants, or lines) may be screened again and reselections made by the same procedure described above.

Techniques for screening and testing alfalfa for resistance are described below for specific insects. These methods were abstracted from the literature. Pertinent details are given in hope that they will be useful to the student or investigator.

5.1.   Spotted Alfalfa Aphid

Field Screening

Field screening for spotted alfalfa aphid (SAA) resistance was common during the decade following the introduction of the pest in the United States in 1954. Seed of adapted varieties was planted in rows in field plots, preferably at a time when peak populations of the aphid would coincide with the seedling stage. Plants that survived aphid attack were caged directly in the field and infested with 10 early instar nymphs. Ten days later the plants were rated on the basis of aphid mortality. Plants that had the highest aphid mortality were moved to an isolated nursery or caged for development of seed. Other effective ways of field screening have been described by Howe and Pesho (1960a, 1960b), Hackerott et al. (1958), and Klement and Randolph (1960).

Direct field selection of resistant plants is also highly desirable whenever sudden outbreaks of new biotypes of newly introduced pests occur in large field plantings. 'Moapa' alfalfa, the first cultivar released with spotted alfalfa aphid resistance, was developed in this fashion (Smith et al., 1958).

Greenhouse Screening

In recent years screening for resistance to SAA has been done in the greenhouse. Populations of the aphid in the field have been reduced by the use of resistant varieties (See Section 8) to such an extent that field screening is no longer practical. Standardized tests have been described by

Nielson (1974) following the general procedures of Howe and Pesho (1960b).

Seed is planted in rows in galvanized flats, 55 × 37 cm, filled with 7.5 cm of soil. Each of 13 rows per flat is planted with 50 seeds and covered with pure silica to reduce damping off. Sterilized soil is recommended wherever *Pythium* is a major problem. Seedlings in each flat in the unifoliate stage are manually infested once with 4 ml of a mixed population of aphids. Tests are conducted at temperatures between 20 and 30°C. Aphids are reared on caged potted plants representing cuttings from a single aphid susceptible 'Caliverde' mother plant or from a seed line of 'Caliverde' alfalfa. Other aphid susceptible varieties will work as well, as long as they are nondormant or semidormant types.

In all screening tests, it is important to control infestation rates so that population pressure does not destroy or obscure potential germplasm sources. Greenhouse screening permits greater control in selecting plants but restricts the amount of material that can be screened over a given period of time. Resistance in alfalfa cultivars developed from greenhouse screening appears to be less durable than that in cultivars from field-selected plants. The reasons for this difference are not fully understood.

Antibiosis Tests

After the screening in the field or greenhouse, selected plants are given an antibiosis test by caging 10 early instar nymphs per plant for 7 days. A known susceptible plant is included with the test. On the eighth day, the plants are examined for live aphids and rated according to the following system: No aphid survival = HR^{++++} (highly resistant); 1 to 5 live nymphs, no adult = HR^{+++} (moderately resistant); 1 or 2 adults, no reproduction = HR^{++} (resistant); 3 to 5 adults, 1 to 5 nymphs = HR^+ (low resistance); 5 to 10 adults, 5 to 100+ nymphs = Susceptible. The rating scale is skewed to favor selection of highly resistant plants.

Evaluation Tests

A necessary component of breeding for resistance is the progeny test, with seed obtained from plants selected in the antibiosis test. This test determines whether resistance has been transferred from the parent plants to the progeny and, if so, the degree or level of resistance that will characterize that potential cultivar or usable germplasm source.

The evaluation techniques are mechanically similar to screening techniques described above. Proper experimental design is important to ensure that data analyses will be valid and useful. Alternate rows of the test entries are planted with a susceptible entry, 'Caliverde,' to maintain uniformity of the population after manual infestation and to adjust for differences in resistance among progenies by permitting equal exposure or access to the aphid populations. Susceptible and known-resistant test entries are also included in the test for comparative purposes. A plant germination count is made prior to infestation. The performance of the progeny is based on the percent survival/mortality of the seedlings. Termination of the test occurs when the susceptible check entries are completely killed and the counts of surviving test plants are made.

5.2. Pea Aphid

Resistance in alfalfa cultivars to field populations of the pea aphid was first reported by Blanchard and Dudley (1934), and Painter and Granfield (1935), with subsequent reports by Dahms and Painter (1940), Albrecht and Chamberlain (1941), and Emery (1946). However, these reports indicated resistance was variable or difficult to use and was not combined into resistant varieties for many years.

In 1965 Howe et al. (1965) presented an appraisal of pea aphid resistance in alfalfa that was based on a review of the literature and their own work. They concluded that (1) Flemish and Turkestan germplasm appeared to have better resistance; (2) resistance could be measured by rate and volume of regrowth of alfalfa under aphid stress; (3) antibiosis as measured by cage tests was correlated with low aphid populations on plants obtained from mass infestation tests; (4) field selections of plants made for pea aphid resistance continued to grow under high aphid populations and had increased clonal and progeny survival; and (5) pea aphid resistance was complicated by phenotypic variation.

Field Screening

Field selections are made by evaluating individual mature field or nursery plants that have been exposed to severe and prolonged infestations of the pea aphid. Evaluations of resistance are based on the lack of plant injury (stunting, killing of top growth, etc.) and low aphid numbers.

Greenhouse Screening

Selection methods and resistance criteria have been discussed by many authors (Sorensen, 1974; Carnahan et al., 1963; Hackerott et al., 1963; Harvey et al., 1972; Ortman et al., 1960; Smith and Peaden, 1960; Howe et al., 1965). Seedling selections can be made in either the greenhouse or growth chamber by planting test rows in a sequence with a susceptible row on at least one side, usually in a pattern of one susceptible row, two test rows, one susceptible row, and so forth. Seedlings are infested in the early cotyledon stage with large numbers of aphids. Aphids are added as needed to maintain population pressure on the seedlings. The test is terminated when most of the plants in the susceptible rows are dead or severely injured. Aphids used in resistance tests are increased on broad beans, *Vicia faba*, by some workers, but aphids obtained in this manner should be used with caution since they may be conditioned to broad bean, which could influence the reaction on alfalfa.

Antibiosis Tests

Antibiosis tests are conducted by caging three to eight first-instar nymphs on a healthy plant terminal for 5 to 9 days. Resistance of the plant is based on survival and aphid reproduction.

5.3. Blue Alfalfa Aphid

Field Screening

Methods for screening resistance to the blue alfalfa aphid (BAA) are similar to those described for the spotted alfalfa aphid and pea aphid. Field screening is preferred since it is difficult to mass rear the blue alfalfa aphid for extended periods in the greenhouse for indoor or greenhouse screening. Such tests can only be done during the cool months of the year.

Large field cage tests have been used by the senior author wherein rows of a BAA susceptible entry are planted and infested with the aphid when the plants are 15 to 20 cm high. When the BAA population has reached the desired level, test entries are planted in alternate rows. After the seedlings reach the first trifoliate leaf stage, the infested rows of the BAA susceptible entry are cut back. This forces the insects to feed on the seedlings. Repeated infestation may be achieved by cutting back alternate rows. This maintains population pressure on the seedlings while allowing buildup on regrowth of the susceptible entry. Selections are made of the most vigorous survivors.

Greenhouse Screening

Greenhouse screening for BAA resistance can be accomplished if an adequate source of aphids is available. Mass rearing in the greenhouse is difficult and at best can only be done during the cool months of the year. Methods are similar to those described for the pea aphid.

Antibiosis Tests

These tests have not yet been developed but presumably those described for the pea aphid will apply here.

Evaluation Tests

Principles described for the spotted alfalfa aphid apply to the evaluation of BAA resistance in progenies of resistant alfalfas.

5.4. Alfalfa Weevil

Field Screening

The first report of weevil resistance in alfalfa was made by Webster (1912). No cultivars were developed from this program and no resistance studies on the alfalfa weevil were reported until work was started in the eastern United States in the 1960s. The studies included oviposition and feeding trials in the field, greenhouse, and/or laboratory.

Early research on alfalfa weevil resistance in the United States involved field evaluations of varieties and introductions for larval feeding response (Dogger and Hanson, 1963; Campbell and Dudley, 1965; Norwood et al., 1967a; and Busbice et al., 1967). Plants were found that appeared to have resistance. Ovipositional nonpreference was also found in certain lines. Emphasis later shifted toward selection for resistance by using controlled laboratory tests such as seedling, leaf disk, and larval tests (Barnes et al., 1970; Busbice et al., 1978).

Germplasm, designated the Starnes strains, was selected for larval feeding resistance from cultivars studied by Dogger and Hanson (1963) (Busbice et al., 1977). This germplasm was improved by using phenotypic recurrent selection for larval feeding in the field under natural infestations of the alfalfa weevil. The cultivars 'Team,' 'Arc,' and 'Liberty' were developed from this germplasm source. Busbice et al. (1977) showed that field selection was the only effective method of several used in the development of these cultivars. After six generations, the Starnes germplasm showed one-third less defoliation than the check cultivars. Heritability of resistance was moderately high on the basis of the response to recurrent field selection in the Starnes strains.

Greenhouse Screening

Seedling Test. Seedling or cotyledon tests were used to screen large numbers of plants (Barnes et al., 1969a, 1969b, 1970). Seed was planted in greenhouse flats and infested with weevil adults when the plants were in the cotyledon stage. Feeding was stopped when about 95 percent of the plants were destroyed. The least preferred plants (about 2 percent) were saved. Adult weevils rather than larvae were used in these trials because of their greater mobility and durability, and because of the ease with which they can be collected, stored, and handled. Plants surviving this test were subsequently tested by using larvae or oviposition and/or leaf disk tests to eliminate escapes.

Plants selected in the seedling test had significantly less adult feeding and produced smaller larvae than susceptible plants from the same variety (Barnes et al., 1969a). Differences, however, were small.

Leaf Disk Test. The leaf disk test was developed by Barnes and Ratcliffe (1967) and Barnes et al. (1969a, 1970) to test large numbers of plants under controlled conditions. Disks were cut from leaves of test plants, placed on moist paper in petri dishes, and fed to weevil adults. The uneaten leaf area was estimated at the end of the feeding period and adjusted according to dry weight of an uneaten sample. Resistance was based on the amount of leaf material uneaten. Estimated weight of a uniform leaf area was used because this was more rapid than using dry weight of damaged and undamaged leaves. Basing the uneaten portion on the weight of a sample eliminated any influence leaf thickness might have. A positive relationship was found between leaf disk feeding and seedling and larval feeding tests. However, correlations between leaf disk and larval feeding tests were nonsignificant. In related work, VanDenburgh et al. (1966) reported that adult oviposition was lower on plants that showed less adult feeding.

Resistance to Larval Feeding. Resistance to larval feeding was measured in laboratory tests (Barnes et al., 1969a, 1970; Byrne and Rittershausen, 1970). Newly hatched to 2-day larvae were caged on plants or placed on detached leaves for 5 to 8 days; then the larvae were counted and weighed. A positive relationship was found between larval feeding and plants selected by the cotyledon test. However, correlation coefficients between adult leaf feeding and larval weight were nonsignificant, and some plants classified as resistant to adult feeding were susceptible to larval feeding and vice versa. On the other hand, Koehler (1971) found larval growth response was correlated with adult feeding response.

Oviposition Tests. Dogger and Hanson (1963) classified 294 cultivars for oviposition nonpreference and concluded that all were susceptible but that significant differences in reaction occurred among cultivars and within cultivars. To isolate resistant plants, Campbell and Dudley (1965) developed a testing method that was precise and capable of eliminating the larval migration and variability of larval feeding found in bulk field plantings. This was an oviposition test in which relatively large numbers of weevil adults were introduced into cages containing alfalfa transplants. After 8 to 11 days the plants were classified for presence of egg masses and damage. Two strains of *Medicago sativa* var. *gaetula* with resistance to oviposition were found in these experiments. Various workers (Norwood et al., 1967a, 1967b; VanDenburgh et al., 1966; and Busbice et al., 1967) confirmed these results by using natural infestations of field plots, and also demonstrated that this type of resistance was associated with small stem diameter. Norwood et al. (1967a) found that oviposition preference was associated with decumbent growth and wide crown, and accounted for only 26 percent of the variability among clones. Norwood et al. (1967a) and Busbice et al. (1967) concluded it might be possible to obtain resistance by selecting for low oviposition preference and larger stems. In related work VanDenburgh et al. (1966), Norwood et al. (1967a), and Barnes et al. (1969a) found an association between low oviposition preference and low leaf feeding. In related work, Byrne (1969) found a chemical stimulus that promoted oviposition.

Evaluation Tests

Barnes and Ratcliffe (1969) evaluated 16 annual species of *Medicago* for weevil resistance by using oviposition, larval development, and adult feeding tests. *M. rugosa* had good resistance in all tests. *M. minima, M. polymorpha, M. scutellata,* and *M. truncatula* contained plants with resistance. Shade et al. (1975) studied 21 species for larval resistance. All larvae died on *M. scutellata* and *M. disciformis*; high larval mortality occurred on *M. marina* and *M. laciniata*; delayed development occurred on *M. echinus, M. intertexta,* and *M. minima*; and delayed prepupal-pupal development occurred on *M. echinus*. Techniques for evaluating alfalfa weevil resistance are described by Ratcliffe (1974).

5.5.   Egyptian Alfalfa Weevil

Screening Techniques

Research on the development of cultivars with resistance to the Egyptian alfalfa weevil was conducted primarily in California, using information and techniques developed in work on alfalfa weevil in the eastern United States (Lehman and Stanford, 1971, 1972, 1975; Summers and Lehman, 1976). However, since this weevil species, its behavioral patterns, and other factors are different in California, certain modifications had to be made. In addition, the germplasm used in the eastern United States could not be used in California because it was winter dormant and susceptible to the spotted alfalfa aphid.

Egyptian alfalfa weevil adults were collected through the spring and summer at their aestivation sites, stored at room temperature until December, and than held in a cold box at 4.5°C for testing from January to May.

The selection methods used by Lehman and Stanford (1972, 1975) were the field, leaf disk, and seedling techniques described earlier under alfalfa weevil. Unlike the seedling or cotyledon test described by Barnes et al. (1969b), these tests were conducted under greenhouse conditions with older plants. For the leaf disk test, all leaf disks were weighed before and after the tests.

The germplasm ('UC 67,' 'UC 63,' 'UC 68') selected for resistance under laboratory methods proved to have little field resistance to the Egyptian alfalfa weevil (Lehman and Stanford, 1972; Summers and Lehman, 1976). 'UC 73,' which was synthesized from field selections, appeared to show a small amount of resistance. Lehman and Stanford (1975) concluded that cultivars made from field selection showed the most promise. Summers and Lehman (1976) found that resistance in alfalfas selected for resistance to *H. postica* and *H. brunneipennis* were similar. They found no correlation between number of eggs or egg clusters per stem and concluded resistance to oviposition by *H. brunneipennis* was not a factor. Since *H. brunneipennis* used debris or dead stems for oviposition the importance of stem size was not a factor in resistance.

In laboratory studies Summers and Lehman (1976) found that mortality among first and second instars was greater on plants from 'UC 73' than on 'Moapa.' Development time was also prolonged. The data indicated some form of antibiosis may have been operating in one of the experimental varieties.

5.6. Potato Leafhopper

Field and Greenhouse Screening

Screening of plants infested with potato leafhoppers in the field and greenhouse has been based on scored criteria. The degree of yellowing caused by leafhopper feeding was adopted as a standard by rating plants from 1 to 9 (*19th Alfalfa Improv. Rep.*, **CR-54-56:** 63, 1964). Schillinger et al. (1964) screened plants by measuring their effect on the reproductive biology of the leafhopper, and Webster et al. (1968b) used seedling survival and leafhopper damage as the basis for screening. Newton and Barnes (1965) exposed potted plants to field populations to determine antibiosis. Similar methods were used by Kindler and Kehr (1974). Seedlings, 6 to 8 weeks old, were transplanted in the field in the spring. Four replications of 25 plants each were used. Single row plots with plants spaced 30 to 90 cm apart were recommended. The plants were then scored from 1 to 9 based on the degree of yellowing. Time of planting was crucial so that maximum effect of high populations might be utilized. Jarvis and Kehr (1966) found a high correlation between population count and nymphs-per-gram of plant material when they evaluated 75 alfalfa clones over a 2-year period. This method appears to have considerable value since long-term effects were measured.

Antibiosis Tests

Newton and Barnes (1965) tested for antibiosis by caging eight newly hatched nymphs on individual plants and recording the time of development and survival of the leafhoppers. Schillinger et al. (1964) conducted similar tests for antibiosis. Apparently this mechanism of resistance was difficult to assess by these methods but refinement of the technique may produce significant antibiosis effects.

5.7. Meadow Spittlebug

Field Screening

Cultivars, seed lines, and plants can be screened for resistance to meadow spittlebug in field trials, seedling screening tests, and insectaries (Wilson and Davis, 1953, 1958).

Screening for meadow spittlebug in large broadcast fields of alfalfa was conducted by interplanting the test materials with small grains the previous summer. Eggs were then laid in the grain stubble by natural infestations of spittlebugs in the fall of the year. As the eggs hatched in the spring, the nymphs moved to the alfalfa. Use of broadcast stands reduced desiccation from wind and rain. Spittlebug population counts were obtained by counting three 1-foot samples per plot. These counts were later converted to spittlebugs per square foot. Damage was rated on a scale of 1 to 9.

Greenhouse Screening

Hill and Newton (1972) developed a greenhouse test for selection purposes. In the fall adult spittlebugs were forced to lay eggs on grain stubble in wooden flats. Flats were stored in the coldroom until about January. Vegetative propagules of alfalfa that were to be tested for resistance were then planted in the flats with the stubble. Resistance was based on number of spittle masses per propagule.

Antibiosis Tests

Antibiosis studies were made on cuttings of individual plants in an open screened insectary. Because success of these tests depended on optimum environmental conditions and good infestation, observations of the test plants were made hourly during the first day and daily thereafter for 3 weeks. Counts of nymphs at the end of the trial were used to classify the plants.

Nonpreference Tests

Nonpreference tests were conducted by Wilson and Davis (1966) by using seedling plants in flats. First-instar nymphs were released in the flats to seek the most preferred lines or plants. Least preferred plants were saved.

5.8. Alfalfa Seed Chalcid

Field Screening

The screening of plants for resistance to the alfalfa seed chalcid is described by Nielson (1967, 1974). Samples should be taken during a period between July and September, depending on locality and climate. Seed is planted in rows 90 cm apart in large, field-size borders. After they germinate, the seedlings are thinned to one plant every 40 to 60 cm, selecting those plants that are most vigorous. As plants mature the last cutting should be made so that pod development will coincide with the period of high chalcid populations.

Samples of 25 racemes having green, fully expanded pods are selected from each plant and placed in rearing cartons fitted with a glass emergence vial. After 1 month the adult chalcids are counted from the emergence vials and recorded. Plants that score less than 1 chalcid/raceme are retained for greenhouse cage testing.

Greenhouse Cage Test

Selected plants are grown in pots in the greenhouse. After the plants flower they are pollinated by hand tripping. Ten racemes are caged individually, each cage containing 10 female chalcid that have been reared from infested seed screenings. After exposure for 2 weeks to chalcid oviposition, the racemes are removed from the plants and placed in rearing cartons. Plants that score less than one chalcid per raceme in three separate cage tests are classified as resistant.

5.9. Lygus Bugs

Development of lygus resistant germplasm or cultivars has been limited. Evaluation, screening, and selecting of material in the past 10 years has produced much information on promising sources of resistance, but this work has not yet resulted in release of resistant cultivars or germplasm sources. Possible genetic sources for resistance have been reported by Aamodt and Carlson (1938); Malcolm (1953); Lindquist et al. (1967); and

Nielson et al. (1974) in their evaluation of a large number of cultivars and lines for resistance to *Lygus hesperus, L. lineolaris, L. elisus,* and *L. desertinus.* A complete review of lygus resistance with concomitant technique problems was reported by Tingey and Pillemer (1977).

Field Screening

Large-scale field plantings of individually spaced plants, similar to the plantings described under the alfalfa seed chalcid, can be used for screening plants for resistance to *L. hesperus.* Plants are allowed to flower and set seed during natural field infestations. Selections are made on the basis of freedom from "blasting" and degree of fullness of seed set.

Greenhouse Screening

Techniques for screening seedlings of alfalfa for resistance to *L. lineolaris* and *L. hesperus* have been described by Lindquist et al. (1967) and Nielson et al. (1974), respectively. Seeds are planted in radial rows in flats and infested with a mixed population of males and females at a rate of 1:2 or 1:3 lygus to seedling. The flats are caged to confine the bugs. Plants surviving the feeding effects are saved for polycross combination and subsequent reselection.

6.   MECHANISM OF RESISTANCE

The mechanisms of resistance in alfalfa to insects are not well understood. Many studies have shown that resistant plants adversely affect the biology and/or behavior of the insect when the plant is used for feeding or oviposition. These studies are discussed below.

6.1.   Spotted Alfalfa Aphid

Several studies of the mechanism of resistance in alfalfa to the spotted alfalfa aphid show the operation of antibiosis, nonpreference, and tolerance. These mechanisms may be expressed singly or *in toto* in the same plant or cultivar. Some early studies by Harvey and Hackerott (1956), Dobson and Watts (1957), Howe and Smith (1957), and Nielson and Currie (1959) demonstrated detrimental effects of resistant plants on various components of the aphid's biology. The results suggested that antibiosis was the principal mechanism of resistance. Later, more definitive studies, particularly those of McMurtry and Stanford (1960), Kishaba and Manglitz (1965), and Kindler and Staples (1969) showed that aphids starved to death on resistant plants because of their apparent inability to ingest sap from the plant. There was no evidence of toxic effects, which suggests that the mechanism of resistance was nonpreference. Studies by Nielson and Don (1974b) on the probing behavior of four biotypes of the spotted alfalfa aphid on resistant and susceptible alfalfa clones revealed that all biotypes probed both types of plants equally well, but that some biotypes were unable to ingest sap when the stylets reached the phloem of resistant plants. Differences between biotypes in their ability to ingest were attributed to the presence or absence of detoxifying mechanisms. The reaction between the aphid and plant was specific and occurred at a specific site in the plant. The mechanism of resistance was described as similar to the phytoalexin concept of interaction between pathogens and plants.

Tolerance was demonstrated in several clones by Jones et al. (1968). Aphid reproduction and survival was greater on these clones than on susceptible checks. Damage was also significantly less on the tolerant plant.

Compounds in resistant and susceptible plants that elicited certain behavioral responses in the aphid were investigated by Maxwell and Painter (1959, 1962) and Kircher et al. (1970). Auxin levels decreased in tolerant plants after aphid infestation. The level of certain nutrients was too low in resistant plants to support aphid population. These investigations were directed more to the causes than to the mechanisms of resistance.

6.2.   Pea Aphid

Several factors have been studied in an effort to determine the mechanism of pea aphid resistance or to isolate factors affecting or influencing resistance. In an early study Emery (1946) attributed plant immunity to an acid condition in the alfalfa plant that was associated with environmental factors. Resistance to pea aphid in peas was associated with higher sugar-nitrogen ratios by Maltais and Auclair (1957); lower concentrations of amino acids at stages of growth corresponding to pea aphid infestation by Auclair et al. (1957); a deep green color in early stages of growth and increased antibiosis at higher temperatures by Cartier (1963). In other color preference tests Cartier and Auclair (1964) found biotypes of the pea aphid that preferred orange or yellow, or both.

Pederson et al. (1976) found that high saponin content was associated with pea aphid resistance after they selected six alfalfa lines for high and low saponin. On the other hand, they found no evidence that breeding for pea aphid resistance had changed foliage saponin content; it was lowered in the roots.

6.3. Alfalfa Weevil

In studies of *Medicago* species Shade et al. (1975) found that the exudate from secretory trichomes of *M. disciformis* affected larval survival. Larvae died either when placed on a plant or when the exudate was applied topically to the first instar larvae. Low concentrations of the exudate resulted in reduced larval development. The mode of action appeared to be primarily mechanical, that is, the stickiness of the exudate. However, the slow development of the larvae appeared to be caused by a toxin.

Smith et al. (1974) investigated the effect of boron on weevil feeding and oviposition activity. Weevils that fed on alfalfa stems containing 0.001 to 0.1 M H_3BO_3 laid fewer eggs, and fewer of these eggs hatched than weevils that fed on alfalfa with no boron.

6.4. Alfalfa Seed Chalcid

Resistance mechanism in alfalfa to the alfalfa seed chalcid was studied by Tingey and Nielson (1974, 1975). Nonpreference for oviposition was detected for a period of 8 days in seed pods 10 to 18 days old in resistant plants. No effect on the reproductive biology or antibiosis was found in any of the resistant clones.

7. INSECT BIOTYPES

Biotypes are natural products of a survival mechanism for the perpetuation of insect species. They are often selected out of a population by a cultivar developed for resistance and grown in areas where exposure to the insect is common.

Many biotypes have been recognized and several have been described and named. They are extremely important as tools for developing insect resistant cultivars, studies of the genetics of inheritance of virulence in insects and inheritance of resistance in plants, and resources for developing concepts of insect/plant relationship and mechanism of insect resistance.

7.1. Spotted Alfalfa Aphid

The first recognized and described biotype of the spotted alfalfa aphid was reported by Pesho et al. (1960). This biotype was designated as ENT-A to distinguish it from ENT-B, the original population introduced in the United States in 1954. Since 1960, seven additional biotypes have been reported, five in the southwestern United States (Nielson et al., 1970a, 1970b; Nielson and Don, 1974a), and two in Nebraska (Manglitz et al., 1966). Most of the biotypes were differentiated on the basis of their biological response to the nine parent clones of 'Moapa' cultivar.

Table 2 shows a composite response of several biotypes on the clones of 'Moapa' over a period of 20 years of bioassay. These biotypes developed on resistant cultivars ('Moapa,' 'Mesa-Sirsa') by the selective pressure of the cultivars themselves which resulted in a progressive increase of virulence in the spotted alfalfa aphid population. This phenomenon is commonly referred to as the gene-for-gene concept; that is, for every gene for resistance in the plant there is a corresponding gene for virulence in the insect.

Breeding methods to prevent the development of biotypes are unknown. However, techniques and sources are available to discourage or reduce the rate of biotype development. For example, breeding for or incorporation of multiple mechanisms of resistance will greatly enhance the longevity of the cultivar in terms of genetic stability for resistance. Moreover, polygenic systems of resistance, consisting of genes at different loci, have a decided protective advantage over monogenic systems or single genes at the same loci. Polygenic resistance is virtually unknown between insects and plants but is not uncommon among pathogens and their hosts.

Table 2.   Response[a] of 'Moapa' Clones to Some Biotypes of the Spotted Alfalfa Aphid, Tucson, Arizona

Moapa Clones	Aphid Biotype					
	B	A	E	F	H	N[b]
C-903	R	S	S	S	S	R
C-904	R	R	S	I	I	R
C-905	R	R	S	S	S	R
C-906	R	R	R	S	S	R
C-907	R	R	R	R	S	R
C-908	R	S	S	S	S	R
C-909	R	R	R	I	I	R
C-910	R	R	R	I	S	R
C-911	R	S	S	S	S	S

[a] R = resistant; I = intermediate; S = susceptible.
[b] N = population from Nebraska.

7.2. Pea Aphid

Biotypes of the pea aphid were first shown to exist on peas by Harrington (1941, 1945) and Cartier (1959), who measured body size, reproduction, and other characters. The existence of pea aphid biotypes on alfalfa was reported by Cartier et al. (1965) by testing 10 alfalfa clones at three temperatures and at two widely separated locations, St. Jean, Quebec, and Manhattan, Kansas.

Frazer (1972) later reported differentiating five pea aphid populations (biotypes) by survival and rates of population increases on alfalfa, broom, white clover, and broad beans. On the basis of this work he suggested that biotypes of pea aphid result from annual adaptation to various species of host plants and that characteristics of the biotypes may vary from year to year. To overcome selecting for annual adaptation to various legume species in resistance tests, he suggested using only fundatrices and immediate offspring that have not been subjected to this selection. In similar work, Hubert-Dahl (1975), working with a green and red biotype on five host species, found that after the insects were reared on *Trifolium pratense* there was a somewhat increased colonization on red clover, compared with insects that had developed on alfalfa.

7.3. Alfalfa Weevil

Local or regional biotypes of the alfalfa weevil that should be considered in breeding or control programs have not been reported. However, two introductions of the alfalfa weevil have been made into the United States and have been commonly referred to as strains, forms, or populations (Blickenstaff 1965, 1969; Armbrust et al., 1970; and Schroder and Steinhauer, 1976). The western strain was found in Utah in 1904 (Titus, 1910) and the eastern strain in Maryland in 1952 (Poos and Bissell, 1953). Most researchers believe these two introductions represent two different but morphologically indistinct strains (Blickenstaff, 1965). Since the differences between the two strains were minor, the biology and control practices depended more on climatic and geographic factors (Metcalf and Luckmann, 1975).

Blickenstaff (1965, 1969), Armbrust et al. (1970), and White et al. (1972) conducted cross mating studies between the eastern and western strains of the weevil and generally agreed that these strains are distinct. Schroder and Steinhauer (1976) crossed the two United States strains with populations from France and Germany. They confirmed the previous work on the United States strains and also determined that the western United States and west European strains probably came from the same source and that the eastern United States strain came from some other area of the Old World.

8. RESISTANT ALFALFAS IN PEST MANAGEMENT

The development of insect resistant alfalfa cultivars through selection and breeding methods had unprecedented success in the United States during the past quarter century. Table 3 is a partial list of cultivars that have been released.

8.1. Spotted Alfalfa Aphid

The spotted alfalfa aphid from India was first described in 1899 by Buckton (1899); the insect is apparently native to that country or nearby areas of southern Asia. Russell (1957) listed many other countries where the pest now lives, including France, Germany, Hungary, Italy, Yugoslavia, Morocco, Egypt, Cyprus, Turkey, Israel, Iraq, Iran, Pakistan, China, Mexico, and the United States, Gentry (1965) also reported the aphid from Sudan, Libya, Ethiopia, and Afghanistan. Several serious outbreaks have occurred in these countries in the Mediterranean area. Harpaz (1955) reported that three devastating outbreaks occurred in Israel since the insect was first found in that country in 1930.

In the United States, the introduction and spread of the spotted alfalfa aphid is well documented (anonymous, 1956, 1957; Smith et al., 1956). No other insect pest has spread as rapidly over as wide an area as has this aphid. The insect occurs in all but two states of the continental United States and eventually it may move into Canada. Thus far it has not been reported in the Neotropical Region. The major sources of germplasm for resistance to the spotted alfalfa aphid in the early cultivars can be traced to three areas of the world, India, Arabia, and Turkestan. The first cultivar developed specifically for resistance was 'Moapa,' which consists of nine clones selected from African, a nondormant cultivar introduced into the United States from Egypt in 1924. 'Lahontan,' a semidormant cultivar developed for resistance to the stem nematode and bacterial wilt, has natural resistance to the spotted alfalfa aphid in the five parent clones. The parent material was selected from 'Nemastan,' a semidormant cultivar whose pedigree can be traced to Turkestan in southern Russia. Spotted alfalfa aphid resistance in 'Lahontan' was fortuitous, an apparent result of accumulation of resistant germplasm over many years of natural selection. The nondormant cultivar 'Mesa-Sirsa' was developed from 13 clones selected from 'Sirsa No. 9' ('PI235,' '736'), an introduction from Sirsa, a

Table 3.   Alfalfas Resistant to Insects[a]

Insect Species	Cultivar
Therioaphis maculata[b]	Lahontan
	Moapa
	Zia
	Cody
	Sonora
	Caliverde 65
	Mesa-Sirsa
	Washoe
	Bonanza
	Dawson
	Mesilla
	WL-504
	UC-Cargo
	CUF 101
Acyrthosiphon pisum	Apex
	Washoe
	Dawson
	Mesilla
	Kanza
	WL 512
	PA-1[c]
	CUF 101
	Paine
A. kondoi	CUF 101
	WL 514
Hypera postica	Team
	Weevlchek
	Arc
Philaenus spumarius	Culver
Empoasca fabae	Cherokee

[a] Reported location: Argentina for Paine, United States for all others.
[b] More than 30 cultivars have been developed for resistance to this insect.
[c] Germplasm release. Commercial seed not available.

small village in India. Other important cultivars included 'Zia,' which has germplasm from 'Lahontan' and 'Turkistan,' and 'Cody,' which was developed from 'Buffalo,' a cultivar that traces back to Chile via the early Spanish conquest of Mexico and Peru.

Resistance in 'Moapa' alfalfa in large field plantings in Arizona was consistent over a 3-year study period from 1959 to 1961 (Barnes, 1963). Populations were 5 to 13 times higher on susceptible varieties than on 'Moapa.' Damage to foliage was 15 to 22 times greater on susceptible varieties. These studies carried out during the period of high aphid abundance clearly demonstrated the advantage of resistant alfalfas for control of a pest.

8.2. Pea Aphid

The first varieties developed for resistance to the pea aphid were 'Washoe,' 'Apex,' 'Dawson,' and 'Mesilla.' All were released in 1966 and 1967, about 32 years after the first report of resistance. However, Howe et al. (1965) felt that work on pea aphid resistance may have been impeded by the more serious problem of the spotted alfalfa aphid. In any case, use of the methods discovered in the spotted alfalfa aphid work eventually were of considerable help in speeding development of pea aphid resistant varieties.

Howe et al. (1965) reported that Flemish type germplasm ('Socheville,' 'Alfa,' 'Du Puits,' etc.) showed good resistance and that the Turkestan-type germplasm ('Ranger,' 'Ladak,' 'Lahontan,' etc.) was intermediate. However, resistant selections have also been reported from many other sources such as numerous introductions, common Chilean and African types, and Grimm (Howe et al., 1965; Jones et al., 1950; Blanchard and Dudley, 1934).

8.3. Blue Alfalfa Aphid

The blue alfalfa aphid, *Acyrthosiphon kondoi,* is a recent pest of alfalfa in the New World. It was first found in California in 1974 (Sharma et al., 1975) and by 1977 it had spread to nine other states in the United States. The pest was reported in Argentina, Australia, and New Zealand in 1976, its first appearance in the Southern Hemisphere.

The insect is native to the Far East (Manchuria, Mongolia, and Japan). It has been a minor pest of alfalfa in these areas. In the United States it is a pest early in the spring and usually occurs in outbreak numbers before the pea aphid population builds up. Both species coexist in the same fields, but the blue alfalfa aphid is the dominant population early in the season; the pea aphid dominates later in the spring. Shinji and Kondo (1938) described the species from Manchuria where it occurs on legume species of *Medicago, Trifolium,* and *Melilotus.*

Sources of resistance to the blue alfalfa aphids were found in UC Cargo, a cultivar resistant to the spotted alfalfa aphid and pea aphid. From this germplasm source the cultivar 'CUF 101' was developed and released. This cultivar has resistance to three aphid species (Nielson et al., 1976; Nielson and Lehman, 1977; Lehman et al., 1977, 1978). Other sources are available that show a low frequency of resistant plants. These included 'Kanza,' 'Dawson,' and 'Washoe,' which have resistance to the spotted alfalfa aphid and pea aphid (Nielson and Lehman, 1977). Promising sources in Argentina include 'Paine' and 'Fortin Pergamino' (C. Bariggi, personal communication). Recent studies on four new experimental alfalfas developed in New Zealand were equally encouraging.

Multiple aphid resistance developed in 'CUF 101' provides protection against the three insect species throughout the growing season and, to a certain extent, also allows access for parasites and predators to feed on aphids that are usually present in subeconomic levels.

8.4. Alfalfa Weevil

Four cultivars have been reported to have some resistance to the alfalfa weevil, 'Team,' (Barnes et al., 1970), 'Weevlchek' (Var. Rev. Bd., Dec. 1970), 'Arc' (Devine et al., 1975) and 'Liberty' (Var. Rev. Bd., Dec. 1975 as 'NCW 20'). Reaction of 'Team' and 'Weevlchek' has been variable (Busbice et al., 1978; Barnes et al., 1970). These cultivars exhibited resistance in some locations or tests but they appeared similar to susceptible check cultivars in other tests. However, each cultivar developed and released from the Starnes germplasm seems to have progressively better resistance (Busbice et al., 1978). No other cultivars or lines tested appeared to be sufficiently resistant to control the weevil, but some plants of certain strains appeared to have resistance (Dogger and Hanson, 1963; Campbell and Dudley, 1965; Norwood et al., 1967a; and Busbice et al., 1967).

8.5. Meadow Spittlebug

Wilson and Davis (1953) showed that alfalfa may be selected for degrees of resistance (nonpreference) to the meadow spittlebug. Plants were selected for resistance and later combined into a synthetic released as the cultivar 'Culver.' It was concluded that 'Culver' was not immune but had a high degree of resistance which appeared to be of three types: antibiosis, tolerance, and, possibly, unattractiveness (Wilson and Davis, 1966). 'Culver' is the only cultivar developed specifically for resistance to meadow spittlebug. Resistance to meadow spittlebug can be found in most alfalfas but may be more frequent in the Flemish cultivars (Sorensen et al., 1972).

8.6. Potato Leafhopper

The potato leafhopper, *Empoasca fabae* is a native pest of alfalfa and other crops in the United States and Canada. The species is restricted to the central and eastern United States and Canada. *Empoasca fabae* was described in 1841 by Harris from specimens collected on bean (Harris, 1841). Although reported as a pest of alfalfa and other crops in many areas of the North and South Americas, *fabae* was actually a complex of eight species that caused nearly identical damage symptoms on alfalfa (Ross and Moore, 1957). These include *fabae* in the eastern half of the United States; *mexara* in Arizona, California, and Mexico; and the six species in Central and South America.

'Cherokee' was developed from seven cycles of selections for leaf yellowing tolerance to the potato leafhopper. However, several other sources have been reported by Davis and Wilson (1953), Hanson et al. (1963), Kindler and Kehr (1970), and Webster et al. (1968a, 1968b). These workers reported that *Medicago falcata* showed more resistance than *M. sativa*. Higher tolerance was reported in 'Culver,' 'Rambler,' 'Rhizoma,' 'Teton,' and 'Vernal,' all of which contain *falcata* germplasm (Sorensen et al., 1972).

8.7. Alfalfa Seed Chalcid

The alfalfa seed chalcid, *Bruchophagus roddi,* is a worldwide pest of alfalfa seed. The species was described from central Asia by Gussakovskii in 1933, but the insect has long been a pest of alfalfa in the United States and other areas of the world under the specific names of *gibbus* and *funebris.* Major areas of the world where the pest is present include southern Russia, Europe, central Asia, western Siberia, North America, and Australia.

Cultivars with resistance to the alfalfa seed chalcid have not been developed. Germplasm has been selected and tested, but the stability of resistance was difficult to maintain in studies in Arizona. Cultivars and *Medicago* species were evaluated for resistance by Howe and Manglitz (1961), Nielson and Schonhorst (1965, 1967), and Strong (1962). 'Hairy Peruvian,' 'Lahontan,' 'Ranger,' 'Sirsa No. 9,' and 'Zia' were sources from *sativa,* and *M. tianschanica* var. *agropyretorum* from the *Medicago* species group. Nielson and Schonhorst (1967) reported low populations in clone 'M-56-11' of 'Mesa-Sirsa' and in clone 'C-89' of 'Lahontan.' Lehman (1967) also reported low populations in four clones that trace their percentage to 'African,' 'Sirsa No. 9,' and 'Lahontan.'

Leaves of cassava plants sus
ceptible to the Cassav
greenmite, *Mononychellu*
tanajoa (upper photo), ar
contrasted with leaves o
resistant plant (lower photo
Photos courtesy of Interna
tional Institute of Tropica
Agriculture, Ibadan.

14

BREEDING APPROACHES IN CASSAVA

Anthony Bellotti

Kazuo Kawano

CIAT, Cali, Colombia

1. THE CASSAVA CROP

Cassava (*Manihot esculenta*) is a perennial shrub of the Euphorbiaceae. It is grown throughout the tropical regions of the world and is a major energy source for 300 to 500 million people. Cassava originated in the Americas, was later taken to Africa, and more recently introduced into Asia (Leon, 1977). Common names include mandioca, yuca, manioc, and tapioca. Cassava is cultivated mainly in developing countries on small farms with little modern technology. Consequently it has received limited attention from research scientists.

Cassava is a root crop that is usually vegetatively propagated by planting stem pieces. Leaves are formed at active apices and consist of an elongated petiole and a palmate blade. The main apex is normally dominant, producing a single stem, and the petioles are borne on raised structures that give the stem a characteristic knobby appearance. When the main apex becomes reproductive, apical dominance is broken and two to four shoot buds immediately below the reproductive structure become active. Carbohydrates are accumulated in the parenchyma to form swollen storage roots. Cassava is one of the highest producers of soluble carbohydrate per unit land area per unit time. Depending on ecological conditions, the growing period is 8 to 24 months.

Since cassava is a long-season crop, often grown by subsistence farmers with a low profit margin, the continual use of pesticides to control insects and mites is economically prohibitive. The most feasible alternative methods of control are host–plant resistance, biological control, and cultural practices, or any combination of these.

2. GENETIC BACKGROUND

The chromosome number of *M. esculenta* is 36 and the species is generally regarded as an allotetraploid (Umanah and Hartmann, 1972). Recent studies (CIAT, 1975) reveal that cassava is a highly heterozygous species whose heterozygosity is easily maintained through vegetative propagation. Both cross-pollination and self-pollination occur naturally in cassava. The proportion of cross-pollination in a particular population depends on the flowering habit of the genotypes and the physical arrangement of the population (CIAT, 1976). Cassava is a monoecious species with the stigma and anthers usually separated in different flowers within the same plant. The male and female flowers almost never open simultaneously within the same branch. However, it is common for the male and female flowers of different branches within the same plant to open at the same time.

Strong inbreeding depression has been observed in characteristics such as root yield and total plant weight (CIAT, 1975), and evidence indicates that selfing may be deleterious. The inbreeding depression, in addition to the vegetatively propagated nature of the species, is the biological mechanism through which the high heterozygosity of the species is maintained. Male sterility is common and this is effective in preventing self-pollination. Hence the efficient use of male sterility is extremely important if the breeding program is based on open pollination.

The vegetatively propagated nature of the species is advantageous to breeders. Once a superior type is obtained, be it for yield or for insect and disease resistance, it can be multiplied indefinitely.

Recent studies have revealed that important characteristics such as harvest index and root dry matter content are highly heritable, and that the role of additive genes in determining these characteristics is significant (CIAT, 1975, 1976). In addition, resistance to diseases such as cassava bacterial blight and Cercospora leaf spot are transmitted with relative ease to the progenies if a resistant genotype is included in the hybridization (CIAT, 1976).

The vegetative propagation of the crop and the additive manner of inheritance of major characters simplify the hybridization and selection schemes for cassava. The identification of good parents and screening of progeny are of a more important role than the details of breeding methods. The accumulation of favorable genes without provoking inbreeding depression is probably the most critical problem that breeders have to deal with. It is only when the desired character is known to be controlled by a recessive gene that the scheme becomes complicated. The possible polysomic inheritance of the character and the heterozygosity of the species are the complicating factors.

3. THE CASSAVA MITE AND INSECT COMPLEX

Cassava is often considered a rustic crop and therefore generally free of arthropod pests. Studies now show that cassava is not free from insect and

Table 1. The Cassava Mite and Insect Complex

Common Name	Important Genus/Species	Reported From	Damage and Plant Parts Attacked
Thrips	*Frankliniella williamsi* *Corynothrips stenopterus* *Caliothrips masculinus*	Mainly in Americas but also in Africa	Deformation of foliage and outer stem tissue
Mites	*Mononychellus tanajoa* *Tetranychus urticae* *Oligonychus peruvianus*	Americas, Africa All regions Americas	Leaf yellowing, necrosis and death of buds
Cassava hornworm	*Erinnyis ello*	Americas	Foliage, buds, and tender stems
Cassava fruitfly	*Anastrepha pickeli* *A. manihoti*	Americas	Fruit (seed) and stem
Cassava shoot fly	*Silba pendula* *Lonchaea chalybea*	Americas	Death of apical buds
Whiteflies	*Bemisia tabaci* *Aleurotrachelus* sp.	Africa, Asia, Americas	Foliage deformation, necrosis, and virus transmission

Stem borers	*Coelosternus* sp.	All regions, but mainly Americas	Stem and possibly swollen roots
White grubs	*Leucopholis rorida*	All regions, but mainly Americas and Indonesia	Planting material roots
	Phyllophaga sp.		
Cutworms	*Prodenia litura*	Americas and Madagascar	Planting material, stem girdling, and foliage
	Agrotis ipsilon		
Gall midges	*Jatrophobia brasiliensis*	Americas	Leaf galls
Lace bugs	*Vatiga manihotae*	Americas	Leaves
Grasshoppers	*Zonocerus elegans*	Mainly Africa but also Americas	Foliage
	Z. variegatus		
Mealybugs	*Phenaococcus gossypii*	Americas and Africa	Foliage and stems
	Pseudococcus spp.		
Scales	*Aonidomytilus albus*	All regions	Stems
	Saissetia sp.		
Leafcutter ants	*Acromyrmex* sp. *Atta* sp.	Mainly the Americas	Foliage
Crickets		The Americas and Africa	Young plants cut off
Termites	*Coptotermes voeltzkowi*	All regions, but mainly Africa	Planting material, roots, stems
	C. paradoxis		

mite attacks, and that these pests are limiting factors in production. Cassava pests represent a wide range of arthropod fauna, with about 200 species recorded. This complex has been extensively reviewed by Bellotti and Schoonhoven (1978).

Insects can damage the plant by attacking the leaves, reducing photosynthetic area and efficiency; by attacking the stems, weakening the plant, and inhibiting nutrient transport; and by attacking planting material, leading to microbial invasion that reduces germination and yield. Some pests, such as whiteflies and fruitflies, are vectors or disseminators of diseases; others attack the roots, leading to secondary rots.

The greatest diversity of insects reported attacking cassava is from the Americas. The 17 general groups of pests described in Table 1 are all found in the Americas, 12 are reported from Africa, and 6 are found in Asia (Bellotti and Schoonhoven, 1977). This is expected since, wherever there is great genetic variation of the host plant, there is also great variability in the organisms that attack the plant or are in symbiotic relationship with it (Jennings and Cock, 1977). Our present knowledge indicates that mites, thrips, stem borers, hornworms, whiteflies, scale insects, and mealybugs cause yield losses.

Insects that attack the plant over a prolonged period, such as mites, thrips, scales, mealybugs, whiteflies, and stem borers, reduce yield more than those that defoliate or damage plant parts for a brief period, that is, hornworms, fruitflies, shoot flies, and leafcutter ants. The cassava plant recuperates from this type of damage under favorable environmental conditions. Adequate rainfall and soil fertility are the critical factors. Cassava is often grown in regions with prolonged dry seasons because it tolerates water stress. However, populations of thrips, mites, lacebugs, and scales increase during dry periods and compound the damage to the crop.

There is a paucity of data available on cassava pest biology, ecology, distribution, seasonal occurrence, and economic damage. Yield losses are reported for several pests but often these reports are not supported by sound scientific studies. Losses due to *Mononychellus tanajoa* are reported to be as high as 46 percent in Africa (Nyiira, 1976), and experiments at CIAT with a complex of four mite species (*M. tanajoa, M. mcgregori, Tetranychus urticae,* and *Oligonychus peruvianus*) resulted in 20 to 53 percent losses, depending on plant age at the time of attack and its duration (CIAT, 1977).

Yield losses due to thrips range from 6 to 28 percent depending on varietal susceptibility (CIAT, 1976; Schoonhoven, 1976). Yield reductions from hornworm attack have been estimated at 10 to 50 percent. Field studies in Colombia show a 15 to 20 percent yield reduction after a single attack. Repeated infestation over the prolonged cassava growing season undoubtedly results in greater losses. Scale attacks at the CIAT farm have reduced yield by 20 percent on susceptible varieties. Similar infestation by scales under less favorable environmental and soil conditions would result in greater reduction.

No yield losses because of the direct feeding by whiteflies have been reported. However, the whitefly, *Bemisia tabaci*, is particularly important as the vector of cassava mosaic disease in Africa and India. Yield losses owing to shoot flies have been measured on an individual plant basis and resulted in a 15 to 34 percent reduction, depending on the age of the plant when the attack occurred. Affected plants were shorter and may have been shaded by healthy neighbors; hence these yield losses may be overestimated (CIAT, 1977).

Losses due to insects such as fruitflies, stem borers, mealybugs, lacebugs, grasshoppers, and others are suspected but unproved.

4. CRITERIA FOR DEVELOPING A CASSAVA INSECT RESISTANCE PROGRAM

Concentrated team-oriented research in cassava is recent. Entomological research is still in its infancy. Cassava pest control programs are being initiated with only fragmentary knowledge about many pests attacking the crop. The entomologist must make a judgment as to the most efficient method of control, based on existing knowledge about cassava, as well as knowledge of other crops and their pest complexes.

Several criteria should be considered before deciding to establish a program that will utilize host–plant resistance for specific cassava pests:

1. The level of economic damage being caused by a particular pest should be significant. For a crop like cassava, where the yield potential is great (i.e., several times actual farm yields), priority should be given to those insects that significantly reduce yield.
2. Resistance should be sought to those pests for which it is considered feasible to find resistance. For example, it would be unlikely to find resistance to such pests as the cassava hornworm, cutworms, leafcutter ants, or grasshoppers; limited resources should not be utilized in this direction.
3. The availability of adequate and low-cost alternative methods of control of certain pests could negate the need to enter into an extensive resistance-breeding program. If pest populations can be suppressed below economic injury levels through biological control or simple cultural practices, these should be employed, especially if resistance

levels are not adequate or are difficult to incorporate in hybrids without sacrificing yield potential.

4. The level of resistance needed to reduce pest populations should be considered. Some cassava varieties have a high economic threshold to pests and can lose considerable foliage (40 percent or more for certain varieties) without reducing yields (CIAT, 1976). Therefore, high levels of resistance to some pests may not be necessary.

5. Low levels of resistance may be combined with other methods of control, such as biological control or cultural practices, to maintain insect populations below economic damage levels. For example, only low levels of resistance have been found for the mite *Tetranychus urticae*. By combining this resistance with a strong biological control program it may be possible to adequately control the mite.

6. The planting system in which cassava will be grown can dictate the level of resistance that might be needed. Studies show that when cassava is grown in multicropping systems, or in association with other crops, such as beans, insect populations are reduced. If a variety of cassava is being developed for intercropping systems the levels of resistance needed may be lower.

5. STATUS OF RESISTANCE TO INSECTS IN CASSAVA

At present, cassava is grown mostly on small plots by small farmers throughout the tropical growing regions of the world. The genetic variability in this system is enormous, because each area or zone often is sown to a distinct variety. The genetic variability in this system constitutes, in essence, a geographic multiline which is a genetic safeguard against major epidemics of pests and diseases.

As new, high-yielding hybrids are developed, released, and sown to extensive areas, genetic uniformity will increase and much genetic variability may eventually disappear. The new hybrids will be well suited to modern agronomic practices, but such genetic uniformity is an invitation to disaster from epidemics of pests and diseases. In subsistence agriculture, in which much cassava is now grown, there is a reasonably stable equilibrium between pests and genotypes. Integrated control programs built around plant resistance are needed to maintain this equilibrium in modern agricultural systems, where extensive areas are planted to uniform genetic material.

Resistance to insects or mites attacking cassava is not extensively reported in the literature. Many of the reports deal only with field observations and, until recently, there was little systematic evaluation of germplasm. Until the CIAT collection was assembled, extensive germplasm was not available in one site to cassava researchers. This germplasm bank is being evaluated for resistance to thrips (Schoonhoven, 1974), the mites *Tetranychus urticae, Mononychellus tanajoa,* and *Oligonychus peruvianus,* scales (*Aonidomytilus albus*), mealybugs (*Phenacoccus gossypii*), whiteflies (*Aleurotrachelus* sp.) and lacebugs (*Vatiga manihoti*) (CIAT, 1975, 1976, 1977). These pests satisfy the criteria outlined in Section 4.

5.1. Mites

Several species of mites attack cassava (Bellotti and Schoonhoven, 1978) but the three most important appear to be *M. tanajoa, T. urticae* (= *T. telarius*), and *O. peruvianus*. Bennett and Yaseen (1975), observed large differences in population levels of *M. tanajoa* on different varieties. Nyiira (1972) reported the lowest *M. tanajoa* population on the varieties 'Kru,' '46301-15,' and 'K-Kawanda.' Reports from Brazil (Universidad Federal de Bahia, 1973) and Venezuela (Barrios, 1972) have identified varieties resistant to *M. tanajoa* and *T. urticae*. The CIAT germplasm bank has been evaluated for all three species. Results show the existence of only low levels of resistance to *T. urticae* and intermediate or moderate levels of resistance to *M. tanajoa* and *O. peruvianus*. Based on our rating scale, nearly 98 percent of the varieties were highly susceptible to *T. urticae,* as compared to 45 percent for *M. tanajoa* (CIAT, 1976, 1977). Approximately 14 percent of the varieties were in the intermediate resistance range for *M. tanajoa,* whereas only 0.4 percent of the varieties were in a similar range for *T. urticae*. This indicates a higher level of resistance to *M. tanajoa* than to *T. urticae* in cassava germplasm, and that few collections are resistant to the two species. These results should be expected since *T. urticae* is a major agricultural pest with more than 400 known hosts. *M. tanajoa* appears host-specific for *Manihot* sp., indicating a possible parallel evolution between the pest and the host.

5.2. Thrips

Part of the CIAT germplasm bank has been evaluated for resistance to the thrips *Frankliniella* sp. and *Corynothrips stenopterus*. Approximately 20 percent of the varieties are highly resistant to thrips attack and an additional 29 percent show only minor damage. Resistance is related to the pubescence of leaf buds and unexpanded leaves (Schoonhoven, 1974). This gross morphological character for resistance appears to be very stable and biotypes are not expected to develop.

5.3. Whiteflies

The whitefly *B. tabaci* has been identified as the vector of African cassava mosaic, a disease not present in the Americas. Varietal resistance that reduces whitefly populations could reduce disease incidence (Costa, 1969). Varietal differences in number of whitefly adults and immature stages have been observed. The varieties with the lowest whitefly counts were also the least infected with mosaic (Golding, 1936). The CIAT cassava collection is being systematically evaluated for resistance to *Aleurotrachelus* sp. an important whitefly species in the Americas. Initial results indicate that moderate levels of resistance are available (CIAT, 1976, 1977).

5.4. Scales

The scale *Aonidomytilus albus* attacks cassava throughout most of the cassava-growing regions of the world (Commonwealth Institute of Entomology, 1957). Varietal resistance has not been reported. Screening of germplasm for resistance to this scale has been initiated at CIAT, and preliminary results indicate there are varietal differences in reaction to scale attack.

5.5. Mealybugs

Mealybugs have recently become important as a pest on cassava, causing defoliation in Africa and the Americas. Incidence appears to increase when cassava is planted in monoculture or continually on the same land. No varietal resistance has been reported. All 150 varieties initially screened in Brazil (Albuquerque, 1976) were susceptible. Evaluations at CIAT indicate that some varieties are attacked more severely than others (CIAT, 1977).

5.6. Stem Borers

Numerous species of stem borers attack cassava, especially in Brazil (Bellotti and Schoonhoven, 1978). Resistance to *Coelosternus* sp. has been reported in '103 Brava de Itu' and '192 Itu' (Normanha and Pereira, 1964). No resistance evaluations are being carried out at present.

5.7. Shoot Flies

Distinct varietal differences in susceptibility to shoot flies have been observed, but no extensive screening has been done (Normanha, 1970). The varieties 'Petit Bel Air 4,' 'Rais Blanc,' 'Campestre 10,' and 'Gabela' were most resistant to *Lonchaea chalybea* (Institute de Recherches Agronomiques Tropicales, 1966), and 'IAC 1418' and 'Ouro do Vale' showed some resistance to *Silba pendula* (Brinholi et al., 1974).

6. TECHNIQUES FOR EVALUATING CASSAVA GERMPLASM

Any breeding program, including one that involves host-plant resistance to insects, must begin with an extensive, working germplasm bank. The CIAT cassava germplasm bank contains more than 2,400 accessions, with considerable genetic variability available. Since cassava does not reproduce true from seed, this collection is grown continuously in the field.

Because breeding plants resistant to insects involves an interaction between the plant and the insect, an intimate knowledge of the biology and feeding habits of the insect is essential. Studies of the insect life cycle, its population fluctuations, ovipositional habits, and seasonal occurrence may be needed before resistance studies can commence. This is especially true when dealing with a crop like cassava, whose insect complex has not been extensively investigated by entomologists.

Techniques for evaluating large amounts of genetic material for pest resistance will differ for each insect of interest. However, there are several standard procedures that must be considered when a germplasm evaluation program—involving hundreds or thousands of varieties—is initiated. These procedures are discussed briefly and are followed by their status for specific insects.

1. Any evaluation program must ensure large and uniform insect populations to guarantee adequate selective pressure. This is extremely important when germplasm is being screened under field conditions using natural populations. There is always the danger that varieties selected as resistant are actually escapes, that is, plants that display no damage because insects did not feed on them. When natural populations are used, it is best to accumulate data over at least two seasons and to test several host cultivars.

2. If field populations are inadequate, two alternatives can be implemented.

   a. Varieties grown in the field can be infested manually with the insect being evaluated. This ensures that every plant will receive an initial population of the pest being screened in addition to whatever natural populations may occur. These pests can be either laboratory- or greenhouse-reared for release in the field, or a field culture of the pest can be maintained and used to infest those varieties being screened. The latter procedure is preferred because it ensures that pests already adapted to field conditions will suffer no environmental shock, which might be the case if artificially reared insects were released in the field. If the laboratory culture is reared on an artificial diet the pest population would require an additional adaptation and the varieties evaluated might appear to be more resistant than they actually are.

b. Germplasm can be initially screened under controlled conditions with artificial infestations. The environment can be regulated to favor either the host or the pest by controlling temperature or humidity. This is especially helpful in Colombia since some of the most important cassava pests (mites) are not naturally abundant where the cassava germplasm is maintained. Natural mite populations (especially *M. tanajoa*) are adequate for field screening in parts of Venezuela and Brazil. However, cassava must be propagated by vegetative cuttings, and quarantine regulations make it difficult to move large amounts of germplasm from country to country. The initial screening of germplasm can be done under controlled greenhouse conditions; a smaller number of varieties showing some resistance can then be forwarded for field screening in other areas.

3. The initial germplasm screening program can be geared to eliminating susceptible germplasm, rather than seeking resistant varieties, the objective being to reduce massive amounts of germplasm to a manageable number of varieties. If a program has 2,500 varieties, 90 percent can be eliminated and more detailed studies done on the remaining 250 varieties. This can be done by exerting heavy selective pressure, that is, maintaining a higher-than-normal insect population on varieties evaluated. The danger in this procedure is that excessive selective pressure may result in the discard of varieties with low levels of resistance. To reduce this possibility, germplasm can be screened in 100-variety units, and the 10 to 20 varieties showing the most resistance can be advanced to the next screening cycle. The use of heavy selective pressure lessens the chance of selecting escapes.

4. When field screening is practiced, pest releases or infestations should coincide with the season or climatic conditions that favor the development of the pest to ensure adequate selective pressure on the test material. If the heaviest attack of a pest occurs during the dry season, as is often the case in the tropics, then infestation of the varieties should coincide with the onset of the dry season. If, for example, cassava plants were infested with mites during the rainy season, populations would never increase sufficiently for symptoms to be manifested for accurate evaluation. Evaluations should be made over a period of time and especially when damage is most severe.

5. When natural populations are used in a field screening procedure, and no artificial infestation is supplemented, numerous rows of known suscepti-ble varieties should be planted throughout the varieties being screened to produce a more uniform pest population.

6. A host reaction scale should be developed in a screening program to describe accurately the damage levels. This scale should define highly resistant, intermediate, and susceptible plants in terms of plant damage. Concise rather than broad terms should be used since workers located in different areas will use the same scale to evaluate germplasm. Basically two types of damage keys can be developed: the first damage scale may be used to evaluate a large number of varieties when the major objective is to eliminate susceptible material. This scale is usually on a 0 to 5 basis. A rating of 0 to 2 = some resistance and suggests further testing; 4 to 5 = highly susceptible for discard; and 3 = an intermediate range. Here the scientist must judge whether a variety merits future testing or should be discarded. A second scale can be used for those varieties identified for further evaluation. In this case a scale is needed with more classes, for example, 1 to 10. The differences between the classes, in terms of damage levels, are smaller than in the 0 to 5 scale. This allows a more accurate definition of varietal reaction and is very important when dealing with low levels of resistance or when trying to increase resistance by combining low or intermediate levels through crossing. In the latter case small increases in resistance must be detectable. (Examples of such scales are given in the following section.)

7. Scales should be devised to take into account the existing insect popula-tion when symptoms are not sufficiently pronounced to evaluate resistance accurately. They are useful if the insect is easily detected so that rapid field evaluation can be made. For example, the female *O. peruvianus* mite spins a small, white web on the underside of leaves, under which the eggs are deposited and the immatures develop. Each web is about 2 mm in diameter, and can be seen easily when a leaf is turned over. Each web, in essence, represents a single mite colony, and the number of webs is an indication of resistance to the mite. An insect population scale has also been used for evaluating resistance to whiteflies, scales, and mealybugs.

8. Methods should be designed to evaluate rapidly large numbers of plants for insect damage or infestation. This can be done in the field or greenhouse. Mass screening of seedlings offers a useful time saving tool, especially with crops such as cassava that have a long growing cycle. These techniques can be developed, but resistance in the seedling should be correlated to that in the older plant.

7. PROCEDURES FOR SCREENING CASSAVA GERMPLASM FOR RESISTANCE TO INSECTS

Procedures for screening cassava germplasm have been, or are being, developed for thrips, mites, whiteflies, scales, and mealybugs.

7.1. Thrips

The symptoms of thrips damage are much more pronounced during the dry season, although the insects are present throughout the year. The procedure for evaluating resistance to thrips in cassava was developed by Schoonhoven (1974). Part of the CIAT germplasm collection (1,254 lines) was evaluated under natural infestation during two successive dry seasons for thrips damage. Plants were evaluated at 4 and at 8 months, and an average of these two assessments was used as a resistance classification. Symptoms of thrips damage were classified into six reaction classes:

0 = No symptoms.
1 = Yellow irregular leaf spots only.
2 = Leaf spots, light leaf deformation, parts of leaf lobes missing, brown wound tissue in spots on stems and petioles.
3 = Severe leaf deformation and distortion, poorly expanded leaves, internodes stunted and covered with brown wound tissue.
4 = As above, but with growing points dead, sprouting of lateral buds.
5 = Lateral buds also killed. Plants greatly stunted, with witches' broom-type appearance.

The nature of thrips resistance was studied on 8-month-old nonflowering clones representing each of the resistance levels. Thrips populations were determined by collecting three terminal buds from single plants in a plastic bag, immersing them in 30 percent alcohol, and counting the insects under a microscope. Plant pubescense was determined by counting the number of hairs on the undersurface of one side of an unexpanded leaf lobe. Two leaves per plant were sampled when leaves measured about 1 cm in length. It was found that the leaves of susceptible clones had few or no hairs whereas the leaves of resistant clones had many. Thrips were found on all clones regardless of resistance, but fewer were found on resistant ones. No correlation was found between thrips resistance and plant cyanide content, thus enabling a combination of thrips resistance and low cyanide content.

7.2. Mites

Procedures were developed to evaluate and screen germplasm to three species of mites: *M. tanajoa, T. urticae,* and *O. peruvianus.* Each species requires a different procedure. As previously indicated, natural or field populations of *T. urticae* and *M. tanajoa* normally are not large enough or sufficiently uniform for field screening at CIAT. Therefore, the initial screening for these two mites is done under screenhouse and greenhouse conditions. The screening for *O. peruvianus* is done with natural field infestations. The procedure for each is described.

Mononychellus tanajoa

In the initial screening phase the main objective is to eliminate about 80 percent of the varieties and reevaluate the remainder. This mite species primarily feeds on the upper leaves of the plant, especially on leaves emerging from the bud. It causes yellow to white speckling and deformation of leaves.

Stem cuttings 2 inches long are planted in 4-inch diameter plastic pots. Approximately 1 month after germinating they are removed to the greenhouse (30 to 34°C) and placed in large (1 x 2 m) plastic screening cages, 60 plants to a cage. Two weeks later, they are infested with mites. Each pot represents one variety, and the variety may be repeated several times in one cage or different cages (CIAT, 1976).

Infestation is done by placing one or two lobes of a mite-infested cassava leaf (~50 to 100 mites) on the upper leaves of each test plant. Mites from the field are regularly reintroduced into the colony. Damage evaluations are made beginning the second week after infestation, and each week thereafter for four consecutive weeks. Second and third inoculations are made if the initial one is not successful. A 0 to 5 damage scale based on these symptoms is used during this initial phase:

0 = No mites or symptoms.
1 = Mites on bud leaves, some yellow to white speckling of leaves.
2 = Many mites on leaves, moderate speckling of bud leaves and adjacent leaves.
3 = Heavy speckling of terminal leaves, slight deformation of bud leaves.
4 = Severe deformation of bud leaves, reduction of bud, mites on nearly all leaves, with whitish appearance and some defoliation.
5 = Bud greatly reduced or dead, defoliation of upper leaves.

The lines that are selected as promising (15 to 20 percent) are reevaluated several times to eliminate the more susceptible ones. Those lines selected as the most resistant are reevaluated using a 0 to 10 damage scale:

0 = No mites or symptoms.
1 = Plants with one or two bud leaves with light brown or whitish speckling, located on a few lobes or dispersed over whole leaf. Average of less than 50 specks.

2 = Light speckling distributed over all bud leaves, average 50 to 100 specks.

3 = Moderate speckling of bud and terminal leaves. Attacked leaves begin to fade in color.

4 = Severe speckling of bud and adjacent leaves. One or two bud leaves show slight deformation. Leaves have whitish appearance.

5 = Severe speckling of apical and middle leaves. Light deformation on leaf margin, leaves turn whitish.

6 = Moderate deformation of leaf margin with indentation almost reaching mid veins, and curling of apical leaves resulting in mosaic-like appearance. Basal leaves also show speckling. Slight bud reduction.

7 = Bud deformed and reduced, apical leaves with intense mottling.

8 = Total plant affected, severe reduction of bud and few new leaves developed, general yellow to white appearance with some apical leaf necrosis.

9 = Bud completely reduced, no new leaves developing, defoliation beginning with apical leaves.

10 = Bud dead and severe defoliation.

This scale uses three distinct damage symptoms—leaf speckling, leaf deformation, and bud reduction—to define damage symptoms and detect small differences in damage. This not only is useful in classifying germplasm but becomes more important when crosses are made and slight increases in resistance must be detected. This scale has been used only in greenhouse screening, whereas the 0 to 5 scale has been employed both for greenhouse and field screening.

Collections selected in the greenhouse phase as promising for resistance are planted in the field in Maracay, Venezuela, where high and uniform natural populations occur. Mite populations usually reach their peak about 3 to 4 months after the initiation of the dry period. Twelve plants of promising lines are set out into two replicates of 6 plants each, interspersed with rows of known susceptible varieties. Evaluations on a 0 to 5 scale are made monthly from the onset of the dry season through the rainy season, to measure the ability of these lines to recover from mite damage.

Field results should coincide with greenhouse screening results. However, some lines have not reacted equally under both screening conditions, possibly because of differences in environmental conditions, the ability of the collection to withstand drought, level of mite infestation, or the ability of the plant to recover from mite attack.

Tetranychus urticae

Screening begins with 2-inch stem cuttings planted in screenhouse floor beds, which are surrounded with plastic to raise the temperature to 32 to

34°C. Each bed contains two plants of each of 100 lines spaced 8 inches apart. Mite inoculation and damage ratings on a 0 to 5 scale are similar to those used for *M. tanajoa*. However, *T. urticae* damage symptoms begin on the lower leaves and there is little or no deformation of leaves. Mite populations and symptoms progress upward, and severe infestation causes webbing on upper and lower leaves.

Screening for these two mite species must be done in separate areas, since *T. urticae* is more successful under greenhouse conditions and will mask a *M. tanajoa* colony and its damage. In addition, the work sequence must start with *M. tanajoa* and go on to *T. urticae*, and *never* the reverse.

Oligonychus peruvianus

The evaluation of germplasm for resistance to this mite is done using natural outbreaks of the pest during two dry seasons. The webs of the female are seen easily. Germplasm is evaluated by counting the webs on a sample of three leaves from the lower half of the plant, where the population is greater.

7.3. Whiteflies

The pupal stage of *Aleurotrachelus* sp. is oblong and black, with a white waxy excretion around the outer edge, and is easily seen on the leaf undersurface. Cassava lines are screened in an area having heavy natural infestations. Ten plants of each line are sown in two replicates of five plants each, and rows of susceptible varieties are dispersed throughout the field. Evaluations are made every 2 months beginning when the plants are 2 months old.

Resistance evaluation uses three 0 to 5 scales for (1) the number of pupae per leaf; (2) the percentage of leaves infested with pupae; and (3) damage symptoms caused by whitefly feeding. The number of pupae per leaf is recorded by sampling three leaves per plant:

First scale

  0 = No pupae.
  1 = Less than 5 pupae per leaf.
  2 = 5 to 10 pupae per leaf.
  3 = 10 to 25 pupae per leaf.
  4 = 25 to 50 pupae per leaf.
  5 = More than 50 pupae per leaf.

The percentage of infested leaves per plant is determined by examining several leaves at various plant levels:

Second scale
   0 = No infestation.
   1 = Less than 20 percent infested.
   2 = 20 to 40 percent infested.
   3 = 40 to 60 percent infested.
   4 = 60 to 80 percent infested.
   5 = 80 to 100 percent infested.

Symptoms are recorded as follows:

Third scale
   0 = No damage.
   1 = Slight speckling of lower leaves.
   2 = Heavy speckling of lower leaves.
   3 = Mosaic-like symptoms on leaves but little wrinkling, sooty mold on lower and central leaves.
   4 = Wrinkling and yellowish mottling of lower and apical leaves, some leaf necrosis, considerable sooty mold.
   5 = Severe wrinkling of apical leaves, leaf necrosis, and death of plant.

These three scales permit correlation of damage symptoms with whitefly numbers. Large whitefly populations with few damage symptoms could indicate that a tolerance mechanism is involved. In this case a damage symptom evaluation alone would not necessarily indicate the whitefly population. Tolerant varieties would not reduce whitefly populations, which would be the main goal of a resistance program aimed at reducing virus transmission.

7.4. Scales

A. albus primarily attacks stems and branches, and rarely the leaves. Leaves of attacked plants become yellow and drop. Plants may be stunted and stems desiccated, causing plant mortality. Since scale infestations move slowly and nonuniformly through a cassava field, natural infestations are not sufficient for germplasm screening. A simple procedure has been developed for the infestation and evaluation of varieties.

A scale colony is maintained in the field on the susceptible variety 'M Col 22.' When the varieties to be evaluated are 3 to 4 months old, an infested 8-cm stem cutting is tied to the stem of the test plant about 30 cm above the ground. Inoculation material should be taken from an active colony at the onset of the dry season. Evaluations are made at monthly intervals using the following rating system:

0 = No scales present on stems.

1 = Some scale transfer to stem and located around buds near the infestation site, 3 to 4 scales per bud.

2 = Scales located around buds and some scales on internodes.

3 = Scales around most buds, especially on lower half of stem and covering about ¼ of internodes. Some lower leaf yellowing and leaf necrosis.

4 = Scales around all buds and most of stem, more than ½ of internodes covered, defoliation of lower leaves.

5 = Scales covering all of stem, drying of stem tissue, defoliation of about ⅔ of plant, death of growing point.

If field conditions are not adequate, this rating method, modified slightly, can also be used in the screenhouse or greenhouse.

7.5. Mealybugs

Mealybugs attack both the leaves and the stem so that there are two methods of infestation. A colony of mealybugs is maintained on a susceptible variety. A nymph-infested leaf is removed from the colony and paperclipped or stapled to a middle or lower leaf of the test variety. Mealybug egg masses can also be placed in the leaf axil. These egg masses contain a sticky substance and usually adhere to the plant. A second or third egg infestation should be made at 3-day intervals, especially if it rained after the first infestation.

8. BREEDING APPROACHES AND SYSTEMS

The systematic evaluation of cassava germplasm for insect resistance as a component of comprehensive breeding programs is comparatively recent. Studies in insect resistance were recently initiated at the International Institute of Tropical Agriculture (Leuschner, 1975) as well as at CIAT. Recommendations for breeding approaches and systems are based on limited background information when compared to more extensively studied crops such as rice, maize, or alfalfa.

As noted earlier, cassava is a vigorous, highly heterozygous, naturally cross-pollinated, woody perennial. It has a long cultivation cycle and is easily propagated by stem cuttings. It is grown in a scattered cultivation pattern with many landrace and traditional varieties having various degrees of susceptibility to insects and diseases. These characteristics indicate that in cassava cultivation there is a minimum of selective pressure exerted by pests. Vertical resistance, in terms of the gene-for-gene theory, probably would not evolve within this system. Therefore, resistance is probably multigenically inherited and of the horizontal or field type. Present research indicates that there is no immunity, except for *Sphaceloma manihoticola*. Resistance to most cassava insects and diseases (Lozano, personal communication) is at low or intermediate levels in existing cultivars.

The stability of horizontal resistance is considered to be much greater than that of vertical resistance (Robinson, 1976) and entails less risk that biotypes will develop (Pimentel and Bellotti, 1976). Given the nature of the cassava crop, its ability to withstand drought and recover from insect damage (CIAT, 1976), and its high economic threshold, field resistance based on numerous genes should be adequate to maintain insect populations below economic injury levels.

A scheme for developing field resistance to cassava pests is outlined in Figure 1. Germplasm is evaluated for resistance to different pests and resistant populations are identified. High levels of resistance may not be found in any single cultivar. If different additive genes are involved, crosses between genotypes may increase the level of resistance. Once resistance is identified for different pests, and improved if needed, genotypes can be intercrossed. Breeding with a base of about 20 genotypes should result in a plant population containing horizontal resistance to several pests. High-yielding germplasm can then be introduced into this population, resulting in high-yielding varieties with stable horizontal resistance.

Clearly, this type of program requires a multidisciplinary approach, involving the entomologist and plant breeder as well as the pathologist, since disease resistance is also required in any commercial variety released.

It must be remembered, when cassava is bred for resistance to insects, that cassava is vegetatively propagated and major characters are inherited in an additive manner. Therefore, once a desirable type is obtained, the genotype can be multiplied indefinitely. If the additive effect is equally important for resistant characters as it is for yield characters, it can be an effective tool to increase resistance where only low levels exist in a single genotype. By crossing cultivars containing low resistance levels, the presence of additive genes could result in increased resistance.

Cassava allotetraploids tend to behave as functional diploids in inheritance. Since the allotetraploid combines the gene content of two different diploid species, its potential capacity for variation is greater. Controlled cross-pollination is recommended as the most effective method for producing desirable recombinations when the characteristics of parental genotypes are known. Therefore, a large number of pollinations involving desirable genotypes will most likely give additive recombinance.

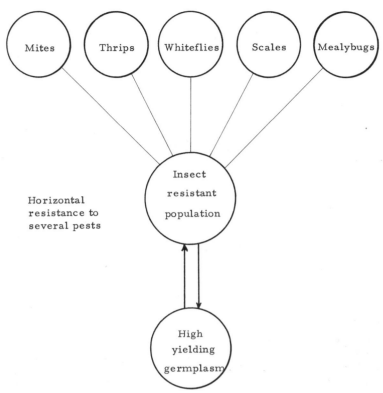

Figure 1. Germplasm improvement scheme for developing field or horizontal resistance to cassava pests. Germplasm improvement pools identified for resistance to insects.

The types of breeding systems used follow:

1. If the resistant characteristic is controlled by dominant genes, then a pedigree breeding method is suggested. Two genotypes are crossed, one possessing the desirable yield characteristics and the other the resistance factor. From this cross, single plant selection in the F_1 generation is made of those individuals possessing the two desirable characteristics, yield and resistance. Further crosses are not needed because desirable genotypes can be propagated vegetatively indefinitely and no further segregation will occur.

2. If the resistant character is controlled by recessive genes, the breeding procedure is more complicated. A double cross is made between hybrids or high yielding varieties that are resistant (Figure 2). Since the genes are recessive this resistance is not expressed in the F_1. The highest-yielding F_1 plants are crossed to give a high-yielding hybrid with resistance. If the

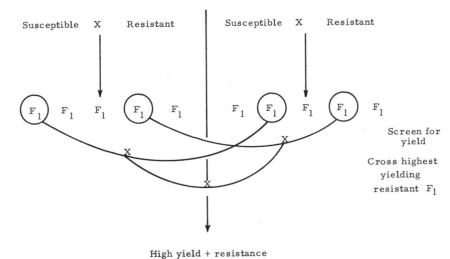

Figure 2. A double-cross program between high yielding susceptible varieties and other varieties having resistance controlled by recessive genes.

inheritance is transmitted in a normal diploid manner (3:1) this method is more successful than if it is tetrasomic (35:1).

3. The details of breeding methods with additive genes are not crucial. The source of resistance must first be identified and an effective selection procedure developed to identify the resistant progeny. Thus it is important to have an accurate and differential scale to select resistant progeny and to identify resistant progeny superior to the two parents.

When several moderately resistant genotypes are available, double or multiple crosses, including resistant genotypes and high-yielding genotypes, are recommended. The scheme is a modified, recurrent, mass-selection process. Effective selection is enhanced in any segregating population by the use of vegetative propagation. It is important to avoid inbreeding depression.

4. If only low levels of resistance are available, for example with the *T. urticae* mite, it may be increased by crossing two distinct varieties that are each moderately resistant. If there are additive genes involved and the resistance genes of each variety are different (i.e., at different loci) then resistance can be increased by making the proper combinations. It is important to make sufficient crosses and to develop an effective damage rating scale for identifying progeny that display more resistance than either of the parents. Selection is for resistance, not yield. This resistance can be incorporated into desired high-yielding hybrids by one of the aforementioned methods.

9. SUMMARY

There is substantial potential for breeding cassava plants resistant to insects. Work at the international level has barely begun, and there is a lack of available information about insect resistance. Since cassava is normally a low value crop, pesticides or other control methods are often prohibitive in cost. Therefore, crop resistance offers an economical alternative.

Numerous mite and insect pests attack cassava, causing yield reductions. Since cassava has a long growing season (8 to 24 months), insects that attack the crop over a prolonged period appear to cause more yield loss than those that attack the crop for brief periods. Thus mites, thrips, scales, mealybugs, and whiteflies can cause substantial yield reductions. Evaluation of host-plant resistance is directed toward these persistent pests. High levels of resistance have been found for thrips and moderate resistance to the mites *M. tanajoa* and *O. peruvianus* and to whiteflies. Resistance studies have recently been initiated for scales and mealybugs.

The following are important guidelines for locating resistance to insects in cassava: (1) collect ample and variable germplasm; (2) screen with high populations of the pest being evaluated and; (3) develop an accurate scale to evaluate damage. Screening can be done under field conditions, using natural and artificial infestation, or under controlled conditions in the greenhouse or screenhouse.

Cassava is generally regarded as an allotetraploid and is highly heterozygous. It exhibits strong inbreeding depression and has a high degree of male sterility. Important characteristics such as harvest index, root dry-matter content, and resistance to various diseases are highly heritable. The role of additive genes in these characteristics is significant.

The vegetative propagation of the crop and the additive inheritance of major characteristics indicate that once a superior type is obtained it can be multiplied indefinitely. If resistant characteristics are controlled by dominant genes, a conventional pedigree method will be successful. If the resistant characteristic is controlled by recessive genes, the procedure is more complicated and requires a double cross between high yielding hybrids or varieties that each have resistance. When working with additive genes the procedure is to first identify the resistance source and then, through an effective selection method, identify the progeny with increased resistance.

Cotton is graded at cotton-development project in Nepal. Photo by Ray Witlin, courtesy of United Nations.

15

BREEDING COTTON FOR RESISTANCE TO INSECT PESTS

G. A. Niles

Department of Soil and Crop Sciences, Texas A & M University, College Station, Texas

1. INTRODUCTION

Of the textile fibers produced and utilized in the world, cotton is the most valuable. Data compiled by the Japan Cotton Trader's Association (1976) show that cotton accounted for 50 percent of the total world production of textile fibers in 1975, compared with 44 percent for man-made fibers, 6 percent for wool, and less than 0.05 percent for silk. Per capita consumption of cotton in 1973 was estimated as 3.4 kg, whereas consumption of the man-made fibers and wool were 3.1 and 0.4 kg, respectively.

In the period 1975 to 1976 cotton was harvested from an estimated 74.4 million acres in 46 countries (anonymous, 1977). As the data in Table 1 indicate, eight nations produced more than 1 million bales, and three of those accounted for more than 50 percent of total world output.

Cotton is a warm-weather plant, and its production generally is limited to the tropics and the temperate zone. In the Northern Hemisphere, the crop is produced as far north as latitude 47 degrees in the Ukraine and latitude 37 degrees in the United States. Production in the Southern Hemisphere extends to about 32 degrees in Australia and South America. Although cotton generally is considered a long-season crop, requiring a minimum growing season of 180 to 200 days, certain combinations of improved varieties and cultural practices permit effective production in areas where the frost-free period is less than 180 days. Agricultural varieties show great flexibility in adaptation to differing soil, rainfall, and cultural systems. To a considerable degree, this broad adaptation is due to a relatively indeterminate fruiting habit which enables it to regulate fruiting activity during periods of stress and to resume fruiting when conditions are more favorable for setting a crop.

Table 1. World Cotton Acreage and Production, 1975 to 1976. (*Cotton International*, 4th Annual ed., 1977)

Country	Harvested Acreage (1,000 acres)	Production (1,000 bales)[a]
El Salvador	185	273
Guatemala	208	460
Honduras	11	14
Mexico	580	902
Nicaragua	335	510
United States	8,796	8,302
Argentina	1,022	611
Brazil	4,485	1,800
Colombia	620	554
Ecuador	85	44
Paraguay	260	150
Peru	290	295
Venezuela	151	101
Greece	335	595
Italy	12	9
Spain	185	170
Yugoslavia	15	9
Albania	70	35
Bulgaria	94	75
USSR	7,220	12,100
Afghanistan	200	150
Australia	73	123
Burma	490	70
China	12,000	11,000
India	18,500	5,600
Iran	720	690
Iraq	150	50
Israel	100	225
South Korea	27	14
Pakistan	4,600	2,360
South Yemen	40	30
Syria	514	727
Thailand	150	100
Turkey	1,655	2,205
Angola	130	50
Cameroon	181	87
Central Africa Republic	250	50
Chad	740	300
Kenya	175	25
Morocco	43	18
Mozambique	700	200
South Africa	168	115
Sudan	1,170	500
Tanzania	575	193
Egypt	1,400	1,750
Uganda	1,370	140
World total	74,387	55,188

[a] 480-pound bales.

Cotton is grown almost exclusively for its fiber, which is used to make numerous textile, pharmaceutical, and industrial products. Cotton lint is a major item in international commerce; in 1975 more than 18 million bales were exported by producing countries. Cottonseed, although considered a by-product, has significant economic value as a source of oil, meal, hulls, and linters. Considerable research is underway to develop improved low-gossypol ("glandless") types which offer great potential as a protein source. The market value of a cotton crop can be apportioned, roughly, as 85 percent for the lint and 15 percent for seed.

2. HISTORY OF COTTON

We do not know when cotton was first used as a textile fiber, but cotton fabrics dating back to about 3000 BC were found in excavations in the Indus River Valley of present-day Pakistan (Gulati and Turner, 1928). Chowdhuri and Buth (1971) found cotton remains in Egyptian Nubia which were estimated to be 4500 years old, and concluded that cotton was cultivated for its seed, which was used as cattle feed, rather than as a fiber crop. Specimens of cotton dating back to about 2500 BC were found in excavations made in Peru (Bird and Mahler, 1951). These discoveries provide evidence for an ancient culture of cotton in both the Old World and the New World. However, the earliest written record of its textile use is in the ancient digest ascribed to Manu, written in 800 BC, which indicates that the Hindus had long known cotton both as a plant and as a textile. Later references by such writers as Pliny and Marco Polo indicate that India was the center of the cotton industry in the Old World until well into the Christian era.

Cotton also was known on the eastern coast of Africa, Asia Minor, Ethiopia, and Arabia. Apparently both *Gossypium arboreum* and *G. herbaceum* were represented in these regions. Cotton cultivation was introduced into Sicily and Spain by the Saracens. The Moors appear to have spread cotton culture and manufacturing to other areas of the Mediterranean region, where a strong commerce in cotton existed for many centuries. Cotton apparently was known in China in the seventh century AD, but extensive cultivation was not made until the eleventh century. Its introduction into Japan took place in the eighth century, but systematic cultivation was delayed until the sixteenth century, probably until day-neutral annual types had evolved. It appears that cotton was not cultivated in Egypt until the thirteenth or fourteenth century, although it was known before that time. Serious cultivation in Egypt dates to the early nineteenth century when Jumel, a French engineer working in Egypt, introduced a type of *G. barbadense*. This first cultivated cotton, named 'Mako-Jumel,' was the predecessor of the Ashmouni type of present-day Egypt. Historical references to the early Indian and Egyptian cottons indicate that they were perennial, whereas the cottons disseminated by the Moors were annual forms of *G. herbaceum*, probably procured from Arabia.

When Columbus visited the West Indies in 1492, he found cotton growing abundantly. Spanish explorers found cotton being cultivated on the mainland of the New World in Peru, Mexico, Brazil, and other tropical regions of the Americas. In the Indies and within the Aztec, Mayan, and Inca civilizations, cotton was widely used for bedding, clothing, household items, and handicrafts, and cotton goods were important trade items. Excavations in New Mexico yielded cotton threads dated around 300 BC, and Spanish explorers found cotton being cultivated in the areas now known as Arizona and New Mexico. Early colonists to the southeastern and midsouth sections of the United States did not find any use of cotton by Indians of the area, and there is no evidence of cotton in these areas until after 1600.

The first effort by Europeans to grow cotton in the United States probably was made in the Jamestown Colony in 1607, but systematic culture was not established until 1621. The source of the cotton seedstocks used in the early American colonies is unclear. Records indicate that seed was obtained early from the West Indies, the Levant, and Siam, and included both Asiatic and *G. hirsutum* types. The Asiatic types soon disappeared, and by 1784, when the first cotton was grown for export, the green-seeded cottons (*G. hirsutum*) introduced from the Caribbean and Central America predominated. Cultivated types of *G. barbadense* were introduced from the West Indies about 1786. These cottons became adapted to the coastal islands of Georgia and the Carolinas, and became known as Sea Island cottons. By contrast, stocks of *G. hirsutum* were best adapted to the interior growing areas, and came to be identified as Upland cottons.

Cotton culture in early colonial times was not extensive, and textile requirements largely were satisfied by use of wool and flax and raw cotton imported from the West Indies. Invention of the cotton gin by Eli Whitney in 1793 made cotton production practical and profitable. Cotton growing rapidly spread westward to the Mississippi Valley and to Texas, and by 1861 the United States was producing 4.5 million bales of cotton annually.

Between World Wars I and II cotton became an important crop in the southwestern states of New Mexico, Arizona, and California, and acreages declined substantially in the southeast. Cultivation of *G. barbadense* types began about 1910 in the Southwest, utilizing the 'Yuma' variety which was selected from the 'Mit Afifi' variety of Egypt. In succession, the 'Pima,' 'S x P,' 'Amsak,' and 'Pima 32' varieties were established from the Egyptian germplasm. 'Pima S-1,' first grown commercially in 1951, was developed from a complex parentage which included both *G. barbadense* and *G. hirsutum* sources. A series of subsequent improvements have been made in the American-Egyptian varieties, culminating in the release of 'Pima S-5' in 1977.

The foregoing brief historical review of cotton was developed from numerous sources, including Mayers (1868), Handy (1896), Watt (1907), Johnson (1926), Ware (1936), Hutchinson (1938, 1962), Feaster and Turcotte (1962), and Lewis and Richmond (1968).

3. TAXONOMY OF COTTON

The true cottons belong in the genus *Gossypium* and are related to several other genera which are included in the tribe Gossypieae (Fryxell, 1968). The Gossypieae are related to the tribe Hibisceae, and these two groups, along with several others, belong to the Malvaceae. In the past, opinion was divided on family classification of Gossypieae and Hibisceae. Certain authors included the groups in the Bombacaceae, but recent evidence establishes that the two tribes should be retained in the Malvaceae (Fryxell, 1968).

Thirty-seven species of *Gossypium* are listed in Table 2, modified from Fryxell (1969). Of these, 33 are wild and 4 cultivated. The cultivated species are distinguished by their seed fibers, which can be spun. The wild species produce short, sparse seed hairs that are not spinnable. Two of the cultivated species, *G. arboreum* and *G. herbaceum*, are distributed throughout Africa and Asia and are termed the Old World cultigens. Both are diploids and possess 13 pairs of chromosomes. The New World cultigens, *G. hirsutum* and *G. barbadense*, are native to the Western Hemisphere and are allotetraploids with 26 chromosome pairs. The origin of the tetraploids through hybridization between Asiatic diploid and American wild species was confirmed by the cytological studies of Beasley (1940). With the exception of *G. tomentosum*, which is a tetraploid, all the wild species of *Gossypium* are diploids.

The genus *Gossypium* is widely distributed in tropic and subtropic regions of the world. According to Hutchinson (1962), the contemporary commercial cottons are almost all short-term annual plants, derived from perennial forms under selection pressures that encouraged early maturity and eliminated photoperiod sensitivity. The truly wild species of *Gossypium* are mostly shrubby perennials, and are confined to frost-free areas.

Centers of variability have been established generally for the four cultivated species of *Gossypium*. In *G. arboreum*, several centers apparently gave rise to annual types; these include India, Southeast Asia, China, and Africa. For *G. herbaceum*, centers of variability have been identified in

Table 2. The Species of *Gossypium* (adapted from Fryxell, 1969)

Species	Distribution	Genome Designation
G. sturtianum	Australia	C_1
G. sturtianum var. *nandewarense*	Australia	C_{1-n}
G. robinsonii	Australia	C_2
G. australe	Australia	C_3
G. bickii	Australia	C_4
G. costulatum	Australia	C_5
G. populifolium	Australia	C_6
G. cunninghamii	Australia	C_7
G. pulchellum	Australia	C_8
G. nelsonii	Australia	Not established
G. pilosum	Australia	Not established
G. thurberi	Mexico, Arizona	D_1
G. armourianum	Mexico	D_{2-1}
G. harknessii	Mexico	D_{2-2}
G. davidsonii	Mexico	D_{3-d}
G. klotzschianum	Galapagos Islands	D_{3-k}
G. aridum	Mexico	D_4
G. raimondii	Peru	D_5
G. gossypioides	Mexico	D_6
G. lobatum	Mexico	D_7
G. trilobum	Mexico	D_8
G. laxum	Mexico	D_9
G. anomalum	Africa	B_1
G. triphyllum	Africa	B_2
G. barbosanum	Africa	B_3
G. capitis-viridis	Cape Verde Island	B_4
G. herbaceum	Old World cultigen	A_1
G. arboreum	Old World cultigen	A_2
G. stocksii	Arabia	E_1
G. somalense	Africa	E_2
G. areysianum	Arabia	E_3
G. incanum	Arabia	E_4
G. longicalyx	Africa	F_1
G. tomentosum	Hawaii	$(AD)_3$
G. hirsutum	New World cultigen	$(AD)_1$
G. barbadense	New World cultigen	$(AD)_2$
G. darwinii	Galapagos Islands	$(AD)_2$

India, Central and East Africa, Iran, and Afghanistan. In the New World, southern Mexico and Central America are considered the centers of variability for *G. hirsutum*; northern South America, the West Indies, and Central America are centers of variability for *G. barbadense*.

4. SOURCES OF GENETIC VARIABILITY

Except for Egypt, Sudan, and Peru, where varieties of *G. barbadense* predominate, and India, where *G. arboreum* and *G. herbaceum* represent the major portion of the crop, the great bulk of the cottons grown in the world today are American Uplands or their derivatives. The Uplands are represented primarily by four varietal types: 'Acala,' 'Stoneville,' 'Coker,' and 'Deltapine.' A study of the origin and development of prominent American Upland varieties indicates that the germplasm is narrowly based. In reviewing the history of cotton variety development in the United States, Ramey (1966) suggested that 17 sources provided the parental germplasm of the modern-day American Uplands. Of the four principal Upland types cultivated worldwide, the 'Coker,' 'Deltapine,' and 'Stoneville' types have a common ancestor in the Bohemian variety which dates back to 1860. The Acala cottons have a different ancestry, tracing back to direct introductions from Mexico into the United States in 1907. Certain varieties of the Acalas developed in recent years represent some infusion of germplasm from "eastern" types.

Despite the apparently narrow germplasm base of the Uplands, they retain a surprising degree of genetic plasticity. Most of the current improvement efforts with the Uplands involve direct selection or intervarietal crossing, and little of the available exotic materials have been incorporated successfully into practical breeding programs. Yet there is increasing interest by breeders in utilizing a broader range of germplasm in improvement programs, especially where improved fiber properties and pest resistance are major objectives.

In cotton, five categories of genetic variability are available for use in breeding programs. These are discussed below in order of general availability and ease of utilization.

1. *Contemporary commercial varieties and strains.* These types are the most convenient to procure and generally do not present any difficulties in handling in the breeding process. Controlled pollinations are easily made and the segregation products are reasonably predictable. Obviously, the range of variability in segregating generations depends on the diversity of the parent stocks from which hybrids are derived.

2. *Obsolete varieties and strains.*   Old cultigens and predecessors of contemporary types may possess useful genetic characteristics which were not brought forward in the development of succeeding varieties. It is unlikely that this residual variability would be useful in improvement of yield or yield components, since production potential has been under constant selection pressure over many years. However, the old sources of germplasm may contribute specific traits for which little or no selection preference has been applied, or for which selection criteria have changed over time. These might include plant conformation, fruiting habit, maturity, fiber properties, seed characteristics, and other morphological and physiological traits.

Many obsolete forms have been maintained, to some extent, by private breeders and public research workers in various countries. In the United States a collection of approximately 800 obsolete agricultural varieties, inbred lines of Upland cotton, and the principal commercial varieties is maintained by the Mississippi Delta Branch Experiment Station. The Agricultural Research Service at the University of Arizona Cotton Research Center has primary responsibility for maintenance and distribution of the regional collection of *G. barbadense* stocks. This collection contains more than 250 stocks, including Sea Island and Egyptian types and historical Pima varieties. Many of the stocks in the two collections also are on deposit at the National Seed Storage Laboratory, Fort Collins, Colorado.

3. *Foreign Uplands.*   Dispersion of the tetraploid species of *Gossypium* from the New World presumably occurred in two ways. Cottons from Central America, South America, and the Caribbean region were carried directly to the Old World by early European traders and missionaries (Hutchinson, 1949, 1962). The major introduction of American cottons, especially the *G. hirsutum* types, was made from the United States cotton belt following the Civil War. In the Old World environment, selective pressures for local adaptation have resulted in forms quite distinct from their progenitors and from the modern commercial varieties of the cotton belt.

For American cotton workers, the Old World forms of *G. hirsutum,* commonly termed 'Foreign Uplands,' represent a unique source of germplasm for improvement of the species. Although their growth habit and maturity characteristics generally are quite different from American Upland types, many of the Foreign Uplands provide potentially useful sources of improved fiber properties, resistance to bacterial blight (*Xanthomonas malvacearum*), and resistance to certain insects, notably the jassid (*Empoasca* spp.).

4. *Primitive stocks of* G. hirsutum.   Central American and Mexican *G. hirsutum* accessions have provided, directly or indirectly, the genetic

sources from which the major portion of the world's cottons have come. Numerous collections have been made in the present century to accumulate a wide diversity of germplasm from this center of variability, and to catalog and preserve it. The collections of Richmond and Manning in 1946, Manning and Ware in 1948, and Stephens in 1946 to 1947 are particularly important, for these provided a prominent part of the existing conservatory of primitive stocks of *G. hirsutum*. Accounts of the Richmond-Manning and Manning-Ware expeditions are included in the catalog of *Gossypium* germplasm (anonymous, 1974).

The living collection of primitive stocks of *G. hirsutum* is maintained at Texas A & M University, College Station, Texas, and seed samples are on deposit at the National Seed Storage Facility, Fort Collins, Colorado. The accessions from Mexico and Guatemala have been studied and classified in geographic races, in accordance with the criteria enumerated by Hutchinson (1951).

Use of the geographic races, or primitive stocks, of *G. hirsutum* has increased in recent years, especially by breeders and geneticists working on resistance to insects. The majority of the accessions are short-day types, which flower sparingly or not at all during the normal growing season in the temperate zone. With most of these types, flowering and boll set can be facilitated by growing them in a greenhouse or in subtropic or tropic regions during the winter season. The primitive stocks generally do not present any substantial problems in respect to fertility or aberrant segregation products, and they readily can be utilized in cotton breeding programs.

5. *Wild species of* Gossypium.   The noncultivated or wild species are the most difficult sources of genetic variability to utilize for cotton improvement. However, they offer probably the greatest diversity of potential genetic resources for improvement of pest resistance and stress tolerance in cultivated cottons. Difficulties in use of the wild species arise from structural differences among chromosomes of different species, as well as in differences in chromosome complement between the wild diploids and the tetraploids. Cytogenetic studies have demonstrated that fertility relationships among the species are highly variable; only about two-thirds of the interspecific crosses studied have produced fertile F_1 plants (anonymous, 1968). Although interspecific hybridization presents substantial difficulties in breeding programs, the use of exotic species offers intriguing potential. Meyer (1973) has developed a series of breeding stocks that combine the nuclear gene complement of two Upland cottons in various combinations with cytoplasms of seven other species of *Gossypium*, providing an unusual and valuable source of germplasm for cotton improvement. Introgression of foreign species germplasm has been utilized effectively in several instances to improve genetic properties of *G. hirsutum*. A collection of wild diploid

cottons is maintained at Texas A & M University, as is a collection of varieties and marker stocks of *G. herbaceum* and *G. arboreum*. Varieties, strains, and marker stocks of *G. barbadense* are maintained at the University of Arizona Cotton Research Center, Phoenix, Arizona.

In addition to the germplasm sources maintained in the United States, valuable collections are held in several parts of the world. Among the better known collections are the following:

1. *Shambat, Sudan*. This collection, formerly maintained by the Empire Cotton Growing Corporation, includes materials from the Trinidad collection; it is now maintained by the Sudanese Department of Agriculture.

2. *Tashkent, USSR*. This collection, maintained by the USSR government, includes accessions of Mauer and Bukasov from Central and South America.

3. *Presidencia Roque Sáenz Peña, Argentina*. The Argentinian collection includes accessions of *G. barbadense* from Northern Argentina, Paraguay, and Bolivia.

5   INSECT RESISTANCE IN COTTON

In all areas of the world where cotton is grown, insect pests constitute a major factor in production. In recent times insect control has been based more on the use of chemical insecticides than on cultural practices. Until recent years, little emphasis was placed on plant genetic resistance as a means of suppressing insect pests. However, research on host plant resistance to insect pests has accelerated in recent years, and cotton workers in many countries now consider this a major objective in breeding programs.

Aston and Winfield (1972) list 46 groups of insects known to occur in cotton throughout the world; 42 are classified as economically important in one or more of the cotton-producing nations. Just as the insect pest complex differs among various regions, so also does the emphasis given to the development of plant resistance to insect pests. An examination of the literature on cotton plant resistance to insects indicates that the bulk of the research has been done within the last 15 to 20 years. Prior to the mid-1950s, research on insect resistance was quite limited; Painter (1951) noted that, except for the breeding of varieties resistant to certain leafhoppers, no extended efforts had been made to develop genetic resistance to cotton insect pests.

Resistance literature of the past two decades is extensive, and no attempt is made here to provide a complete review. Rather, the intent is to assess the present status of insect resistance in cotton, including recent developments with respect to the several types of cotton insect pests.*

5.1. Jassids

The genus *Empoasca*, to which jassids belong, occurs throughout the world and inflicts damage on numerous crops. In cotton, three species are of greatest concern, particularly on American Upland cottons and their derivatives: *E. facialis* in Africa, *E. devastans* in India and Pakistan, and *E. terrareginae* in Australia. An excellent discussion of the development of jassid resistance up to 1951 is provided by Painter (1951). Genetic resistance to the jassid, initially developed more than 50 years ago in South Africa, evidently was the first success in utilizing resistance to control a cotton pest (Parnell, 1925). The growing of hairy varieties has essentially eliminated the jassid as a major cotton pest in tropical Africa (Reed, 1974). In the Sudan, strains of *G. barbadense* have been developed which carry resistance to *E. lybica*, and selection for resistance to jassids has been successful in Australia.

Development of jassid resistance apparently has been less successful in India and Pakistan, possibly because of differences in species of *Empoasca* and because four species of cotton are grown, two of which (*G. herbaceum* and *G. arboreum*) are highly resistant to the pest. Indian researchers have done extensive work to screen and evaluate cotton germplasm for jassid resistance, and to identify mechanisms of resistance. Pandya and Patel (1964) reported that *G. tomentosum, G. armourianum,* and *G. raimondii* possess resistance to jassids; resistance from the first two species was transferred into *G. hirsutum* types.

The consensus among most research workers is that plant hairiness and resistance to infestation are associated; some differences of opinion have been expressed about the degree to which other plant characteristics influence resistance. Joshi and Rao (1959) cite several authors who variously reported that not all hairy varieties are jassid resistant, that glabrous *G. hirsutum* types may be resistant under certain conditions, that presence of hairs does not necessarily confer resistance, and that toughness of leaf veins is the source of resistance. Tidke and Sane (1962) concluded that thickness of the leaf lamina is more determinant of resistance than are number of hairs on veins or lamina, length of hair, and angle of hair insertion. A certain diversity of opinion notwithstanding, the conclusions of

---

* Comprehensive reviews of literature through 1971 are provided by Dahms (1943), Painter (1951), and Maxwell et al. (1972).

Parnell et al. (1949) appear to have universal application. These authors considered the relationship of hairiness (properly measured) and jassid resistance to be one of direct cause and effect, and not due to any genetic linkage between hairiness and some other factor conferring resistance. They also concluded that length of hair is most important and, provided length is maintained, increased hair density increases resistance.

5.2. Plant Bugs

This general group includes the cotton fleahopper (*Pseudatomoscelis seriatus*), the tarnished plant bug (*Lygus lineolaris*), and *Lygus hesperus*.

Host-plant resistance studies with the fleahopper mostly began in the 1960s after several experiments had demonstrated that the glabrous character substantially reduces populations of fleahoppers (Cowan and Lukefahr, 1970; Lukefahr et al., 1970). Subsequent studies indicated that, although glabrousness effectively reduces fleahopper numbers, many nonhairy genotypes are highly sensitive to even low populations of the pest (Walker and Niles, 1973; Walker et al., 1974; Schuster et al., 1976). Lukefahr et al. (1976) suggested that the problem with glabrous cottons is their sensitivity to leafhoppers (*Empoasca* spp.), rather than fleahoppers. Although the question has not been resolved fully, it appears that glabrous cottons vary in their sensitivity, according to genetic background, and this trait may yet prove useful as a source of plant resistance to the fleahopper.

Nectarilessness and high bud gossypol are other plant characteristics that provide resistance to the cotton fleahopper. Schuster and Maxwell (1974) showed that fleahopper populations in a nectariless cotton were reduced by 58 percent, in comparison with a nectaried counterpart. Schuster and Frazier (1976) suggested that nectarilessness confers resistance through nonpreference and antibiosis, the latter as a result of nutritional deficiency in the absence of nectar. The effect of high gossypol on reducing populations of fleahoppers was demonstrated by Cowan and Lukefahr (1970) and Lukefahr and Houghtaling (1975). Population reductions of 50 to 60 percent were measured in Texas, and corresponding effects have been noted in Mississippi and Louisiana.

The tarnished plant bug, *Lygus lineolaris*, is an economic pest of cotton in the eastern half of the United States cotton belt; *L. hesperus* is important primarily in the irrigated areas of the western United States. Nectarilessness generally has been effective in reducing populations of *L. lineolaris* (Meredith et al., 1973; Schuster and Maxwell, 1974; Laster and Meredith, 1974b; Schuster and Frazier, 1976). California workers reported no apparent effect of nectarilessness on *L. hesperus* (Tingey et al., 1975). High levels of bud gossypol likewise have an adverse effect on behavior and development of Lygus bugs (Schuster and Frazier, 1976; Tingey et al., 1975). The latter authors presented data showing significant suppression of insect growth rates on four stocks of *G. barbadense* and on several stocks possessing cytoplasm from wild diploid species of *Gossypium*.

5.3. Pink Bollworm

The pink bollworm, *Pectinophora gossypiella*, occurs throughout the cotton-growing regions of the world; it is of major economic importance in the western United States, Mexico, India, Pakistan, Egypt, China, and several other countries in Africa and South America.

Brazzel and Martin (1956) summarized numerous previous reports dealing with host-plant resistance to the pink bollworm in cotton. Those prior to 1950 indicated that the Old World cultivated species, *G. arboreum* and *G. herbaceum*, are resistant to the pest, and that certain other species, including *G. thurberi, G. trilobum, G. armourianum,* and *G. somalense* show appreciable resistance. The extensive research of Brazzel and Martin (1956), and the reports of Reed and Adkisson (1961), Adkisson et al. (1962), and Wilson and Wilson (1975c) indicate that races of *G. hirsutum* offer considerable promise as sources of resistance to the pink bollworm.

In one of the earlier studies with the nectariless trait, Lukefahr and Rhyne (1960) were unable to demonstrate that nectaries had any appreciable effect on pink bollworm in small field plots. In a later cage study, Lukefahr et al. (1965) reported that 39 to 50 percent fewer mines were formed by pink bollworm larvae in bolls of nectariless cotton; Davis et al. (1973) reported that infestations were slightly but not significantly lower in a nectariless type. Wilson and Wilson (1976) field tested glabrous and nectariless strains for resistance to pink bollworm and found that both the nectariless and glabrous traits significantly reduced insect numbers and amount of seed damage, compared to a normal nectariless-pubescent variety. The combination of the nectariless and glabrous traits had an additive effect in lowering larvae numbers and seed damage. On the other hand, Smith et al. (1975) concluded that high plant hair density is a promising mechanism for resistance to the pink bollworm in cotton.

Early-rapid blooming and early maturity are effective mechanisms of escape from late-season infestations by pink bollworm (Noble, 1969; Singh and Butani, 1963). Used in conjunction with suitable cropping practices, early maturing varieties may prevent development of diapausing pink bollworm and allow for early crop residue disposal, thereby reducing overwintering populations of the pest. In Texas, these general production practices and regulatory control of planting date and crop termination have reduced the pink bollworm to the status of a minor pest.

5.4. Bollworms

In addition to the pink bollworm, several other lepidopterous insects, classed as bollworms, are economic pests of cotton. These include *Heliothis zea* (bollworm), *H. virescens* (tobacco budworm), *H. armigera* (American bollworm), *Earias* spp. (spiny bollworm, spotted bollworm) and *Cryptophlebia leucotrea*. *H. zea* and *H. virescens* are important pests in the Western Hemisphere; the others are important primarily in Africa, Asia, and Australia.

Most research on host-plant resistance to the bollworms has been done with *H. zea* and *H. virescens*. Numerous reports made in the 1960s indicated that four plant traits effectively impart resistance—glabrousness, nectarilessness, increased gossypol in flower buds, and the so-called X factor. The most comprehensive studies with the glabrous trait are those reported by Lukefahr et al. (1971, 1975), working with bollworm and tobacco budworm. The glabrous character consistently reduced numbers of eggs, larvae, and damaged squares and bolls. In discussing these studies, Lukefahr (1977) suggested that cotton having less than 200 trichomes per square inch on the leaf surfaces should effect a 50 percent reduction in egg deposition and larval population. Comparable effects of glabrousness were reported by Camplis et al. (1976) from studies in Mexico.

The value of the nectariless character as a source of plant resistance to bollworm and tobacco budworm is unclear. Several investigations have demonstrated the efficacy of nectarilessness in reducing egg deposition and square damage (Lukefahr et al., 1965; Davis et al., 1973; Laster and Meredith, 1974; Schuster and Maxwell, 1974). More recently, Lukefahr et al. (1975) reported that, although the nectariless character conferred some resistance in cage tests, the trait showed no benefit in field plots. In a recent discussion of host-plant resistance in cotton, Lukefahr (1977) stated that nectarilessness apparently does not suppress *Heliothis* spp. in the field, possibly because of the flight habit of the adult moth.

Extensive studies have been made on antibiosis as a source of resistance to *Heliothis* spp., and Lukefahr et al. (1966) discussed utilization of high bud gossypol as a source of resistance. Data from several bioassays and other studies indicate that a bud possypol content greater than 1.2 percent (dry weight basis) will effect a 50 percent mortality of *Heliothis* spp. larvae. The initial source of high square gossypol was a wild cotton endemic to Socorro Island, and most of the breeding work with high gossypol involved this accession, a good source of high gossypol, but relatively poor in agronomic properties. Dilday and Shaver (1976a, 1976b) attempted to identify other sources of the trait, to expand the range of genetic diversity in high gossypol types. They indicate that several of the primitive race stocks of *G. hirsutum* possess relatively high levels of gossypol in flower buds, and they suggested that these might be better than the Socorro Island stock for breeding. Effects of high bud gossypol can be enhanced in combination with glabrousness, resulting in larval population suppression of 60 to 80 percent (Lukefahr et al., 1975).

Other natural plant constituents also have antibiotic effects on *Heliothis* spp. Shaver and Lukefahr (1971) reported screening primitive stocks of *G. hirsutum*, utilizing three bioassay techniques. Several of the accessions were shown to possess antibiosis mechanisms other than gossypol, and Lukefahr et al. (1974) showed that the larval growth inhibitor (X factor) is heritable. Two biologically active components appear to account, at least in part, for the inhibitory properties of X factor (Lukefahr et al., 1977). These compounds, classed as terpenoids, include hemigossypolone and heliocide H_1. Of the terpenoids assayed thus far, gossypol is the most active biologically, followed by heliocide H_1 and hemigossypolone, in that order.

5.5. Boll Weevil

The boll weevil, *Anthonomus grandis*, is distributed in the United States cotton belt from west-central Texas to North Carolina; in Mexico, Central America, Cuba, and Hispaniola; and in the northern extremities of Venezuela and Colombia (Cross et al., 1975). Cultivated American Upland cotton, *G. hirsutum*, and its wild and semiwild forms are the primary host of the insect; *G. barbadense* and several wild diploid species (*D*-genome) are involved to a lesser extent.

Of several plant characteristics that exhibit resistance to the boll weevil, the frego-bract trait is the most promising at the present time. Frego bract originated as a mutant trait in a commercial planting of 'Stoneville 2B' cotton; it is characterized by relatively narrow, elongated, and twisted bracts which flare away from the flower bud and boll. It is controlled by a single pair of recessive genes whose expression may be modified somewhat by genetic background.

Boll weevil resistance associated with frego bract was first reported by Jones et al. (1964), who demonstrated the strong degree of nonpreference conferred by this trait. Hunter et al. (1965) subsequently reported that the abnormal bract inhibited feeding and oviposition by weevils until population pressure became heavy. In certain frego lines, abcission of damaged squares was incomplete and flared bracts remained attached to the plant where desiccation and high temperatures appeared to increase mortality of larvae and pupae. The resistance properties of frego bract have been substantiated by several other studies including field experiments in Mississippi reported by Jenkins (1976). In the largest of the tests, the frego-bract cotton showed a 50 percent reduction in oviposition-damaged squares, compared with a normal type, and required 46 percent less insecticide. Lint yields of the frego and normal cottons were the same.

The efficacy of frego bract in weevil resistance appears to be due mainly to modifications of weevil behavior. Studies reported by Mitchell et al. (1973) compared female weevil activity on frego-bract and normal-bract cottons. On the frego type, weevils showed eight times as much plant-to-plant movement, required twice as much time per feeding and oviposition puncture, and spent much less time in squares. The restless movement and dispersal of weevils over terminals, leaves, and stems may have other advantages in increasing the probability that the weevil will contact insecticide residues on plant surfaces. In this connection, Jenkins (1976) summarized a study showing that buds of frego cotton hold seven times more insecticide deposit than do buds of normal-bract cotton.

An obvious drawback of the frego-bract trait, in certain genetic backgrounds, is the susceptibility of frego types to plant bug damage. This sensitivity can cause delayed flowering and maturity, and result in reduced yield. Numerous reports (Scales and Hacskaylo, 1974; Tingey et al., 1975; Jones, 1972) have shown this effect for *Lygus hesperus, L. lineolaris,* and cotton fleahopper.

Red plant color, determined by the dominant gene R_1, was one of the first morphological characteristics recognized as conferring resistance to the boll weevil. Isely (1928) worked with an intense red genotype and showed a distinct nonpreference by the weevil for red plants when both red and green plants were present. Isely's findings have been substantiated by several other studies. More recent studies have shown that medium and light intensities of red plant color also confer significant degrees of nonpreference (Reddy and Weaver, 1975; Glover et al., 1975). The so-called 'Ak Djura' red, a medium-red type, and 'NC Margin,' a light-red type, were described by Jones (1972) as nonpreferred by the boll weevil. The red trait is cumulative in effect, the degree of nonpreference being highest in the R_1 type and least in the light-red 'NC Margin' stock. Preference for each of the three red phenotypes is significantly less than for normal green types.

High degrees of plant pubescence, controlled by the H_1 and H_2 genes, long have been known to confer significant resistance to the boll weevil (Stephens and Lee, 1961; Hunter et al., 1965). Resistance generally is considered to be mechanical, due to protection provided to the enclosed flower bud by interlocking of bracteoles by the trichomes. Stephens and Lee (1961) also presumed existence of an antibiotic effect, because larvae developed less rapidly in hairy than in nonhairy buds. Although high degrees of pubescence and pilosity confer good levels of resistance to the boll weevil, potential for utilization of the traits appears limited for four principal reasons: (1) plant hairiness intensifies the problem of trash in lint, especially in machine-harvested cotton; (2) increased trash levels may accentuate the problem of byssinosis (brown lung disease); (3) the gene conditioning the pilose trait (H_2) apparently is pleiotropic for short, coarse, and weak fiber; (4) increased populations of *Heliothis* spp. are associated with high pubescence.

Okra leaf (L^o) and super-okra leaf (L^s) are leaf shape mutants characterized by palmately lobed leaves, which have the effect of opening up the plant canopy. Studies conducted in Louisiana and reported by Jones (1972) showed that mature okra leaf plants had 40 percent less foliage than normal leaf counterparts and permitted a 70 percent greater penetration of sunlight (Andries et al., 1969; Major, 1971; Reddy, 1974). Similarly, super-okra leaf plants produced 60 percent less foliage and increased sunlight penetration by 190 percent (Andries et al., 1970; Reddy, 1974). Weevil mortality in weevil-punctured squares under okra leaf canopies was greater than under the canopy of normal leaf cotton, presumably because of a desiccation effect of higher temperature after the soil surface became dry (Reddy, 1974).

Certain types of male sterility show a pronounced nonpreference interaction with the boll weevil. Weaver (1974) first observed a strong nonpreference for cytoplasmic male-sterile cotton (*G. harknessii*); this resistance effect was confirmed by the later reports of Reddy and Weaver (1975) and Glover et al. (1975). In addition to reducing oviposition, the male-sterile types also resulted in decreased adult emergence from oviposition-punctured squares. From Louisiana experiments, Glover et al. (1975) concluded that the boll weevil resistance of the *G. harknessii* was due to the male sterility factor, rather than the cytoplasm per se. McCarty (1974), who worked with lines developed with cytoplasms of *G. anomalum, G. arboreum,* and *G. herbaceum,* noted a reduction in anther numbers, relative to 'Deltapine 16,' and suggested that the significant reduction in boll weevil oviposition probably was associated with reduced anther production. There also is evidence that cytoplasmic male sterility, as well as Ms_4 genetic male sterility, exerts an antibiosis effect on boll weevil development, because both adult emergence and adult weights are greatly reduced (Weaver, 1971; Glover et al., 1975).

Maxwell et al. (1969) described the discovery and utilization of an oviposition suppression factor that effectively reduces egg laying by the boll weevil. This unidentified factor, located in Sea Island 'Seaberry' (*G. bar-*

badense) has been transferred into Upland lines in which 25 to 40 percent reductions in oviposition have been measured, compared with standard Upland commercial varieties. A number of primitive stocks of *G. hirsutum* effectively reduce boll weevil oviposition (Earnheart, 1973; Jenkins, 1976). The latter author reported oviposition suppression up to 84 percent; in several stocks egg deposition was reduced more than 50 percent. Selected stocks were introduced into a backcross breeding program from which three lines were selected which are day-neutral in flowering response and carry the oviposition suppression factor.

The phenomenon of escape or pest evasion as a resistance characteristic should not be overlooked. The use of fast-fruiting, early-maturing varieties has long been advocated as an effective tool for boll weevil management. Currently, the escape potential provided by "short-season" genotypes and the cultural systems used for growing them is the most effective and generally available tool for minimizing crop damage by this pest in certain areas of the United States. The principles of the "short-season" approach were discussed by Walker and Niles (1971), and several studies have demonstrated the advantages of "short-season" production systems where boll weevils are a major hazard (Sprott et al., 1976; Heilman et al., 1977). The data reported by Walker et al. (1977) on boll development and age distribution demonstrate that susceptibility of the developing boll to the boll weevil decreases markedly after about 12 days and that this phenological stage reduces the probability of damage from second-generation weevils. Rapid-fruiting "short-season" cottons offer distinct advantages in addition to their in-season boll weevil evasion potential. Early maturity, generally associated with early and rapid fruiting, allows early harvest and crop residue disposal, thereby reducing food and breeding sites and limiting overwintering potential of adult boll weevils. Reduction of diapausing populations may provide a carryover effect into the following season in delaying the buildup of damaging boll weevil infestations, allowing effective early-season fruiting with little or no use of insecticides.

The relatively high determinacy of "short-season" cottons also may be an effective component in the management of bollworm, tobacco budworm, and pink bollworm. Rapid fruiting and early maturity often provide escape from late-season damage by bollworm and tobacco budworm, and early harvest and residue disposal reduce overwintering potential of the pink bollworm.

5.6. Spider Mites

The most comprehensive research in host-plant resistance to spider mites in cotton has been conducted in Mississippi. Schuster et al. (1972b, 1973) and

Schuster and Maxwell (1976) reported comparative resistance to the two-spotted spider mite (*Tetranychus urticae*) in a series of tests that included Upland cultivars, wild races of *G. hirsutum* and *G. barbadense*, and interspecific hybrids and species of *Gossypium*. Most of the primitive and varietal stocks of *G. barbadense* were resistant, but none of the Upland varieties and stocks were scored as resistant. In addition to *G. barbadense*, the species *G. lobatum* and *G. australe* showed high levels of resistance. Neither the glandless nor high-gossypol characteristics had any appreciable or consistent effect on plant damage by spider mites, although fecundity was significantly greater on the glandless lines (Schuster et al., 1972b).

The mechanism of resistance to spider mites is unknown. Kamel and Elkassaby (1965) suggested that resistance in the variety 'Bahtim 101' is due to its heavy pubescence, but the Mississippi evaluations provide no indication that either hairiness or glabrousness is related to mite damage. Schuster et al. (1972a) demonstrated antibiosis to the two-spotted mite in the 'Pima S-2' variety, reflected in reduced fecundity and lengthened life cycle.

5.7. Thrips, Aphids, Whiteflies, and Cotton Leaf Perforators

Host-plant resistance research with these insect pests has been limited. Various studies have been made to evaluate genotype responses to thrips (*Frankliniella* spp. and *Thrips* spp.), but little is known about mechanisms of resistance. Ballard (1951) speculated that dense pubescence of juvenile leaves is associated with resistance but is not the only factor. Gawaad and Soliman (1972) in Egypt concluded that plant hairiness is not correlated with thrips resistance; from subsequent work, Gawaad et al. (1973) suggested that resistance to thrips is associated with a relatively thick epidermis on the lower leaf surface. Unpublished data from Louisiana, cited by Lukefahr (1977), indicate that high levels of bud gossypol result in greatly increased thrip and banded-wing whitefly (*Trialeurodes abutilonea*) populations.

The association between plant hairiness and resistance to aphids (*Aphis gossypii*) is contradictory. Dunnam and Clark (1939) found that aphid populations increased in direct proportion to the number of hairs on lower leaf surfaces; Pollard and Saunders (1956) substantiated the relationship. Gawaad and Soliman (1972) cited reports from India which indicate that susceptibility to aphids is associated with smooth leaves. Their own studies showed that hairiness and resistance to aphids are not correlated. In a laboratory cage test, Bottger et al. (1964) found gossypol to be highly toxic to aphids; they suggested that removal of pigment glands through breeding would make the plant susceptible to aphids.

In the Sudan, Mound (1965) found that highly pubescent, jassid-resistant cottons bear larger populations of cotton whitefly (*Bemisia tabaci*) than do glabrous types, substantiating similar previous findings by Pollard and Saunders (1956). However, there appear to be no reports that glabrousness would provide useful resistance to the cotton whitefly. Modification of leaf shape seems to be an effective means of increasing tolerance to the banded-wing whitefly, for Jones et al. (1976) reported that okra and super-okra leaf types confer high degrees of resistance.

Wilson and Wilson (1975) investigated the effects of varying plant pubescence on the cotton leaf perforator (*Bucculatrix thurberiella*) in Arizona. In both field and greenhouse tests they found that deviations in both directions from normally hirsute cottons reduced leaf perforator populations; the reduction in numbers was greatest on the most pubescent genotypes. Apparently, degree of hairiness is of greater importance than is source of hairiness.

6. GENETICS OF INSECT RESISTANCE TRAITS

At least a dozen plant characteristics have been evaluated for their effects on insect pest response in cotton. Many of the plant-insect interactions have been corroborated by different investigators, but evidence is sketchy in some instances and completely lacking in others. Table 3 provides a summary of several plant traits and insect responses reported in the literature. The table is incomplete and indicates that most of the plant resistance research reported has involved only a portion of the economically important insect pests of cotton. The mode of inheritance has been established for some, but not all, of the characteristics listed.

Frego bract (*fgfg*) is an Upland mutant inherited as a monofactorial recessive, and results in elongated, twisted bracts which flare away from the square, flower, and boll. Phenotypic expression of the *fgfg* genotype varies appreciably, depending on genetic background. In most instances the heterozygote is recognizable as an intermediate phenotype, but expression varies somewhat in different genetic backgrounds.

The nectariless trait ($ne_1ne_1ne_2ne_2$) is controlled by two pairs of recessive genes transferred to *G. hirsutum* from *G. tomentosum*. The double recessive genotype lacks the extrafloral nectaries which, in the normal type, are present on the lower leaf midrib and basal to the involucral bracts, and the inconspicuous floral nectaries located inside the bracts. The extrafloral

Table 3. Insect Resistance Characteristics in Cotton[a]

Trait	Boll Weevil	Heliothis spp.	Lygus spp.	Cotton Fleahopper	Spider Mites	Pink Bollworm	Empoasca spp.	Thrips	Aphids	Cotton Leaf Perforator	Whitefly
Frego bract	R	N	S	S	N		S	N	N		
Nectarilessness	N	R	R	R		R	R	N	N		
Glabrousness	N	R	(?)	R(?)		R	S	S	(?)		S
Terpenoids (high square gossypol, heliocides)	N	R	R	R			R	S	R(?)		
Heavy pubescence	R	S	(?)	R	N	R	R	(?)	(?)	R(?)	(?)
Red plant color	R	N	N	N					N		R
Okra leaf	N										
Oviposition-suppression factor	R			R							
Plant bug suppression factor			R								
Early-rapid fruiting	E	E				E					

[a] R = resistance; S = susceptible; E = escape; N = neutral; (?) = conflicting evidence or not verified.

nectaries are the most important in insect resistance. Five phenotypes recognizable in segregating populations are described by Holder et al. (1968), as follows:

$Ne_1Ne_1Ne_2Ne_2$	Full leaf nectaries, full bract nectaries
Any 2 dominant alleles	Full leaf nectaries, reduced bract nectaries
$ne_1ne_1Ne_2ne_2$	Full leaf nectaries, no bract nectaries
$Ne_1ne_1ne_2ne_2$	Reduced leaf nectaries, no bract nectaries
$ne_1ne_1ne_2ne_2$	All nectaries absent

Phenotypic expression of these classes may vary somewhat, depending on background genotypes.

The glabrous or smooth-leaf character is governed by at least four alleles distributed at three loci. Three alleles are located at the first locus, sm_1 being the normal. The Sm_1 allele removes trichomes from the stem but has little effect on leaf trichomes. The $Sm_1{}^{sl}$ allele results in highly glabrous leaves and also removes most of the stem trichomes.

The Sm_1 gene was transferred from 'Orinoco,' a wild strain. Sources of $Sm_1{}^{sl}$ are available in cottons descended from 'Meyer's D$_2$' smooth in which the glabrous trait was transferred from *G. armourianum*.

At the sm_2 locus there is one smooth-leaf allele (Sm_2), which, in a homozygous condition, imparts a degree of smoothness similar to that imparted by $Sm_1{}^{sl}$. Sm_2 was introduced from a Nicaraguan feral cotton, and is present in the 'North Carolina Smooth' strains, NCS-1 and NCS-2.

At least three alleles are present at a third locus. The normally pubescent Upland varieties carry $sm_3{}^h$; the recessive sm_3 allele has been widely used in development of commercial smooth-leaf varieties such as 'Deltapine Smooth Leaf' and 'Rex Smoothleaf.' The dominant allele at this locus, Sm_3, gives a phenotype similar to that produced by sm_3. The sm_3 allelomorphs are native to Upland cotton, and Sm_3 originated in the wild or dooryard cottons of Central America. The inheritance of trichome distribution has been reported by Lee (1968), who also provided an excellent discussion of breeding smooth-leaf cottons (Lee, 1971).

Increases of plant hairiness above the degree normally found in the pubescent cultivated varieties are governed by two major genes and a complex of modifier genes. In *G. hirsutum, G. barbadense, G. herbaceum,* and *G. arboreum* this key major gene is designated as H_1. A second major gene, H_2, controls the finely dense pubescence in an Upland mutant designated as 'Pilose,' and the same gene (or its allele) is responsible for the densely pubescent characteristic of *G. tomentosum*. The H_2 gene apparently has a pleiotropic effect on lint length, resulting in fiber length of 13 to 19 mm, too short to be of commercial use. In F_1 populations, both H_1 and H_2 show incomplete dominance.

The nature of biochemical resistance ascribed to "high gossypol" and X factor recently has been clarified and reported in papers by Seaman et al. (1977) and Bell and Stipanovic (1977). Pigment glands of cotton contain a variety of terpenoid chemicals, of which gossypol and heliocide H_1 are the most biologically active constituents. Generally, increasing gland density in the cotton plant results in increasing concentrations of these toxic compounds. Gland density is regulated by genes at six loci designated as gl_1 through gl_6. At each locus, the alleles that increase gland formation are identified Gl; those that reduce gland density are designated gl.

The Gl_1, Gl_2, and Gl_3 alleles are the principal determinants of gland density. The Gl_1 allele affects gland formation only in stems, petioles, and carpel walls, whereas the Gl_2 and Gl_3 alleles affect gland formation in cotyledons and leaves, as well as those organs affected by Gl_1. For the tissues noted, Gl_2 has a stronger effect than Gl_3. In flower tissue, the relative contribution of the two alleles is reversed—Gl_3 is the stronger determinant of terpenoid content. The effects of Gl_2 and Gl_3 in flower parts are largely additive with some epistatic interactions.

A full complement of glands is provided by the joint action of the Gl_1, Gl_2 and Gl_3 alleles, that is, the $Gl_1Gl_1Gl_2Gl_2Gl_3Gl_3$ genotype. Gland density and terpenoid content in Upland cottons can be increased by substitution of Gl alleles from other species or from wild forms of *G. hirsutum*. Bell and Stipanovic (1977) suggested that, in addition to breeding for increased gland density, it also may be possible to alter genetically the relative content of terpenoids in the glands and to select for "delayed gland morphogenesis," thereby alleviating the dilemma of gossypol in the seed.

Okra leaf is a deeply lobed leaf shape that is a monogenic trait governed by the L^o gene which is incompletely dominant to normal l^o. A more extreme leaf shape, termed super-okra, is produced by the L^s allele at the L^o locus. L^s is incompletely dominant; the heterozygote produces a phenotype very similar to okra leaf. Expressions of the L^o and L^s alleles are modified somewhat by genetic background.

Red plant color is available from at least five sources, but only three have been utilized to an appreciable extent in resistance research. R_1 is an incompletely dominant gene that results in red coloration of the aerial plant parts; it was the first qualitative character known to confer resistance to the boll weevil. Lesser intensities of red color are produced by the R_2 allele in the 'Ak Djura' and 'North Carolina Margin' stocks.

Inheritance of the "plant bug suppression factor" has been studied in Mississippi (Jenkins, 1977). In a cross of 'Stoneville 213' with 'Timok 811,' resistance to *Lygus lineolaris* (as measured by yield components) showed highly additive gene action in two early harvests and dominant gene action for total yield. Inheritance of the boll weevil "oviposition suppression factor" noted in the wild races of *G. hirsutum* has not been established.

7. BREEDING TECHNIQUES

The breeding methods used for cotton improvement are based on its mode of reproduction. Cotton is considered a normally self-pollinating species, although natural crossing may range from zero to more than 50 percent, depending on local conditions. A normal cotton flower produces an abundance of pollen, of which little or none is disseminated by wind. Dispersal of pollen among different flowers and plants is effected by insects, primarily wild bees, bumblebees, and honeybees.

Control of pollination in cotton is relatively easy and is facilitated by relatively large flower parts. Self-pollination can be ensured by sealing the flower bud before it opens and anthesis takes place. Selfing can be done with paper or cloth bags, malleable wire, string, cellulose acetate, paper clips, or other devices that will prevent flower opening and insect transfer of pollen. Crossing involves two steps—emasculation and pollination. Emasculation of the female or seed parent is accomplished by removal of the anthers from the flower bud, either the day preceding flower opening or in early morning before opening occurs. Anthers can be removed with tweezers, various mechanical emasculating devices, or with the fingers. Either the anthers themselves can be removed, or the staminal sheath can be peeled from the style of the pistil. After emasculation the pistil may be covered with a short length of soda straw to prevent desiccation and to exclude unwanted pollen. When anthesis occurs in flower buds of the male parent, pollen is transferred to the seed parent, the pistil again is covered for protection, and the peduncle is tagged to identify the crossed boll. With good growing conditions and techniques, a cross will yield 15 to 30 hybrid seeds.

Obviously, artificial enforcement of self-pollination is not always necessary in a breeding program. In areas of intensive insecticide use, populations of pollen vectors may be very limited and the incidence of natural crossing may be negligible. Likewise, when cotton is grown in a greenhouse or similar enclosures, pollen vectors may be so few that enforced self-pollination is not needed.

The breeding methods utilized for cotton are based on those considered appropriate for self-pollinated species, and on various adaptations of methods used with cross-pollinated crops. Selection for pure lines is not common in cotton variety development, and most commercial varieties represent either a mixture of relatively homozygous lines or are developed from single lines having at least a modest level of genetic variability. Mass selection is often used in variety-maintenance programs or to establish improved lines from a heterogeneous base population.

Backcrossing is used mainly to transfer specific genetic traits into a selected recurrent background. This has been done extensively for transfer of disease resistance, insect resistance, glandlessness, and other morphological and physiological characteristics that are inherited in a relatively simple manner and for which heritability is high.

Most breeding efforts in cotton involve hybridization and some adaptation of either the pedigree or bulk methods of selection. Several modifications of these selection methods have been proposed to slow approach to homozygosity, to increase opportunities for genetic recombination, and to allow for natural selection to operate more effectively on genetically variable populations. Recurrent selection techniques have been utilized to a limited extent in cotton, and genetic male sterility and various systems of random intermating have been useful in developing germplasm pools for selection.

Thus far, hybrid varieties have not been used much in cotton, at least not to the same manner as in grain sorghum and maize. Both cytoplasmic male sterile and fertility restorer lines have been developed in cotton, and research is in progress to determine the practicability of hybrid breeding programs for cotton (Weaver, 1977). A major obstacle to the development of hybrid varieties is the problem of pollen transfer from a fertile pollen parent to a sterile seed parent. Since this requires an effective insect pollen vector, it substantially precludes the use of insecticides to protect the crop against insect pests.

Utilization of species hybrids for cotton improvement presents certain unique problems, and breeding techniques vary according to the species involved. *G. barbadense* and *G. tomentosum*, both tetraploid species, cross readily with *G. hirsutum* and produce vigorous F_1 plants. However, the F_2 generations produce a variety of aberrant plants with poor reproductive ability. With continued selection of the more normal plants in succeeding progenies, it is possible to transfer exotic chromosomes or chromosome segments into a stable, essentially all-Upland background.

Development of hybrids of Upland with diploid species is more difficult than with tetraploids, although they can be achieved in several ways. Diploid species that cross directly with Upland produce sterile triploid F_1

hybrids, which may be treated with colchicine to produce fertile hexaploids. Breeder selection within the hexaploids rather quickly will reduce chromosome number to the tetraploid condition. It is hoped that during this process of chromosome elimination appropriate selection will result in the incorporation of a portion of the genetic material from the diploid species into a stable Upland background. Not all diploid species will cross directly with Upland; some will produce hybrids with hexaploids derived from crossing Upland with other diploid species as described above. These trispecies hybrids theoretically should be tetraploids which should cross directly with Upland.

A third method of developing Upland × wild diploid hybrids involves crossing of two diploid species, doubling the chromosome number in the hybrid, and then crossing the tetraploid hybrid with Upland cotton. This results in a trispecies hybrid. Obviously, the interspecific approach in cotton breeding is likely to be quite difficult. As Meyer (1974) has pointed out, early generations of interspecific hybrids frequently do not fit the usual plant breeding assumptions and techniques, and with a paucity of viable seed from a cross, we should not assume a random sample of genes or that there has been no differential selection in the transmission of chromosomal material.

8. SCREENING TECHNIQUES

The evaluation of cottons for resistance to insects and mites generally is based on either of two criteria: (1) quantitative measurements of size, weight, numbers, fecundity, mortality, or other biological characteristics of insect populations; and (2) estimates of damage inflicted on plants by insects.

The following is a brief summary of certain techniques used to evaluate insect resistance in cotton. Numerous modifications may be needed to fit specific situations.

8.1. Boll Weevil

Field evaluations have been used very successfully to assess plant resistance to the boll weevil. A widely used method involves the collection of squares and an examination to determine frequencies of feeding and oviposition punctures. Since the two types of punctures can be differentiated, it is fairly easy to show relative differences in resistance among the cotton lines being tested. Both natural infestations and introduced populations of weevils may be used in field screening trials. Insecticide treatments can be employed effectively to regulate natural infestations and synchronize insect buildup with specific plant phenological stages. Various kinds of cages or screened enclosures also are effective for screening for weevil resistance, in much the same way as in field plots. Enclosure tests offer the advantage of more positive control of boll weevil populations, but the number of plants that can be evaluated is limited. Egg-punctured squares from either field or cage tests may be incubated in the laboratory, and emerging weevils counted, weighed, or measured to obtain information on antibiosis effects.

Workers in Mississippi developed a seedling screening technique for evaluation of resistance to the boll weevil (Maxwell et al., 1969). Boll weevils are confined in cages and allowed to feed on seedling plants for 2 to 3 days, or until most plants are killed or severely damaged. Surviving plants are transplanted to the greenhouse and seed are harvested for progeny testing. This technique may be especially useful, for it permits screening of large plant populations in a short period of time.

Bailey et al. (1967) reported a technique for evaluating cotton lines for antibiosis to the boll weevil. Eggs were implanted into intact flower buds, and infested buds were collected after they were shed. Data were collected on percent emergence, weevil weight, and elapsed time between implantation and adult emergence. Jenkins et al. (1964) used a similar method of evaluating antibiosis in which boll weevil eggs were extracted from live squares and placed in vials containing an artificial diet prepared from lyophilized square powder. Determinations of days to emergence and weight of emerging adult weevils provided good assessment of antibiosis in the cotton strains tested. An advantage of this technique is that the amount of food and morphology of the square are not limiting factors.

Fresh squares also are used frequently to measure antibiosis. These can be collected in the field and examined for oviposition punctures, after which they are held in a suitable environment until adult emergence.

8.2. *Heliothis* spp.

Various procedures are used for evaluating cotton strains for resistance to *Heliothis* spp. These generally involve counts of larvae, eggs, or damaged fruit forms. Comparisons of treated and untreated plots of different genotypes frequently are utilized to provide a basis for estimating crop damage. As with most insect pests, natural populations are unpredictable with respect to either time or density, and resistance is often difficult to evaluate under field conditions.

Cage and laboratory tests, which provide a more consistent screening method with *Heliothis* spp., have been utilized extensively in the USDA bollworm-budworm research program at Brownsville, Texas (Lukefahr et al., 1975). In large cage tests, replicated plantings of cotton are made and insecticides are used to remove all insects until plants have reached the squaring state, when specified numbers of adult moths are released in the cages. Fixed sampling points are used for monitoring, and counts are made of fruit forms, eggs, larvae, and damaged forms throughout two generations. With use of a plastic netting, insects can be confined on individual plots to create a no-choice situation.

Small screen bags or cages may be used to infest individual squares with early instar larvae. Larvae are transferred every second day to provide fresh squares, and after confinement to four different squares, the larvae are removed and weighed to determine growth rate (Lukefahr and Houghtaling, 1969).

Satisfactory bioassay methods have been devised to evaluate aspects of antibiosis to *Heliothis* spp. Oliver et al. (1967) used lyophilized square powder incorporated in an artificial diet and dispensed into 2-ounce plastic containers. Late first- or early second-instar larvae are weighed individually and then placed in the diet cups. After 6 days the larvae are reweighed and returned to the cups until pupation. Determinations are made of larval survival, larval growth, and percent pupation. The test can be extended to obtain data on fecundity and longevity of emerging adults. A modification of this technique was used by Shaver and Lukefahr (1971) to screen accessions of the *G. hirsutum* race collection for resistance to *Heliothis* spp.

8.3. Pink Bollworm

Where natural populations of pink bollworm exist, screening and evaluation for resistance can be done adequately in the field. Green bolls can be collected from field plots and held in emergence containers until adult emergence. Boll collections should be made periodically over several weeks, and held in emergence containers 2 to 3 weeks. Laboratory dissection and examination may be used to determine degree of damage to boll contents and insect mortality. Workers in Arizona have followed this general scheme, and also use X-ray methods to determine seed damage (Wilson and Wilson, 1975a, 1975b).

Infesting green bolls with pink bollworm eggs and larvae has been effective in screening cottons for resistance. Brazzel and Martin (1956) described various greenhouse and cage experiments in which eggs were placed onto bolls 10 to 20 days old. Following a 2 to 3 day oviposition period, the bolls were removed, dissected, and examined to determine the number of entry holes and the number and condition of larvae within the bolls. Refinements were later made to provide for the use of newly hatched larvae (Reed and Adkisson, 1961), which were confined to the bolls with tanglefoot applied to the peduncle.

Bioassay techniques can also be used effectively with the pink bollworm to screen cottons for resistance to this insect. Wilson and Wilson (1974) described the use of artificial diets containing lyophilized carpel walls or boll contents. Freshly hatched larvae are placed in diet cups and held until pupation, at which time they are weighed. Pupae are returned to the cups and allowed to develop to adults. Cotton genotypes can be evaluated for effect on larval development time and larval weight, percent pupation, pupal weight, pupation time, and percent adult emergence.

8.4. *Lygus* spp.

Field screening of cottons for *Lygus* spp. reactions is complicated by mobility of the pest, its variable dispersion, and its preference for other crop species. However, field evaluations may provide useful information if tests are properly designed and sufficient observations are made to minimize the probabilities of plant escape. Cotton genotypes to be tested may be planted in a split-plot arrangement to provide for comparison of the entries with and without insecticide treatment. Comparative data on flowering, fruiting, and plant development can be used to characterize plant resistance.

In field studies conducted in Mississippi, mustard (*Brassica juncea*) is interplanted in a cotton nursery to develop populations of *L. lineolaris* (Laster and Meredith, 1974a). The mustard is cut, forcing plant bugs to move to cotton plants in adjacent rows. By including an insecticide treatment to one set of plots grown without interplanted mustard, the effects of lygus bugs on lint yield can be determined. Cotton plants also are evaluated for terminal damage as an indicator of resistance.

Numerous variations of cage tests are utilized in screening for resistance to *Lygus* spp. Bugs can be introduced into yield cages where the caged plants are monitored for fruiting response and growth characteristics. Uninfested cages of the lines being evaluated should be included in the experiment. Tingey et al. (1973) compared three methods of screening for ovipositional preference, including individual caging of whole plants, caging of excised leaves, and bagging terminals of field-growing plants. The bagged terminal technique was judged to be best for mass screening. It is relatively inexpensive and comparatively easy, since females are exposed to both foliar and fruiting stimuli and nymphs are easily recoverable.

8.5. Cotton Fleahopper

Screening for resistance to cotton fleahopper is complicated because cotton is not a preferred host plant. Natural populations are highly unpredictable and rearing the insect is rather difficult. For most effective screening, plants should be infested in the early squaring period when they are about 40 to 60 days old. Natural infestation of cotton at this time usually requires that the wild host plants (primarily croton, *Croton* spp., and horsemint, *Monarda* spp.) have matured sufficiently to be relatively unsuitable for fleahopper habitation. Preference characteristics can be determined by counting adult and nymphal fleahoppers in the plant terminals; these counts are made periodically for a specified time period and may be summed for a total seasonal count (Lukefahr et al., 1968). Relative determinations of tolerance can be made by taking blooming counts and yields in sets of both treated and untreated plots of the cotton genotypes being tested.

Cages can be used in much the same way as field planting except that they provide more control of infestations. Fleahoppers can be collected from wild hosts and used for infestation of cage tests.

8.6. Spider Mites

Screening for resistance under field conditions is difficult because of the irregular dispersion of mites in field cotton. The appearance of mite populations often is associated with hot, dry weather or with the use of certain insecticides used for treatment of other pest species. Field plantings may be evaluated by some form of damage index reflecting the degree of leaf injury, necrosis, or defoliation. Assessment of relative resistance also may be made on the density and types of infestation colonization (Leigh and Hyer, 1963).

As with most insect pests of cotton, screening for resistance in cages or the greenhouse is more reliable than field testing. A method of seedling screening suitable for mass evaluation of cotton germplasm (Schuster et al., 1973) involves mite cultures maintained on bush beans under conditions of controlled light, temperature, and humidity. Seeds of the cotton lines to be evaluated are planted in the greenhouse, and 5-day old seedlings are infested by placing bean leaves on top of the small plants. Mites respond to sunlight and desiccation of the bean leaves by moving onto the cotton seedlings.

Cotton plants are rated 24 to 45 days later using a damage scale of 1 to 5. An injury index for each entry is calculated by multiplying the damage score by number of plants in each category and dividing by total number of plants for that entry.

8.7. Jassids

Both field and cage tests are used to screen cotton plants for resistance to jassids. Where natural jassid populations are appreciable, counts of nymphs may be made, either in one-time or periodic samplings. Counts are made on a specified number of leaves of randomly selected plants in each plot or entry (Parnell et al., 1949). Jassid nymphs also may be used to infest plants confined in individual cages (Batra and Gupta, 1970). The caged insects are permitted to feed, develop into adults, and reproduce for a chosen period of time. At the end of the test period, the individually potted and caged plants are fumigated and the dead insects are counted.

8.8. Thrips and Aphids

Evaluation for thrips resistance usually involves use of an injury index based on plant damage inflicted by the insect. These injury or damage indexes can be done with relative ease under field conditions, providing infestations are adequate to produce visible plant response (Hawkins et al., 1966).

Aphid resistance evaluations also can be made in the field if natural populations are uniform and sufficiently great. Greenhouse or cage screenings provide more accurate evaluation of plant resistance. Cultured aphids can be released onto cotton plants and plant ratings made 10 to 14 days after infestation, when the most susceptible entry shows severe injury. Entries are rated visually on degree of injury sustained by seedling cotton plants. Adult female aphids also may be confined to cotton leaves in suitable mini-cages or cloth bags, and the number of progeny determined at selected times after infestation (Kamel and Elkassaby, 1965).

8.9. Cotton Leaf Perforator

Techniques for field and greenhouse screening for plant resistance to cotton leaf perforator (CLP) have been described by Wilson and Wilson (1975). A selected number of leaves are collected from field test plots, weighed, and examined for "horseshoe" (resting fourth-instar) larvae. Data are recorded as number of CLP per gram of leaf tissue.

For the greenhouse tests, leaf disks including fourth-instar "horseshoe" larvae are removed, with a cork borer, from leaves of culture plants. The leaf disks are placed in leaf cages and secured to undamaged leaves of greenhouse test plants. The emerged fifth instar is allowed to feed until pupation. After pupae are removed, the leaf feeding area is removed, dried, and pressed. The dry leaf areas are then examined microscopically, using a dot grid to estimate the area of leaf tissue consumed by the CLP larvae. A micrometer is used to measure leaf thickness and permit calculation of volume of tissue consumed. Infestation with fifth-instar larvae is repeated several times over a 2-month period.

8.10. Whitefly

Reports of evaluations of cotton plant resistance to the whitefly are scarce, and apparently no critical techniques of screening have been developed. There are two methods for screening with natural field populations (Butler and Muramato, 1967). These involve physical counts of whitefly larvae and pupae on selected leaves of random plants, and use of vacuum sampling devices to collect adult whiteflies from field plots. Both the physical counts and vacuum collections are made at several dates to ensure an adequate determination of infestation levels.

Placing pollen on the female influorescence must be done with care to prevent contamination by other pollen floating in the air. Photo courtesy of CIMMYT.

16

BREEDING FOR INSECT RESISTANCE IN MAIZE

Alejandro Ortega
Surinder K. Vasal
John Mihm
Clair Hershey

International Maize and Wheat Improvement Center CIMMYT,
Mexico D.F., Mexico

1. THE CROP

Zea mays, or maize from the Arawak-Carib word "mahiz" (Hatheway, 1957), is the world's third leading cereal after wheat and rice (Table 1). Maize origin and evolution are still matters of controversy and speculation. Its cultivation probably started in Mexico and/or South America about 7,000 years ago (Mangelsdorf, 1974).

Maize is the most widely distributed cereal and generates 8.9 percent of the total world food production (Table 1). Its genetic diversity and plasticity allow it to grow in very diverse environments. It is cultivated in irrigated

Table 1. World Food Production 1974

Food	Production (Million Metric Tons)	Percent Total Food Production
From the land[a]		
All cereals	1,333.9	40.70
Wheat	360.2	10.90
Rice	323.2	9.90
Maize	293.0	8.90
Barley	170.9	5.30
Other cereals	186.6	5.70
Root crops (cassava, yams, etc.)	266.2	8.20
Potatoes	293.7	8.90
Pulses (beans, peas, chickpeas, etc.)	44.1	1.30
Vegetables, melons, fruits, and nuts	331.5	10.09
Vegetable oils, sugar, coffee, cocoa, tea	237.4	7.20
Meat, milk, and milk products, eggs, honey	683.5	20.94
Total from land	3,190.3	97.37
From the ocean and inland waters[b]	86.1	2.62
Total from land and water	3,276.4	99.99

[a] *FAO 1974 Production Year Book.*
[b] *FAO 1974 Year Book Fishing Statistics.*

Table 2.   World Changes in Maize Production 1961–1965 to 1971–1975[a]

Production	1961–1965	1971–1975	Percent Increase
Average annual area harvested (million hectares)	99.6	112.0	12
Average annual yield (kg/hectare)	2170.0	2749.0	27
Average annual production (million metric tons)	216.0	308.0	42

[a] Source: *FAO Production Year Books.*

deserts under very high temperatures; in the subtropical and temperate regions during the summer, usually under favorable rainfall conditions; in the tropical highlands up to 3300 m above sea level with variable rainfall; and in the lowland wet tropics, year round. Such diversity has been stratified into hundreds or thousands of narrowly adapted land varieties serving a multitude of specific uses for human food, animal feed, and industry (Inglett, 1970).

Maize is grown on about 112 million hectares in 134 countries. Collectively these countries produce 300 million tons of grain with an average yield of 2749 kg/hectare (Table 2). Across the world, yields average from about 5.9 metric tons/hectare in the highly developed countries in the temperate regions to 0.5 tons/hectare in many developing countries in the tropical belt. Only 11 percent have yields above 4.0 tons/hectare (Table 3).

Maize varieties may be classified (1) according to *maturity range* as early, intermediate, and late; (2) by *seed texture and appearance* as flint, dent, floury, and wrinkled; (3) by *color* into white, yellow, and others; and (4) by *climatic adaptation* as lowland tropical, subtropical, temperate, and tropical highland (Table 4).

Table 3.   Average Maize Production per Hectare[a]

Average Production (Tons/Hectare)	Number of Countries	Percent of Total Countries	Cultivated Area (Million Hectares)	Percent of Total Area
5.0 to 5.9	12	8.9	28.8	25.7
4.0 to 4.9	3	2.2	3.5	3.2
3.0 to 3.9	8	5.9	15.3	13.7
2.0 to 2.9	16	11.9	16.0	14.2
1.0 to 1.9	57	42.5	34.6	30.8
0.3 to 0.9	38	28.3	13.0	12.3
Total	134	99.7	112.0	99.9

[a] Source: *FAO Production Year Book 1975.*

In the African countries where maize is a staple human food, there is a preference for white dent types of tropical and subtropical adaptation, of early, medium, and late maturities to fit different cropping systems. In Asia, yellow and white flints and dents of tropical, subtropical, and temperate adaptation are used. In Europe the area devoted to maize both for silage and grain for animal feed has increased considerably in recent years. Temperate, cold resistant yellow flints of early to medium maturity are grown. In temperate regions of the Americas, early, medium, and late orange and yellow dents and flints are used. The greatest genetic diversity occurs in the tropical regions of the Americas (Mexico and Guatemala in the north and the Andean region in the south), where a multitude of color, texture, and maturity types are grown, mainly for human consumption. Throughout the world, there is a great demand for improved early maturing maize varieties.

Table 4.   Agroclimatic Characteristics Considered in Classifying Maize Germplasm

Maturity Class	Altitude (Meters above Sea Level)	Latitude N–S (Degrees)	Mean Temperatures of Growing Season (°C)			Days to Physiological Maturity
			Min.	Max.	Av.	
Tropical lowland						
Early	Below 1000	Within 23	22	32	28	±80
Medium	Below 1000	Within 23	22	32	28	±100
Late	Below 1000	Within 23	22	32	28	±120
Tropical highland						
Early	Above 1800	Within 23	7	22	16	±150
Medium	Above 1800	Within 23	7	22	16	±180
Late	Above 1800	Within 23	7	22	16	±220[a]
Subtropical						
Early	Below 1800	Within 34	17	32	25	±100
Medium	Below 1800	Within 34	17	32	25	±130
Late	Below 1800	Within 34	17	32	25	±160
Temperate						
Early	Below 500	Outside 34	14	24	20	±110
Medium	Below 500	Outside 34	14	24	20	±130
Late	Below 500	Outside 34	14	24	20	±160

[a] South American Andean cultivars may take up to 13 months.

The transfer of new maize technology to farms has met with limited success in the developing world. On-farm testing programs are needed, and the value of the so-called "improved technology" must be demonstrated to the farmer in his own fields. During the last three decades maize production in a few temperate countries has doubled. In the United States, where 50 percent of the world's maize grain is produced (most of it for animal consumption), such success can be attributed to mechanization, the widespread application of chemical fertilizers, extensive use of improved maize hybrids carrying genetic resistance to some key insect pests and appropriate use of effective pesticides (anonymous, 1972). In contrast, production in most developing regions has barely kept pace with demand. Most of the gains made have been obtained through an increase in area planted, rather than an increase in productivity.

In developing countries hybrids are grown on about 15 percent of the area devoted to maize production. Hybrids produced by crossing of inbred lines are productive but their effective distribution to farmers demands a degree of sophistication lacking in many of the developing nations. Recent international testing of open pollinated experimental varieties (anonymous, 1976) suggests that some of them compete favorably with the best available commercial hybrids. The "population improvement" approach holds promise for many less technologically advanced countries. Genetic improvement—in this particular case, plant resistance to insects—will reach the farmer only via the seed, and efficient mechanisms of improved seed distribution must be developed and implemented.

The slow progress in improving maize production in developing regions is due primarily to (1) the gap between research and extension activities, (2) socio-economic constraints, coupled with inadequate production policies, (3) difficulties in establishing successful improved variety-seed distribution systems, and (4) difficulties in effective control of pests with pesticides at the farm level.

Maize genetic improvement in the temperate growing areas has resulted in the development of many inbred lines. A few of these have proved to be outstanding and are widely used. These are characterized by: (1) good standability; (2) reduced plant height; (3) uniform ear height and maturity for mechanical harvest: (4) a high grain to stover ratio; (5) plant types suitable to high densities; and (6) genetic disease resistance. With the exception of European corn borer and corn earworm, only within the last decade has insect resistance received significant attention in some developed countries. In most of the developing world, work on insect resistance has not yet reached the conceptual stage. Breeding efforts in the tropics have emphasized yield, disease resistance, and more recently reduction in plant height and days to maturity.

2. THE PEST COMPLEX

Biology, habits, distribution, and control measures of insects and pathogens injurious to maize have been described in several publications (Dickson, 1956; Metcalf et al., 1962; Shurtleff et al., 1973; Sprague, 1977; Ullstrup, 1974).

In most maize growing environments only a few pests threaten production. Table 5 lists the most widespread maize insect pests and diseases, with an indication of their relative prevalence and importance in the temperate, subtropical, and tropical environments.

The seedling stage of the maize plant is attacked by seed maggots, rootworm larvae or adults, flea beetles, thrips, and cutworms. Leafhoppers, armyworms, and borers in successive broods may damage the leaves from the early whorl stage (four leaves) to the late whorl stage (pretassel period). Borers also damage the growing point ("dead heart"), leaf collars, sheaths and midribs, and stems. Rootworms attack the root system. Earworms, borers, armyworms, rootworm adults, and ear maggots can damage the developing ear, stems, and tassels. Finally, field infestation by stored-grain insects during the grain maturation period may be carried to storage, reducing the quality and usable quantity of grain.

The most important maize insect pests and the areas where they occur are *Ostrinia nubilalis* in the temperate region of the Northern Hemisphere; *O. furnacalis* in eastern Southeast Asia; *Chilo partellus* in northeast Africa and western Southeast Asia; *Busseola fusca, Sesamia calamistis, Eldana saccharina, Spodoptera exempta,* and *Ciccadulina* spp. throughout Africa south of the Sahara; *Diatraea* spp., *Spodoptera frugiperda,* and *Dalbulus* spp. in southern United States and Latin America; *Heliothis* spp., and the *Diabrotica* complex in the Americas; and *Rhopalosiphum maidis, Sitophilus* spp., and *Sitotroga cerealella* throughout the world (Ortega and DeLeon, 1974). In order of prevalence and importance, lepidopterous, coleopterous, and homopterous pests govern priorities in terms of insect resistance in maize.

Losses from insect feeding are compounded by many common insect–pathogen associations: aphids with viruses; leafhoppers with viruses and spiroplasmas; corn rootworms with *Fusarium* root rots; borers with corn smut, *Diplodia* and *Gibberella* stalk rots; earworms with *Fusarium* ear rot; and flea beetles with bacterial leaf blight (Dicke, 1975). In some instances the picture is further complicated by the development of insect and pathogen biotypes, particularly with aphids and viruses.

Crop losses from the pest complex vary considerably. On a world basis, annual maize grain losses from insects alone have been estimated at about

12 percent (Crammer, 1967) (Table 6). Crop loss assessment and methods for their estimation are being compiled by FAO (Chiarappa, 1971).

The following sections are concerned primarily with insect pests. However, any attempt to improve the genetic resistance of maize varieties to insects and pathogens must be an integral part of the overall crop improvement process. All too often, isolated efforts by breeders, entomologists, or pathologists are unproductive. Scientists trained in different disciplines increasingly are working together in interdisciplinary programs of crop improvement at national, regional, and international levels, providing viable production alternatives from which farmers may choose. An improved variety has no impact until it is distributed and utilized by the farmer.

3.   MANAGEMENT OF INSECT PESTS

3.1.   Artificial Infestation and Mass Rearing

Progress in identifying or building levels of genetic resistance to insect pests depends on the ability to distinguish, in each cycle of selection the most resistant genotypes. Uniform infestation levels at appropriate stages of plant development are required for selecting resistant genotypes, reducing or eliminating "escapes," and accumulating genetic resistance.

Natural populations of maize pests seldom occur at a time (stage of plant development), intensity, or uniformity adequate for the improvement of genetic resistance. The limitations and advantages of natural and artificial infestations, indicated in Table 7, should be understood by workers in crop improvement programs.

Success in developing resistant materials have emphasized that without the capability of artificial rearing for mass production, and artificial infestation of donor parents and segregates, (Guthrie, 1974; Ortman, et al., 1974). For many of the lepidopterous pests there is much new information relating to mass production with artificial diets. For host plant resistance work it matters little whether the diet is holidic (chemically pure), meridic (one or more chemically undefined materials, such as yeast or wheatgerm), or xenic (host plant materials plus supplemental nutrients); what is important is that the insects produced retain their ability to attack and damage plants being evaluated so that differences in resistance become apparent. This can be assured only by a combination of a suitable diet and a periodic renewal of the laboratory insect colony with wild stock.

Table 5. The Relative Prevalence and Importance of the Most Damaging Insect and Disease Pests in the World[a]

Insects and Diseases	Temperate	Subtropical	Highland	Tropical	
				Winter	Summer
Root pests and seedling blights					
Rootworms (*Diabrotica* spp.)	+++3	++2	+1	++2	+++2
Seedling blights (*Pythium, Fusarium* etc.)	+1	++2	+1	+1	+1
Leaf feeders and leaf diseases					
Armyworms (*Spodoptera* spp.)	+1	+++3	+1	+++2	+++3
Northern leaf blight (*Helminthosporium turcicum*)	+++3	+1	+++3	++2	+1
Southern leaf blight (*H. maydis*)	+++3	+++3	+1	++2	+++3
Curvularia leaf spot (*Curvularia lunata*)	0	+1	+1	+1	+++3
Common rust (*Puccinia sorghi*)	+++2	++2	+++2	++1	+1
Southern rust (*P. polysora*)	0	++2	+1	+1	+++3
Downy mildews					
Sorghum downy mildew (*Sclerospora sorghi*)	0	++2	0	+++3	+++3
Philippine downy mildew (*S. philippinesis*)	0	++2	0	+++3	+++3
Sugarcane downy mildew (*S. sacchari*)	0	++2	0	+++3	+++3
Viruses, spiroplasmas vectors					
Sugarcane mosaic–maize dwarf mosaic virus (*Rhopalosiphum maidis*)	+++3	+++3	+1	++2	++2
Maize chlorotic dwarf virus (*Graminella nigrifons*)	+++3	0	0	0	0
Corn stunt (*Dalbulus* spp.)	+1	+++3	+1	+++3	+++3
Maize streak (*Ciccadulina* spp.)	0	++2	+1	++2	+++3

Stalk borers and stalk rots

European corn borer (*Ostrinia nubilalis*)	+++3	+1		0	0
Asian maize borers (*Chilo partellus*); (*Ostrinia* spp.)	0	++2	0	++2	+++3
Pink maize borer (*Sesamia* spp.)	0	++2	0	++2	+++3
African maize borer (*Busseola fusca*)	0	++2	0	++2	+++3
Sugarcane borer (*Diatraea saccharalis*)	0	++2	0	++2	+++3
Neotropical corn borer (*Diatraea lineolata*)	0	0	0	++1	+++3
Diplodia maydis	+++3	+++3	+1	0	0
Gibberella zea	+++3	++2	+1	0	0
Fusarium (*Gibberella fujikuroi*)	+++2	+++2	++1	++1	+++2
Charcoal rot (*Macrophomina phaseolina*)	++1	++1	+++2	+++2	+++2
Black bundle	++1	++2	+1	++1	++2
Late wilt (*Cephalosporium maydis*)	0	+++3	0	+++3	+++3
Pythium aphanidermatum	++1	+++3	0	++1	+++3
Bacterial (*Erwinia chrysanthemi*)	+1	+++3	+1	++1	+++3
Anthracnose	0	+1	0	++1	+++3

Earworms and ear rots

Corn earworm (*Heliothis zea*)	++2	+++2	+++3	++1	+++2
Diplodia maydis	++2	+++2	+1	0	0
Kernel rot (*Gibberella fujikuroi*)	+++2	++1	+1	0	0
Botryodiplodia theobromae	0	+1	0	0	+++2
Fusarium graminearum	+++2	+++2	+++2	++2	++2

Stored grain insects

Angoumois grain moth (*Sitotroga cerealella*)	++1	+++2	+2	+++3	+++3
Maize weevil (*Sitophilus* spp.)	++1	+++2	+2	+++3	+++3

[a] Sources: *CIMMYT, World Wide Maize Improvement in the 70's and the Role for CIMMYT*, 1974; B. L. Renfro and A. J. Ullstrup, *PANS* **22**(4): 491–498 (1976).

PREVALENCE: +++ = abundantly present; ++ = commonly present; + = occasionally present.

IMPORTANCE: 3 = major importance; 2 = moderate importance; 1 = minor importance; 0 = absent or rare.

Table 6. Estimated Annual World Losses in Maize Based on Monetary Value, 1967 (Cramer, 1967)

Region	Production Actual	Potential	Losses Insects	Diseases	Weeds	Total
North America and Central America						
Million tons	102.9	144.7	16.9	13.6	11.3	41.8
Million $	4733,4	6657,4	778,9	625,8	519,3	1924,0
Percent			12	9	8	29
South America						
Million tons	19.1	31.8	6.3	3.2	3.2	12.7
Million $	800,4	1334.0	266,8	133,4	133,4	533,6
Percent			20	10	10	40
Europe						
Million tons	28.3	32.9	1.7	1.0	1.9	4.6
Million $	2096,0	2437,2	121,9	73,1	146,2	341,2
Percent			5	3	6	14
Africa						
Million tons	14.9	47.5	9.5	6.5	16.6	32.6
Million $	746,0	2375,8	475,1	323,1	831,6	1629,8
Percent			20	14	35	69
Asia						
Million tons	16.5	26.2	2.6	3.1	3.9	9.6
Million $	1320,8	2096,5	209,6	251,6	314,5	775,7
Percent			10	12	15	37
Oceania						
Million tons	0.2	0.2	0.01	0.01	0.02	0.04
Million $	9,2	11,2	0,6	0,6	0,7	1,9
Percent			5	5	7	17
USSR and China						
Million tons	36.5	56.1	7.0	5.2	7.3	19.5
Million $	1681,3	2578,8	319,8	242,3	335,3	897,4
Percent			12	9	13	34
World totals						
Million tons	218.4	339.4	44.0	32.6	44.2	120.8
Million $	11387,1	17490,9	2172,7	1649,9	2281,0	6103,6
Percent			12	9	13	34

The state of diet refinement varies among species, and usually depends on a given species' importance as a pest, the length of time it has been investigated, and the effort expended in attempts to rear it. There are several comprehensive compilations and reviews of various diets, their composition, preparation, and utilization (Chippendale, 1972; Dadd, 1973; House, 1976; Smith, C. N., 1966; and Vanderzant, 1974). As examples, the ingredients of several diets successfully utilized in producing millions of insect larvae are given in Table 8.

Walker et al. (1966) have suggested the following criteria for a suitable diet: (1) high larval survival; (2) vigorous adults with high reproductive capacity; (3) normal rate of larval development; (4) low-cost ingredients; (5) easy preparation from readily available ingredients of uniform quality; and (6) good keeping quality.

Several diets have been found to be adaptable to rearing different species (George et al., 1960; Shorey and Hale, 1965). A given diet may have to be modified over time to increase production, including changes in proportion of ingredients, adding or deleting microbial inhibitors, adjusting consistency, and/or including feeding stimulants (crude plant fractions or defined chemicals). To reduce costs, machines may be devised to prepare diets and dispense the food into rearing containers (Burton et al., 1966; Burton and Harrell, 1966; Sparks and Harrell, 1976). As costs of diet ingredients increase, alternatives of equal nutritional or gelling quality must be found (Brewer and Martin, 1976; Burton and Perkins, 1972; and Raulston and Shaver, 1970). Though producing insect material at minimum cost is important, when the total costs of a successful insect resistance program are considered, diet ingredient costs are a small component. Many factors affect the efficiency of mass production and testing (Table 9), which is best evaluated as the cost per plant infested.

3.2. Infestation Techniques

Early work on the resistance of maize to several of its insect pests depended upon natural infestations. Painter (1951) has given a comprehensive review of these studies. Artificial infestations were initiated with European corn borer (ECB) egg masses (Patch and Pierce, 1933), and corn earworm (CEW) larvae (Blanchard et al., 1942) produced from insect stock. Results of both studies were encouraging. Later workers adopted similar systems giving controlled, repeatable infestations that enhanced success in breeding for resistance.

Prior to the "diet era" the insect material was produced in the laboratory from field-collected insects; in time, improved collection techniques and

Table 7. A Comparison of Infestation Techniques for Improving Maize Insect Resistance

Characteristic	Natural Infestations	Artificial Infestation	
		Traditional Methods	Improved Methods
Inputs necessary			
Manpower	Low	High	Intermediate
Funds	Low to intermediate	High	Intermediate
Materials, equipment	Low or none	High	Intermediate
Land requirements	High (many reps.)	Intermediate (fewer reps.)	Intermediate (fewer reps.)
Egg mass handling	None	High	Low
Level of infestation/plant	Variable; uncontrollable	Controlled	Better controlled
Number of plants infested	Uncontrollable	Intermediate	High
Uniformity of infestation	Variable	Controlled; as uniform as infestation technique	
Speed of infestation	Variable, unpredictable	Laborious, slow	Rapid
Timing of infestation	Variable, unpredictable	Controlled; can infest all plants at same stage of maturity	
Number of escapes	Variable, high	Low	Very low
Probability of success for:			
Identifying tenative resistance	Low	Intermediate to high	Intermediate to high
Improving levels of resistance	Low	Intermediate to high	Intermediate to high

Table 8. Approximate Quantities of Ingredients to Prepare about One Kilogram of Artificial Diet for European Corn Borer (ECB); Corn Earworm (CEW); Fall Armyworm (FAW); Sugarcane Borer (SCB); and Southwestern Corn Borer (SWCB)[a]

Ingredient	ECB (Guthrie, 1974)	CEW (Burton, 1969)	FAW–CEW (CIMMYT, 1977)	SCB, SWCB (CIMMYT, 1977)	SWCB (Davis, 1976)
1. Water	850.0 ml	660.0 ml	735.0 ml	825.0 ml	860.0 ml
2. Agar	18.0 g	13.0 g	16.0 g	16.0 g	17.0 g
3. Wheat germ	34.0 g	53.0 g	4.0 g	—	—
4. Dextrose	29.0 g	—	—	—	—
5. Casein	29.0 g	—	—	—	—
6. Cholesterol	2.0 g	—	—	—	—
7. Salt mixture # 2	9.0 g	—	7.0 g	—	—
8. Vitamin supplement	6.0 g	—	15.0 g	15.0 ml	5.0 g
9. Ascorbic acid	8.0 g	3.0 g	4.0 g	2.0 g	2.0 g
10. Aureomycin	3.0 g	—	5.0 g	5.0 g	—
11. Fumidil B	0.5 g	—	—	—	—
12. Methyl hydroxybenzoate	5.0 g	2.0 g	2.5 g	1.5 g	1.5 g
13. Propionic acid	5.5 g	—	—	2.8 g	—
14. Formaldehyde	40% 0.5 g	10% 8.0 g	10% 2.5 ml	2.5 ml	—
15. Sorbic acid	2.5 g	1.0 g	1.25 g	0.5 g	0.5 g
16. Pinto beans	—	227.0 g	—	—	—
17. Yeast	—	34.0 g	40.0 g	—	—
18. Soybean flour	—	—	50.0 g	—	—
19. Opaque-2 maize flour	—	—	96.0 g	—	—
20. Ground maize tassel[b]	—	—	20.0 g	20.0 g	—
21. Choline chloride	—	—	2.0 g	—	—
22. Streptomycin	—	—	0.1 g	0.1 g	—
23. Vanderzant–Adkisson Wheat Germ Diet®	—	—	—	85.0 g	88.0 g
24. Corn cob grits (60 mesh)	—	—	—	25.0 g	26.0 g

[a] For detailed instructions in procedures for preparing the diets consult the original references.
[b] Green (before pollen shedding), dried, ground, and autoclaved.

Table 9. Factors that Influence the Efficiency of Insect Mass Production in a Host-Plant Resistance Program

Mass Production Components	Insect Factors
1. Permanent facilities and equipment.	1. Number of species being reared.
a. Planning and construction for efficiency of laboratory operation.	2. For each species reared, general requirements for life (physical factors—optimum and range of temperature, humidity, light).
b. Equipment complexity, durability, sensitivity, and maintenance.	3. Biological characteristics.
2. Laboratory personnel, characteristics.	a. Larval characteristics.
a. Training.	Nutrition.
b. Experience.	General or specific feeding habits.
c. Inventiveness.	Cannibalism.
d. Dislike of unnecessary labor.	Susceptibility to diseases.
3. Recurring labor-consuming operations.	b. Pupal characteristics.
a. Diet preparation.	Pupation site.
b. Diet infestation.	Pupation within cocoon, or not.
c. Pupal removal.	Fragility or durability.
d. Adult handling.	c. Adult characteristics.
e. Egg handling.	Mating habits.
f. "Dishwashing" and general sanitation procedures.	Feeding or nonfeeding.
g. Collecting initial and replacement insect materials for colonies.	Oviposition habits.
4. Recurring consumptive supplies.	Longevity.
a. Diet ingredients.	Activity.
b. Oviposition substrate.	Fecundity.
c. Rearing containers.	d. Egg characteristics.
d. Pupation rings.	Sensitivity to temperature.
e. Utensils, glassware, etc.	Sensitivity to humidity.
f. Sanitation supplies.	Susceptibility to microbes, pathogens.
5. State of knowledge on rearing techniques.	
a. Initial experimentation on rearing procedures.	
b. Development of more efficient techniques.	

laboratory production evolved (Guthrie et al., 1960, 1965). Even though diets were used for several lepidopterous species, infestation methods changed little until recently. Studies for CEW and ECB have defined the best (1) plant stage to infest, (2) number of eggs or larvae per application, (3) number of applications, (4) time of day to infest, and (5) egg versus larval infestation and sources of insect material (Josephson et al., 1966; Showers and Reed, 1972; Wiseman et al., 1974a).

Infestation is done (1) by manual distribution of egg masses, either dropped into the whorl of the plant or affixed (usually pinned) to the plant; or (2) by manually placing eggs or larvae on various plant parts, usually with a camel-hair brush. Most of the work in other countries with other lepidoptera have utilized or adapted the techniques developed in the United States (Anglade, 1961; Chatterji et al., 1968; Dolinka et al., 1973; Rangdang, 1971).

For one lepidopterous maize pest an improved technique is to infest field corn with an agar suspension of CEW eggs (Widstrom and Burton, 1970; Wiseman et al., 1976). This reduces the time and labor of infestation, as well as the number of escapes, compared to earlier techniques.

An agar solution technique has been used to distribute the eggs of fall armyworms (FAW) onto diets in the laboratory and to plants in the greenhouse (McMillian and Wiseman, 1972; Wiseman et al., 1974b). There have been no reports of successful use of the technique in the field.

A manual larval dispenser, calibrated to deliver a uniform amount of a larvae/corncob grits mixture to each plant (Table 10) (Mihm et al., Unpublished), has been used successfully for four maize pests (*S. frugiperda, D. saccharalis, D. lineolata,* and *D. grandiosella*). This has several advantages over traditional techniques, eliminating laborious egg mass cutting or punching and laboratory or field pinning. An ovipositional substrate of waxed paper sacks permits rapid collection of egg masses. With optimal conditions for egg hatch provided in the laboratory, a large number of plants can be infested with a given egg production. A highly uniform infestation is achieved with first-instar larvae (Figure 1). The escape problem is minimized by eliminating predator attacks on eggs and by the protection of the larvae within the grits during early establishment. This technique may be adapted for other lepidopterous species.

Development of resistance to maize soil insects is difficult. The corn rootworm (CRW) complex (*Diabrotica* spp.), is the most important soil pest in terms of world maize production and the insects' distribution. The ecology of the northern and western corn rootworms, *Diabrotica longicornis* (NCR) and *D. virgifera* (WCR), has been reviewed by Chiang (1973). To assure infestations, researchers have used a trap crop of late-planted maize to attract egg-laden adults to an area planted the following season with the evaluation

Table 10.   Variation in Number of Fall Armyworm Larvae per Plant Based on Simulated Infestation of Plants in Laboratory

	0.3 cm³ Bazooka			0.5 cm³ Bazooka		
	Mean	S.D.	C.V.	Mean	S.D.	C.V.
Row-to-row variation with one application per plant						
Row 1	12.6	2.6	20.6	22.3	4.6	20.6
Row 2	12.7	2.8	22.0	18.3	4.7	25.7
Row 3	11.6	3.2	27.6	19.0	4.4	23.1
Row 4	12.5	3.5	28.0	17.5	3.9	22.3
Over	12.3	3.0	24.4	19.3	4.7	24.4
Plant-to-plant variation within row with 1 to 4 successive applications						
Appl. 1	12.6	2.6	20.6	22.3	4.6	20.6
Appl. 2	25.2	3.1	12.3	40.6	6.0	14.8
Appl. 3	36.8	4.0	10.9	59.5	7.8	13.1
Appl. 4	49.3	5.4	11.0	77.0	9.0	11.7

[a] Mihm et al., unpublished data, 1976.

nursery. This usually provides sufficient infestation to evaluate inbred lines and varieties for tolerance. However, the variability in egg and larval distribution in natural field populations makes it difficult to screen for the antibiosis type of resistance in segregating populations (Ortman et al., 1974).

Laboratory rearing and mass production of eggs (Branson et al., 1975; Chiang et al., 1975) have improved identification of and breeding for antibiosis to rootworms. Techniques were developed to evaluate segregating materials in the seedling stage for antibiosis in the laboratory (Ortman and Branson, 1976), after which the seedlings are transplanted in the field for further testing and selection. Artificial field infestation techniques have been developed (Chiang et al., 1975; Palmer et al., 1977) that promise to enhance resistance breeding.

There are two components of maize resistance to stored-grain insect pests such as *Sitophilus* spp.: resistance to field infestation and resistance to infestation of grain in storage. Studies of field resistance have utilized natural infestations when populations were high (Eden, 1952; Giles and Ashman, 1971). The irregularity of natural populations (Kirk, 1965) has required supplementing field weevil populations with weevil-infested corn screenings from grain elevators (Kirk and Manwiller, 1964) or with other infested materials (McMillian et al., 1968).

Work on resistance of stored maize grain is done in the laboratory. Grain samples are exposed to insects from laboratory cultures in free choice and

Figure 1. Procedures involved in egg production and larval infestation using the "bazooka" larval infestation technique. (a) Waxed paper bags containing 15 pairs of FAW adults for oviposition. (b) Rapid removal of FAW egg masses eliminates need for cutting or punching. (c) A metal plate with sharpened edge and collecting tray for rapid removal of sugar cane, neotropical, and southwestern corn borer (SCB, NCB, and SWCB) egg masses. (d) Removal of borer egg masses from plastic oviposition sheets eliminates need for cutting or punching. (e) Sieving larvae-corn cob grits mixture to remove foreign particles. (f) 500 ml. plastic bottle

Figure 1. (*Continued*)

containing larvae-grits mixture screwed onto bazooka. Note larvae on sides of bottle. Contents must be periodically mixed by gentle swirling to apply uniformly on test plants. (g) Field infestation of maize seedlings with FAW larvae using bazooka. (h) Plant damage after infestation with FAW, illustrating from left to right, resistant, intermediately resistant, susceptible. (i) Field infestation of mid-whorl maize with SCB larvae using bazooka. Other species, FAW,

Figure 1. (*Continued*)

NCB, SWCB, and CEW may also be infested with this technique at this plant stage. (j) Placement of larvae-grits mixture into maize whorl. (k) Field infestation with SCB larvae at silking stage. Larvae-grits mixture is dispensed into the axil of the ear leaf. (l) Larvae-grits mixture placement with SCB. To study ear damage by FAW or CEW, larvae-grits mixture is applied directly onto silks.

no-choice situations (Diaz, 1967; Van der Schaaf, et al., 1969; Villacis et al., 1972). A more precise measure of maize susceptibility to weevils, proposed by Dobie (1974), involves the number of emerged adults from the first generation and the length of the developmental period of the weevils.

Little work on resistance to insect vectors of pathogens has been done. Most efforts concentrate on improving resistance to the pathogens. Materials to be screened may be planted to coincide with maximum exposure to natural populations of the insect vectors, in areas where the disease is prevalent, or planted in large screenhouses where the insect vectors are released. Seedlings of progenies of plants rated as resistant in the field are then reevaluated in the greenhouse by infesting them with known viruliferous insects. Susceptible plants are discarded. The remaining seedlings are then transplanted to the field for evaluation of disease-symptom development and other characteristics.

3.3. Evaluating Plant Reaction to Insect Damage

The reactions of plants exposed to insect attack must be measured at the proper stage of growth. This can be done by visual observations or by actual measurements of the effects of insects on plants or the effects of plants on insects. It is also important to recognize the extent to which environmental factors influence the expression of resistance.

Specific evaluation techniques will depend on (1) the insect species; (2) the type of resistance sought; (3) the range of genetic variance for resistance in the host and for virulence in the insect, (4) the stage of plant development, and (5) the amount of material available for evaluation. Table 11 lists several evaluation criteria and the resistance components they identify. Techniques used to evaluate host resistance have been discussed by several authors (Painter, 1951; Horber, 1972; Dahms, 1972).

Since large amounts of material must be evaluated in resistance breeding programs, especially in the initial stages, visual rating scales are used in place of the more laborious actual measurements. Most rating scales include five or more classes that describe the insect damage or the plant response. Useful rating criteria should include both gross and small differences in insect damage and plant reaction, particularly where existing levels of resistance are low and need to be selected and concentrated. The ratings should be repeatable over time, in different locations and by different observers; permit easy and rapid classification; and be usable by different researchers working on a given species, or for different species that cause similar kinds of damage.

Table 12 presents some common rating scales for important maize insects. The diversity of scales for some insects indicates that many workers

Table 11. Criteria Commonly used in Evaluating Resistance to Maize Insects and the Components They Help Identify

Criteria	Anti-biosis	Toler-ance	Nonpref-erence
Direct methods of evaluation on insects			
1. Surviving insect population (various life states) (A)	X		X
2. Size, volume, or weight of surviving insects (A)	X		X
3. Progeny produced by surviving insects (A)	X		X
4. Time to complete life cycle (A)	X		X
5. Amount of food consumed or utilized (leaf area, cavity number, and length) (A, N)	X		X
6. Level of infestation in various life stages (N)			X
Direct methods of evaluation on plants			
1. Number of surviving plants at various intervals (A, N)	X	X	X
2. Yield loss between infested and protected plots (A, N)		X	
3. Proportion of plants infested (or showing disease symptoms, vector transmitted) (N)			X
4. Attractance of plants, plant parts, or extracts (N, Lab.)			X
5. Correlation of plant morphological characters with injury (A, N)	X	X	X
Visual estimates and indirect methods on insects			
1. Percent parasitism or number of predators present (aphids) (A, N)	X		X
2. Counts of cast skins (aphids)	X		X
3. Exreta production (A, N)	X		X
Visual estimates and indirect methods on plants			
1. Amount of damage to plant (estimation of number of lesions or percent of plant parts affected (A, N)	X		X
2. Type of feeding (size and shape of lesions and cavities) (A, N)	X		X
3. Number of pest exit holes (A, N)	X		X
4. Simulated or actual insect damage with observations on amount and speed of recovery (A, N)		X	
5. Estimates of standability, ears dropped, pulling resistance, stunting (A, N)		X	

^a A = artificial infestation; N = natural infestation; Lab. = laboratory.

Table 12. Commonly used Damage Rating Scales for Evaluation and Development of Resistance to Some Important Maize Insect Pests

Insect Group	Factor Evaluated: Insect Damage	Plant Response	Plant Part Damaged[a]	Damaging Insect Stage[b]	Type of Infestation[c]	Categories of Damage — Least Damaged ..... / Excellent ..... Very Good ..... / Highly Resistant ..... Resistant		
Corn rootworms (*Diabrotica* spp.)	×	×	R	L	Nat.	(1) No damaged plants	(2) 4–12% damaged plants	
	×		R	L	Nat.	(1) No feeding	(2) Feeding, no roots pruned to 4 cm long	
		×	R	L	Nat.	(1) Acceptable root generation		(3)
		×	Lf	A	Nat.	(0) No feeding injury	(1) Slight feeding injury	(2) Moderate feeding injury
		×	R	L	Nat.	Little resistance to pulling		
Leaf feeders (budworms) (*Spodoptera* spp.)	×		Lf	L	Nat.	(1) No damage or very slight		
	×		Lf W	L	Nat.	(0) 0–10% leaf area damaged	(1) 11–20% leaf area damaged	
	×		Lf W	L	Art.	(0) No damage (1) Few pinholes (2) Several to many pinholes (3) Few shot holes and 1 or 2 elongated lesions (4) Several shot holes and a few elong. lesions		
	×		Lf W	L	Art.	(0) Slight pinhole damage (1) Pinholes on 2+ leaves (2) Shot holes and few elongate lesions (3) Shot holes and several elong. lesions		
	×		Lf N	L	Nat.	.... Number of leaf nodes (0.6 per plant)		

Table 12. (*Continued*)

Resistance or Susceptibility Indicated by the Rating Classes

				Reference[a]
....Most Damaged				
......Good.............Fair...........Poor...................Unacceptable				
......Intermediately Resistant...................................Susceptible				
(3) Some slightly damaged; 12+% severely		(4) Many plants severely damaged and lodged	(5) All plants severely damaged and lodged	1
(3) 1–several roots pruned to 4 cm long	(4) 1 ring roots destroyed	(5) 2 rings, roots destroyed	(6) 3+ rings, roots destroyed	2
(4) Marginal root regeneration	(6)	(7) Unacceptable root regeneration	(9)	3
(3) Heavy feeding injury	(4) Some damage to whorl	(5) Much damage to whorl	(6) Leaves ragged from feeding injury	4
(Pounds of force needed to pull plants vertically)		Plants resistant to pulling		5
(2) Moderate, but general foliar damage		(3) Heavy foliar damage, and stunted plant growth		6
(2) 21–40% leaf area damaged	(3) 41–60% leaf area damaged	(4) 61–80% leaf area damaged	(5) Beyond recovery	7
(5) Several shot holes and elong. lesions (6) Many shot holes, several elong. lesions, and few portions eaten away (7) Several lesions and portions eaten away and areas dying		(8) Elong. lesions, portions eaten away, and areas dying (9) Whorl almost eaten away and several lesions and areas dying (10) Dying or dead plant		8 9
(4) Many elong. lesions (5) Many elong. lesions and few portions eaten away (6) Many elong. lesions and several portions eaten away		(7) Many elong. lesions portions eaten away and damage in whorl (8) Many elong. lesions, portions eaten away, and whorl destroyed (9) Plant dying or dead		10
per plant having lesions longer than 1.3 cm at leaf node .... (4.0 per plant)				11

Table 12. (*Continued*)

Insect Group	Insect Damage	Plant Response	Plant Part Damaged[a]	Damaging Insect Stage[b]	Type of Infestation[c]	Categories of: Least Damaged .... Excellent .... Very Good .... Highly Resistant ..... Resistant ....
Leaf feeders/ stalk borers (*Ostrinia, Chilo, Diatraea*)	×		Lf	L	Art.	(1) No damage or few pinholes (2) Few shot holes on few leaves (3) Shot holes on several leaves
	×		Lf	L	Art.	(1) Pinholes rare, sporadic (2) Pinholes intermediate (3) Many pinholes
	×		Sh & C	L	Art.	.... Cumulative number of (1 lesion per plant)
	×		St	L	Art.	.... Cumulative number of
Leafhopper vectors of maize chlorotic dwarf virus and maize dwarf mosaic virus	×	×	Lf	A, N	Nat.	(0) No disease symptoms (1) = 10% incidence; no stunting
Ear feeders *Ostrinia*	×		E	L	Art.	(1) 0–6 tunnels/husk; 0–10% damaged ear tips (5 cm) (2) 7–12 tunnels/husk; 11–60% damaged ear tips
Heliothis	×		E	L	Nat.	(0–10%) (11–20%) ....
Spodoptera	×		E	L	Art.	
Centimeter scale						(1) No damage (2) Silk feeding only (3) Tip damage only (4) Kernel damage to 1 cm
Survey scale						(...0...) (1) ....
Visual scale						(.....1—Resistant.....)
Revised centimeter						(0) No damage (1) Silk feeding only (...2...kernel dam. to 1 cm)
Sitophilus	×		E	A, L	Nat. + Art.	(1) 0–5% kernels infested/ear (2) 6–15% kernels infested/ear
	×		E	A, L	Nat. + Art.	(1) 0–10% kernels damaged (2) 11–20% kernels damaged ....

Table 12. (*Continued*)

Resistance or Susceptibility Indicated by the Rating Classes

			Reference[a]
....Most Damaged			
......Good............Fair...........Poor...................Unacceptable			
......Intermediately Resistant.....................................Susceptible			

(4) Several leaves with shot holes and few elong. lesions	(7) Long lesions common on half of the leaves		12
(5) Several leaves with elong. lesions	(8) Long lesions common on ½ to ⅔ of leaves		13
(6) Several leaves with lesions = 2.5 cm	(9) Most leaves with long lesions		14, 15

(4) Match-head sized holes rare or sporadic	(7) Holes bigger than match head size rare or sporadic		16
(5) Match-head holes intermediate	(8) Intermediate		
(6) Many match-head holes	(9) Many holes bigger than match head		

lesions, where a lesion = feeding of ⅓ distance around stalk ....

(10 lesions per plant)

"cavities," where a cavity = 2.5 cm of tunneling in stalk 17

(2) 10–25% incidence; mild stunting	(3) 25–50% incidence; moderate stunting	(4) 50+% incidence; 50% stunted	18

(3) 13–18 tunnels/husk; 61–100% damaged ear tips	(4) 19+ tunnels/husk; severe damage extending more than 5 cm beyond ear tip	19

.... Percent of ear length damaged.................................(91–100%) 20

.... Additional unit of rating for each additional cm of ear damage ....

(5)   (6)   (7)   (8)   (9)   (10)   (11)   (12)...*n*   21

.... Additional rating unit for each additional cm ear damage ....

(...2 ...)   (...3 ...)   (.....4.....)   (.....5.....)

(.....2—Intermediate.....)   (.....3—Susceptible)

.... Additional rating unit for each additional cm of ear damage ....

(3)   (4)   (5)   (6)   (7)   (8)   (9)   (10).....*n*

(3) 16–40% kernels infested/ear	(4) 41–70% kernels infested/ear	(5) 71–100% infested/ear	22

		(10) 91–100% damaged kernels	23

....

See next page for notes.

have developed and modified them to permit identification of degrees of resistance in their materials, under their conditions. The similarity of the scale used by workers in the United States for estimating ECB resistance and that utilized in the international effort of the International Project on *Ostrinia* (Dolinka, 1975) indicates scope for standardization of techniques. This would be a significant contribution to the improvement of resistance to maize pests.

Insecticides can be a useful tool in evaluating plant reaction, especially where natural infestations by several pests obscure damage by the species of primary interest. In tropical environments, systemics such as carbofuran and mephosfolan granules applied to the soil provide "clean" seedlings for uniform artificial infestations 2 to 4 weeks after plant emergence. Application of insecticides is a key component in the "yield differential technique," which provides information on both antibiosis and tolerance components of resistance. It consists of replicated paired rows of each entry, one continuously protected and the other exposed to artificial and/or natural infestations. Granular formulations or sprays of nonpersistent insecticides also are used in the "sequential infestation technique," where a set of plants of each entry can be infested at successive stages of plant development by first, second, or third broods, or by different pests such as armyworms in early and borers in late stages (CIMMYT, unpublished data).

Standard resistant and susceptible entries should be interspersed regularly within the evaluation nursery. Such standards are necessary as benchmarks of progress in developing resistance. Susceptible standards usually are easy to identify and multiply for this purpose; resistant standards may not appear until later in the improvement program.

4. MANAGEMENT OF MAIZE GERMPLASM

4.1. Mechanisms and Inheritance of Resistance

Insect resistant genotypes were developed in the past without knowledge of either the mechanisms of resistance or the mode of inheritance. This

---

[a] R = root; Lf = leaf; W = whorl; N = node; Sh = sheath; C = collar; St = stalk; E = ear.
[b] L = larva; A = adult; N = nymph.
[c] Nat. = natural; Art. = artificial.
[d] Fitzgerald and Ortman, 1964; 2 = Wilson and Peters, 1973; 3 = Ortman et al., 1974; 4 = Sifuentes and Painter, 1964; 5 = Ortman et al., 1968, Rogers et al., 1976; 6 = Brett and Bastida, 1963; 7 = Hormchong, 1967; 8 = Wiseman et al., 1966; 9 = Widstrom et al., 1972b; 10 = Mihm et al., unpublished; 11 = Wiseman et al., 1967; 12 = Guthrie et al., 1960; 13 = Starks and Doggett, 1970; 14 = Davis et al., 1973; 15 = Mihm et al., unpublished; 16 = Dolinka et al., 1973; 17 = Guthrie, 1974; 18 = All et al., 1977; 19 = Carlson and Andrew, 1976; 20 = Ditman and Ditman, 1957; 21 = Widstrom, 1967, 22 = Kirk and Manwiller, 1964; 23 = Wiseman et al., 1970.

information increases the efficiency of a breeding program. Knowledge of the resistance mechanism makes it easier to select plant characteristics that confer resistance. Resistance traits should be studied to understand their relationship to other selection criteria. For example, a plant that tolerates insect damage by producing excess foliage may be physiologically inefficient in terms of yield.

Inheritance-of-resistance studies estimate the number of genes controlling resistance and determine the type of gene action. This information indicates the best breeding system to increase resistance levels. The breeding methods best suited to given objectives are described in a later section.

Relatively few comprehensive studies have been made on the genetics of resistance to maize insect pests. In most cases information about mechanisms and inheritance of resistance is often contradictory and inconclusive. The difficulties in getting reliable information are compounded by the need to control two interacting organisms. Reliable natural infestations are seldom attainable in the field (Elias, 1970; Torres et al., 1973; Parisi et al., 1973), but artificial mass rearing of insects is difficult, and rearing techniques are lacking for some important maize insects. These weaknesses have hindered the study of the mechanisms and inheritance of resistance. Early reviews comprehensively cover information published on European corn borer and other insects (Brindley and Dicke, 1963; Brindley et al., 1975; Gallun et al., 1975; Horber, 1972; Maxwell et al., 1972; Painter, 1951; Sprague and Dahms, 1972).

European Corn Borer

ECB may have one or more broods each season, but it develops two broods throughout most of the United States corn belt. First-brood larvae damage the leaves at the early to mid-whorl stage, and the second-brood infestation starts at the pretasseling stage. Larvae feed on leaf blades, sheaths, ribs, collars, pollen, stalks, shanks, and ear shoots. The primary economic damage of the first brood is in reducing kernel filling; dropped ears and broken stalks result from shank and stalk tunneling by the second brood. There is little or no correlation between resistance to the first and the second brood in most maize materials. Genes controlling resistance to the two broods are mostly independent (Guthrie et al., 1971), but some genes may contribute resistance to both broods (Jennings et al., 1974). There are several reports on the genetics and mechanisms of resistance to the first brood, but only limited data are available on second-brood resistance.

Resistance to first-brood leaf feeding has been found in several inbreds. Resistance may result from one gene pair (Penny and Dicke, 1957), two (Mohamed et al., 1966), or more. Gene action controlling resistance has been variously reported as dominant or partially dominant (Ibrahim, 1954), primarily additive (Chiang et al., 1973; Jennings et al., 1974; Scott et al.,

1964), or having a significant epistatic component (Scott et al., 1967). The evidence of additive variance as a major type of gene action for first-brood resistance indicates that a breeding system aimed at accumulating desirable genes should be effective.

The biochemical basis of resistance to first-brood ECB has been identified as 2,4-dihydroxy-7-methoxy-1,4-benzoxazin-3-one, commonly known as DIMBOA. DIMBOA levels are high in most seedling stage maize plants, but decrease as plants mature. At the mid-whorl stage of development, some lines retain high DIMBOA levels and are resistant to ECB. DIMBOA levels are low in all lines at the time of second-brood infestation; hence DIMBOA is not related to stalk boring resistance. Additive gene effects are of primary importance in controlling high DIMBOA levels (Gallun et al., 1975; Reed et al., 1972).

Resistance to second-brood ECB has been found in few inbred lines (Andrew and Carlson, 1976; Guthrie et al., 1970). Inbred line 'B_{52}' has a high degree of antibiosis and has been used as a source of second brood resistance in several breeding programs. Resistance of 'B_{52}' is dominant or partially dominant (Guthrie et al., 1971). In other lines resistance to second-brood ECB is predominantly additive, but dominance may also play an important role (Jennings et al., 1974).

Some Latin American germplasm has been screened and selected for both first- and second-brood resistance. DIMBOA is not responsible for resistance in these varieties, since DIMBOA content is as low as that of the susceptible check. Resistance levels to both leaf feeding and stalk boring in these materials are equal or superior to those found in United States inbreds (Sullivan et al., 1974; Scriber et al., 1975).

Asian Maize Borer

Maize varieties resistant to AsMB have been reported (Chatterji et al., 1973; Sarup et al., 1974). Different parts of the plants such as the whorl, stem, cob, and tassel show varying degrees of antibiosis. Resistance may be due to differential nutrient compositions which act as feeding depressants or suppressors of toxins. Little is known of the genetics of resistance to AsMB. Inheritance seems to be multigenic (Sharma and Chatterji, 1972) and both additive and nonadditive effects are involved (Singh, 1967).

Other Borers

Information on variety resistance to other borers (*Busseola, Sesamia, Diatraea*) is either lacking or scarce. Based on natural incidence, the response of 57 tropical varieties to *Diatraea* spp. was determined in several environments in Mexico. The less damaged entries were recombined and later reported as sources for ECB resistance (Sullivan et al., 1974; Scriber et al., 1975). Several sources for resistance to the southwestern corn borer (*D.*

grandiosella) have been released (York and Whitcomb, 1966; Davis et al., 1973).

Corn Earworm

CEW is widely distributed and attacks many different crop species. Ear damage is serious in maize. Much of the work on CEW resistance has been done with sweet corn, for which consumer acceptance of insect damaged ears is very low. Published information on resistance to CEW in maize is extensive; several comprehensive review articles have been presented (Dicke and Guthrie, 1975; Maxwell et al., 1972; McMillian and Wiseman, 1972; Painter, 1951; Wiseman et al., 1974b). Nonpreference for oviposition, silk and kernel antibiosis, and tolerance through length and tightness of husk cover are associated with CEW resistance in maize.

Silk and kernel antibiosis to CEW larvae have been studied in the field, with extracts incorporated into artificial diets, and by comparisons of biochemical constituents of resistant and susceptible varieties. Silks of resistant varieties may lack a particular nutrient; contain a feeding inhibitor, a feeding deterrent, or a larval growth inhibitor; or incorporate a combination of these factors (Knapp et al., 1967). Specific chemicals involved in conferring resistance have not been identified (Luckmann, et al., 1964). The role of nutritional components of silks is unclear. Differential concentrations of free amino acid fractions (Knapp et al., 1965), ascorbic acid, and reducing sugars (Knapp et al., 1967) have been reported in resistant, intermediate, and susceptible lines. Kernel carbohydrate content in sugary versus starchy genotypes is of little importance during early development but may confer resistance in later stages (Cameron et al., 1969). Studies of larval feeding on extracts of leaves, silks, and kernels indicate that larvae respond, in order of preference, to kernels, silks, and leaves (McMillian et al., 1967; Starks et al., 1967). In some resistant strains, silk age is not a factor, whereas in others differential injury occurs up to 1 week after silking (Straub and Fairchild, 1970). Natural selection seems to have synchronized larval feeding response with maturation of maize kernels; early instars prefer silks and immature kernels and later instars the more mature kernels (McMillan et al., 1970).

The importance of a long, tight husk cover in providing protection from CEW damage has been studied extensively. Good husk cover and long silk channels generally reduce CEW injury (Wiseman et al., 1977). However, husk cover is affected greatly by moisture stress and does not give consistent protection over years and in different locations (McMillian et al., 1972; Starks and McMillian, 1967; Del Valle and Miller, 1963).

Field maize sources from Mexico, Cuba, and Guatemala provide considerable resistance to CEW damage (Walter, 1962). A Mexican strain, 'Zapalote Chico,' has excellent resistance to earworm, probably because of a silk chemical factor which acts as a feeding deterrent or growth inhibitor, and to long, tight husks (Keaster et al., 1972; Walter, 1962; Widstrom et al., 1972a). However, its tight husk, a drawback in sweet corn, limits its use as a resistant donor for commercial (canning and freezing industry) varieties.

The inheritance of resistance to CEW in maize has been worked out, but the specific characteristics associated with the genes have not been well documented. Each mechanism of resistance (husk cover, lethal factors, nutritional deficiencies) may have its own mode of inheritance. Additive, dominance, or epistatic effects are involved in distinct germplasm sources (Blanchard et al., 1941; Dungan et al., 1938). A minimum of one to as many as six gene pairs have been credited with a role in resistance (Poole, 1936; Sifuentes, 1964). Chromosomes 1, 3, 4, 5, and 8 are known to carry genes for CEW resistance and other chromosomes may also carry resistance factors (Widstrom and Davis, 1967; Widstrom and Hamm, 1969; Widstrom et al., 1972b; Widstrom and McMillian, 1973; Widstrom and Wiseman, 1973).

Fall Armyworm

FAW attacks corn in all stages, although the most serious damage occurs as leaf feeding from seedling to mid-whorl stages. This insect is widely distributed in Mexico and Central and South America. Despite the large economic losses resulting from this pest, little improvement of genetic resistance to FAW has been accomplished.

A number of materials, many with 'Antigua' germplasm, are reported to be resistant (anonymous, 1966; Wiseman, 1967a). Northern United States lines are more damaged by FAW than southern maize inbred lines (Sprague, 1977). Maize is highly preferred over *Tripsacum dactyloides* by first-instar larvae (Wiseman et al., 1967b). Additive gene effects seem to be important in the inheritance of resistance of maize inbreds to FAW, whereas dominance effects are relatively unimportant (Widstrom et al., 1972b).

Corn Rootworms

Three main species of corn rootworms (CRW), western corn rootworm (*Diabrotica virgifera*) (WCR), northern corn rootworm (*D. longicornis*) (NCR), and southern corn rootworm (*D. undecimpunctata*) (SCR) cause varying degrees of damage to maize in the United States corn belt (Chiang, 1973; Luckmann et al., 1974). *D. balteata, D. speciosa* and others damage maize in Latin America. Larvae damage subterranean parts and adults feed mainly on leaves and silks.

Larval feeding of CRW on maize is difficult to evaluate because damage occurs beneath the soil surface. Nevertheless, several criteria can be used to measure the tolerance of maize lines, single crosses, and varieties to larval damage. These include root size, ability to regenerate secondary roots following insect damage, lodging, and the capacity to survive under rootworm attack. Comparisons of plant response to CRW, with and without insecticide treatment, were used to identify tolerant lines (Zuber et al., 1971a). In addition, indirect methods have been developed for tolerance evaluations (Ortman et al., 1968; Ortman and Gerloff, 1970; Rogers et al., 1976).

The best source of resistance to CRW larval feeding is inbred line 'SD 10,' which has high potential for root regeneration and growth despite moderate rootworm attack. Other moderately tolerant lines are 'B_{54},' 'B_{67},' and 'B_{69},' (Fitzgerald and Ortman, 1964; Ortman and Gerloff, 1970; Ortman et al., 1974; Russell et al., 1971; Wilson and Peters, 1973; Zuber et al., 1971a). Materials of different maturity groups react differently to rootworm infestation; early genotypes suffer more damage than late ones (Wilson and Peters, 1973). Resistance to CRW larval feeding is heritable, and many resistant lines have the ability to transfer this characteristic to their hybrids (Ortman and Gerloff, 1970; Walter, 1965).

Rootworm resistance may be encountered in area where the insect is indigenous (Melhus et al., 1954). Germplasm possessing antibiosis may exist in Central and South America and the West Indies because of the prevalence of severe rootworm infestation (Ortman and Gerloff, 1970). It may be possible to transfer resistance to the larvae from relatives of maize such as *Tripsacum dactyloides,* which either possesses antibiosis or is extremely nonpreferred (Branson, 1971). Adequate levels of hydrogen cyanide or cyanide-related metabolites in the roots may provide CRW resistance (Ajani and Lonnquist, submitted for publication).

Reports on the reaction of maize strains to adult feeding damage are conflicting (Ortman and Fitzgerald, 1964; Reissig and Wilde, 1971). Pubescence may act as a barrier to CRW adult feeding (Hagen and Anderson, 1967). Differential concentrations of feeding stimulants have been found in different corn plant parts, with the highest concentrations in the kernels (Derr et al., 1964). The inheritance of leaf-feeding resistance to CRW adults in maize is probably monogenic and recessive (Sifuentes and Painter, 1964).

Corn Leaf Aphid

Large colonies of the corn leaf aphid (CLA) (*Rhopalosiphum maidis*) may occur in the leaf whorl, upper leaves, and tassels of scattered maize plants. Heavy infestations usually reduce pollen shedding and weaken plants by withdrawing nutrients. The sugary excretions of aphids may result in growth of certain fungi which produce black spores such that the corn plant has a "sooty" appearance. This condition would be expected to have a detrimental effect on photosynthesis and therefore productivity.

Maize genotypes react differently to infestation by CLA (Rhodes and Luckmann, 1967). Leaves of strains that expose the tassels quickly and completely have the lowest infestation. A correlation has been reported between susceptibility to aphids and damage by ECB (Huber and Stringfield, 1942). DIMBOA was identified as an active agent in resistance to CLA (Long et al., 1977).

Three species of *Tripsacum* have been reported to be resistant to CLA of which *T. floridanum* is highly resistant. The resistance mechanism could be either antibiosis or extreme nonpreference. It appears that some water-soluble substance either interferes with feeding behavior or is toxic to CLA (Branson, 1972).

Inheritance of resistance to CLA has not been extensively studied. Susceptibility was reported to be dominant (Huber and Stringfield, 1940) or recessive (Chang and Brewbaker, 1975; Snelling et al., 1940; Walter and Brunson, 1940).

Maize and Rice Weevils

Adult maize and rice weevils (*Sitophilus zeamais* and *S. oryzae*, respectively) eat small cavities in the grain into which they deposit their eggs. Larvae feed on both germ and endosperm. A tight husk cover provides protection against field infestation by weevils (Singh et al., 1972; Wiseman et al., 1970; Wiseman et al., 1974). Kernel hardness may be important in decreased weevil infestation (Eden, 1952; Mills, 1972). The pericarp may act as a physical barrier to oviposition, lack feeding stimulants, or seal off feeding stimulants from the endosperm (Schoonhoven et al., 1974, 1975, 1976). Low sugar content (Singh and McCain, 1963) and high amylose content (Rhine and Staples, 1968) of the kernel adversely affect larval nutrition and other chemical components of the kernal may influence weevil infestation (Eden, 1952; Mills, 1972; Schoonhoven et al., 1972).

Studies on the relative importance of cytoplasmic, maternal, and endosperm resistance to injury by weevils show that maternal tissues developed during seed formation confer seed resistance to weevils. Resistance is attributable to both additive and dominance effects (Widstrom et al., 1975). Cytoplasm of the lines used was not important in determining resistance.

Angoumois Grain Moth

The Angoumois grain moth (AnGM) (*Sitotroga cerealella*), causes heavy grain losses both in the field and in storage. Differential response of maize strains to AnGM indicates the possibility of breeding for resistance (Warren, 1956). High amylose content in kernels results in reduced adult size (Peters et al., 1972; Rhine and Staples, 1968).

4.2. Breeding Methods

Maize is a highly heterogeneous species, with 10 pairs of chromosomes. The flowering habit is monoecious. The ear bears from 150 to 500 or more seeds (caryopses, kernels, or grains). Because maize is naturally cross-pollinated, artificial control of pollination is necessary both for self-fertilization and for crossing of selected plants.

Detasseling of female parents is commonly practiced in commercial hybrid seed production fields and in half-sib selection programs. Cytoplasmic male sterility has also been used in hybrid seed production to avoid the need of detasseling. Isolation from contaminating pollen sources is required. Precise control of pollination is achieved by covering both tassel and ear shoot with bags and manually transferring the pollen as desired. Because the silks (stigmas) are receptive to pollen immediately upon emergence, the ear shoot must be pre-covered, usually with a glassine or parchment bag. Stigmas are receptive for about 14 days. Anthers normally begin to shed pollen 1 to 3 days before silks appear and continue to shed pollen for up to 10 days. When most of the silks have emerged, a pollen bag is placed over the tassel of the desired male parent and fastened with a paper clip. Pollen is collected about 5 to 20 hours after bagging. During this interval any foreign pollen that may have been on the tassel will have died. The ear shoot bag is removed and the pollen bag immediately placed over the ear shoot. The pollen is distributed over the silks by vigorous shaking of the pollen bag. The pollen bag is then pulled down over the ear shoot, fastened with a stapler and marked to identify male and female parents and the date of polination.Fertilization of the ovule takes place about 24 hours after the stigmas have been pollinated. Selfing is accomplished by collecting pollen from the same plant as that to be pollinated. Crosses are made by transferring pollen of one plant to the silks of another.

Breeding methods used for improving insect resistance are similar to those used for improvement of other characters. The breeding methods and the statistical and genetic theory that supports them are presented in several books (Allard, 1960; Jugenheimer, 1976; Sprague, 1977).

Through the long association of certain maize populations with insect pests and pathogens, natural selection has contributed, with some human participation, to evolution of types that resist or tolerate damage to various degrees. Most often their resistance levels are not adequate for modern, intensive crop production. The favored approach to combat some of the most important insect problems is through improved varietal resistance.

Breeding for insect resistance requires knowledge of the host plant and the insect pest, their interactions, and how these may be modified by breeding and by environmental factors such as tillage practices and planting dates. The most important pests of a given region govern the direction of the breeding efforts. In Southeast Asia borer resistance must be accompanied by downy mildew resistance; in equatorial Africa borer resistance needs to be coupled with streak virus and leaf rust resistance.

A range of environments is needed to wide test maize materials. Multilocational tests assist the monitoring of levels of resistance, their buildup or breakdown, and shifts in biotypes of races of pests. Systematic exposure of progenies selected under artificial infestation and/or inoculation in a diversity of environments is basic to the development of widely adapted, stable, pest resistant varieties.

Agronomic management of the breeding material is important. Factors to be controlled include soil fertility, moisture, weed competition, and plant density. Optimum growing conditions should be provided, to identify true resistance and minimize pseudoresistance. This is particularly important when exotic and poorly adapted materials are being evaluated.

The standard breeding procedure for hybrid maize consists of development of inbreds and their subsequent use in hybrid combinations. The chances of obtaining inbred lines resistant to a particular insect pest depend on the frequencies of genes conditioning resistance. If gene frequency is low, there is a low probability that lines extracted from the source population will be of value for resistance. Source populations thus influence the level of incorporation of desired traits in inbred lines and subsequent use in hybrid combinations.

The simplest procedure for incorporating resistance into hybrids is evaluation of available inbreds and direct use of the most resistant lines in hybrid combinations. When resistant lines are available, the production of a resistant commercial hybrid is relatively easy. If resistance is controlled by additive effects, involving as many resistant inbred lines as possible in the hybrid will result in a higher level of resistance. When resistance is governed by recessive genes, it is necessary to have them in a homozygous recessive condition in all lines. Where resistance is governed by dominant genes all lines need not necessarily be resistant.

Resistant lines can also be used to transfer resistance to susceptible but agronomically desirable lines by crossing a susceptible and a resistant line and extracting from the progeny the most resistant and agronomically best lines. This procedure has been used to combine resistance to both first- and second-brood European corn borer. One F_6 line originating from the cross of 'B_{52}' $\times$ 'Oh_{43}' was slightly better than 'Oh_{43}' for first-brood resistance and equal to 'B_{52}' for second-brood resistance (Russel et al., 1974).

Resistance can be transferred through a backcrossing program when only one or two genes are involved. When many genes are involved the backcross procedure may result in reduced resistance with each successive backcross. Various modifications of the backcross program can be used to increase the level of resistance. One method is to advance each backcross to F_2 or F_3 by intermating among resistant plants to increase the resistance level before making the next backcross. A modified backcross program was used to improve the agronomic characteristics of 'Oh$_{45}$' having resistance to first brood of ECB (Guthrie and Stringfield, 1961).

Interest is growing in several hybrid-producing countries to develop germplasm source populations. These materials are subjected to various environmental stresses, including insects, either naturally or artificially. Source populations are improved for several generations under insect pressure. Inbred lines derived from such source populations are also selected under insect infestation to develop acceptable resistance.

Gene action conditioning resistance to most insect pests in maize appears to be additive. This indicates that population improvement procedures, such as mass selection and various recurrent selection schemes, are effective in accumulating desirable genes for resistance to various maize insect pests. Ten cycles of mass selection for corn ear worm resistance in two synthetics, C and S, resulted in average reduction of earworm damage per generation of 2.75 and 2.81 percent, respectively (Zuber et al., 1971b).

Progress from recurrent selection for CEW resistance has been achieved while maintaining desirable agronomic traits (Widstrom et al., 1970). Two cycles of recurrent selection in five synthetic varieties were sufficient to shift gene frequencies for ECB resistance to a high level in all varieties, and three cycles produced essentially ECB resistant varieties (Penny et al., 1967). The selection technique consisted of selfing 100 desirable plants. These were planted in progeny rows the following season. Plants were infested with corn borers and the 10 most resistant S_1 lines were selected. In the third year the selected S_1's were intermated in a partial diallel system $(n(n-1)/2)$ producing 45 combinations to start the second cycle of selection. Plants were infested again and 100 of the best plants were selfed. The following year the 100 S_1's were again planted and infested. The 10 most resistant families were selected and again the 45 possible crosses were made. The procedure could be repeated as many times as desired and as progress continued.

A modified recurrent selection scheme was applied to a 24-line synthetic and Cash synthetic (Guthrie, 1974). Half-sib ears were planted ear-to-row and selfed. In the next cycle the S_1 ears were recombined in a half-sib crossing block with each S_1 ear planted as a female family and detasseled. The pollinator rows consisted of a composite of selected S_1 ears. The female rows were infested with ECB egg masses and inoculated with leaf blight and stalk rot organisms. This system utilized a broad gene base and concurrently concentrated genes for resistance to corn borers, stalk rots, and leaf blight.

In schemes of recurrent selection for general combining ability (Jenkins, 1940) and reciprocal recurrent selection for general and specific combining ability (Comstock et al., 1949) topcrosses can be evaluated for resistance in an insect nursery at the same time yield trials are conducted. Thus the ability to transmit resistance to either the broad based tester (recurrent selection for general combining ability) or the other source population (reciprocal recurrent selection) is evaluated.

There is a need for breeding schemes designed to capitalize on all components of resistance (antibiosis, nonpreference, and tolerance). Selection of resistant plants is usually based on plant damage ratings or on insect response. The primary purpose of breeding for pest resistance is to reduce yield losses from a particular pest organism or pest complex. It follows then that selection for resistance should be based on the criterion of yield loss, or a closely correlated parameter. Any selection criterion should be evaluated for its relationship to yield loss. Damage rating scales and insect response often predict only a part of yield loss; other undetermined factors may contribute as well to overall resistance.

Various damage rating techniques were related to yield loss caused by ECB (Penny and Dicke, 1959; Scott et al., 1967; Guthrie et al., 1975). Generally, neither leaf feeding ratings for first brood nor stalk boring ratings for second brood accurately predicted yield loss. Differences in tolerance of the hybrids were postulated.

The differential response of plants under infested and insecticide protected conditions has been studied for western corn rootworm (Zuber et al., 1971a). A low differential in root volume or in pulling resistance between infested and protected plots was found to be useful to identify maize strains with a high level of tolerance. The technique was not extended to include the measure of other resistance components.

The array of resistance mechanisms (involving morphological, physiological, and biochemical factors) within a population could be simultaneously selected for and integrated into a combination that minimizes yield loss by selecting for least yield loss under infestation. In a practical field situation, this measurement is made by comparing yields of a given genotype under infested and noninfested conditions. Comparisons of percent yield loss due to pest damage for various genotypes provide a relative measure of their resistance.

For some insects extension of the yield differential technique permits relative measures of the contributions of the tolerance and antibiosis components of resistance, and the variability for each in the population. For example, antibiosis to the fall armyworm is indirectly manifested in the degree of leaf feeding damage. Tolerance is measured as the plants' ability to yield well despite feeding damage. A regression of percent yield loss due to FAW damage on leaf-feeding ratings for full-sib families from a maize population illustrates the principle (Figure 2). That part of the yield loss which can be predicted from leaf-feeding damage is an estimate of the antibiosis contribution to resistance. However, all points do not fall on the

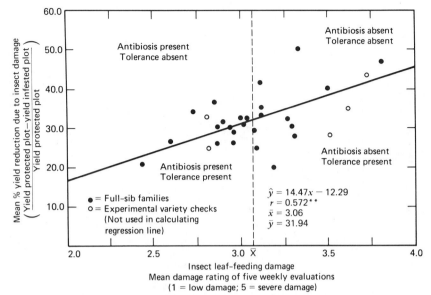

Figure 2. Identification of components of resistance to *S. Frugiperda* (FAW) using leaf-damage ratings (antibiosis indicator) and yield loss (tolerance indicator) in "TUXPENO CARIBE" maize population, Poza Rica, Veracruz, Mexico, 1977 (Hershey, unpublished data).

regression line, so differences in leaf-feeding resistance among families are not wholly responsible for the variation to yield loss. The deviation from the regression line reflects the tolerance component, that is, the sensitivity to a given level of leaf damage. Some families may incur little yield loss despite high damage levels, whereas others show a relatively high yield loss with only slight damage. The "antibiosis" and "tolerance" measurements are relative to the population, and do not imply a distinct separation of the categories. Progenies falling within the category "antibiosis, tolerance" that are satisfactory for other criteria should be selected.

Different populations can be compared on the bases of slope and intercept of the linear regression line, population means for damage rating and percentage yield loss, and the correlation coefficient (r) (Figure 3). The slope of the regresssion line indicates the sensitivity of the population to feeding damage. A steep slope ('Mezcla Amarilla') shows that a high yield loss is incurred per unit increase in feeding damage; conversely, a gradual slope ('Tuxpeño 1') indicates a low sensitivity to feeding damage. Population means for leaf-feeding damage and for percentage yield loss provide a ranking at the population level for these two parameters, and may be most useful in selecting a more resistant population for an area where several populations are adapted and acceptable. The correlation coefficient reflects the closeness of the fit to a straight line for the points on the graph. The r^2

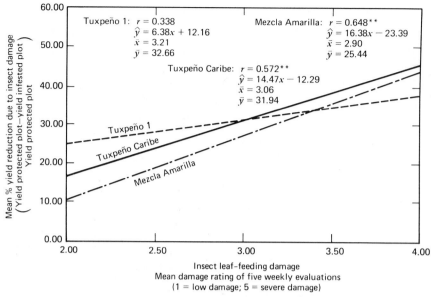

Figure 3. Comparisons of leaf-feeding damage (antibiosis indicator) by fall armyworm and resulting yields loss (tolerance) indicator) for three tropical maize populations (Hershey, unpublished data).

value is also informative; it indicates the proportion of the variation in y that can be predicted from x. For example, the 'Mezcla Amarilla,' $r = 0.648$ and $r^2 = 0.42$. Thus, although r is significant at the 1 percent level, only 42 percent of the variation in percentage yield loss can be predicted from leaf-feeding ratings. This underscores the need to select families by a comprehensive measure such as yield loss.

In summary, the following points are important in breeding for insect resistance in maize:

1. Knowledge of gene action conditioning resistance is available for few maize insects, and more work is required to provide a clear understanding of the modes of inheritance. Although this information is potentially useful in structuring an efficient breeding program, most programs have made little use of inheritance studies because they have been inconclusive.

2. To the extent possible, resistance should be improved in adapted genotypes. Introduced germplasm that may have resistance can also be used. Transfer of resistance from related species should be the last resort because of the long process required in recovering good agronomic characteristics.

3. The levels of insect resistance are generally low. This requires the evaluation of resistance in an array of germplasm sources. Plant population size is extremely important for accumulating levels of resistance or transferring resistance to susceptible material.

4. Escapes occur under both natural and artificial infestation. If the proportion of escapes is not great, they can be eliminated in the successive cycles of evaluation and selection. Breeding procedures that rely on some form of progeny or family evaluation are extremely useful, because progeny rows escape much less than individual plants.

5. Insect population and infestation level, whether natural or artificial, should be regulated depending on the level of resistance in plant materials. If resistance levels are low, expression of existing genes for resistance may be masked by heavy infestation and all genotypes may appear equally susceptible. As resistance genes are accumulated, infestation level can be increased accordingly.

6. When natural insect levels are adequate and uniform, effective selection is possible if appropriate plot techniques are used. Environments known to have relatively high insect levels over seasons may be utilized for this purpose. If natural insect levels are inadequate or nonuniform they may be augmented in the field by timely planting of a susceptible variety as plot borders for plant materials to be evaluated. Alternatively, appropriately distributed staggered plantings may provide several generations of pest population augmentation and overlapping broods for evaluating experimental plant materials.

7. If rapid and sustained progress is desired, successful artificial infestation is necessary to identify plant reaction. However, limits are imposed by the insect mass production capabilities of a given program.

8. Rating scales to measure differences in resistance should be practical, rapid, and reliable. However, when the general level of resistance is low and range of variability narrow, genotypic differences are difficult to distinguish. More laborious and precise measurements may then be necessary to identify the most resistant genotypes, which may have a rating only slightly better than the susceptible ones.

9. Through the use of appropriate breeding procedures the components of resistance (nonpreference, antibiosis, and tolerance) should be exploited singly or in combination. Desirable plant morphological traits that contribute to resistance should be given due importance in selection.

10. Breeding progress depends on the cooperation of scientists specializing in different disciplines. Isolated efforts by breeders, entomologists, and pathologists will produce varieties deficient in one or more aspects. Therefore an integrated, multidisciplinary approach is of the utmost importance.

4.3   Maize Improvement in an Integrated System

Seventeen institutions in different parts of the world maintain substantial maize germplasm banks; they collectively approach 80,000 items (Gutierrez, 1974). Additional collecting has been programmed by the International Board for Plant Genetic Resources (IBPGR). Unfortunately, preservation of these collections is difficult and costly. Only a few institutions have relatively complete collections, and quantities of seed on hand are often insufficient to provide expedient service to plant breeders.

Experience indicates that individual entries or collections per se do not provide immediate solutions to the problem of developing insect resistant varieties. Level of resistance for most entries is low. Where it is high, a lengthy breeding program for agronomic improvement is often necessary. However, maize growing countries should have source populations in their programs, or have access to gene pools with considerable genetic diversity, within which insect resistance can be improved while retaining those agronomic traits needed by their farmers. Enhancement of such genetic diversity to reduce pest vulnerability through planned introduction programs should be encouraged.

Germplasm resources and environmental diversity must be carefully managed to minimize the loss of genetic diversity during utilization and improvement. Figure 4 illustrates the approach followed at CIMMYT. In an attempt to provide appropriate genetic diversity for the different environments where maize is grown (Table 4), a set of 34 gene pools has been developed (Table 13), with emphasis on adaptation to highland tropical, tropical lowland, and temperate–subtropical regions. Within each of these categories, materials have been classified by maturity as early, medium, and late types. Within each maturity group, materials are further subclassified by endosperm color (white and yellow) and grain texture (flint, dent, and floury) (anonymous, 1973). Additional pools may be created according to needs in different parts of the world. Each pool eventually consists of about 500 half-sib progenies. These pools are grown in Mexico for recombination and selection in two or three different environments to widen the range of adaptation and maintain substantial genetic variability. Frequently, new germplasm bank accessions and introductions from national programs (previously evaluated in "introduction nurseries") are added as female rows only. Handling of the new entries crossed to the pool will depend on the specific objective for making the cross. It may undergo several backcrosses to the pool, or an alternation of sibbing and backcrossing may be used. During sibbing or backcrossing the cross is constantly evaluated for its potential contribution to the pool. A gradual merging process is followed. A pool thus has a core of progenies that are regularly incorporating new genetic variability. Outstanding progenies from the pools are promoted to the appropriate advanced popu-

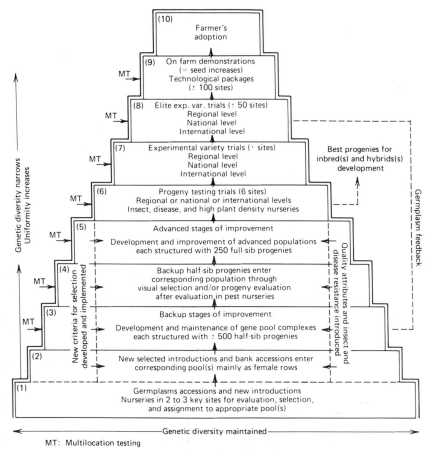

Figure 4. Stages in maize germplasm management and improvement. MT = multilocation testing.

lations. Table 13 gives the present advanced populations and corresponding gene pools. This merging constitutes the genetic bridge between backup and advanced stages of improvement.

The populations in the advanced stages are continuously improved through a full-sib recurrent selection procedure. From each population 250 full-sibs are tested in the international progeny testing trials conducted at six locations. On the basis of site specific performance, about 10 outstanding families are recombined to generate experimental varieties. Also, 100 to 120 full-sib progenies are selected to continue the improvement of the population as described in Figure 5. In this approach, for each cycle of improvement two seasons are devoted to evaluation and selection for pest resistance with the best 100 to 120 progenies from a given advanced population(s).

When priority is selection for insect resistance in the two successive

Table 13. CIMMYT's Gene Pools and Advanced Maize Populations, 1977

Pool No.	Pool Designation	Abbreviation	No. H.S. Prog.[a]	Corresponding Advanced Populations[b]	Number
		Backup Stages		Advanced Stages	
Highland pools					
1.	Highland early white flint	HEWF	230	—	
2.	Highland early white dent	HEWD	512	Blanco Dentado Precoz de Altura	52
3.	Highland early white floury	HEWFL	1010	Blanco Harinoso Precoz de Altura	53
4.	Highland early yellow flint	HEYF	432	Amarillo Cristalino Precoz de Altura	54
5.	Highland early yellow dent	HEYD	457	Amarillo Dentado Precoz de Altura	55
6.	Highland intermediate white flint	HIWF	156	—	
7.	Highland intermediate white dent	HIWD	471	—	
8.	Highland intermediate white floury	HIWFL	471	Blanco Harinoso Intermediate de Altura	58
9.	Highland intermediate yellow flint	HIYF	533	—	
10.	Highland intermediate yellow dent	HIYD	1072	Amarillo Dentado Intermedio de Altura	60
11.	Highland late white flint	HLWF	278	—	
12.	Highland late white dent	HLWD	575	—	
13.	Highland late yellow flint	HLYF	390	—	
14.	Highland late yellow dent	HLYD	552	—	

Lowland tropical pools

No.	Pool	Code		Source	
15.	Tropical early white flint	TEWF	425	—	
16.	Tropical early white dent	TEWD	518	—	
17.	Tropical early yellow flint	TEYF	576	—	
18.	Tropical early yellow dent	TEYD	416	—	
19.	Tropical intermediate white flint	TIWF	461	—	
20.	Tropical intermediate white dent	TIWD	362	—	
21.	Tropical intermediate yellow flint	TIYF	448	Mezcla Amarilla; PD(MS)6 H.E. o_2[c]	26; 38
22.	Tropical intermediate yellow dent	TIYD	481	Antigua × Republica Dominicana	35
23.	Tropical late white flint	TLWF	544	Blanco Cristalino 1; ETO Blanco; W.H.E.o_2	23; 32; 40
24.	Tropical late white dent	TLWD	576	Tuxpeño 1, Mezcla Tropical Blanco; (Mix. 1 × Col. Gpo. 1)ETO;	21; 22; 25
25.	Tropical late yellow flint	TLYF	436	Tuxpeño Caribe; La Posta, Tuxpeño o_2	29; 43; 37
				Amarillo Cristalino 1; Y.H.E. o_2	27; 39
26.	Tropical late yellow dent	TLYD	537	Antigua × Veracruz 181; Amarillo Dentado; Cogollero	24; 28; 36

Temperate-subtropical pools

No.	Pool	Code		Source	
27.	Temperate early white flint	TmEWF	352	—	
28.	Temperate early white dent	TmEWD	224	—	
29.	Temperate early yellow flint	TmEYF	320	—	
30.	Temperate early yellow dent	TmEYD	418	Templado Amarillo o_2	41
31.	Temperate intermediate white flint	TmIWF	351	Compuesto de Hungria	48
32.	Temperate intermediate white dent	TmIWD	448	Blanco Subtropical	34
33.	Temperate intermediate yellow flint	TmIYF	480	AED × Tuxpeño	44
				Amarillo Subtropical	33
34.	Temperate intermediate yellow dent	TmIYD	369	ETO × Illinois; Templado × Tropical	42; 45

[a] Numbers of half-sib progenies.
[b] 250 full-sibs in each population.
[c] H.E.o_2 = Hard endosperm opaque—2.

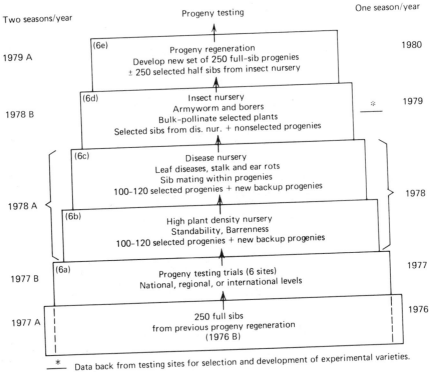

Figure 5. Maize germplasm management and improvement: pest resistance improvement cycles. These steps are involved in stage 6 of maize germplasm management and improvement (Figure 4).

seasons, (first season—improvement breeding nursery; second season—family recombination nursery), if possible instead of one row, each progeny may be grown in paired rows, one protected and the other unprotected and artificially infested (yield differential technique). Differential response in terms of leaf-feeding ratings, growth recovery, and yield may be used to capitalize on both antibiosis and tolerance components. Plants selected under insect pressure are recombined and, in the following season, generate half-sib progenies for subsequent development of 250 new reciprocal full-sib progenies for national, regional, or international testing. In addition, if desirable, about 10 best families selected after the two improvement seasons may be recombined to develop an experimental variety, presumably with higher levels of genetic resistance. Data from the multilocation trials of progeny and experimental varieties may also assist in monitoring pest problems.

Experimental varieties are formed by making all possible crosses among the 10 best progenies at each location. Sufficient seed is produced to test such experimental varieties at 30 or 40 locations in replicated trials. The best varieties are promoted to the elite category and are reevaluated. Thus at each stage, from progeny testing through the identification of elite germplasm, national, regional, or international programs are involved.

Inbred lines and their hybrids may be developed in this system of continuous flow from unimproved germplasm to elite experimental varieties. Such hybrids are likely to have a broader genetic base, wider adaptation, and higher stability of performance than if developed from narrow genetic base sources. The systematic improvement of maize germplasm in the manner described above is aimed at realizing gains in several agronomic aspects while reducing genetic vulnerability to pests.

5. IMPROVEMENT OF PEST RESISTANCE AS AN INTERNATIONAL EFFORT

Establishment of "international testing centers for crop pests and pathogens (Stakman, 1968) or "regional centers for plant pest research" (Ling, 1974) has been suggested to cope with pest problems of worldwide importance. Utilization of the present network of commodity-oriented international centers seems an alternative possibility. These centers are strategically located and could participate in the improvement and monitoring of genetic host-plant resistance to pests, as an integral part of the overall crop improvement process. The infrastructure of facilities and services is already established at each center's headquarters. Links with national programs exist and key environments are available for progeny evaluation and selection. Many close links already exist between the international centers and universities throughout the world. These contribute to bridging the gaps between basic and applied research. An example of the potential of such cooperation is the recent finding of ECB resistance in exotic tropical maize progenies derived from a composite that included entries with low *Diatraea* spp. damage (Sullivan et al., 1974; Scriber et al., 1975).

In addition, the centers are establishing cooperative informal agreements with national programs to improve resistance to widespread pests. An example of such effort, illustrated in Figure 6, is discussed below.

Downy mildews are a major maize disease in Southeast Asia. Sorghum downy mildew is endemic in equatorial Africa and is a potential threat to the rapidly expanding maize production areas. In the Americas sorghum downy mildew has been reported from the southern United States to Argentina. Corn stunt, disseminated by American maize leafhoppers, is a limiting factor in Central America, the Caribbean region, and northern South America. Streak, disseminated by African maize leafhoppers, limits maize production in Africa south of the Sahara.

The initial objective is to improve the level of resistance to downy mildew

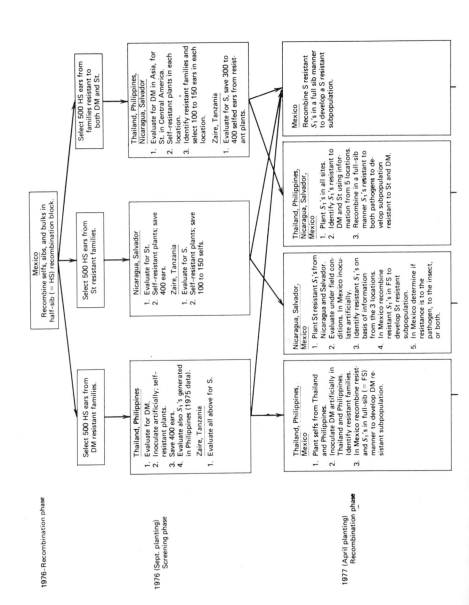

1976-Recombination phase

Mexico
Recombine selfs, sibs, and bulks in half-sib (= HS) recombination block.

1976 (Sept. planting)
Screening phase

Select 500 HS ears from DM resistant families.

Select 500 HS ears from St resistant families.

Select 500 HS ears from families resistant to both DM and St.

Thailand, Philippines
1. Evaluate for DM.
2. Inoculate artificially; self-resistant plants.
3. Save 400 ears.
4. Evaluate also S_1's generated in Philippines (1975 data).

Zaire, Tanzania
1. Evaluate all above for S.

Nicaragua, Salvador
1. Evaluate for St.
2. Self-resistant plants; save 400 ears.

Zaire, Tanzania
1. Evaluate for S.
2. Self-resistant plants; save 100 to 150 selfs.

Thailand, Philippines, Nicaragua, Salvador
1. Evaluate for DM in Asia, for St. in Central America.
2. Self-resistant plants in each location.
3. Identify resistant families and select 100 to 150 ears in each location.

Zaire, Tanzania
1. Evaluate for S, save 300 to 400 selfed ears from resistant plants.

1977 (April planting)
Recombination phase

Thailand, Philippines, Mexico
1. Plant selfs from Thailand and Philippines.
2. Inoculate DM artificially in Thailand and Philippines. Identify resistant families.
3. In Mexico recombine resistant and S_1's in full-sib (= FS) manner to develop DM resistant subpopulation.

Nicaragua, Salvador, Mexico
1. Plant St resistant S_1's from Nicaragua and Salvador.
2. Evaluate under field conditions. In Mexico inoculate artificially.
3. Identify resistant S_1's on basis of information from the 3 locations.
4. In Mexico recombine resistant S_1's in FS to develop St resistant subpopulation.
5. In Mexico determine if resistance is to the pathogen, to the insect, or both.

Thailand, Philippines, Nicaragua, Salvador, Mexico
1. Plant S_1 s in all sites.
2. Identify S_1's resistant to DM and St using information from 5 locations.
3. Recombine in a full-sib manner S_1's resistant to both pathogens to develop subpopulation resistant to St and DM.

Mexico
Recombine S resistant S_1's in a full-sib manner to develop a S resistant subpopulation.

416

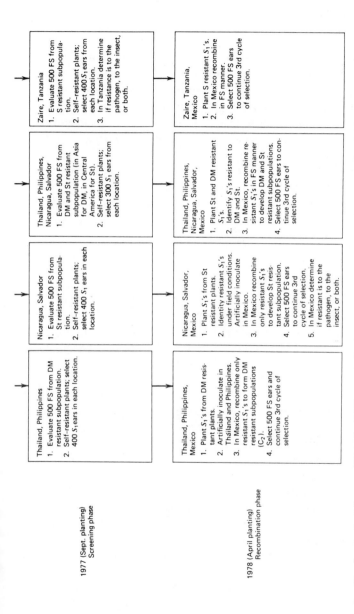

Figure 6. Development of downy mildew (= DM), streak (= S), and stunt (= St) resistant populations in CIMMYT's collaborative research program.

417

and streak virus and to downy mildew and stunt mycoplasma. For this purpose, otherwise well adapted maize populations with a wide genetic base are being used. Appropriate studies will determine whether selected progenies are resistant to the pathogen, to the insect, or both. Steps may be taken in the near future to include improvement for borer and armyworm resistance.

Other international programs exist. An example is the project conducted by the International Working Group on *Ostrinia*, which has been in operation since 1968 and now involves 15 countries. In its current phase the program is concerned with the development of composites to combine borer resistance and local adaptation. Their recent testing has included *Chilo zonellus, Sesamia inferens, Oscinella frit, Oulema melanopus, Diabrotica* spp., and *Helminthosporium* (Chiang, 1978).

Systematic and well organized efforts, coupled with reliable multilocation testing, should contribute to the development of yield-stable, widely adapted, and pest resistant varieties that will increase world maize production.

6. SUMMARY

Maize is one of the main commodities that contributes to the sustenance, directly or indirectly, of the human population. Among the major insect pests that limit its production are the borers, armyworms, earworms, rootworms, and insect vectors of pathogens (stunt, streak, maize dwarf mosaic diseases). These are the target pests that govern priorities in breeding for improved levels of resistance.

Considerable efforts have been made to preserve maize genetic diversity. However, its systematic utilization for overall maize improvement has only recently begun. Breeding for insect resistance has not been a component of most maize improvement programs. In others, progress has been limited and slow because of reliance on attempts to breed maize solely under natural insect incidence.

The development of artificial diets in the last decade has made possible mass production of lepidopterous insects. Timely, uniform manual infestations of maize progenies with eggs or larvae, while possible in only a few well established insect resistance programs, has permitted sustained progress in breeding for resistance to insects. Both mass rearing and infestation techniques are being improved to permit the evaluation and selection of larger numbers of progenies in established insect resistance programs.

Visual injury rating scales of 1 to 5 or 1 to 9 have been widely used for evaluating leaf-feeding injury by lepidopterous pests. For stalk damage, the number of injured internodes or cavities per plant is commonly used. Damage by earworms is estimated by a 1 to 5 scale or in centimeters of feeding. For rootworms, damage and recovery ratings (1 to 6 or 1 to 9 scale) and pulling resistance are common evaluation methods.

Gene frequency for insect resistance is low in most maize germplasms. Therefore, screening of the thousands of available maize collections is not a sound approach. Rather, the objective should be to build up gene frequency by recombination of the least damaged (apparently resistant) progenies within populations in successive cycles of evaluation and selection.

Antibiosis and tolerance are the most commonly reported components in borer, earworm, armyworm, and rootworm resistance improvement. DIMBOA, an ECB resistance factor, is the only chemical positively identified as associated with insect resistance in maize.

Inheritance-of-resistance studies have been conducted primarily for ECB and CEW. The gene action that conditions ECB resistance to first-generation leaf feeding is predominantly additive. Results of CEW resistance studies have been inconclusive. Mass selection and recurrent selection have been effective breeding techniques for improving the levels of insect resistance. Backcrossing techniques have been used to transfer resistance to desirable commercial types when relatively few genes are involved.

Host-plant resistance is a component of pest management. Pest management loses its significance if it is not intimately associated with crop improvement and production practices. Therefore interdisciplinary teams of agronomists, breeders, entomologists, pathologists, physiologists, and economists are better equipped to face the demands of increased production than are isolated workers.

The complexity in improving levels of pest resistance requires collaborative efforts at national, regional, and international fronts. Fortunately there is an increasing linkage of such efforts and more rapid progress may be made in the future.

ACKNOWLEDGMENTS

The authors wish to express deep appreciation to Rosa Maria Cruz for her perseverance in typing this chapter, and the following maize workers who kindly provided reference lists, reprints and unpublished information: H. C. Chiang, F. F. Dicke, W. D. Guthrie, V. E. Gracen, and B. Wiseman. Thanks are also expressed to our colleagues at CIMMYT for their constructive suggestions and to the CIMMYT Library staff for their assistance. Special recognition and deep gratitute are expressed to H. C. Chiang, F. F. Dicke, W. D. Guthrie, J. H. Lonnquist, and B. R. Wiseman for their thorough review of the manuscript. Their comments, suggestions, and corrections have greatly contributed to a better presentation.

A scientist at IIRI cross-pollinates strains of rice to develop improved varieties. Photo by Edwin G. Huffman, courtesy of World Bank.

17

BREEDING APPROACHES IN RICE

M. D. Pathak
R. C. Saxena

The International Rice Research Institute, Manila, Philippines

1. INTRODUCTION

1.1. Rice

Rice feeds more people than any other food crop in the world. It is the staple food of 60 percent of mankind and is grown on about 140 million hectares, one-fifth of the total cereal area. Nine-tenths of the world's rice is produced and consumed in Asia.

Rice is grown from latitudes 55 degrees north to 55 degrees south and from sea level to altitudes of 3,000 m. It is grown either from seed—broadcast or drilled—or by transplanting, and under diverse water regimes: as an upland crop where there is no standing water and rain is the sole source of moisture, or under lowland conditions where water from rain or irrigation systems is impounded in the fields. On slopes, rice is cultivated in terraces, and in low lying sites, deep-water rice is grown in as much as 5 to 6 m of standing water. Temperature is a major factor in rice cultivation. The optimum is about 30°C, but temperatures of about 20°C induce sterility, particularly in the flowering stage. Thus only one crop a year may be possible in regions of cool winters, but in warm areas as many as three crops are common.

The rice plant belongs to the Gramineae and the genus *Oryza*. There are 20 valid species of *Oryza* but almost all cultivated rice belongs to *O. sativa,* which originated in Asia several thousand years ago (Chang, 1976). The African rice *O. glaberrima* developed later. It differs from *O. sativa* in lacking secondary branches of the panicles and pubescence of lemmas, and has much longer ligules. *O. glaberrima* accounts for less than 4 percent of the world's total rice hectarage. Both Asian and African rices are diploid and have 24 chromosomes ($2n$ = 24). Essentially all of the work on breeding for insect resistance has been on *O. sativa*.

1.2. Breeding for Insect Resistance: General Considerations

More than 100 insect species attack rice. Of these, about 20 are major pests. Together they infest all parts of the plant at all growth stages, and a few transmit virus diseases (Pathak, 1977). In addition a number of insect pests attack stored rice.

The warm and humid environment in which rice is grown is conducive to proliferation of insects. Heavily fertilized, high-tillering plants and the practice of growing rice throughout the year favor the buildup of pest populations. Thus in the tropics the rice grown with modern technology often suffers more severe pest infestations than the poorly managed fields of conventional varieties. Average yield losses due to various pests have been estimated at 35 percent in Asia (excluding China and Japan) and 21 percent in North and Central America (Cramer, 1967).

The use of insect resistant varieties is an ideal method of controlling rice pests. Apart from the undesirable effects of pesticides, many farmers in southern and Southeast Asia, where most of rice is grown, have limited access to capital, pesticides, and application equipment. Nevertheless, until recently there was little sustained effort to develop insect resistant varieties because of a general lack of qualified personnel and facilities, limited availability of germplasm collections, and lack of simple insect mass rearing and varietal screening techniques. However, outstanding progress in resistance breeding was made during the last decade, and varieties resistant to several insect species are now being grown over millions of hectares (Table 1). This chapter reviews the work on breeding for resistance for some of these insects.

During its long history of cultivation a large number of rice varieties and strains evolved to suit various agroclimatic conditions and the grain quality preference of various peoples. It is estimated that there are about 80,000 to 100,000 strains of rice in the world. These strains show immense variation in agronomic and physiological characteristics and are expected to have vast potential as sources of resistance.

Many of these strains are maintained at various national rice research or germplasm resource centers such as the Central Rice Research Institute, Cuttack, India; the Bangladesh Rice Research Institute, Dacca; the Central Research Institute for Agriculture, Bogor, Indonesia; Hiratsuka, Japan;

Table 1. Major Insect Pests of Rice against which Varietal Resistance has been Recorded

Common Name	Scientific Name	Progress in Breeding for Resistance[a]	Selected References
Striped rice borer	*Chilo suppressalis*	A, B	Pathak et al. (1971); IRRI (1976)
Yellow rice borer	*Tryporyza incertulas*	A, B	Manwan (1975); IRRI (1976)
Green rice leafhopper	*Nephotettix virescens*	A, B, C	Athwal et al. (1971); Athwal and Pathak (1972); Siwi and Khush (1977)
Zigzag rice leafhopper	*Recilia dorsalis*	A	Pongprasert (1974); IRRI (1976)
Brown planthopper	*Nilaparvata lugens*	A, B, C, D	Athwal et al. (1971); Athwal and Pathak (1972); Lakshminarayana and Khush (1977)
White-backed planthopper	*Sogatella furcifera*	A, B, C	IRRI (1976); Pablo (1977)
Rice delphacid	*Sogatodes orizicola*	A, B	Jennings and Pineda (1970)
Rice bug	*Leptocorisa varicornis*	A, C	Sethi et al. (1937)
Rice stinkbug	*Oebalus pugnax*	A	Nilakhe (1976)
Rice gall midge	*Orseolia oryzae*	A, B, C, D	Venkataswamy (1968); Fernando (1972); Shastry et al. (1972); Hidaka et al. (1977); Khush (1977)

Rice whorl maggot	*Hydrellia phillippina*	A, B	Viajante and Herrera (1976); Pathak and Khush (1977)
Rice stem maggot	*Chlorops oryzae*	A, B	Koyama and Hirao (1971)
Stalk-eyed borer	*Diopsis thoracica*	A	Soto and Siddiqi (1976)
Rice leaf folder	*Cnaphalocrosis medinalis*	A	Lippold (1971); Gonzales (1974); Velusamy et al. (1975); Soehardjan et al. (1975)
Rice hispa	*Dicladispa armigera*	A	Lippold (1971); Venkata Rao and Muralidharan (1977)
Rice water weevil	*Lissorhoptrus oryzophilus*	A	Bowling (1973)
Rice thrips[b]	*Baliothrips biformis*	A	Velusamy et al. (1975); Kudagamage (1977)
Angoumois grain moth	*Sitotroga cerelella*	A	Cohen and Russell (1977); Russell and Cogburn (1977)
Rice weevil	*Sitophilus oryzae*	A	Bishara et al. (1972); Rout et al. (1976)
Maize weevil[b]	*S. zeamais*	A, C	Rossetto et al. (1973)
Lesser grain borer	*Rhyzopertha dominica*	A	McGaughey (1973); Cogburn (1977)

[a] A = sources of resistance identified; B = resistant varieties released; C = genes for resistance identified; D = insect biotypes encountered.

[b] Generally a minor pest.

and Fort Collins, Colorado. The International Rice Research Institute (IRRI) in the Philippines is assembling the largest collection of rice. At present it has about 38,000 accessions of *O. sativa*, 1,372 strains of *O. glaberrima*, 866 wild rices, and 637 genetic testers and mutants from different parts of the world. In collaboration with various national programs and by appropriate direct acquisitions, it expects eventually to include most of the rice strains in its collections.

Resistance to a particular pest is usually found in materials from areas where the pest occurs. Exceptions include 300 collections of *O. glaberrima* rice highly resistant to the green leafhopper, *Nephotettix virescens,* although this species is not known to occur in Africa (Pathak, 1977), as well as many Asian varieties of *O. sativa* highly resistant to the delphacid *Sogatodes orizicola,* which is found only in the Americas (Jennings and Pineda, 1970). However, rice strains grown in areas indigenous to the pest should receive priority in screening for resistance. Diverse sources and types of resistance are important as protection against the development of insect biotypes.

The preferred procedure is to discard the obviously susceptible germplasm using mass screening, field, or greenhouse tests. The selected strains are then retested intensively to confirm their resistance and to discard the pseudoresistant plants, including cases of escape and host evasion (Painter, 1951). Genetic studies indentify distinct resistance genes. These genes can also be sought among closely related species of rice and in rice mutants. Although information on the nature of resistance or its inheritance is important, experience has shown that it is not essential for the development of rice varieties resistant to insect pests.

Once sources of resistance have been identified, the resistant factor(s) must be incorporated into rice plants of good agronomic characters. Also, because the rice plant has several insect pests, diseases, and other problems, resistance to a particular pest is only a part of the overall strategy required for breeding insect resistant varieties. Varieties lacking resistance to any of these critical problems will have limited use. This consideration led IRRI to develop an interdisciplinary varietal improvement program called Genetic Evaluation and Utilization (GEU) in which scientists of different disciplines work and share the results. The IRRI program works simultaneously with resistance to five major insect pests and seven diseases, and tolerance to four different kinds of problem soils, drought, flood submergence, and suboptimal temperatures. The example of procedures for breeding for brown planthopper resistance is in Figure 1.

The stability of resistant varieties under different agroclimatic conditions and pest (biotype) situations is tested by growing a set of these varieties at selected locations. IRRI coordinates these activities and provides seeds and booklets containing information on methodology. Also, it arranges for the

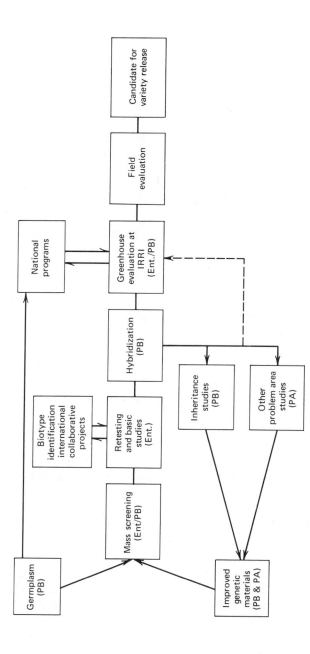

Figure 1. The building of brown planthopper resistant varieties. Pest resistance and superior agronomic characteristics are bred into all varieties released through the Genetic Evaluation and Utilization Program (GEU). Entomologists (Ent.), plant breeders (**PB**), and other problem area scientists (**PA**) work together to obtain these objectives (**IRRI**, 1973).

427

participating scientists to visit the test sites, and analyzes and disseminates the resulting data.

2. BREEDING TECHNIQUES AND PROCEDURES

2.1. Floral Morphology and Flowering

The rice inflorescence is a terminal, branched, compound panicle, having numerous pediceled flowers or spikelets (Chang and Bardenas, 1965). Each spikelet has two sterile glumes, a lemma and palea, two lodicules representing the perianth (calyx and corolla), six stamens, and a monocarpellary superior ovary with a short style and bifid feathery stigma.

Normally, flowering in a panicle continues for 4 to 7 days. Anthesis takes place from about 7 to 11 AM, depending on the season and weather conditions. The opening of the lemma and palea is followed by anther protrusion and the release of pollen. Spikelets remain open for 5 to 10 minutes and completely close in about 30 minutes. Stigmas remain receptive for a few days after anthesis. Pollen remains viable for about 5 minutes. Essentially all flowers are self-pollinated.

2.2. Techniques and Procedures

Parental material is usually grown in large pots in a greenhouse or a screenhouse. To make crosses, spikelets of the female parent are emasculated by clipping spikelets with scissors above or through the anthers, which are removed by hand with forceps or with a vacuum emasculator.

Because anthesis normally begins in the forenoon and lasts about 2 hours, it is better to emasculate late in the afternoon. The emasculated panicle is covered with a small glassine bag and identified. The next morning, the emasculated spikelets are pollinated by dusting them with pollen from a blooming panicle of the male parent. The date of pollination and the male parent are identified and the pollinated panicle is reenclosed with the glassine bag.

The crossed seed matures about 25 days after pollination. The seed is removed, threshed by hand, and the remnant glumes adhering to the seeds are removed. The naked seeds are counted and kept in small envelopes marked to identify the cross and number of seeds obtained. Seed dormancy is broken by air-drying the seeds for 7 days at 50 to 55°C.

In terms of appearance and the subsequent expression of segregation ratios, it is irrelevant which of the two parents is emasculated and which provides the pollen. The single cross F_1's are normally backcrossed to the more desirable parent or top crossed to a third parent.

The number of crossed seeds to produce for each single cross depends on the fertility of the F_1 and the desired size of the F_2. When the F_1 is fertile each plant normally produces a minimum of 10 panicles each bearing about 100 seeds. Thus 10 to 20 crossed seeds allow the breeder to keep a portion in storage for future use and to plant the remainder, giving sufficient F_1 plants to establish a large F_2 population. Proportionately more seeds are required when the F_1 is partially sterile or when backcross or three-way F_1 populations are needed.

The F_1, F_2, and the following segregating generations are grown in the field without insecticide protection. Each selection made in the segregating populations is screened against diseases and insect pests and other stresses depending on the parental combination of the cross. Promising uniform cultures from the F_6 or F_7 are advanced to observational yield nurseries or to replicated yield trials.

3. PROGRESS IN BREEDING FOR RESISTANCE

3.1. Stem Borers

The common stem borers include 17 species of Pyralidae and 3 species of Noctuidae. Of these, the striped rice borer, *Chilo suppressalis,* dark-headed rice borer, *Chilotraea polychrysa,* yellow rice borer, *Tryporyza incertulas,* white rice borer, *T. innotata,* and pink borer, *Sesamia inferens,* are the most widely distributed and the most destructive. Progress has been made in breeding for resistance to the striped and yellow rice borers. A few rice varieties carry resistance to more than one borer species (Das, 1976a).

Striped Rice Borer

More than 14,000 rice varieties have been screened for resistance to the striped borer (Pathak, 1977). Such high volume screening requires simple and efficient field, greenhouse, and laboratory tests.

Screening of Varieties. Initial varietal screening is conducted in paired rows 5 m long. The planting of test varieties is timed so that their maximum tillering coincides with the harvest of neighboring rice crops. This generally brings about heavy infestations as borer moths emerging from maturing fields migrate to oviposit on the younger plants. Thus representative data on varietal susceptibility can be obtained in unreplicated experiments and up to 3,000 varieties can be tested in one planting.

The borer incidence is recorded as dead hearts 60 days after transplanting and as white heads near harvest. The percentage of damage is calculated as follows (Onate, 1965):

$$x = P\bar{x}_{nz}\ 100$$

where $P = \dfrac{\text{Number of affected hills}}{\text{Number of hills in the plot}}$

$\bar{x}_{nz} = \dfrac{\text{Number of dead hearts}}{\text{Number of tillers in the affected hills}}$

x = Percentage dead hearts or white heads

Stems of selected few varieties are dissected to record larval populations.

Varieties that show low borer infestation in preliminary screening tests are retested to confirm their resistance. The procedure is the same as for mass screening except that each variety is planted in four-row plots with three to four replications. Resistant and susceptible check varieties are planted every other tenth plot. Finally, all varieties noted as resistant in different groups of retested materials are tested in one batch in a replicated field experiment. The most resistant of all are further tested in the greenhouse or laboratory using uniform borer infestations and one or more of the following methods.

1. *Infesting potted plants.* Four 20-day-old seedlings of the test varieties are planted in 12-inch clay pots. After 30 days, each plant is infested with 10 first-instar borer larvae and placed in a water-filled tray to prevent larvae from leaving the plant. Discoloration of leaf sheaths, dead hearts, and white heads are recorded at 5-day intervals. At 20 to 30 days after infestation, the plants are harvested and individual tillers are examined for counting the surviving borer larvae and pupae which are weighed separately.

2. *Rearing larvae on rice stalks.* Stalks 7 cm long are cut from 60-day-old test varieties grown in a greenhouse. Two such stalks and 10 first-instar larvae are enclosed in a 7.5 × 2 cm glass vial having a screened cap. Fresh stalks of the same variety are replaced periodically (usually every 5 days) until most larvae on the susceptible check variety have pupated. The surviving larvae are counted and weighed whenever the stalks are replaced.

3. *Rearing larvae on seedlings.* Seeds (30 g) of each test variety are germinated in a 8 × 17 cm glass jar having a screened cap. After 5 days, 100 first-instar larvae are placed on the seedlings in each jar and transferred to fresh seedlings at 10-day intervals until pupation.

Differences in Insect Survival and Damage Caused. The above studies have identified a number of varieties that are moderately resistant to stem borers, as well as various characteristics associated with this resistance. The larvae caged on resistant varieties suffer higher mortality and have slower rates of growth, smaller body size, and smaller percentages of pupation than those caged on susceptible varieties (Israel, 1967; Pathak, 1969; Pathak et al., 1971). Many varieties, particularly japonica types, are moderately resistant to the striped borer during the vegetative stage but become susceptible after panicle initiation. In field experiments, these varieties have a generally low incidence of dead hearts but frequently suffer heavily from white head damage. This change in susceptibility appears to be due to differences in susceptibility of different plant parts (Pathak et al., 1971). A few indica varieties show uniformly low stem borer infestation at all stages of plant growth.

Nature of Resistance to Stem Borer. The striped borer moths strongly prefer certain varieties for oviposition. Under low borer populations, they oviposit heavily on certain varieties while others remain virtually egg-free (Patanakamjorn and Pathak, 1967; Pathak, 1969; Pathak et al., 1971). However, even under heavy infestations, many varieties receive only a few egg masses. Generally, the varieties that receive more eggs also have more dead hearts, indicating that oviposition is a critical factor in determining borer damage. However, a few varieties, despite receiving a larger number of eggs, suffer less borer damage than others. These varieties have been found to have adverse effects on survival of young borer larvae. Thus antibiosis is another major factor of resistance of rice varieties to the striped borer. On the resistant varieties 'Taitung 16' and 'Chianan 2,' only half as many larvae survived as on the susceptible varieties 'Rexoro' and 'Sapan Kwai' (Figure 2). The larvae also pupated earlier and in greater numbers on the susceptible varieties. Furthermore, the larvae caged on the susceptible varieties weighed about twice as much as those caged on the resistant varieties (Figures 3 and 4).

A general association between several morphological and anatomical characteristics of the rice plant and resistance to stem borers has been recorded (Table 2). Each of these characteristics appears to contribute to borer resistance and none by itself appears to be the main cause of such

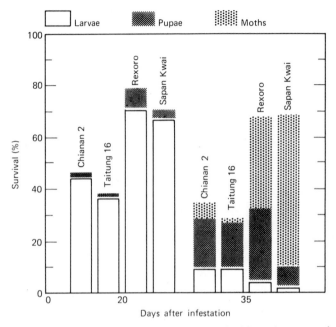

Figure 2. Survival and development of 600 first-instar striped borer larvae caged on each of four rice varieties (Pathak et al., 1971).

resistance. This relationship was evident in several varieties that reacted as susceptible to the borers even when one of the characteristics they possessed was positively correlated with resistance.

Tall varieties, because of their height, might be more attractive to ovipositing moths. The length and width of the flag-leaf blade were positively correlated with borer susceptibility. In separately conducted ovipositional preference tests these characteristics were positively correlated ($r = 0.743$ and 0.924, respectively) with the number of egg masses laid.

A hairy leafblade surface might act as a physical deterrent for the egg-laying moths. However, even the removal of the hairs from the leaf surface of the resistant variety 'TKM6' did not make it more attractive for oviposition by the borer moths. Varieties whose resistance was attributable to biophysical characteristics were generally more resistant to more than one species of the stem borers.

Resistance to the striped borer has been recorded in rice varieties having leaf sheaths tightly wrapped around the stem, closely packed vascular bundles, thick sclerenchyma, and high silica content. These characteristics probably interfered with the boring activity of larvae in the stem as larvae feeding on silicious rice varieties exhibited typical antibiosis effects and had worn out mandibles (Djamin and Pathak, 1967; Patanakamjorn and Pathak, 1967).

A rice plant biochemical, "oryzanone" (p-methylacetophenone), attracted ovipositing moths, which laid more eggs on the treated than on the untreated surface (Munakata and Okamoto, 1967). The odor of "oryzanone" also attracted borer larvae.

Incorporation of fresh plant extracts of resistant and susceptible varieties into artificial diets showed that the borer larvae preferred the susceptible 'Rexoro' variety over resistant 'TKM6,' and grew poorly on the other resistant variety 'Taitung 16' (Das, 1976b).

Recent studies at IRRI showed that striped borer moths oviposited heavily on surfaces treated with the odor of the susceptible 'Rexoro' variety, very little on untreated surfaces, and none on surfaces treated with the odor of 'TKM6' variety. The effect of plant odor on oviposition was so strong that the moths laid three times more eggs even on the resistant 'TKM6' plant when it was sprayed with the odorous extract of the susceptible 'Rexoro' variety, but oviposition was inhibited when the latter was sprayed with the odor of 'TKM6' (Figure 5).

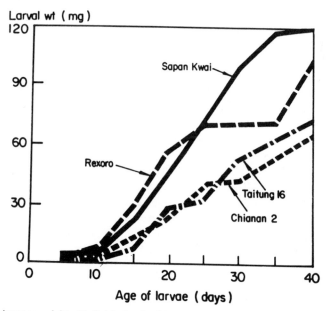

Figure 3. Average weight of individual striped borer larvae reared on resistant and susceptible varieties of rice (Pathak et al., 1971).

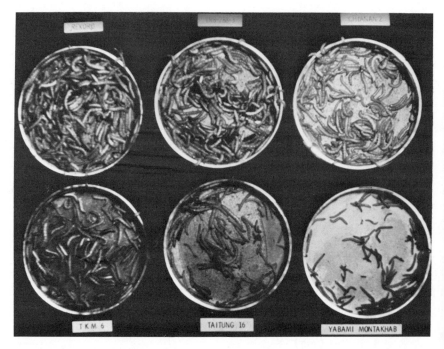

Figure 4. Survival and development of striped borer larvae are better on susceptible than on resistant lines (Pathak et al., 1971).

Effects of Resistant Varieties on Striped Borer Populations. The cumulative effects of varietal resistance on striped borer populations were demonstrated by caging an identical number of borers for several generations on two resistant ('Taitung 16' and 'Chianan 2') and the susceptible variety 'Rexoro' (Pathak, 1972). At 120 days after infestation, two egg masses and 21 larvae were recovered from resistant 'Chianan 2,' whereas susceptible 'Rexoro' yielded 83 egg masses and 1,583 borer larvae. 'Rexoro' had 56 percent dead hearts but 'Chianan 2' suffered only 1 percent damage. Beside the low survival and slow growth rate of larvae, the uneven emergence of male and female moths on resistant varieties may fail to synchronize mating and therefore reduce oviposition. This latter effect is illustrated in data presented in Figure 6.

Breeding for Striped Borer Resistance. Selected resistant varieties have been utilized in a hybridization program to improve their level of borer resistance and to incorporate their resistance in plants with desirable agronomic characters. 'TKM6' has been used extensively in several coun-

Table 2. Correlations between Rice Plant Characters and Percentages of Tillers Infested with Striped Borer[a]

Plant Character	Correlation Coefficient[b]
Elongated internodes, number	0.632
Third elongated internode, length	0.715
Flag leaf, length	0.798
Flag leaf, width	0.836
Culm height	0.796
Culm, external diameter	
At half its length	0.672
At one-fourth its length from the base	0.785
Culm, internal diameter	
At half its length	0.671
At one-fourth its length from the base	0.790
Tillers per plant, number	−0.756
Stem area occupied by vascular bundle sheaths (percentage)	−0.756

[a] Data from Pathak et al. (1971).
[b] All values highly significant.

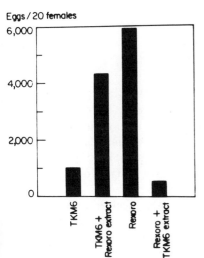

Figure 5. Oviposition by striped borer moths on plants treated with ether extract of steam distillates of resistant ("TKM6") and susceptible ("Rexoro") rice plants (IRRI, 1976).

435

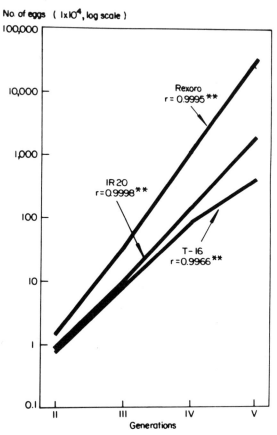

Figure 6. Population buildup of striped borers on three varieties. An initial population of 350 first-instar larvae was caged on each variety (IRRI, 1972).

tries in breeding for borer resistance. 'IR20,' the first borer resistant improved plant-type variety was developed by crossing 'TKM6' with 'Peta³ × TN1.' It has moderate resistance to striped and yellow borers and resistance to green leafhopper, tungro virus, bacterial leaf blight, and several races of rice blast. Its cultivation expanded rapidly, to more than several million hectares in southern and Southeast Asia. However, it succumbed recently to another pest, the brown planthopper.

Recent breeding for resistance to striped borer involves making diallel crosses and evaluating progeny lines. Seven varieties moderately resistant to the striped borer have been used in diallel crosses, and the resistant progenies were selected and intercrossed. Three generations of this selective crossing systems have produced progenies distinctly more resistant than any parent (Figure 7).

Yellow Rice Borer

Evaluation of rice varieties for resistance to the yellow borer has been rather difficult because of the problems in mass rearing of this insect and the extreme sensitivity of its larvae to handling. Therefore, the insect is reared on plants grown inside a large screenhouse and the emerging moths are allowed to oviposit on the test varieties grown in adjacent nursery beds inside the same screenhouse.

Screening for resistance to yellow borer is enhanced considerably in the tropics by adjusting the planting date of the test varieties so that they are at

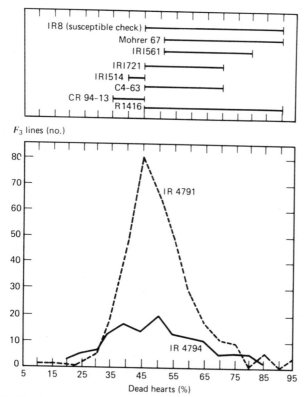

Figure 7. Distribution of dead hearts among F_3 lines and parents of the diallel crosses "IR4791" and "IR4794" (IRRI, 1975).

an appropriate stage of infestation when most of the rice crop in the area is ready for harvest. The moths emerging from the maturing crop oviposit heavily on this young crop. Using this procedure, several thousand breeding lines and selections were evaluated for resistance and 14 lines or varieties were identified as moderately resistant (Manwan, 1975). The resistance of these lines appeared to be due primarily to antibiosis. Larvae feeding on resistant varieties were smaller, had lower survival, and caused lower percentages of dead hearts than those feeding on susceptible varieties (Figure 8). Larval weight was greatly influenced by the stage of plant growth, fertilizer application, herbicide treatment, and soil salinity. Larval weights and survival were least on plants infested at 15 days before harvest. Plants receiving high doses of nitrogen fertilizer or 2,4-D herbicide favored oviposition, larval growth, and establishment. Also, more larvae survived on plants growing in saline soils. Thus plant age and agronomic practices influence borer infestation.

Differences in nonpreference for oviposition were not distinct in the screenhouse tests. This could be due to the fact that the yellow borer moths freely oviposited on plant species other than rice and even on such objects as iron poles and wooden frames.

Figure 8. The line "IR1820-52-2" has the highest level of resistance to the yellow borer of any rice yet screened at IRRI. "Rexoro" is susceptible. Screenhouse trials. (IRRI, 1975.)

3.2. Leafhoppers and Planthoppers

Several species of leafhoppers and planthoppers damage the rice plant by sucking sap and by transmitting virus diseases. The most damaging species are the green leafhoppers, *Nephotettix* spp., the zigzag leafhopper, *Recilia dorsalis*, the brown planthopper, *Nilaparvata lugens*, the smaller brown planthopper, *Laodelphax striatella*, the white-backed planthopper, *Sogatella furcifera*, and the rice delphacid, *Sogatodes orizicola*. Several recent studies have demonstrated that natural resistance to leafhoppers and planthoppers exists in rice varieties, and such resistance is easily transferred to the high yielding rice varieties (Pathak, 1972; Pathak and Khush, 1977). Breeding for leafhopper and planthopper resistance has now become a major research objective in most of the rice-growing countries of Asia.

Mass Rearing of Insects

Leafhoppers and planthoppers are reared on 8- to 9-week-old plants of a susceptible rice variety. However, for rearing *Nephotettix nigropictus* a mixture of rice and barnyard grass, *Echinochloa crusgalli* is better than rice alone (Sajjan, 1972). Four to five hundred adults are collected from the stock culture and released for feeding and egg laying on potted plants kept inside rearing cages. The insects can also be reared on young seedlings growing in glass, aluminum, or plastic trays. Fresh plants are replaced every third day. The used plants are transferred to other rearing cages for emergence of nymphs from eggs laid by the hoppers. This method provides a continuous supply of insects of uniform eggs.

Screening of Varieties

The test varieties are sown in small pots or seed boxes filled with soil. A susceptible check and a resistant check variety are also sown in random rows in each seed box. After one week, each variety is thinned to 15 or 20 seedlings per row and the seed boxes are transferred to a water-filled tray. Several thousand second- and third-instars of the test insect are uniformly scattered on the seedlings. This infestation is sufficient to kill the susceptible varieties. The number of insects on each variety and the damage are recorded at 5-day intervals according to the standard scoring (Table 3). Final grading is done after all susceptible check rows are killed.

The selected varieties from this mass-screening procedure are retested by using a similar procedure but by replicating it three or four times.

Table 3. Standard Rating for Damage by Leafhoppers and Planthoppers

Grade of Damage	Rating[a]	Seedling Damage by			
		Green Leafhopper	Brown Planthopper	White-Backed Planthopper
0	HR	No damage	No damage	No damage
1	R	First leaf yellow	First leaf partially yellow	First leaf yellow-orange
3	MR	50–70% of all leaves yellow	First and second leaves partially yellow	50% of leaves or their tips yellow-orange, slight stunting
5	MS	All leaves yellow	Marked yellowing, stunting	Most leaves or their tips yellow-orange, stunting
7	S	50% of plants dead	Severe wilting and stunting	50% of plants dead, severe wilting and stunting
9	HS	Plants dead	Plants dead	Plants dead

[a] HR = highly resistant; R = resistant; MR = moderately resistant; HS = highly susceptible; S = susceptible; MS = moderately susceptible.

Differences in Insect survival and Damage Caused

The varieties rated as resistant or moderately resistant are further evaluated by determining the survival of adult or nymph insects on individual potted plants.

The results of one such experiment are in Figure 9. Few nymphs of the brown planthopper survived on the variety 'Mudgo' and they died within 10 days after caging; many survived on 'Taichung Native 1' (TN1) and 'Pankhari 203.' Survival of green leafhopper, however, was low on 'Pankhari 203,' but high on 'Mudgo' and 'TN1.' This pattern demonstrated that resistance to the brown planthopper is different from resistance to the green leafhopper. In repeated similar experiments, insects caged on resistant varieties had slower growth and higher mortality than on susceptible varieties. Moreover, even a large population of the insects caged on resistant plants caused barely noticeable symptoms whereas the susceptible variety was killed (Figure 10).

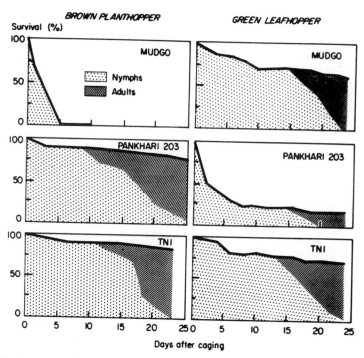

Figure 9. Survival and development of first-instar nymphs of brown planthopper and green leafhopper on 60-day-old plants of resistant and susceptible varieties (Pathak et al., 1969).

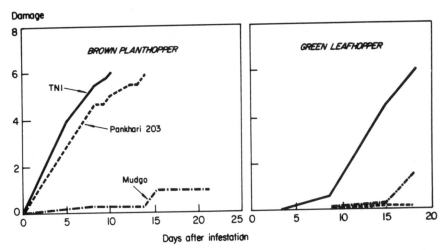

Figure 10. Damage caused by caging 100 first-instar nymphs on resistant and susceptible rice varieties (Pathak et al., 1969).

In all cases investigated so far, varieties resistant at juvenile stages are also resistant at later stages of plant growth, except for 'ARC 575,' which is highly resistant to the white-backed planthopper at the vegetative stage of growth although its panicles are as susceptible as those of susceptible varieties (Rodriguez-Rivera, 1972; Pablo, 1977). Normally white-backed planthopper does not feed on panicles.

Nature of Resistance to Leafhoppers and Planthoppers

The insects exhibit distinct nonpreference for certain varieties. This response is gustatory rather than visual, mechanical, or olfactory, since the insects show no differences in alighting upon different varieties but they do not stay on resistant varieties for sustained feeding (Sogawa and Pathak, 1970; Pura, 1971; Cheng and Pathak, 1972; Pongprasert, 1974; Rezaul Karim, 1975; Pablo, 1977; Saxena and Pathak, 1977). The brown planthopper biotype 1 feeds much less on resistant 'Mudgo' and 'ASD7' varieties than on the susceptible 'TN1' variety. On the other hand, brown planthopper biotypes 2 and 3 ingest an almost equal amount of food from 'Mudgo' and 'ASD7' varieties, respectively, as those taken from the susceptible 'TN1.' 'Mudgo' is susceptible to biotype 2, while 'ASD7' is susceptible to biotype 3.

Generally, weight gain of insects is low on resistant varieties as compared to that on susceptible varieties (Figure 11). This gain in weight reflects the amount of sap sucked by the insects and is shown by measuring the amount of honeydew excreted by the insects (Sogawa and Pathak, 1970; Saxena and Pathak, 1977). The green leafhopper also excretes more honeydew on the susceptible variety 'TN1' than on the resistant varieties 'Pankhari 203' and 'IR8' (Cheng and Pathak, 1972), but differences are not as marked as for the brown planthopper.

The possibility of a mechanical barrier preventing the stylets of the insects from reaching proper feeding sites in resistant varieties was investigated by scoring the feeding sites in plant tissues. Adult brown planthoppers made two to three times more feeding punctures on the resistant 'Mudgo' than on susceptible 'IR8' and 'TN1' varieties (Sogawa and Pathak, 1970; Sogawa, 1971). Furthermore, there were more stylet punctures through the fiber tissues, which are harder than parenchymatous tissues. Also, a higher percentage of the salivary sheaths terminated in the vascular bundles of "Mudgo' than in 'IR8' or 'TN1' plants. Thus hardness of the tissues is apparently not a factor in planthopper resistance. Similar results have been obtained for the green leafhopper (Cheng and Pathak, 1972), zigzag leafhopper (Pongprasert, 1974) and the white-backed planthopper (Rodriguez-Rivera, 1972; Pablo, 1977) caged on resistant and susceptible plants.

Thus the resistance to brown planthoppers, which do little feeding on resistant varieties, appears to be due to either lack of feeding stimulants or

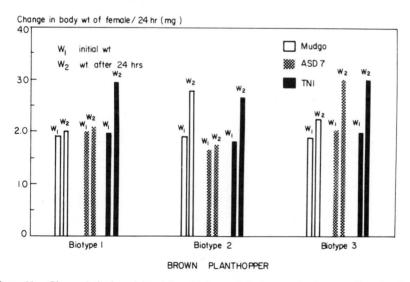

Figure 11. Change in body weight of three biotypes of the brown planthopper allowed to feed on "Mudgo," "ASD7," and "TN1" rice varieties (IRRI, 1976).

presence of feeding deterrents for these insects (Sogawa and Pathak, 1970; Sogawa, 1971). However, the green leafhopper, zigzag leafhopper, and white-backed planthopper do some feeding on resistant varieties and therefore the resistant plants either contain toxins or lack nutrients vital to these insects.

There were no significant differences in the number of eggs laid by the gravid biotype 1 females on resistant and susceptible rice varieties, but on the resistant barnyard grass a much smaller number of eggs was laid by the hoppers (Table 4). However, a significantly smaller percentage of eggs hatched on the resistant varieties 'Mudgo,' 'ASD7,' and 'IR26' than on the susceptible 'IR20,' 'IR8,' and 'TN1' varieties. Hatching on barnyard grass was even poorer than on the resistant rice varieties (Table 4). Similar observations have been made on egg laying by biotype 2 (Md. Iman, personal communication) and biotype 3 hoppers (Alam, 1978) and hatching of their eggs on respective resistant and susceptible varieties.

Examination of the unhatched eggs showed that early embryonic development, recognized by the onset of eye pigmentation, proceeded normally, but hatching was apparently prevented because of the failure of developing larvae to split the chorion. Eggs laid in the susceptible variety 'TN1,' and those artificially transplanted in the leaf sheaths of another 'TN1' plant, hatched normally, whereas those transplanted in barnyard grass had very low hatching (Saxena and Pathak, 1977).

Table 4.   Oviposition and Hatching of Brown Planthopper (Biotype 1) Eggs on Resistant and Susceptible Rice Varieties and on Barnyard Grass, IRRI, 1975–1976 (Saxena and Pathak, 1977)

Test Plants	Eggs Laid/10 Females/24 Hours[a,b] (Number)	Eggs Hatched[a,c] (Percent)
Mudgo	328 ab	68 b
ASD7	348 ab	78 b
IR26	292 ab	74 b
IR20	325 ab	95 a
IR8	309 ab	90 a
TN1	400 ab	91 a
Barnyard grass	205 a	18 c

[a] In a column, any two means followed by a common letter are not significantly different at 5 percent level.
[b] Based on values transformed to log X, base 10.
[c] Based on values transformed to $\arcsin^2 (x)$.

The resistance of 'Mudgo' to brown planthopper biotype 1 is attributable to its lower asparagine content than in the susceptible variety 'TN1' (Sogawa and Pathak, 1970). In addition, the free amino acid concentration in whole plant extract and xylem exudates is three to four times lower in 'Mudgo' than in 'TN1' (Sogawa, 1971). Asparagine is highly phagostimula- tory to brown planthopper biotype 1 at the highest (4 percent) concentration tested (Saxena and Pathak, unpublished). The asparagine content of the rice plant is believed to be greatly influenced by the amount of nitrogenous ferti- lizer applied. However, application of high rates of nitrogenous fertilizers did not alter the resistance of 'Mudgo' (Kalode, 1971).

A brown planthopper antifeedant, *trans*-aconitic acid, was isolated and identified from barnyard grass, which is nearly immune to the brown plant- hopper (Kim et al., 1975, 1976).

Breeding for Resistance to Leafhoppers and Planthoppers

Breeding for resistance to leafhoppers and planthoppers has become one of the main objectives of the breeding programs at IRRI and many other countries. The IRRI program started in 1967 with a cross between 'Mudgo' (resistant) and 'IR8' (improved plant-type) which produced brown plant- hopper-resistant progeny but had poor grain quality. Since then, progress in breeding for resistance to leafhoppers and planthoppers has been quite dra- matic and several high-yielding resistant varieties have been released. The resistance to the brown planthopper of one such breeding line is illustrated in Figure 12. These varieties are also resistant to a number of other insect pests and diseases. In 1975, two new genes were identified for resistance to brown-planthopper, bringing the total to four (Lakshminarayana and Khush, 1977). Similarly, two new genes were identified for resistance to the green leafhopper (Siwi and Khush, 1977). These genes are being incor- porated into improved plant-type backgrounds. At present, many countries are using local as well as IRRI improved lines or varieties in breeding pro- grams for resistance to these insects. Because the development of biotypes of the brown planthopper endangers the stability of resistant varieties, the present breeding endeavors are aimed at utilizing major as well as minor genes for resistance.

Breeding for resistance to the white-backed planthopper started in 1975. About 450 resistant lines were found among more than 4,000 entries tested. 'N22,' 'ARC6003,' and 'Dharia' were used as donor parents in a backcross- ing program. Breeding lines with resistance to white-backed planthopper and multiple resistance to diseases and other insects are available now.

Among more than 500 varieties tested by CIAT for resistance to the rice delphacid in Colombia, about 100 indicas from Asia, where the insect does

Figure 12.   Both replications of a resistant line showed no visible damage, whereas two other varieties (susceptible) were killed by the brown planthopper (IRRI, 1972).

not occur, were found resistant (Jennings and Pineda, 1970). All varieties from the Western Hemisphere were susceptible. The insects caged on resistant varieties suffered high mortality. The resistance was highly heritable and easily recombined with other agronomic traits. All varieties released by CIAT are highly resistant, with no indication of biotype development. This resistance was also independent of resistance to the hoja blanca virus, transmitted by the rice delphacid. Varieties resistant to the vector and susceptible to hoja blanca, however, show little virus disease in the field.

3.3.   Rice Gall Midge

The rice gall midge, *Orseolia (Pachydiplosis) oryzae,* is a serious pest of rice in southern and Southeast Asia and a few African countries. The young larvae feed at the growing points of the rice plant causing development of onion leaf-like galls (onion shoots or silver shoots). The tillers thus affected produce no panicles. Yield losses are total if the infestation is high. Chemical control of this insect has been difficult, but use of resistant varieties is practicable. Most of the work on breeding for gall midge resistance has been done in India, Sri Lanka, and Thailand.

In greenhouse screening for gall midge resistance, the adults are first reared from field-collected, pest-infested plants. Ten-day-old seedlings are infested with these insects for oviposition and then transferred to cages kept in a moist place. Galls elongate 20 to 25 days after infestation. Adults emerge at night, copulate, and are ready to start new infestations. The test varieties and a susceptible check such as 'IR8' are randomly planted and replicated in eight blocks of three rows each, seven plants per row, in seed boxes. Twelve to 15 gravid females are released in each seed box. Damage, rated when gall elongation is complete in the susceptible check, is based on the percentage of galls formed (Fernando, 1972, 1975).

Field screening is done when the gall midge incidence is high. Electric lighting (one 100-watt bulb per 25 m², lit from 7 PM to 4 AM) over 40-day-old test varieties, until the crop is 80 days old, has been reported to increase gall midge incidence by 1.5 to 2.5 times (Prakasa Rao, 1975).

Varietal screening has identified several sources of resistance. Upland rices and early-ripening varieties were generally found to suffer less from gall midge damage in Taiwan (Li and Chiu, 1951). In India, low infestations of the pest were found in scented, low-tillering varieties (CRRI, 1954). In early efforts to breed for resistance, a resistant Indian variety 'Eswarakora' was crossed with 'MTU15.' This resulted in several tall 'Warangal' cultures, such as 'WC1263,' 'WC1257,' which have been used extensively as donor parents for gall midge resistance (Venkataswamy, 1968). Since then, more than 10,000 rice strains comprising local and exotic germplasm have been screened in India and several outstanding sources ('HR42,' 'HR63,' 'Ptb18,' 'Ptb21,' 'JBS446,' 'JBS673,' 'Siam29,' etc.) were identified (Shastry et al., 1972; Israel, 1974). The varieties 'Shakti' and 'Vikram' and two lines, 'RPW-15' and 'CR93-4-2,' have high resistance and good yield (Mishra et al., 1976; Naidu et al., 1976). In Sri Lanka, varying levels of resistance have been found among four 'Eswarakora' selections (Fernando, 1972). But in Indonesia the variety 'RD4,' having resistance from 'Eswarakora,' is susceptible (Soehardjan et al., 1975). This points to the occurrence of biotypes among gall midges (Hidaka et al., 1977). In Thailand, 'RD4' and 'RD9' varieties have good gall midge resistance, but poor cooking qualities and susceptibility to the brown spot disease. Two new lines, 'BR 1031-3-4-3' and 'BR 1031-7-5-4,' have high gall midge and brown spot resistance, good plant type, and high yield potential (Weerawooth Katanyukul et al., 1977). Two IRRI varieties, 'IR32' and 'IR36,' are resistant to gall midge (Khush et al., 1977).

The resistance appears to be primarily antibiosis. The development of larvae is retarded on resistant varieties whereas development proceeds normally on susceptible varieties (Fernando, 1972; Shastry et al., 1972; Wongsiri et al., 1971). No differences among varieties in egg laying by the gall midge or in penetration by its larvae to the growing point have been recorded.

3.4. Rice Whorl Maggot

The rice whorl maggot, *Hydrellia philippina,* infests the rice plants from the seedling stage to the beginning of panicle initiation. The maggots generally feed on the developing central whorl leaves (Andres, 1975). The damage symptoms are whitish discolorations of the leaf margin and small holes on the leaf surface caused by larval feeding. In endemic areas, such as the Philippines and Thailand, nearly all rice fields are infested and severe infestation causes stunting of the plants.

Since 1972, about 20,000 rices have been screened for whorl maggot resistance (Viajante and Herrera, 1976). The test varieties are planted at a 25 × 25 cm spacing in a single row. The least damaged entries are retested for resistance in 1 × 5 m plots in replicated trails. Damage is evaluated on a row basis 25 to 30 days after transplanting using a 0 to 9 standard scoring scale. A few varieties have shown moderate resistance but most are highly susceptible. To fortify the existing levels of maggot resistance, a diallel crossing program was initiated using seven moderately resistant varieties. 'IR40,' a moderately resistant variety, has a somewhat higher level of resistance than its parents, 'IR20' and 'CR94-13' (Viajante and Herrera, 1976; Pathak and Khush, 1977).

Resistance to the rice whorl maggot is mainly due to antibiosis. Few maggots survive on the resistant varieties and the survivors are smaller than those reared on the susceptible 'TN1.'

3.5. Rice Stem Maggot

The rice stem maggot, *Chlorops oryzae,* is an important pest of rice in the mountainous regions in Japan (Koyama and Hirao, 1971). The maggot feeds within the plant on developing leaves and unemerged panicles, causing damage to leaves and reduction in number of filled grains.

About 3,000 rice varieties have been screened in nursery beds, and selected varieties were retested with controlled infestations in greenhouse experiments (Koyama and Hirao, 1971). A few varieties such as 'Ou 188,' 'Sakaikaneko,' 'Oha,' and 'Ou 230' were highly resistant. Their resistance was derived from the old resistant variety, 'Joshu.' Resistance is monogenic and incompletely dominant. Although there are differences in egg laying by the fly on different varieties, high mortality of maggots on resistant varieties is the main factor in resistance.

3.6. Other Insect Pests of Rice

Sources of resistance to several other insect pests of rice have been identified, but generally the causes of resistance and breeding for resistance have not been investigated. Greenhouse and field screenings against heavy pest populations of the stalk-eyed borer, *Diopsis thoracica,* at the International Institute of Tropical Agriculture (IITA), Nigeria, identified some high yielding resistant lines such as 'IR589-53-2' and 'IR1561-38-6-5,' (Soto and Siddiqi, 1976).

Varietal screening using natural field populations of the rice leaf folder, *Cnaphalocrosis medinalis,* has been done in several countries. In Bangladesh, local varieties with narrow leaves (less than 0.8 cm wide) and introduced varieties with wider leaves (more than 1.5 cm wide) have shown resistance (0 to 10 percent infestation), whereas those with intermediate leaf width are highly susceptible (Lippold, 1971). In India, progenies of the cross 'TKM6' × 'IR8' and varieties 'IARI 6579' and 'Tetep' were resistant (Velusamy et al., 1975). In Indonesia, out of 188 local and foreign selections, 21 varieties from India and Sri Lanka remained undamaged by the leaf folder (Soehardjan et al., 1975). Of nearly 1,000 rices tested in a screenhouse at IRRI, 46 selections had low leaf folder damage (Gonzales, 1974). The resistance appears to be due to nonpreference for oviposition by moths and poor survival and development of larvae.

Lippold (1971) recorded differences in field incidence of the rice hispa, *Dicladispa armigera,* in Bangladesh on a few hundred foreign introductions. In Andhra Pradesh, India, a local variety 'Bulk H9' ('Molagolukulu') was heavily infested but a few entries in the International Rice Observational Nursery and the National Screening Nursery showed less than 1 percent leaf area damage (Venkata Rao and Muralidharan, 1977).

Field evaluation for resistance to the rice water weevil, *Lissorhoptrus oryzophilus,* has been conducted by planting rice varieties in rows (6 feet long, 27 inches apart) and recording the number of weevil larvae 3 weeks after first flooding. However, such investigations have not shown marked differences in varietal susceptibility (Bowling, 1963). Recently, a procedure was developed for screening varieties in the laboratory. The test varieties are grown in plastic trays filled with soil. Two weeks after planting, trays are flooded and one or two pairs of adult weevils are caged for oviposition on the test plants. After 1 week, the seedlings are removed, washed, and replaced in water-filled test tubes. The larvae are counted when they emerge from the eggs laid on the plants. Thus a number of varieties have been screened and a few showing weevil resistance identified (Bowling, 1973).

Varietal resistance trials have been conducted against rice thrips, *Baliothrips biformis,* using natural field infestations. Damage is assessed by calculating the percentage of affected leaves in 20 randomly selected plants. Out of 100 varieties tested in India, 15 indicas showed resistant reaction while the selection IR2070-747-6-3-2 was highly susceptible (Velusamy et al., 1975). In Sri Lanka, field screening of about 800 indigenous varieties identified 'Dahanala 2220' and 'Dahanala 682' as resistant to rice thrips (Kudagamage, 1977).

Uichanco (1921) recorded bearded or awned varieties of rice as almost immune to attack by rice bugs, *Leptocorisa* spp., for awns interfere with feeding, the hulls are tougher, and the glumes fit more closely. The poor-quality 'Sathia' varieties in India also carry resistance to rice bugs because the panicle remains enclosed in an extension of the leaf sheath (Sethi et al., 1937). A few sixth-generation true breeding hybrids between 'Sathia' and improved rices were found resistant. This resistance was controlled by three pairs of genes.

Recently, of the 228 lines screened for resistance to the rice stink bug, *Oebalus pugnax,* in Louisiana, a few showed moderate levels of resistance (Nilakhe, 1976).

In resistance trials of 500 local varieties in Indonesia against *Spodoptera mauritia* only two varieties, 'Jambrek' and 'Java Serut,' showed reduced damage. The others suffered moderate to total destruction (Soejitno and van Vreden, 1975).

3.7.   Pests of Storage

Nearly 20 species of insects damage paddy or rough rice (kernels with lemma and palea), parboiled rice (soaked and steamed), brown rice (kernels without lemma and palea but with germ), and milled rice (kernels without lemma, palea, or germ) during storage. The most injurious pests of paddy in the tropics are the Angoumois grain moth *Sitotroga cerealella,* the rice weevil *Sitophilus oryzae,* the maize weevil *Sitophilus zeamais,* and the lesser grain borer *Rhyzopertha dominica.* Milled rice is susceptible to more pests than is rough rice. The important pests are the rust flour beetle *Tribolium castaneum,* the saw-toothed grain beetle *Oryzaephilus surinamensis,* and the rice moth *Corcyra cephalonica.*

For evaluating resistance, 6-g samples of each variety are exposed in suitable containers to a specified number of pests for feeding and oviposition. Equilibration of moisture is necessary before evaluation and is obtained by adding water to the grain samples according to the following formula:

$$\text{g of water added} = \left(\frac{100 - \text{present percentage of moisture}}{100 - \text{desired percentage of moisture}} \times 6\,\text{g}\right) - 6\text{g}$$

The samples are placed in a rearing room 4 days beforehand because moisture content in rough rice stabilizes in 2 to 4 days (Juliano, 1964). The test samples may be offered in two ways: (1) when the insects have a free choice of all the samples or (2) when insects are confined to each sample. Resistance is evaluated by counting insects at different intervals, the damaged seeds of kernels, and/or insect progeny; however, Russell (1976) reported that infestation by eggs appeared to be the most dependable method for screening for resistance in rice to the Angoumois grain moth.

Screening of varieties for resistance to the Angoumois grain moth showed that the number of grains with broken or gaping palea and lemma was significantly correlated with the number of emerging moths (Cohen and Russell, 1970). About 1000 varieties originating in China, Cuba, Panama, Pakistan, and Taiwan were screened in the United States. Ten percent of these showed resistance, as moth emergence from them was 10 percent or less (Russell and Cogburn, 1977). Subsequent testing of fresh samples of these varieties confirmed their resistance. Hull morphology was important but testing of homogenates of rough rice indicated that varieties also differed in their nutritive suitability. However, the exact cause of this effect is not known.

The rice weevil causes considerable damage to rough, parboiled, and brown rice. In Egypt, among four varieties tested, 'Giza 159' proved to be the most susceptible, followed by 'Arabi,' 'Nahda,' and finally the 'Hybrid 170' (Bishara et al., 1972). In India, screening of eight high-yielding rices showed that 'Rajeswari' was the most susceptible and 'Vijaya' the most resistant (Rout et al., 1976). A negative correlation existed between grain hardness and susceptibility, but a consistent relationship between grain starch and protein content and susceptibility was lacking.

The maize weevil can fly from grain stores and infest maturing rice in fields. Performance of 1,700 varieties of rough rice from 35 different countries showed that only kernels with husk defects were infested (Rossetto et al., 1973). The most common husk defects were separation between lemma and palea and broken husk or hulled grains. The first defect was common in Korean and Japanese varieties but rare among Chinese varieties. Separated lemma and palea and ease of dehulling were considered to be simple recessive characters and it was suggested that breeders should select against them to obtain resistant varieties (Rossetto et al., 1973).

Resistance to the lesser grain borer was found in the long-grained 'Dawn' and 'Labelle' varieties (McGaughey, 1973; Cogburn, 1977). The insect produced significantly smaller progenies on the rough and milled rices of these varieties than on 'Belle Patna' and 'Bluebelle' (McGaughey, 1973). 'Dawn' was not resistant to the saw-toothed grain beetle (McGaughey, 1974).

Studies on varietal resistance to other grain pests have not received much attention so far. Broken grains in milled rice increase susceptibility to certain species but often it is not known whether this susceptibility is varietal or induced by poor harvesting and threshing methods. It is often claimed that parboiled milled rice is less susceptible to infestation than raw milled rice, since parboiling imparts hardness to grains. On the other hand, an increase in milling is also reported to render stored rice less suitable to several species of insect pests.

4. INSECT BIOTYPES

The development of insect biotypes capable of surviving on resistant varieties limits their full potential in insect control. The forerunners of new biotypes may be a few individuals in natural populations that can survive on resistant plants. If farmers intensively plant resistant varieties, such insects are more likely to survive. In time, the general population can shift to a new biotype. Biotypes may also develop through mutation.

Evaluation of differential host varieties at different locations is a commonly used method of detecting biotypes. The evidence to date suggests occurrence of biotypes in the brown planthopper, green leafhopper, and gall midge. Some rice varieties resistant to the gall midge in Thailand are susceptible in Indonesia (Soehardjan et al., 1975). The gall midges collected in Thailand have a larger body, forewings, halters, tibiae, and longer abdominal hairs than the Indian gall midges, confirming the occurrence of different biotypes (Hidaka et al., 1977). In India, gall midges infesting rice in Cuttack (Orissa) and Hyderabad (Andhra Pradesh) may be different biotypes (Shastry et al., 1972; Israel, 1974).

The rice variety 'Pankhari 203' is highly resistant to green leafhoppers at IRRI and many other places where it has been tested. However, it is susceptible to green leafhoppers in Bangladesh (Rezaul Karim, personal communication) and at Bogor, Java, Indonesia (Harahap, personal communication).

At least three distinct biotypes of the brown planthopper have been isolated and reared in the Philippines (Figures 13 and 14). Besides these, the brown planthopper populations in southern India (Kalode, personal communication) and in Taiwan (Cheng, personal communication) appear to belong to different biotypes. A total of 4 different pairs of genes have so far been identified that impart resistance to brown planthopper (Athwal et al., 1971; Athwal and Pathak, 1972; Lakshminarayana and Khush, 1977). The brown planthopper biotype 1 can survive on and damage only those varieties which do not carry genes for resistance, while biotype 2 survives on

Figure 13. Most of the recent IRRI varieties are resistant to the brown planthopper biotype 1, which is common in the Philippines. But biotype 2 attacks varieties that are resistant to biotype 1. Biotype 3 attacks plants resistant to biotype 2 (IRRI, 1975).

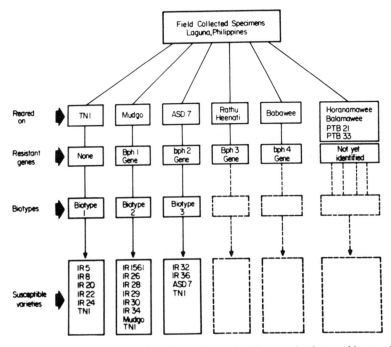

Figure 14. Schematic representation of development of brown planthopper biotypes "(Fil Medrano—personal communication)" (IRRI, 1976).

resistant varieties carrying gene *Bph 1* and biotype 3 on varieties carrying gene *bph2*. However, none of these biotypes survives on varieties with genes *Bph 3* and *bph 4*. These resistance genes have been bred into various rice varieties with improved plant types.

The monogenic resistance of rice to the brown planthopper makes it more vulnerable to the biotype problem. Current breeding efforts are therefore aimed at developing quantitative resistance and moderately resistant lines.

To monitor biotypes and to identify resistant material, the International Rice Brown Planthopper Nursery and International Rice Gall Midge Nursery have been established in many countries through IRTP. Each nursery consists of a uniform set of varieties. If particular rices are resistant in one area but susceptible in another, the insects at the two locations are suspected to be of different biotypes.

5. CONCLUSIONS

Breeding for insect resistant rice varieties has a number of advantages: no significant costs to the farmer are involved, it does not impair the quality of the environment, and it is generally compatible with other methods of pest control. Since costs of agricultural chemicals are increasing in the developing countries, rice varieties resistant to insect pests offer a great potential because they imply a certain amount of independence from chemically dependent pest control systems, and therefore should be emphasized in the years ahead. Genetic resistance in rice plants wherever available should be combined with other desirable plant characters, such as high yields and good quality, and should provide the basic foundation on which to build integrated pest managment systems. This method of insect control is highly practical and should receive greater attention by scientists throughout the world.

Although the exact mechanism of resistance is not always known, the stability of resistance against the development of new biotypes deserves top priority. Identification of sources of stable resistance and breeding for this type of resistance are major research objectives, along with basic studies on insect–host plant relationships for a better understanding of the nature of resistance.

Leaves of greenbug resistant (right) and susceptible (left) sorghums are shown Photo courtesy of George L. Teetes, Texas A&M University.

18

BREEDING SORGHUMS RESISTANT TO INSECTS

G. L. Teetes

Department of Entomology, Texas A & M University, College Station, Texas

Insect and mite pests play a major role in limiting sorghum yield. Interest in the breeding and use of pest resistant sorghum in developed countries has been stimulated by the failures and adversities brought about by heavy reliance on synthetic organic chemicals for pest control. The use of chemical pesticides without regard to the complexities of the ecosystem—especially the population dynamics of sorghum pest species and their natural enemies—has been one of the basic shortcomings of this approach. Much sorghum is produced in less developed areas of the world where it is not economically or logistically possible to use pesticides. In these areas, growers must depend on a more fundamental approach to pest management, such as use of insect resistant sorghums.

Sorghum (*Sorghum bicolor*) is an extremely diverse plant species. Its genetic variability makes it an attractive candidate for breeding programs aimed at developing insect resistant varieties.

This chapter describes the "ways and means" of breeding, screening, and evaluating sorghums for plant resistance to insect and mite pests. It makes no attempt to review all the literature dealing with the subject of insect resistant sorghums, although pertinent references are cited. Several recent reviews and bibliographies on sorghum have included either references or a chapter on the subject of sorghum entomology (Bottrell, 1971; Doggett, 1970; George Washington University, 1967; Rao and House, 1972; Rockefeller Foundation, 1973; Wall and Ross, 1970; Young and Teetes, 1977).

1. SORGHUM

1.1. Origin

Sorghum, originated in the Old World, probably in the northeast quadrant of Africa where the greatest variability in cultivated and wild sorghums is found. Doggett (1970) postulates that cultivated sorghums were developed from the wild sorghum of Africa by disruptive selection, and then by isolation and recombination following its introduction into highly varied habitats via the movements of people throughout that continent. Cultivation of sorghum probably began between 5000 and 7000 years ago (Martin, 1970).

1.2. Areas of Production

World production of sorghum grain currently totals about 52 million metric tons produced on some 42 million hectares (FAO, 1975). Sorghum is grown on six continents, within a zone extending to about 40 degrees latitude on either side of the equator. Based on production of grain, the major sorghum producing countries and their percent of the total are the United States (45 percent), India (17 percent), Argentina (10 percent), Nigeria (6 percent), Mexico (6 percent), Sudan (3 percent), Ethiopia (2 percent), and Yemen (2 percent).

1.3. Methods of Production

Production methods for sorghum vary considerably from small subsistence plots to immense monocultures. The latter are most common in the New World, but are also found in parts of Africa, Asia, and Australia. In the Old World, as in parts of Central and South America, multipurpose varieties are usually grown. Combine-harvestable sorghum hybrids are predominant in North America where they produce about one-half of the world's sorghum grain. The introduction of hybrids in the late 1950s greatly increased grain production for the livestock feedlot industry, but it also resulted in an intensification of pest problems and pesticide use. Pests are also a major yield-limiting factor in Africa and Asia, where about three-fourths of the world's sorghum acreage currently produces one-third of the grain crop.

1.4. Uses

Sorghum is used directly for human consumption and as feed for poultry and livestock in the form of grain and fodder. Sorghum grain is a staple food in many parts of Africa and Asia. It is also an important source of alcoholic beverages in many countries. Sorghum is an important feed grain and forage crop in the United States, where it is produced for both domestic and export markets. New industrial uses are also being found for it. Palatable syrups are made from the pressed juice of sweet sorghum stalks, and the panicles of broom corn are made into brooms.

2. INSECT AND MITE PESTS

In each agro-ecosystem, there are one or two key insect pests that attack sorghum. *Key pests* are serious, perennially occurring, persistent species that dominate control practices. In the absence of human intervention, they commonly attain population densities that exceed the economic-injury level each year, often over wide areas. The sorghum midge, *Contarinia sorghicola,* shoot fly, *Atherigona soccata,* stem borer, *Chilo partellus,* and greenbug, *Schizaphis guaminum,* are examples of key pests of sorghum in various geographic areas.

Secondary pests, although often present in sorghum fields or surrounding areas, rarely occur in economically important numbers. Nevertheless, such pests can exceed the economic injury level as a result of changes in cultural practices or crop varieties, or because of injudicious use of insecticides applied for a key pest. Spider mites, *Oligonychus* spp., are often regarded as secondary pests in sorghum.

Occasional pests cause economic damage only in localized areas or at certain times. Such pests are usually under natural control and exceed the economic injury level only sporadically. Most pests of sorghum are occasional pests.

A summary of information on common sorghum pests worldwide is given in Table 1. It includes data on geographic distribution, pest status (key, secondary, occasional), nature of damage, and whether plant resistance to a particular pest has been reported.

3. SORGHUM GERMPLASM RESOURCES

3.1. Germplasm Collections

Approximately 18,000 lines from sorghum-producing areas of the world have been assigned accession (IS) numbers in a world collection maintained at the International Crops Research Institute for the Semi-Arid Tropics (ICRISAT), Hyderabad, India, and at the National Seed Storage Laboratory, Fort Collins, Colorado. The Fort Collins laboratory also has a collection of several hundred varieties that either originated in the United States or were introduced over a period of about 100 years. In addition, seed of hundreds of introductions are stored at the Regional Plant Introduction Station at Experiment, Georgia.

Varieties in these world seed collections were obtained from tropical and temperate zones, from low and high elevations, and from sorghum grown during different seasons of the year. These varied environments ensure germplasm diversity. If resistance to any sorghum insect pest in the world exists, one can probably find it in these collections, although, much of the existing germplasm has not yet been collected.

3.2. Conversion Program

Many of the varieties in the world collection are from Asia and Africa. Most of these tropical varieties cannot be evaluated in temperate zones of the world because they require long nights to induce flowering. As a result, only a small fraction of the total genetic diversity within the species could be evaluated in its original form and used to produce hybrids in temperate areas.

In 1963 the Texas Agricultural Experiment Station and the U.S. Department of Agriculture, SEA, AR began a cooperative program, referred to as the sorghum conversion program, to convert tropical sorghum into short, early types for direct use in temperate areas (Stephens et al., 1967).

Sorghum is a short-day species, but strains do exhibit differential sensitivity to photoperiod. Four genes influence the photoperiodic reaction which governs the time of floral initiation (duration of growth). Lateness of maturity and tallness are dominant characters, and each is controlled by genes at four independently inherited loci (Quinby, 1974).

In the conversion program, crossing and backcrossing are done during the winter at the Federal Experiment Station, Mayaguez, Puerto Rico, and selection of short, early genotypes in segregating populations is done during long days in Texas. Tropical sorghums are used as male parents in all crosses and backcrosses except the last, when the tropical line is used as the female so that the converted line is placed in its original cytoplasm. The original female (nonrecurrent parent) is an early, 4-dwarf, B-line (male fertile) in normal cytoplasm. Each backcross is made to an F_3 progeny grown from a short, early F_2 selected in Texas. Four cycles of backcrossing and selection are required for conversion; however, partially converted lines

Table 1. Most Common Insect and Spider Mite Pests Injurious to Sorghum

Common Name	Scientific Name	Geographical Distribution[a]	Pest Status[b]	Nature and Symptoms of Damage	Resistance Reported
Soil pests					
White grub	*Phyllophaga crinita*	NW	Occ	Pruning of roots, seedling death, stunting and/or lodging	No
White grub	*Schizonycha* sp.	AF	Occ	Pruning of roots, seedling death, stunting and/or lodging	No
White grub	*Holotrichia consanguinea*	AS	Occ	Pruning of roots, seedling death, stunting and/or lodging	No
Wireworms	Several species of true and false wireworms (*Eleodes, Conoderus, Aeolus*)	NW	Occ	Destroy planted seed, stand loss	No
Rootworms	*Diabrotica* spp.	NW	Occ	Pruning and tunneling of roots, stunting and dead heart	No
Aphids					
Greenbug	*Schizaphis graminum*	C	Key	Suck plant sap, injects toxin that kills leaves, virus vector, disease predisposer	Yes
Yellow sugar- cane aphid	*Sipha flava*	NW	Occ	Suck plant sap, injects toxin that kills leaves	No
Sugarcane aphid	*Aphis sacchari*	AF, AS	Occ	Suck plant sap	No
Corn leaf aphid	*Rhopalosiphum maidis*	C	Occ	Suck plant sap, virus vector	Yes
Shoot fly	*Atherigona soccata*	AF, AS	Key	Injure growing point, causing dead heart	Yes

Common name	Scientific name	Distribution	Status	Damage	Seedling
Armyworms					
Fall armyworm	*Spodoptera frugiperda*	NW	Occ	Leaf feeder in whorl or destruction of seed in head	Yes
Armyworm	*Mythimna separata*	AS	Occ	Leaf feeder in whorl and on leaf margins	Yes
Nutgrass armyworm	*Spodoptera exempta*	AF	Occ	Leaf feeder in whorl and on leaf margins	No
Beet armyworm	*Spodoptera exigua*	AF, NW	Occ	Leaf feeder in whorl and on leaf margins	No
Armyworm	*Pseudaletia convecta*	AF, O	Occ	Leaf feeder in whorl and on leaf margins	No
Stem borers					
Sorghum stem borer	*Chilo partellus*	AF, AS	Key	Some leaf feeding, boring in stalk, dead heart, stalk lodging	Yes
Southwestern corn borer	*Diatraea (= Zeadiatraea) grandiosella*	NW	Occ	Some leaf feeding, boring in stalk, stalk lodging	Yes
Sugarcane borer	*Diatraea saccharalis*	NW	Occ	Some leaf feeding, boring in stalk, stalk lodging	Yes
European corn borer	*Ostrinia nubilalis*	NW, EE	Occ	Some leaf feeding, boring in stalk, stalk lodging	Yes
Sorghum maize borer	*Eldana saccharina*	AF	Occ	Some leaf feeding, boring in stalk, dead heart, stalk lodging	No
Maize stalk borer	*Busseola fusca*	AF	Occ	Some leaf feeding, boring in stalk, dead heart, stalk lodging	No
Sorghum (pink) borer	*Sesamia* spp.	EE, AF, AS	Occ	Some leaf feeding, boring in stalk, dead heart, stalk lodging	No
Lesser cornstalk borer	*Elasmopalpus lignosellus*	NW	Occ	Boring in stalk at soil surface of seedling plants	No
Sugarcane root-stock weevil	*Anacentrinus deplanatus*	NW	Occ	Boring in stalk and roots above and below soil surface causes lodging	No

Table 1. (Continued)

Common Name	Scientific Name	Geographical Distribution[a]	Pest Status[b]	Nature and Symptoms of Damage	Resistance Reported
Corn planthopper	*Peregrinus maidis*	C	Occ	Suck sap from leaves in plant whorl	No
Chinch bug	*Blissus leucopterus leucopterus*	NW	Occ	Suck sap from leaves and stems	Yes
Spider mite					
Banks grass mite	*Oligonychus pratensis*	NW	Sec	Suck plant sap causing discoloration and death of leaves	Yes
Sorghum mite	*Oligonychus indicus*	AS	Sec	Suck plant sap causing discoloration and death of leaves	No
Sorghum midge	*Contarinia sorghicola*	C	Key	Destroys developing seed	Yes
Head caterpillars					
Sorghum webworm	*Celama sorghiella*	NW	Occ	Destruction of seeds in head	No
Webworm	*Stenachroia elongella*	AS	Occ	Destruction of seeds in head	No
Webworms	*Eublemma* spp.	AS	Occ	Destruction of seeds in head	No
Yellow peach moth	*Dichocrocis punctiferalis*	AS, O	Occ	Destruction of seeds in head	No
Head caterpillar	*Cryptoblabes adoceta*	O, AS	Occ	Destruction of seeds in head	No
Corn earworm	*Heliothis zea*	NW	Occ	Leaf feeder in whorl or destruction of seed in head	No
American bollworm	*Heliothis armigera*	AF, AS, O	Occ	Destruction of seeds in head	
Head bugs					
'Jowar' earhead bug	*Calocoris angustatus*	AS	Key	Feed on developing seed causing smaller, lighter distorted seed	No
False chinch bug	*Nysius raphanus*	NW	Occ	Feed on developing seed causing smaller, lighter distorted seed	No
Stinkbugs	Pentatomidae, e.g., *Oebalus pugnax*	C	Occ	Feed on developing seed causing smaller, lighter distorted seed	Yes
Pyrrhocorid bug	*Dysdercus superstitiosus*	AF	Occ	Feed on developing seed causing smaller, lighter distorted seed	No
Leaf-footed plant bug	*Leptoglossus phyllopus*	NW	Occ	Feed on developing seed causing	No

Stored-grain pests

Rice weevil	*Sitophilus oryzae*	C	Key	Consume whole grain in fields and storage	Yes
Maize weevil	*Sitophilus zeamais*	C	Occ	Consume whole grain in storage	Yes
Angoumois grain moth	*Sitotroga cerealella*	C	Key	Consume whole grain in field and storage	Yes
Lesser grain borer	*Rhyzopertha dominica*	C	Occ	Consume whole grain in storage	No
Indian-meal moth	*Plodia interpunctella*	C	Occ	Feed on cracked grain, a secondary feeder	No
Grain mite	*Acarus siro*	C	Occ	Feed on cracked grain, a secondary feeder	No
Red flour beetle	*Tribolium castaneum*	C	Occ	Feed on cracked grain, a secondary feeder	Yes
Confused flower beetle	*Tribolium confusum*	C	Occ	Feed on cracked grain, a secondary feeder	No

[a] C = Cosmopolitan, AF = Africa, EE = eastern Europe, NW = New World, AS = Asia, O = Oceania.
[b] Occ = occasional; Sec = secondary; Key = key pest.

465

(lines with less than four backcrosses) are sometimes released as breeding stocks.

4. BREEDING TECHNIQUES AND METHODS

4.1. Anthesis and Pollination Characteristics

Breeding methods are influenced by the anthesis and pollination characteristics of the species (Quinby and Schertz, 1970). Typically the sorghum panicle has two types of spikelets. One is sessile and contains a perfect flower. The other is pedicellate and contains a staminate floret. When flowering of sorghum commences, the glumes spread and usually the anthers extrude from the glumes, become pendant, and then dehisce the pollen. Flowering of a single spikelet may require as little as 30 minutes. Six to 9 days are needed for the completion of anthesis of a panicle, with maximum flowering occurring usually on the third or fourth day. Flowering begins with the spikelets at or near the tip of the panicle and progresses downward in subsequent days. In many varieties the pediceled spikelets are staminate. These shed considerable pollen and begin flowering about the time the primary spikelets of the panicle complete flowering at the base of the panicle. It is a common practice to begin hand pollinating when pollen is available during early morning and to continue pollinating until noon. Seed sets from pollination made after that time are often sparse, depending on local conditions. However, time of pollen availability varies dramatically.

Sorghum is mostly self-fertilized but occasionally cross-fertilization occurs. Most studies indicate about 6 percent crossing between plants in adjacent rows of grain sorghum (Quinby et al., 1958). By contrast, Sudan grass, *S. bicolor* var. *sudanense,* crosses readily, with up to 34 percent crossing for plants in adjacent rows (Garber and Atwood, 1945).

4.2. Selfing

Sorghum plants have perfect flowers (i.e., both male and female on the same spikelet) and are easily self-pollinated. This is done by placing a paper bag over the panicle after it emerges from the leaf sheath but before anthesis occurs. The bag may either be left on the panicle until the seed is mature or removed after flowering is completed.

4.3. Emasculation

The anthers must either be removed from the florets or the pollen prevented from functioning to make crosses in sorghum (Quinby and Schertz, 1970). If only a few crossed seed are desired, a hand-emasculation technique can be used. Anthers are removed from each floret with a tweezer or pencil point in a region of the panicle which is ready to flower. These emasculated florets are then covered with a small paper or glassine bag. Pollen from the intended male parent is dusted, brushed, or otherwise placed onto the exposed stigmas 1 or 2 days later. Care should be taken to remove all nonemasculated fertile and staminate spikelets to prevent contamination through selfing.

Emasculation can be done by killing pollen with heat when large numbers of seed are needed, as in a backcrossing program or when it is not feasible to use male sterility. Immersing the panicle in water at 48°C for 10 minutes kills the pollen without damaging the ovules (Stephens and Quinby, 1933).

Another technique of pollen control on the intended female parent is to cover the panicles with plastic bags. Moisture accumulation inside the bags prevents the anthers from dehiscing (Schertz and Clark, 1967). Pollen from the intended male parent is dusted onto the exposed stigmas before the anthers of the female parent are dry enough to dehisce and shed pollen.

4.4. Hybridization and Selection

Selection, hybridization and selection, and hybridization for F_1 production are the main sorghum breeding methods employed (Quinby and Schertz, 1970). Because not all the desirable traits exist in any one variety, the hybridization and selection method is used to concentrate as many desirable traits as possible in one strain. These strains may be used as true breeding varieties or as parents of hybrids. The initial breeding methods used to produce parental lines are similar to those used to develop new varieties.

In a hybridization and pedigree selection program, parents are chosen that possess desired characteristics. These two parental varieties are crossed, and first generation (F_1) plants are grown and self-pollinated. Since both parents are pure varieties, all F_1 plants are alike. A segregating or F_2 generation is then grown, and plants that appear promising are self-pollinated. The required size of the F_2 population depends on the numbers of traits to be selected and on the inheritance of those characteristics. F_3 progeny rows are then grown and selections are made within the rows that produce the desired plant type. Progeny-row selection continues through subsequent generations until the best types are recognized and all the plants within a progeny row appear to be similar (F_6 to F_8).

Backcrossing is often used when desirable parents with high combining ability need improvement for other characteristics, such as insect resistance. Backcrossing is effective for transferring one or a few characteristics to a line while nearly duplicating the original line in other attributes.

Radiation and chemical-induced variation has been used very little for sorghum improvement, but mutations have been obtained from these procedures (Kaukis and Webster, 1956; Harris et al., 1965; Franzke and Sanders, 1965).

Most of the sorghum improvement work in the United States is aimed at producing strains suitable as parents of hybrids. Breeding strains consist of A and B-lines (female parent) and R-lines (male parents). Sorghum lines can be identified for B- or R-line reaction by crossing to an A-line. A-lines are male sterile because they possess sterile-inducing cytoplasm and are recessive for the fertility restoration factor. The F_1 of an A-line × B-line will be male sterile. The F_1 of an A-line × R-line will be male fertile. The mechanism of sterility is genetic-cytoplasmic male sterility. Some varieties partially restore fertility and consequently are not suitable for R or B lines, unless improved for this characteristic.

B-lines with potential as a female parent must be male sterilized. B-lines possess male-sterile genes, but are male fertile because they have normal (fertile-inducing) cytoplasm. B-lines are male sterilized to produce A-lines by incorporating the B-line's chromosomes into the sterile-inducing cytoplasm. This is done by paired-progeny selection in a backcrossing process. R-lines are crossed to A-lines in hybrid combination to restore fertility to the F_1, which the producer then plants.

4.5. Populations

Population breeding is done with or as an alternative to the traditional crossing and backcrossing approach that utilizes narrow range of genotypes (Doggett, 1970). The basic difference between the two approaches is that in population breeding methods, the population itself is improved generation by generation, thus steadily increasing the chances of finding individuals with excellent combinations of characters. A great diversity of germplasm can be combined in a single population and many sources of a specific characteristic, for example, insect resistance, can be entered into the population. By this method, cyclic hybrid recombination is followed by selection in a recurrent selection system. The necessary tool to apply recurrent selection to sorghum is available in ms_3, the male-sterile gene from the 'Coes' variety, or in other appropriate genetic male-sterile genes. Male sterility is essential in a sorghum population to enhance random mating in this mainly self-pollinated species. The population breeding methods for sorghum can be similar to those used in maize (Eberhart et al., 1967; Eberhart, 1972; Gardner, 1972; Ross, 1973) and can give rise to varieties or parents of hybrids.

5. SCREENING TECHNIQUES FOR DETECTING RESISTANCE

Procedures used to screen sorghums for resistance to insects and mites are based on the nature of damage and resulting injury symptoms caused by the pests. Progress in selection for resistant plant types depends on uniform and sufficient selection pressure. Although selection can be done under natural infestations, the pest population and/or crop often must be manipulated to create selection pressure and to ensure that susceptible plant types do not escape infestation. Modification of existing screening and evaluation procedures is often required. The general procedures used for pest groups or major pest species are summarized below. The techniques described are a compilation of those reported in the literature, and the amount of detail varies with the amount of emphasis the species or group has received.

5.1. Aphids

Several aphid species infest sorghum. Resistance in seedlings and mature plants can be evaluated successfully under natural pest infestations, provided populations reach uniform, high levels at appropriate times. Because of the fluctuation and seasonality of natural aphid populations, it is often necessary to evaluate sorghums for resistance in the absence of natural infestations. The following techniques are common for evaluating resistance to the greenbug, *Schizaphis graminum*, in seedlings and adult plants (Johnson et al., 1976; Starks and Burton, 1977).

Greenhouse Seedling Screening

Aphids are reared in a greenhouse on culture plants, usually sorghum or a sorghum–barley mixture, which are grown in plastic pots or metal cans. A 3:1:1 mixture of soil, sand, and peat is the preferred growth medium. If soil alone is used, sand should be used to cover planting seed. A small amount of complete fertilizer is added to the soil medium. Soil mixtures are sterilized if plant diseases are a problem. Seed should be treated with a fungicide to control soil-borne diseases. From 30 to 50 seeds per container are planted to a depth of 2.5 cm or covered with sand to that depth. Sand prevents excess water evaporation and aids in keeping cage bottoms clean. Prior to plant emergence, a bottomless, cylindrical cage of clear vinyl or of nitrocellulose film is inserted into the soil or sand to protect emerging plants from premature aphid infestation and to exclude extraneous insects, especially predators and parasites. The ventilation holes and cage tops are covered with fine-mesh nylon cloth glued with rubber cement. Once plants attain a height of 15 to 20 cm (about 2 weeks), they are infested with about 200 aphids. The culture should have a maximum number of aphids two weeks later (Figure 1).

Temperature and humidity requirements are often dictated more by the aphid-culture plants than by the aphids. Greenbugs, for example, reproduce in a temperature range of 15.6 to −32.2°C, although fecundity is greatest at about 22.2°C, at which temperature each ovoviviparous female produces an average of 100 nymphs over a 20-day period.

Breeding lines to be evaluated for resistance are planted in galvanized metal flats 35.6 cm wide × 50.8 cm long × 9.5 cm deep filled with soil to

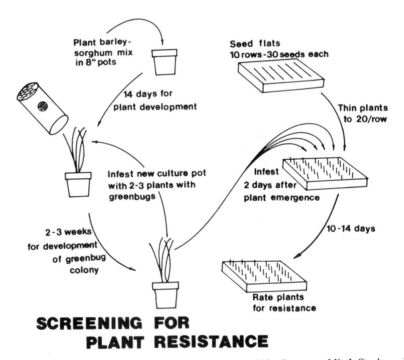

Plant barley-
sorghum mix
in 8" pots

Seed flats
10 rows-30 seeds each

14 days for
plant development

Thin plants
to 20/row

Infest new culture pot
with 2-3 plants with
greenbugs

Infest
2 days after
plant emergence

2-3 weeks
for development
of greenbug
colony

10-14 days

Rate plants
for resistance

SCREENING FOR
PLANT RESISTANCE

Figure 1.   Screening techniques for plant resistance to aphids. Courtesy of K. J. Starks and R. L. Burton.

about 2.5 cm from the top. Ten equally spaced furrows about 2.5 cm deep are made in each flat by pressing a planting board into the top of the soil. Each flat can accommodate 10 entries if one entry is planted per row or 20 entries in rows 17.8 cm long. A test can contain any number of flats, depending on the supply of aphids and available greenhouse space. Approximately 20 to 30 seeds of each entry are planted per row and thinned to about two-thirds that number after plant emergence. Whenever possible, known resistant and susceptible lines should be planted in each flat as controls. If breeding selections from resistant crosses are to be evaluated, the resistant parent used in the cross should be included as the resistant control.

After thinning, plants are infested with aphids of all ages and instars by brushing or shaking them from culture plants fairly uniformly over flats, or by placing uprooted, infested, culture plants between rows and allowing the aphids to crawl to the test plants. Plants are examined about 2 days after being infested and additional aphids are applied to flats which have inadequate infestations. From 4 to 10 greenbugs per plant are considered adequate. Test flats are left uncovered before and after infestation.

Generally, plants in each flat are rated for resistance when plants in the susceptible, control row are near death, usually about 10 to 14 days post-infestation. A visual rating of an entire row is possible for nonsegregating material; in segregating rows, individual plants can be rated. A 0 to 9 rating system for seedling evaluation is feasible, where 0 = no damage, 1 = 1 to 10 percent plant necrosis, 2 = 11 to 20 percent, and so forth, and 9 = 81 to 100 percent necrosis or a dead or dying plant.

Adult Plant Screening

Leaf-damage ratings are fairly easy to make and offer a good measurement of resistance if an adequate, natural, aphid infestation occurs. The following rating system is feasible for field evaluation: 0 = no damage; 1 = red spotting on leaves; 2 = portion of leaf killed; 3 = 1 leaf killed; 4 = 2 leaves killed; 5 = 4 leaves killed; 6 = 6 leaves killed; 7 = 8 leaves killed; 8 = 10 leaves killed; and 9 = dead plant. Data may be collected at any plant growth stage when aphids are present and aphid numbers should be estimated. The plant growth stage should be recorded at the time aphid counts and damage ratings are made. The growth stage descriptions of Vanderlip (1972) are recommended. If aphid populations differ markedly among entries, an indication of the level of infestation on each entry can be made using the following code after the rating: 1 = low incidence, 2 = medium incidence, 3 = high incidence. Yield comparisons of resistant and susceptible sorghums may be used as resistance indicators. In addition, insecticidal treatments may be used to compare yields of infested and noninfested plants.

An alternative to natural infestations of aphids in the field is the use of cages. These can be large enough to enclose groups of plants or small enough to attach to a portion of a leaf. Large cages should be constructed of metal frames covered with small mesh screen to exclude parasites. In large cages, aphid population levels from natural or artificial infestations increase rapidly, often to unnaturally high levels, and damage may be accentuated. Small plastic clip-on cages can be used in the field for evaluation of resistance. Small cages (2.54 cm³) clipped to leaf blades need cloth-covered ventilation holes (1.9 cm in diameter) on at least one side. Five to 10 aphids, usually adults, are put in each cage with a small artist's brush. The cages keep the aphids confined to a small area and exclude parasites and predators. Cages are inspected the day following attachment to the leaves to ensure that all aphids remain alive and feeding on the plant. Additional aphids are added when necessary to ensure equal numbers per cage. Ratings of the damaged leaf area covered by the cage begin about 1 week after infestation and continue at 2-day intervals until the caged areas of the susceptible plants are near death. A useful rating scheme is as follows: 0 = no necrotic plant tissue in the caged area, 1 = 1 to 10 percent necrosis, 2 = 11 to 20 percent, and so on to 9 = 81+ percent necrosis.

5.2.   Shoot Fly

Plant resistance screening for this seedling pest may employ either natural field infestations, artificially created high field populations, or greenhouse (screenhouse) populations cultured on appropriate host plants.

Greenhouse or screenhouse techniques require rearing flies on seedling plants in specially constructed cages or using eggs taken from infested plants in the field. Rearing cages are divided into three compartments, one of which serves as an emergence and collecting unit (Soto, 1974). The adults are attracted to light, which facilitates their collection and transfer among units. The cages are covered with fine 26 mesh wire screen and a metal sheet increases the durability of the bottom of the cage.

Adult flies are fed a 20 percent aqueous solution of sucrose in sponges and a dry mixture (1:1) of Brewer's yeast and glucose in petri dishes (Soto and Laxminarayana, 1971). The sugar solution requires changing daily and the dry mixture twice a week. Two-week old sorghum seedlings grown outdoors in flats are placed in adult-infested cages to serve as oviposition sites and hosts for larvae. Flats are removed from the cage and placed outdoors after plants are infested with eggs (about 24 hours). When larvae reach the third instar, infested seedlings are uprooted and placed in flats containing a layer of moist sand. Flats are then held in the emergence units until adults emerge.

In greenhouse culturing, the pre-oviposition period is about 3 days, the approximate developmental period of the egg and the larval stages is 12 days, and pupa and adult stage is 10 days.

Seedling plants to be screened are placed in one of the units in the three compartment cage. Adult flies are transferred from the rearing compartments and allowed to oviposit on the test plants. Surviving plants may then be transplanted for seed production (Soto, 1972).

Natural populations of the shoot fly can be obtained for field testing by using attractants, trap planting of susceptible varieties, and late-season planting. Flies are attracted to fish meal (Starks, 1970). A fly-susceptible sorghum variety such as 'CK 60' is planted as border rows and/or spreader rows between blocks of the trial about two weeks before test entries are planted (Doggett et al., 1970). Fish meal is also applied to these trap plantings. Shoot fly population levels build up on the earlier-planted, susceptible plants, and as soon as the test varieties reach a susceptible stage, fish meal is spread between rows to attract the flies.

Seedling resistance is generally based on percent infested seedlings or percent "dead hearts." Standard check cultivars, both resistant and susceptible, are useful. Recovery resistance or tolerance should be determined and is generally measured on the basis of successful head production by tillers (Starks et al., 1970). Main plant and tiller infestations should be distinguished. In addition, egg counts are usually taken as a record of oviposition nonpreference.

5.3. Armyworms, Panicle Caterpillars, and Borers

A large complex of caterpillars attack sorghum as foliage and grain feeders or as stalk borers. Different species are encountered in different geographic locations. Most sorghum screening trials involving lepidopterous species have been conducted in the field using natural pest infestations. Greenhouse screening techniques have been reported for a few of these species; for a somewhat larger number, a technique combining greenhouse or laboratory culturing and artificial inoculation in the field has been used (Mayo, 1967). Compared to sorghum, considerably more progress has been made in screening for resistance to lepidopterous pests in corn (see Chapter 16).

Late-planted sorghum is usually more heavily and uniformly infested with lepidopterous species than early-planted sorghum (Chada, 1962). Field screening trials planted where stubble exists from previously planted sorghum help ensure natural infestations of stalk borers. Plant escape is a problem associated with field screenings by natural infestation.

Several caterpillar species can be reared on artificial diet, allowing mass rearing for use in screening when natural populations of the pest species are inadequate. Using diet-produced insects, it is possible to ensure uniform test-plant infestation at desired plant growth stages. The plants are infested with eggs and/or larvae. In greenhouse screening trials, with foliage feeding caterpillars, second- or third-instar larvae are placed on plants grown in flats as described earlier for aphid screening (McMillian and Starks, 1967). Generally, the severity of damage is recorded by visual classification. A rating system of 0 to 9 can be used by slight modification of the system reported by Wiseman et al. (1966) or Hormchong (1967). Resistance to panicle-feeding larvae is based on relative amounts of damaged seed (Buckley and Burkhardt, 1962, 1963).

For several borer species, egg masses in the "black head" stage are sometimes placed on the underside of the top leaves of individual test plants (Dicke et al., 1963). Reaction to borers is generally based on percentage infested stalks (cavities in the penduncle area), length of tunnel, and percentage stalk breakage and lodging. However, early leaf feeding by borers may or may not be highly correlated with stem tunneling, and length of tunneling may not be related to yield.

5.4 Sorghum Midge

Greenhouse techniques for screening sorghums for resistance to the sorghum midge are not available because techniques for artificially rearing the insect have not been developed. At present, workers rely on naturally occurring infestations in field plantings. The unreliability and/or fluctuations of midge population levels and variation of maturity in test plants, as in segregating rows, are problems inherent in field screening sorghums for this pest.

Damaging midge population levels are best attained by delayed planting, by multiple plantings of the same test materials, and/or by the use of earlier planting of susceptible sorghums in which midge populations reach high levels by the time the test plants are blooming. In the latter technique, susceptible sorghums are planted at 10-day to 2-week intervals beginning as early in the season as practicable, thus providing a continuous supply of blooming sorghum over an extended period. Also, bulks (mixtures) of sorghum seed of varying maturity lines may be planted as spreader rows between blocks of test plants.

Midge damage is usually rated as percent "blasted" seed. Generally, plants cannot be rated sooner than 20 days after bloom. Individual heads in a row are rated and a mean damage rating calculated, or the entire row is rated by visual observation. A useful rating scheme is as follows: 0 = no damage, 1 = 1 to 10 percent blasted seed, 2 = 11 to 20 percent, and so on to 9 = 81+ percent blasted seed. A more objective evaluation can be obtained by "protecting" portions of the test plants with pollinating bags or insecticides. Seed yield comparisons of protected and unprotected heads are then made. Standard resistant and susceptible varieties should be included as controls.

Adult midge population levels should be estimated by visual examination of blooming heads or by placing a large-mouth jar or plastic bag over the heads to capture adults. Days to 50 percent flower of each entry should be recorded and correlated with any fluctuation in midge adult abundance. Care should be taken to differentiate midge damage from bird damage and from sterile florets. The presence of midge cocoons at the tips of glumes indicates that the seed was damaged by midges.

5.5.   Chinch Bug

Infestations of chinch bugs in sorghum generally result from mass migration of the pest from winter-grown small grains. Screening for and evaluating chinch bug resistant sorghums is usually done by field trials under natural infestations (Snelling et al., 1937). Field infestations are increased by growing sorghums adjacent to infested small grains. Growing sorghums of mainly milo germplasm in the test may increase the infestation level, and these same sorghums can serve as susceptible checks. Greenhouse screening of seedling plants for chinch bug resistance might be done by slightly modifying the technique previously described for aphids, if the pest is reared artificially (Parker and Randolph, 1972). Field-collected bugs might be used for greenhouse screening, although this method has not been reported.

Chinch bug damage to sorghum is recorded as percentage of plants killed or injured (Snelling et al., 1937). The degree of stunting and yield reduction are additional criteria for measuring relative resistance. Although no damage rating schemes have been reported, one based on the amount of reddish discoloration on sorghum stems could be used. Rating damage to sorghum is complicated by date of planting, earliness, drought, and varietal adaptation (Painter, 1951). Seedlings are more readily damaged than more mature plants, and the latter may not be killed but will show the characteristic stem discoloration.

5.6.   Spider Mites

Screening sorghums for resistance to spider mites is relatively difficult because many factors affect mite abundance and plant reaction to infestation. The more damaging species generally require hot, dry, climatic conditions for injurious populations to develop. Also, mite infestations tend to be most damaging to sorghum that is in the reproductive stage (Ehler, 1974). All these factors hinder greenhouse screening (Owens et al., 1976).

Field evaluations depend on the availability of injurious mite densities. Manipulating planting to coincide with dry environmental conditions during sorghum reproductive stages often increases mite abundance. Moisture stressed plants are generally more heavily infested with mites than nonstressed plants. Plant maturity influences the reaction of plants to mites.

Rating schemes are usually based on percent leaf area damaged by mites (Foster et al., 1977a). Mites generally begin feeding in colonies adjacent to the midrib of the leaf. Feeding is accompanied by webbing, and desiccation of the leaf tissue begins at the tip and margins of the leaves. Mites generally cause damage to sorghum after it has bloomed; however, ratings are made at any stage of plant growth at which there are sufficient mite infestations. Stalk death ratings may be made when severe damage occurs. A useful rating scheme is as follows: 0 = no damage, 1 = 1 to 10 percent leaf necrosis or stalk death, 2 = 11 to 20 percent, and so on to 9 = 81+ percent or dead plant. Plant lodging and differential seed weight reduction among varieties can also be used. To record mite abundance, a population density rating is used where 1 = no mites, 2 = few individuals above midrib only, 3 = colonies along midrib, 4 = mites spreading away from midrib, and 5 = entire leaf covered with mites (Foster et al., 1977b).

5.7.   Stored-Grain Pests

A common practice in searching for resistance to stored-grain insects among varieties of sorghum is to infest a grain sample of each variety with a specific number of insects for a period of feeding and oviposition. Resistance is then evaluated by counting damaged grains and/or first-generation progeny (Stevens and Mills, 1973). Parent insects, in some cases, are given free choice of all samples tested; in others, they are confined to specific samples of grain (Rogers and Mills, 1974a). Balancing of the relative humidity and seed moisture content of the test samples is necessary for valid results (Rogers and Mills, 1974b).

Evaluations of groups of varieties involve small samples, typically 50 to 100 kernels contained in small plastic boxes measuring 48 × 48 × 18 mm

(Mills, 1976). Three to five samples of each variety are included in each test. In no-choice tests six females and three males are used to infest each sample, allowed a 5 to 7 day ovipositioning period, and then removed. The relative resistance ranking of the entries is based on numbers of emerged progeny per sample. For secondary stored-grain pests, eggs or newly hatched larvae are placed in each sample. Resistance is evaluated by recording the numbers of larvae that become adults.

In free-choice tests, samples of grain are arranged in a circle equidistant from the center of a circular test chamber (Stevens and Mills, 1973). An appropriate number of unsexed adult test insects is dropped into the center of the chamber and allowed free access to all the grain samples. Insects within each sample are counted in 4 to 7 days for nonpreference evaluation and then removed. Grain samples are then held to evaluate for progeny emergence. Free-choice tests probably are more practical where large numbers of entries are to be evaluated because insects do not have to be sexed, and obviously susceptible entries can be eliminated quickly.

The most commonly used measure of resistance has been progeny production. Other factors have been used such as developmental periods and weights of progeny. Dobie (1974) proposed an index of susceptibility based on progeny numbers and developmental periods. Soft X-rays have been used in resistance screening trials to detect developing larvae within the grains (Russell, 1962).

6. DETERMINATION OF RESISTANCE MECHANISMS

Plant resistance to insect and mite pests commonly is divided into three basic types or mechanisms: nonpreference, antibiosis, and tolerance (Painter, 1958). Resistance mechanisms involved in host–pest relationships can be determined, but the evaluation techniques differ from those normally used in screening diverse germplasm. Resistance is usually a result of more than one mechanism.

Differential preference reactions of insects and mites to resistant and susceptible sorghums are usually determined by allowing the pest a choice of resistant and susceptible plants. In the laboratory or greenhouse, this is done by randomly planting resistant and susceptible selections in a circular arrangement in pots or metal cans. After emergence, each entry is thinned to one plant. Pest insects are then placed on the soil surface in the center of the circle of plants at a rate five times the number of test plants. Test plants are covered with a cage. Insects present on individual test plants are counted at 24-hour intervals for at least 4 days after initial infestation. Frequently some insects die or are lost during the testing procedure.

Antibiosis is expressed as an adverse effect of the plant on the biology of the insect or mite when the plant is used for food. Techniques are used which allow the experiment to compare the fecundity, size (weight), longevity, and increased mortality of the pest on resistant versus susceptible plants. Some "conditioning" of the insect on the resistant plant is normally required.

Of the three resistance mechanisms, tolerance is perhaps the most difficult to quantify. It basically involves a comparison of pest numbers with subsequent plant damage. Consequently, pest numbers must be determined and related to visible damage, and eventually to yield.

7. RESISTANT SORGHUMS IN PEST MANAGEMENT

Sorghum resistance to insect and mite pests is best viewed as a component of a pest management system which can interact with and influence other components in the system. Some sorghums reported as insect or mite resistant are listed in Table 2. Insect resistant sorghum may inhibit a pest's ability to attain the economic injury level because of nonpreference or antibiosis, or the resistance may increase the damage tolerance level of the crop because of a tolerance mechanism. Similarly, insect resistant sorghums could create a situation in which natural biological control agents are more efficient because of slower rates of increase of the pest. Of course, resistant plants must allow the pest to reach a "threshold" of predation or parasitism. Low densities of the pest species may provide a pivotal reserve food supply for beneficial organisms needed later in the growing season to sorghum or neighboring crops. Also, resistant varieties may contribute to the effectiveness of insecticides or make it possible to omit or reduce treatments.

The role of host-plant resistance in sorghum pest management in the United States and the implications of this vital component can be seen in the examples that follow.

7.1. Chinch Bug

Painter (1951), in discussing insect resistant sorghums, devoted most of his attention to the chinch bug, because it was the most important pest of sorghum at that time. However, during the past 20 years this species has not been a significant factor in sorghum production. It is speculated that chinch bug resistant sorghums were at least partly responsible for the decreased severity of damage by the insect. In some cases, during most years, resistant

Table 2. Sorghums Resistant to Insects and Mites (Abbreviated List)

Insect/Mite Species	Variety/Line	Location Reported
Schizaphis graminum	IS 809 (PI 221613)	U.S.
	KS 30	U.S.
	SA 7536-1 (Shallu)	U.S.
	PI 264453	U.S.
Rhopalosiphum maidis	Piper sudan 428-1	U.S.
	TAM 428	U.S.
Atherigona soccata	Maldani 35-1 (IS 1054)	India
	IS 2123	India
	IS 5604	India
	Serena	Uganda
	Namatera	Uganda
Spodoptera frugiperda	Freed (PI 29166)	U.S.
Chilo partellus	Maldani 35-1 (IS 1054)	India
	BP-53	India
	C-10-2	India
	E 302	India
	E 303	India
Ostrinia nubilalis	Kafirs (several)	U.S.
	Feterita	U.S.
	Shantung Brown Kaoliang	China
Diatraea grandiosella	Y-4 (Tex 63 × Kaura)	U.S.
	NK × 3007	U.S.
Diatraea saccharalis	King's Diamond	U.S.
Blissus leucopterus	Atlas	U.S.
leucopterus	Axtel	U.S.
	Dwarf Kafir 44-14	U.S.
	Redline 60	U.S.
Oligonychus pratensis	Rio (SC 599-6)	U.S.
	IS 12568 (SC 56-6)	U.S.
Contarinia sorghicola	Nunaba	Africa
	Huerin Inta	Argentina
	AF 28	Brazil
	SGIRL-MR-1	U.S.
	TAM 2566 (IS 12666C) (SC 175)	U.S.
	IS 2508C (SC 0414)	U.S.
	IS 2549C (SC 0228)	U.S.
	IS 2579C (SC 0423)	U.S.
	IS 3071C (SC 0237)	U.S.
	IS 8100C (SC 0424)	U.S.
	IS 12612C (SC 0112)	U.S.

Table 2. (Continued)

Insect/Mite Species	Variety/Line	Location Reported
Oebalus pugnax	White Darso (Kans. 33-378)	U.S.
Sitophilus oryzae	Double Dwarf Early Shallu	U.S.
	Sagrain	U.S.
Sitophilus zeamais	Double Dwarf Early Shallu	U.S.
	Early Kalo	U.S.
	Early Sumac	U.S.
Sitotroga cerealella	Double Dwarf Early Shallu	U.S.
	Early Kalo	U.S.
	Early Sumac	U.S.
Rhyzopertha dominica	Double Dwarf Early Shallu	U.S.
	Early Kalo	U.S.
	Early Sumac	U.S.
Tribolium castaneum	Double Dwarf Early Shallu	U.S.

sorghum varieties proved to be a most practical means of chinch bug control and largely replaced the creosote barriers once recommended for keeping populations from migrating from small grains to sorghum.

Prior to the introduction of combine-type sorghum hybrids, most milo varieties of sorghum were found to be highly susceptible to chinch bug damage (Hayes, 1922). Several of the sorgo varieties are relatively resistant, 'Atlas' sorgo being one of the most resistant. Most of the kafir varieties also exhibit resistance whereas the feteritas are susceptible or intermediate in reaction (Dahms, 1943). Data for the period 1870 to 1923 show that considerably more kafir than milo sorghum was grown in the chinch bug-infested areas of Oklahoma, Kansas, and Texas, whereas milo sorghum was grown in areas where the pest was less serious (Snelling et al., 1937).

Factors influencing varietal resistance are tolerance of plants, differences in food values for the bugs, preference of bugs for susceptible varieties, and specific morphological characters of plants, such as loose fitting leaf sheaths which tend to be associated with increased susceptibility. Sorghums that produce a dense canopy tend to be less infested. Hybrid vigor reflected in rapid and lush plant growth also tends to reduce chinch bug damage.

The relative chinch bug resistance of current sorghum hybrids is largely unknown, and recent literature on the subject is lacking. Based on the early literature, sorghum hybrids possessing kafir germ plasm, such as 'Combine Kafir 60,' should exhibit a moderate level of resistance. In contrast, those with milo germ plasm should be relatively susceptible. The same may be true for Wheatland which is also susceptible (Dahms and Sieglinger, 1954).

In very recent years, chinch bugs have caused serious damage to sorghum adjacent to small grains, which underscores the need to assess currently used germ plasm.

7.2. Greenbug

This aphid has been a pest of small grains in the United States for almost 100 years, and a key pest of sorghum since the appearance of biotype C in 1968. Management approaches to dealing with biotype C were developed rather rapidly, using extremely low dosage rates of registered insecticides (Cate et al., 1973). Treatment need was based on knowledge of the pest's seasonal abundance trends (Bottrell, 1971; Teetes et al., 1975a) and economic threshold levels (Teetes and Johnson, 1973, 1974). This integrated control approach reduced greenbug population levels below numbers capable of causing economic yield loss; it also preserved natural biological control agents, which prevented resurgence.

Since many producers continued their heavy use of persistent systemic insecticides, resistance to these chemicals developed in greenbug populations in 1974 (Teetes et al., 1975b; Peters et al., 1975).

Breeding programs to develop greenbug resistant sorghums were initiated in 1969 by several public (Kansas, Oklahoma, and Texas Agricultural Experiment Stations and USDA, SEA) and private agencies. As a result, greenbug resistant sorghum hybrids became commercially available in 1976.

The resistance mechanisms have been identified for sorghums from sources such as 'SA 7536-1,' 'KS 30,' 'IS 809,' and 'PI 264453' (Hackerott and Harvey, 1971; Schuster and Starks, 1973; Teetes et al., 1974). Moderate levels of nonpreference exist as well as some antibiosis. Antibiosis is expressed as an increase in the duration of the developmental stages and a decrease in fecundity, adult longevity, and duration of the reproductive period. The primary resistance mechanism is tolerance, which is ecologically advantageous and increases the value of resistant sorghum as a component in pest management.

Field tests show that greenbug population densities are lower in resistant sorghum than in susceptible ones (Figure 2). As would be expected, predator numbers are similarly lower in resistant sorghum (Figure 3). However, when compared with the ratio of predators to greenbugs, it is about equal and sometimes greater in the resistant sorghum than in the susceptible (Figure 4). Predator–prey ratios would indicate that greenbug resistant sorghum may complement biological control. Biological control is certainly of value in the performance of greenbug resistant sorghum (Teetes et al., 1975; Starks et al., 1972).

The cumulative affect of use of greenbug resistant sorghum hybrids and biological control, when these hybrids constitute a majority of the sorghum

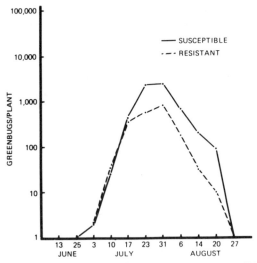

Figure 2.   Seasonal greenbug population levels on resistant and susceptible sorghum hybrids.

grown in a large geographical area, is yet to be determined. If insecticides are not applied to resistant sorghums, thereby preserving natural enemies of the aphid, the cumulative effects in stabilizing greenbug population levels should be considerable. Historically, the result has often been a reduction of insect species to noneconomic levels (Maxwell et al., 1972).

The longevity of greenbug-resistant sorghums is unknown. Plants whose major resistance mechanism is antibiosis exert selection pressure for new damaging biotypes. This happens in the same manner that insecticides select for individuals with the ability to withstand normally lethal dosage

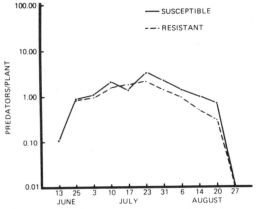

Figure 3.   Seasonal predator population levels on greenbug resistant susceptible sorghum hybrids.

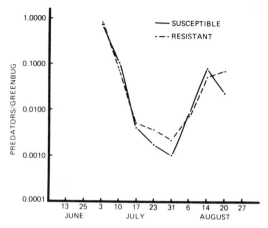

Figure 4. Seasonal predator-to-greenbug ratios on greenbug resistant and susceptible sorghum hybrids.

rates. As noted earlier, greenbug resistant sorghums exhibit all three resistance mechanisms. Therefore, the potential exists for new greenbug biotypes. The fact that the inheritance of greenbug resistance is not fully understood complicates the issue. It appears that resistance is dominant and simply inherited. Possibly only one gene is involved. Should this be the case, then the chances of a new biotype being selected may be increased. Also, new, physiologically distinct biotypes of the species rapidly appear. However, greenbug resistant sorghums primarily exhibit tolerance, which has no inhibitory effect on the insect and does not contribute to the selection of new biotypes.

7.3. Sorghum Midge

This dipterous species is the most cosmopolitan insect pest of sorghum and is a key pest in most sorghum growing regions. In the United States the primary means of reducing midge damage is uniform regional planting of sorghum early in the growing season. Where sorghum is produced late, insecticides have been the only effective control method. Both methods have economical and ecological disadvantages.

Insecticidal control of the sorghum midge is difficult and costly since multiple applications are required during flowering of the crop to reduce numbers of ovipositing adults. Early planting of the sorghum crops is not always possible or feasible, nor even an effective control tactic in some areas.

The use of host-plant resistance in midge management appears promising for the future. Screening of sorghum germplasm for resistance has been underway for some time (Jotwani et al., 1971; Rossetto and Banzatto, 1967). Bowden (1965) and Passlow (1965) found midge resistance in cleistogamous sorghums such as 'Nunaba,' but the resistance breaks down in the absence of a more favorable alternative variety. The sorghum line 'AF 28' shows a high level of resistance in field trials in Brazil according to Rostetto et al., (1975). Sources for resistance in sorghums adapted for temperate zones have been identified and released (Johnson et al., 1973; Wiseman et al., 1973).

At present, midge resistant sorghums such as 'AF 28,' 'SGIRL-MR-1,' and several lines from the sorghum conversion program ('IS 12612C,' 'IS 12666C,' 'IS 2508C,' 'IS 2579C,' and 'IS 3071C') provide sources for the development of commercial midge resistant varieties or hybrids. The inheritance of resistance in converted lines is not well understood, but it does not appear to be a dominant character. Consequently, for hybrids, resistance will be required in both parental lines. To date, sources of midge resistance have been in R-lines, but midge resistant B-lines are being developed.

The mechanisms of resistance of midge-resistant sorghums are not well understood. Ovipositional nonpreference appears to be the responsible mechanism for 'AF 28.' Types adapted to temperate zones appear to be antibiotic, since in some resistant lines fewer adults emerge from spikelets than eggs are laid.

Commercial midge resistant sorghum varieties or hybrids will play a very significant role in management of sorghum pests. In addition to providing another means of controlling the pest, resistant sorghums would provide flexibility in time of planting. In some areas, sorghums could be planted later in the season, thus allowing for more efficient use of natural rainfall. Also, freedom to plant later in the season without fear of midge damage would permit sorghum to flower and mature during the cooler, wetter part of the season. This not only favors yield, but reduces irrigation requirements and suppresses population levels of spider mites, which are more abundant during hot, dry conditions.

7.4. Spider Mites

The Banks grass mite, a New World species, often severely injures sorghum as a secondary pest influenced by weather, plant maturity (Ehler, 1974) and insecticide treatments for earlier occurring pests. Infestations of mites appear to be temporally and spatially separated from populations of effective natural enemies (Ehler, 1974). Chemical control is the one available method of suppressing outbreaks of mites, but insecticide resistance has accounted for control failures (Ward and Tan, 1977).

Sorghum germ plasm resistant to mites has been identified (Foster et al., 1977). 'SC 0599-6,' a partially converted Rio selection from the sorghum conversion program, appears especially promising as a source of mite resistance. This line is a nonsenescing type which maintains green leaves and healthier stalks much longer than most lines. Interestingly, it continues to maintain green leaves even when infested with mites; consequently, the resistance mechanism appears to be of a tolerance type (Foster et al., 1977). The plants of this line are higher in total sugars than standard grain sorghums, and this may be involved in the resistance mechanism. Mite population levels on the resistant sorghum are about equal to those on susceptible ones, and there is little effect on the biology of the pest.

7.5. Corn Leaf Aphid

This common aphid species is sometimes abundant in sorghum, infesting the whorl or panicle of plants. Opinions differ as to the importance of the pest in reducing yield. References published prior to the introduction of combine-type hybrid sorghums indicate that the pest severely injures sorghum (Painter 1951), but more recent work indicates that the pest does not reduce sorghum yield (Wilde and Ohiagu, 1976).

The corn leaf aphid is one of the first insects to infest sorghum in the growing season and is often responsible for maintaining high population levels of several species of natural enemies. Consequently, the aphid is extremely important in the sorghum ecosystem, which would be destroyed by insecticidal treatments.

McColloch (1921) was the first to report on sorghum resistance to the corn leaf aphid which he found in Sudan grass. Howitt and Painter (1956) confirmed this discovery and later selected a highly resistant plant from Piper Sudan grass. Four aphid biotypes have subsequently been identified based on differential responses to resistant and susceptible sorghums (Pathak and Painter, 1958a, b; Cartier and Painter, 1956).

Recently, new sources of corn leaf aphid resistance in sorghum were identified from converted exotic lines (Teetes et al., 1976). Compared to highly susceptible B 'Redlan,' a common parent of sorghum hybrids, several 'Zera Zera' sorghums, especially 'TAM 428' ('SC 0110-9'), exhibit high resistance to the corn leaf aphid and show little damage. The resistance in F_1 hybrids of 'TAM 428' is influenced by the other parental line in hybrid combination.

The utility of resistant sorghums in limiting populations of corn leaf aphid to noninjurious levels would appear beneficial in sorghum pest management. Resistant sorghums assist in providing stability to the ecosystem by maintaining natural biological control agents.

Dr. Obi Okato applies pollination bags on whites of spruce to produce hybrid trees. Photo courtesy of J. W. Hanover.

19

BREEDING FOREST TREES RESISTANT TO INSECTS

J. W. Hanover

Department of Forestry, Michigan State University, East Lansing, Michigan

Progress in breeding trees for resistance to insects has not been commensurate with the great economic losses from insect damage to forests and ornamental trees. This is partly because forest genetics is a young science with few participants, most trees have relatively long generation intervals, regeneration of forests is mainly by natural seeding rather than by planting improved strains, and there has been a dearth of knowledge about host physiology and insect biology. In recent years each of the above limiting factors has been changing to favor the increased use of tree resistance to combat insect pests.

Trees are relatively long-term crops (10 to 120 years), even though short rotations of 5 years or less may be typical of some species in the future. Hence the incorporation of resistance into future planting stock offers the ideal, one-time method for reducing economic and aesthetic losses from insect damage. At the same time the development and use of resistant strains of trees is not without problems. These problems, which will become evident later in this discussion, include (1) the large number of tree species/insect combinations for which breeding can be done; (2) the occurrence of several insect pests on the same species; (3) the necessity of maintaining a broad genetic base in the early phases of a tree breeding program to avoid inbreeding depression in growth traits later; (4) a general lack of existing tests to demonstrate genetic resistance for most host/insect combinations; (5) the potential vulnerability of resistant trees with long rotations to "readaptation" by the insect before the host is harvested; (6) the difficulty in testing for resistance in the case of a tree species that is attacked by an insect late in its life; and (7) the potential effects of site diversity in modifying genetic resistance to an insect.

Any of the above problems may or may not be associated with the development of genetic resistance for a specific insect/host combination. However, if the proper procedures of research, development, and practice are followed, none of the problems should be insurmountable. This chapter describes these procedures and reviews the kinds of background knowledge needed to begin a breeding program with trees. Past treatments of this subject include those of Hanover (1975) and Gerhold et al. (1966).

1.   TREES AS INSECT HOSTS

1.1.   Uniqueness of Trees

Trees differ from most other plants in several respects that influence the strategies used in breeding for insect resistance. First, the vast majority of our tree resource still exists in wild, uneven-aged stands. These stands often are units of extensive forests distributed over diverse environments and have been subjected to little or no management by man. In other words, the domestication of forest trees lags far behind that of our agricultural crops. In recent years there has been a movement toward tree cultivation, the extreme forms of which are plantations that are established and managed as short rotation crops. Thus initial selection for insect resistance in trees must often be done in natural populations or in young plantations.

Second, tree species are naturally outcrossing and highly heterozygous for most genetic traits. Inbreeding, especially selfing, often results in a significant depression of growth rate. For this reason breeding programs with trees must always begin with a sufficiently broad genetic base to avoid the effects of inbreeding in advanced generation populations. The large amount of heterozygosity in tree species also leads to much variability within the species at all levels including races, populations, and families. Thus pure lines are difficult to obtain for trees except through vegetative propagation.

Most tree species can be vegetatively propagated by stem or root cuttings. *Populus* spp., for example, are commonly established by stem cuttings in large-scale plantations. Other species, particularly the conifers, are less readily propagated as clones, and these methods are used only for research or seed orchard establishment. Recent advances in tissue culture methods, involving both conifers and deciduous trees, may stimulate more widespread use of vegetative reproduction of genetically improved strains for insect resistance.

A fourth distinct attribute of trees is their great longevity and massive size compared with crop plants. Because a tree is a perennially enlarging organism, insects adapt to specific host-tissue systems to complete their life cycle. The preferred tissue may be located high above ground, which makes it more difficult to study than in the case of short-lived, small plants. Certain tree insects such as *Dendroctonus* bark beetles do not attack their host species until a certain tree size or age is reached. This makes progeny testing of parental selections a slow process if one wishes to create natural or artificial infestations on offspring. The use of indirect selection is especially appropriate in these situations and will be discussed later.

Another factor associated with the large size and longevity of trees is that breeders often must work in the tree crown to do controlled pollination and collect seed for progeny testing (Figure 1). As mentioned earlier, attainment of advanced-generation breeding stock in trees may take many years because of the time required for some species to reach sexual maturity. On the other hand, some tree species flower in their first year and progress in breeding can be more rapid.

1.2.   Important Insects On Trees

There are thousands of tree species, each of which may be a host for numerous insect pests. However, to be considered for a resistance breeding

Figure 1. Controlled pollination in upper portions of large western white pine (*Pinus monticola*).

490

effort an insect must be responsible for economic losses large enough to jus-
tify the costs of breeding. There are many such insects and according to
Baker (1972) they include (1) species that damage or destroy the flowers
and seeds of trees and that are particularly important pests in seed orchards
and seed production areas; (2) species that stunt, deform, or kill young trees
by damaging or destroying the terminals, laterals, or roots of plantation
trees or natural reproduction; (3) species that cause loss of vitality, growth
reduction, and often death of trees by eating the foliage; and (4) species that
feed under the bark or in the wood of living trees and girdle and kill them or
riddle them with tunnels. In addition to insects that directly damage trees
there are those that indirectly weaken or kill trees by acting as a vector for
fungal, bacterial, and viral diseases. Notable examples of such vectors are
the smaller European elm bark beetle (*Scolytus multistriatus*) and the native
elm bark beetle (*Hylurgopinus rufipes*). These insects transmit the fungus
Ceratocystis ulmi which causes the devastating Dutch elm disease.

Most tree-insect species normally occur in such low densities that they
are of little or no economic consequence. In these cases development of
resistant varieties would be justified only if the costs were minimal or the
effort was incidental to breeding for other purposes. The minority of tree
insects that do cause great damage to forests either consistently or sporadically should be the focus for evaluation and possible implementation of
breeding programs to develop long-term protection. Worldwide, some
indication of which insects do sufficient damage to warrant resistance
breeding is given in review publications such as those of Gerhold et al.
(1966); Toda (1974); Søegaard (1964); Hanover (1976); Davidson and
Prentice (1967). There have been surprisingly few published estimates of the
actual monetary damage done to trees by insects throughout the world. A
few national and regional estimates suggest the magnitude of the losses. For
example, Baker (1972) gives the following estimates for insect damage to
United States forest and shade trees during an average year (1952):

Young, growing trees killed	= 1 billion cubic feet
Mature sawtimber killed	= 5 billion cubic feet
Young, growing trees growth loss	= 1.8 billion cubic feet
Mature sawtimber growth loss	= 8.6 billion cubic feet

Losses during epidemic years would be much greater. On a regional basis
the U.S. Forest Service (1977) reported that during 1977 the Western
spruce budworm (*Choristoneura fumiferana*) defoliated 1.2 million acres of
forest in north-central Washington alone. Total expenditures in 1975 by all
ownerships in the United States for chemical suppression of just the spruce
budworm amounted to $6,219,000. Another damaging insect, the spruce

bark beetle (*Dendroctonus rufipennis*) killed almost all Engelmann spruce (*Picea engelmanni*) over 10 cm in diameter on the White River Plateau in northwest Colorado from 1940 to 1948. This epidemic killed almost 4 billion board feet of spruce in western Colorado and changed the forest succession pattern dramatically (Wood, 1973).

The few selected examples of tree-damaging insects mentioned above are associated primarily with wild or unmanaged forests. Many more such situations exist in the United States and in other countries (Mattson and Addy, 1975; Gerhold et al., 1966). In recent years there has been a steady increase in the establishment of intensively managed forests to help meet the growing world demand for wood. These new forests contrast markedly with natural forests (Shea, 1971). They are usually composed of a single species, uniformly spaced, and often occupying sites which may be convenient but marginal for the species. Establishment and operation of these plantations is highly mechanized and involves fertilization and weed control. Thus they are similar in many respects to agricultural crops and may be even more subject to catastrophic losses to insects unless many of the precautions recommended by foresters (Shea, 1971; Wilson, 1976) and agronomists (Robinson, 1977; Painter, 1966) are heeded. One of these precautions is to preserve rigorously a broad genetic base of materials used to establish the new forests so that genetic resistance to both insects and diseases may be exploited.

2. TREE RESISTANCE TO INSECTS

2.1. Nature of Resistance

Trees and their insect pests coexist in a delicate relationship. Most of these specific host–pest systems are the result of very long evolutionary processes. Both the tree and insect constitute competing, interdependent biochemical and morphological systems. The complexity of these systems is illustrated diagrammatically in Figure 2. Here the main feeding and reproductive stages and stimuli of the insect are shown to interact with the major physiological and morphological characteristics of the host. The net result of these interactions is some degree of utilization of the tree by the insect. Although the responses of both tree and insect that permit this relationship are largely determined by their genotypes, environment may play a significant role in modifying the relationship.

The foregoing description of a host–insect complex serves as the basis for a definition of host resistance to insects. At one extreme in the complex is the complete success of an insect infestation, leading to death of a host

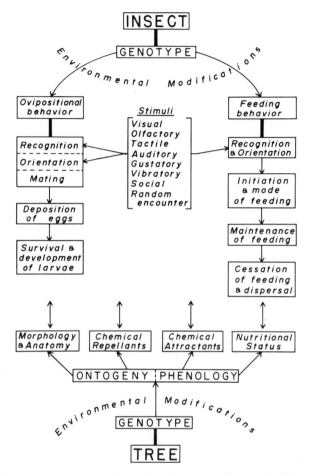

Figure 2. Diagramatic summary of the host-insect complex emphasizing the potential mechanisms for tree resistance to an insect (Hanover, 1975).

species, population, or individual. At the other extreme is immunity of the host species, population, or individual to damage by the insect (Hanover, 1975). Host resistance lies in the broad area between these extremes and has been defined by Beck (1965) as the collective heritable characteristics by which a plant species, race, clone, or individual may reduce the probability of successful utilization of that plant as a host by an insect species, race, biotype, or individual. Excluded in this definition is pseudoresistance (Painter, 1951), which is the escape from insect attack or lack of insect success on a susceptible host resulting from some transient situation. For example, temporary environmental conditions and a decrease in host vigor can result in apparent, nongenetic resistance in a potentially susceptible host tree.

2.2. Causes of Resistance

Because of the great complexity of any given tree–insect system it is helpful to examine the components shown in Figure 2 more closely in terms of insect development, or inhibition thereof, on the tree. It is probably true that an understanding of the actual causal mechanism for insect resistance is unnecessary for its discovery, development through breeding, or practical use. However, a knowledge of the basis for resistance may have considerable practical value by revealing host chemicals that could serve as tools for behavior modification of the insect or as possible new insecticides. In addition, by knowing the chemical, physical, or physiological basis for resistance, a breeder may increase the efficiency of selection and incorporation of resistance genes into a new variety, especially when plateaus for improvement are reached. Finally, in the case of trees where generation intervals are relatively long or the insect attacks beyond the seedling stage, screening parents or progenies for resistance factors by some method other than bioassay could be extremely important (Coppel and Mertins, 1977; Hanover, 1975; Weissenberg, 1976). This could also facilitate methods of indirect selection for traits that may be highly correlated with resistance but not causally related.

Insight into the mechanisms involved, beyond those given by Painter (1958) is provided by the classification scheme shown in Figure 3. This was devised by van Emden and Way (1973) to reflect the dynamic and interacting nature of resistance during the sequence of events from arrival at the host to development of the insect population on the host.

From the various descriptions of causal mechanisms for insect resistance it is clear that resistance to insect attack is often a quantitative rather than a qualitative response for a given host–insect system. This results from the fact that even gene-controlled resistance is continually subject to environmental modification as implied in Figure 2, from the standpoint of both the tree and the insect. This fact must be recognized by tree breeders as they proceed with selection and testing for resistance and by forest managers who will use the resistant strains.

2.3. Genetic Variation In Resistance

The major prerequisite to any breeding effort for insect resistance is the existence and demonstration of heritable resistance to the insect pest.

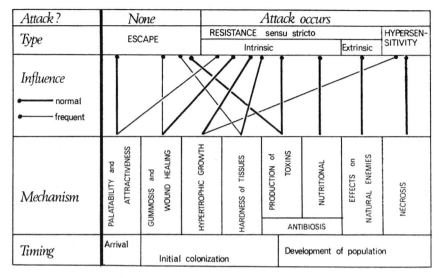

Figure 3. Classification of host-plant responses to an insect population (Van Emden and Way, 1973).

Procedures for revealing genetic resistance in trees are discussed below in Section 3.1. First, let us review some examples of tree resistance and then examine one system, the sawflies (*Neodiprion* spp.), in more detail

Examples Of Tree Resistance

Several publications have summarized the known cases of tree resistance to insects (Gerhold et al., 1966; Søegaard, 1964; Toda, 1974; Hanover, 1975; Roth, 1970). However, it is remarkable how few rigorously demonstrated examples of genetic resistance there are for any of the major insect pests on trees. This probably reflects the relative youth of forest genetics as well as a preoccupation by geneticists with the improvement of growth traits.

Perhaps the foremost example of genetic resistance by a tree species is that of the maidenhair or ginkgo tree (*Ginkgo biloba*). Apparently all individuals of this extremely old species are highly resistant, if not immune, to insects as well as bacteria, viruses, and fungi (Major, 1967). Although there obviously is no need for improved strains of ginkgo, this species offers a good example of the potential stability of genetic resistance to insects once achieved.

There are a number of good examples of genetic variation within species, some of which serve as bases for insect resistance breeding programs. A dangerous pest on spruces in central Europe is the black-marked tussock moth (*Lymantria monacha*) (Schonbörn, 1966). This insect feeds on fresh tree shoots in May. Late-flushing varieties of spruce avoid damage by the larvae of this insect, so tree breeders select and hybridize to produce these resistant strains. The late-flushing spruces are also less sensitive to late frosts but are more susceptible to larvae of the spruce sawfly (*Nematus abietum*). Fortunately the sawfly does far less damage to spruce than the tussock moth.

An early example of tree resistance to an insect pest is that of black locust (*Robinia pseudoacacia*), host for the locust borer (*Megacyllene robiniae*). Hall (1937) reported that two recognized varieties of black locust ('Shipmast' and 'Higbee') were resistant to borer attack. Later, more rigorous tests of these and other clones of black locust verified that several strains are less susceptible to damage than others but the degree of genetic control is still unclear (Boyce and Jokela, 1966). Nevertheless this represents a clear case of selection, breeding, and practical application of genetic resistance in a tree species.

In Japan the testing and breeding of chestnut (*Castanea crenata* and *C. mollissima*) varieties resistant to the chestnut gall chalcid (*Dryocosmus kuriphilus*) has progressed through the selection of three resistant varieties from F_1 crosses which are now being produced on a commerical scale (Kajiura and Machida, 1961).

The eastern spruce gall aphid (*Adelges abietis*) is a serious pest on two ornamental and Christmas tree species, white spruce (*Picea glauca*) and Norway spruce (*P. abies*) in the northeastern United States. It is partially controlled by the use of insecticides, a method that has proved expensive, time-consuming, and ecologically disruptive. Research on plantations in Ohio has recently revealed the existance of gall-free trees growing among heavily infested trees (Thielges and Campbell, 1972). These gall-free trees support overwintering aphids but do not allow maturation of the insect in the spring. There are no differences in host phenology to suggest that an escape mechanism is responsible for aphid resistance. Preliminary studies indicate that resistance to this insect is determined by a single gene and the development of resistant strains through breeding is underway. A chemical marker for aphid resistance in Norway spruce has also been found which should facilitate selection and production of resistant strains (Tija and Houston, 1975).

The American bark beetles (*Dendroctonus* spp. and *Ips* spp.) and their tree hosts have received possibly more research emphasis than any other tree–insect systems in the world. Apparent resistance has often been observed in natural populations (Callaham, 1966; Furniss, 1972; Rudinsky, 1966) and concerted research has been done on possible resistance mechanisms (Smith, 1966a; Person, 1931; Vité and Wood, 1961; Hanover, 1975).

Yet, surprisingly, there is no good documentation of resistance to the bark beetles in terms of total number of attacks per tree, maximum attack density, and rate of attack (Wood, 1973). Nor is there any evidence of genetic resistance to bark beetles among the pines, spruces, or Douglas fir which they attack. One explanation for a lack of demonstrated genetic resistance is that these insects generally attack their hosts only after a considerable age or stem diameter (about 10 cm) has been attained.

Another severely damaging insect for which genetic resistance has been sought is the white pine weevil (*Pissodes strobi*) which attacks the terminal shoots of white pines (*Pinus* spp.), Sitka spruce (*Picea sitchensis*), Norway spruce (*Picea abies*), and other species. Pauley et al. (1955) provided the first evidence of intraspecific variation in white pine resistance to the weevil from results in a seed source test. Wright and Gabriel (1959) also reported variation in resistance to weeviling among geographic sources of *Pinus strobus* in a New York provenance test. A long-term breeding program for the development of weevil resistant white pine is being conducted by the U.S. Forest Service in the northeastern United States (Santamour, 1964; Garrett, 1972; Wilkinson, 1978). This program includes the replicated testing of progenies over a number of sites and years, establishment of clonal tests with promising selections, and investigations as to the existence of chemical markers for resistance. Other research centers have also contributed valuable information on weevil resistance in white pine (Heimburger, 1963; Soles and Gerhold, 1968; Connola and Beinkafner, 1976) and other host species (Mitchell et al., 1974).

The pine reproduction weevil (*Cylindrocopturus eatoni*) offers an example of success in using hybridization between closely related tree species to secure resistance to a serious insect pest. This weevil attacks and kills young trees, especially Jeffrey pine (*P. jeffreyi*) in California. Breeding work led to the development of a resistant interspecific hybrid from crosses with the highly resistant Coulter pine (*P. coulteri*). The backcross of this hybrid to Jeffrey pine has also been mass produced and planted on pine sites where the weevil was a problem (Smith, 1966b; Libby, 1958; Smith, 1960).

Seed source, provenance, or common garden tests with tree species provide effective and efficient means for obtaining information on natural variation in resistance to insects as well as in other traits. Examples of seed source or varietal resistance revealed by examination of such plantations include the eastern pine-shoot borer (*Eucosma gloriala*) attacking Scotch pine (*Pinus silvestris*) (Steiner, 1974); the pine root collar weevil (*Hylobius radicis*) attacking Scotch pine (Wright and Wilson, 1972); and the white pine weevil attacking eastern white pine (Wright and Gabriel, 1959). Instances of genetic variation in the resistance of the southern pines of the United States to several insects have been summarized by Dorman (1976). As the practice of establishing seed source trials increases and as data on insect resistance are collected from those already established, natural sources of resistance will be found and important breeding can commence.

A final source of potential genetic variation in tree resistance to insects is to be found simply by observing natural selection in existing forests. There are remarkably few instances of either documentation or preservation of phenotypically resistant individuals to even the most damaging insect pests. This procedure is most important because destruction of these resistant specimens through salvage harvesting or other forms of mortality represents an irreplaceable loss of germplasm.

Diprionid Sawflies: Genetic Specialization On Tree Hosts

An appreciation of the complexities involved in developing insect resistance through breeding may be gained by briefly examining one major group of insects, the diprionid sawflies. These insects cause much damage to trees of the Pinaceae and Cupressaceae throughout northern coniferous forests and plantations. Their popular name is derived from the sawlike ovipositor in the female, a nonstinging wasp. This "saw" is used to slit or cut the tree needle and for insertion of eggs. The pattern and mode of insertion is highly specific for each species of the genera *Neodiprion* and *Diprion* but nearly all species prefer old needles rather than new tissue on the host tree. Emerging sawfly larvae form feeding colonies and usually consume whole needles of the host (Figure 4). Since defoliation is usually not complete the host tree can sustain repeated attack by the sawfly without succumbing. However, tree growth and appearance are affected, and mortality can result from repeated infestations.

The sawflies have developed highly specific adaptation mechanisms to their host trees which can provide the tree breeder with some insight about the potential stability of resistance and the host factors most likely to be associated with resistance to a sawfly strain.

Knerer and Atwood (1973) provided an excellent description and synthesis of the gene-controlled physiological systems in synchrony between sawflies and their tree hosts. Differences in phenology or seasonal development are a major mechanism by which host selection has been achieved by sawflies. For example, different strains of *N. abietis* have as their normal host plants white spruce (*Picea glauca*), white fir (*Abies concolor*), and two strains ("early" and "late") of balsam fir (*Abies balsamea*) (Figure 5). These strains of sawfly and their hybrids clearly vary in their time of larval hatching and adult emergence when reared under identical conditions (Figure 6). Host physiology is of primary importance in determining both the degree of preference by a sawfly strain and success of an adapted strain to its host. Host toxins and repellents apparently play a major role in sawfly adaptation to its preferred host. This was shown in tests done by Knerer and Atwood (1973) in which strains of *N. abietis* were placed on different

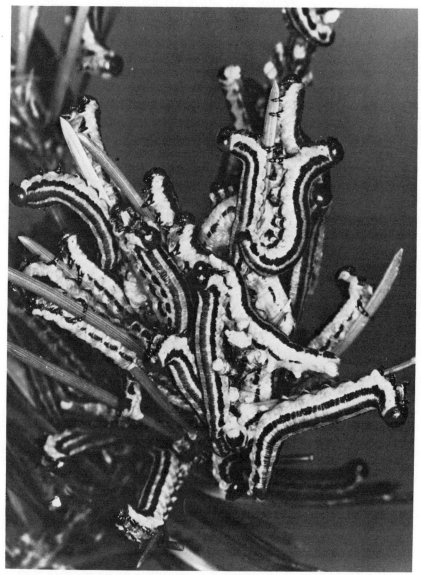

Figure 4. Close-up view of sawfly larvae (*Neodiprion rugifrons*) consuming needles of a jack pine tree (Knerer and Atwood, 1973). Copyright 1973 by the American Association for the Advancement of Science.

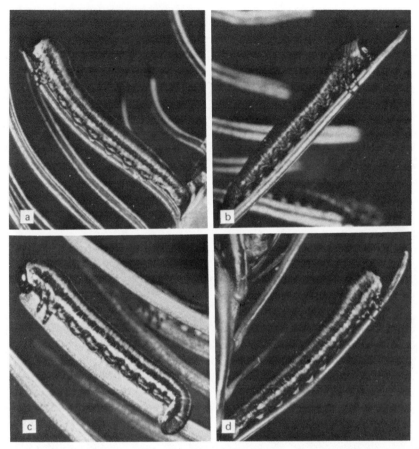

Figure 5. Mature larvae of different strains of *N. abietis* on their normal food plants. (*a*) White spruce strain from Alberta eating whole spruce needle. (*b*) Early balsam strain from Ontario, skeletonizing balsam fir. (*c*) White fir strain from California eating whole needle of white fir. (*d*) Late balsam strain from Nova Scotia, skeletonizing balsam fir. (Knerer and Atwood, 1973. Copyright 1973 by the American Association for the Advancement of Science.

conifer hosts, resulting in striking patterns of larval mortality and feeding success (Figure 7). Moreover, some indication of the strong genetic control over sawfly larvae specialization was provided by hybridization of sawfly strains and application of the hybrid larvae to the same range of conifer hosts (Figure 8). Thus the sawflies illustrate how a group of more than 30 polyphagous insect species or races have evolved rigid systems of host-tree selection which are under simple gene control. Further study of these systems, and especially the mechanisms involved in host resistance to certain sawfly strains, could lead to the development of wider-based

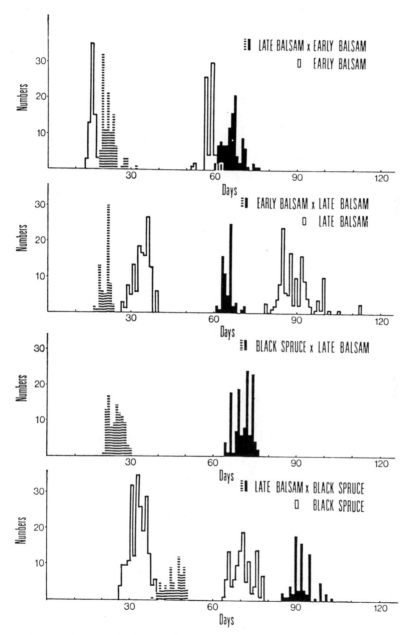

Figure 6. Inheritance of developmental rates in hybrids of *N. abietis*. A reversal of the sexes of the first cross was without effect. A cytoplasmic factor seems to be responsible for the marked shift in emergence of the second pair of crosses (Knerer and Atwood, 1973). Copyright 1973 by the American Association for the Advancement of Science.

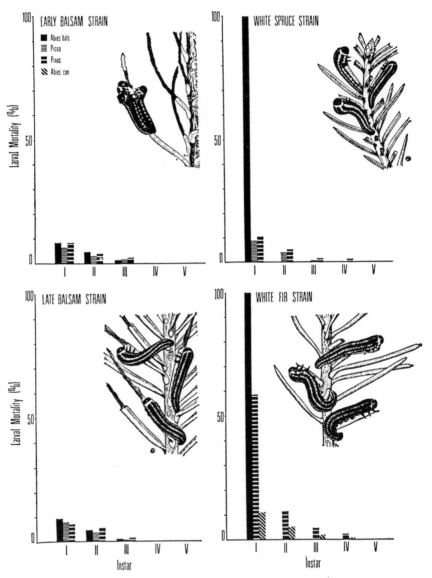

Figure 7. Larval mortality and feeding patterns of four *N. abietis* strains on a variety of conifers (Knerer and Atwood, 1973). Copyright 1973 by the American Association for the Advancement of Science.

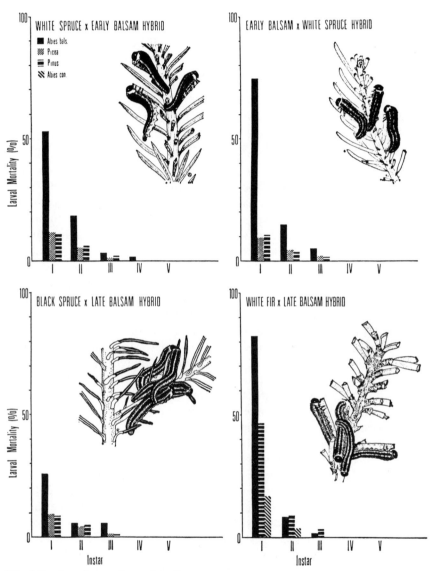

Figure 8. Larval mortality and feeding patterns of some *N. abietis* hybrids (female is first name of the cross). Striking differences exist not only between hybrids of different strains but even between the same cross if sexes are reversed (Knerer and Atwood, 1973). Copyright 1973 by the American Association for the Advancement of Science.

resistance to other strains through breeding. Some progress has already been made toward this end with the discoveries of sex pheromone specificity in the sawflies (Jewett et al., 1976) and also of a chemical (13-keto-8(14)-podocarpen-18-oic acid) deterrent in juvenile foliage which is the basis for larval feeding preference for old jack pine (*Pinus banksiana*) foliage by *N. rugifrons* and *N. swainei* (Ikeda et al., 1977).

3.   BREEDING METHODS AND PROCEDURES

3.1.   Selection for Resistance

The decision to begin a breeding effort for insect resistance in a tree species and all subsequent stages in the development of resistant varieties should involve the cooperation of at least a geneticist and an entomologist. Other disciplines may also be involved in some phases of the total program but the breeder–entomologist team is of utmost importance in achieving the goal.

The first step which the breeder–entomologist team must take, even before the commitment to a program is made, is to determine whether or not natural resistance to the insect exists. As Painter (1966) has pointed out, the likelihood of finding resistance in trees, as in other crops, is very high provided large numbers of species, varieties, or individual plants are examined. Forest trees offer several advantages in this respect because they often grow in large numbers over extensive areas. Moreover, trees are outcrossing organisms and exhibit a high degree of polymorphism. They are also prolific seed and pollen producers, and both their seed and pollen can be stored for years with proper treatment. Each of these factors favors the creation, maintenance, and increase of genetic diversity necessary for large-scale natural or artificial screening for resistance.

To determine the existence of insect resistance in a tree species a search must be made. Searching may be done in natural stands or in already established plantations or progeny tests. Little use has been made of mass selection for resistant phenotypes in natural stands despite its high potential for success and low cost relative to other insect control measures available. Extensive, relatively undisturbed forests still occur in many countries and natural selection for insect resistance continues. For example, control of the western spruce budworm which damages millions of acres of Douglas fir (*Pseudotsuga menziesii*) in the northwestern United States has been solely by use of insecticides. Until recently there were no attempts to find and document host resistance to this pest. When infested stands were finally surveyed, substantial evidence for phenotypic resistance was obtained (Johnson and Denton, 1975). Work is now underway to identify, preserve, and test trees exhibiting resistance in heavily infested stands (Figure 9) (McDonald, 1976). Undoubtedly, similar instances of natural resistance would be found for other important insects on trees if searches were made. It is important that this work begin for all major insect pests before logging, fire, and other factors destroy a valuable source of germplasm for future breeding.

Selection for insect resistance is common in plantations and progeny tests conducted for other commercial or research purposes. Most tree breeding programs include the assemblage of diverse genetic seed sources as standard procedure. Although other traits, especially growth rate, are usually of major concern to breeders, resistance to insects and disease can be conveniently observed and utilized as indicated earlier. In such tests it is important that as wide a genetic base as possible be preserved to allow multiple trait selection without incurring the detrimental effects of inbreeding in future generations.

Selection of trees for insect resistance is usually direct; that is, resistance itself is measured in the selected individuals or populations. However, in cases where the insect attacks a tree at some age beyond the seedling stage, such as most bark beetles, indirect selection methods should be sought. Indirect selection involves finding another trait in the host, causally related or not, which is highly correlated with the resistance response but can be measured at an early age. Some attempts are being made to establish correlated responses for purposes of indirect selection but the practice is not yet established for any insect–host system (Smith, 1975; Hanover, 1975; Weissenberg, 1976). Investigations into the physiological basis for resistance to an insect are important in this regard and should be encouraged for tree species.

3.2. Testing Trees for Resistance

When the intitial selection for phenotypic resistance has been done the breeder–entomologist must then obtain evidence that the resistance has a genetic basis and can be transmitted by clonal or sexual means. For tree species that are commonly planted over a range of site conditions and that receive little cultural treatment after planting, the resistant strains should exhibit some degree of environmental stability in resistance. That is, changes in the environment, whether geographic or temporal, should not disrupt or decrease the resistance to any great extent. Therefore, tests for genetic resistance should include provision for exposure to the insect under varying conditions whenever possible.

Testing of directly selected phenotypes must be done on the parental selections in the wild by artificial or forced attacks using proper control trees in cases where the insect attacks older trees only (Figure 10). This does not ensure that the resistance is genetically controlled but it provides added evidence that escape is not the reason for the apparent resistance.

Figure 9. Phenotypically resistant Douglas fir among trees heavily infested with western spruce budworm in Montana (McDonald, 1976). Courtesy of G. I. McDonald and U.S. Forest Service.

Trees that continue to express resistance after forced attacks may then be used for mass clonal propagation or as parents for hybridization work. Logically, they should also be used to investigate the mechanisms for resistance as well as the existence of correlated traits for use in indirect selection.

It should be emphasized that in testing trees for resistance the goal is not to achieve immunity to the insect. Test procedures should be heavy enough to decrease the probability of escape yet not so heavy as to eliminate tolerance to the insect by the host. This will also encourage accumulation of more genes for resistance in a variety and more environmental stability when it is planted widely. The actual procedures used to screen for resistance will of course vary with different insects and hosts.

If the insect is of the type that attacks small trees, bioassays may be done on either vegetatively propagated mature trees or on their seedling progeny. Here again adequate controls from susceptible phenotypes should be included in the testing scheme. Bioassays can be either artificial, such as cage tests, or natural exposure to normal insect populations in plantations.

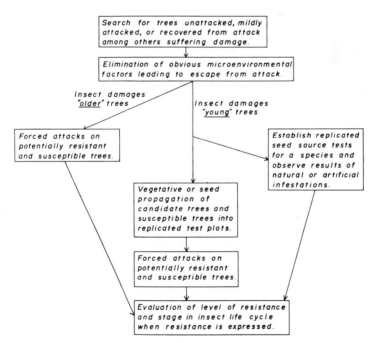

Figure 10.   Procedures for demonstrating resistance of trees to insects (Hanover, 1975).

In either case dependence on data from over- or underattack must be avoided.

Other details of the resistance testing phase of tree breeding against insects can be found in the literature for specific insects and in other reviews of the subject (Gerhold et al., 1966). Of course, the testing phase also requires very close coordination between the tree breeder and entomologist for success.

3.3. Hybridization

If the testing phase of a resistance breeding program reveals substantial phenotypic variation in insect damage, the breeder must next determine if the resistance can be transmitted through seed. There are several ways of doing this, depending on how the initial selection and testing work were done. The simplest and least expensive procedure results from geographic variation in resistance revealed by provenance tests. Seed from sources whose trees suffer significantly less damage than others can be assumed to possess some genetic resistance that is of immediate use to a breeder. In other words, by collecting open-pollinated seed from the same geographic origin or race as tested one can expect to achieve the same level of resistance relative to other sources in the provenance test. In addition, several resistant races of the species may be combined and used as a synthetic mixture to establish new plantations. This procedure should provide some protection against variation in resistance owing to environmental changes over locations and time, or possible changes in the insect itself.

Another efficient and relatively quick method for determining the degree of genetic control over resistance is to collect open-pollinated seed from selections made with the aid of natural selection in wild infested stands. Seedlings must than be produced from this seed and retested to determine their level of resistance. If it is sufficiently high, it would indicate a strong genetic component and allow the limited production of seed from these original selections.

Genetic gains in resistance will be greater than in the two methods described above if controlled pollination is used on either the provenance test trees or the wild parents. Crossing the wild parents may be expensive and dangerous if the trees are large (Figure 1). If the parents are cloned and placed in plantations to flower, or if the trees in young provenance tests are used, hybridization is more feasible and safer (Figure 11).

Hybridization of resistant phenotypes by controlled pollination may be either intra- or interspecific. An advantage of tree species over many other plants is that interspecific crosses are relatively successful in many genera.

This provides the tree breeder with another source of genetic variation to exploit for resistance to pests as well as improving other traits. Thus the tree breeder may employ close or wide (racial) intraspecific hybridization or interspecific hybridization to generate F_1, backcross, or advanced generation materials in an insect resistance program. Little use is made of inbreed-

Figure 11. Plantation of young, early-flowering, white spruce in which pollination is controlled to produce advanced generation hybrid trees.

ing by tree breeders because of the severe depression frequently found in inbred offspring.

Whatever the means by which seed from phenotypically resistant parents is obtained, the resulting progenies must be retested to determine the degree and mode of inheritance. In this phase the same principles of testing for resistance as were stated above will apply. However, experimental designs may vary depending upon the type of genetical analysis desired with the progeny. Further details on experimental design and methods of hybridizing are available in Gerhold et al. (1966), Wright (1976), Hanson and Robinson (1963), and other plant breeding literature.

3.4. Resistant Seed Production

The final stage in the development of insect resistant trees is the production of large amounts of commerical seed. The procedures for this stage also depend somewhat on how the selection, testing, and hybridization were done. Moreover, they depend on how advanced the breeding program is. For example, it would not be uncommon to have several methods of providing seed of differing levels of resistance in a long-term program, using several types of selection and hybridization concurrently.

The most expedient source of large amounts of improved seed would be native stands that proved to be resistant in provenance tests. Seed can be collected in the unimproved stands using conventional seed harvesting equipment or from squirrel caches. If higher seed yields are desired seed production areas can be established. These consist of native stands in which thinning, fertilization, weeding, animal protection, and other cultural treatments are used to maximize seed production.

The provenance test itself may be rogued or thinned heavily to favor the resistant seed sources. Intensive culture would then be applied to the remaining trees to stimulate early, prolific flowering. However, since a provenance test of a tree species is seldom established for the purpose of improving a single trait its use as a seed production area is not recommended. It is possible to use controlled pollination of original parental selections, in the wild or in provenance tests, to mass produce seeds. Although expensive, this procedure may be useful when the seed has a very high value and ample flowers are available.

Resistant seed in large quantities is most efficiently produced in seed orchards. Orchards are established by planting grafts, cuttings, or seedlings from parental selections. Often a seed orchard will be an F_1 or F_2 progeny test that is first established in a systematic design to enable thinning later and conversion for seed production. In this case the initial spacing may be relatively close so that final spacing after roguing will be wide.

Figure 12. Western white pine cone infested with *Conopthorus monticolae*, one of many cone and seed insects that cause heavy damage to tree seed orchards.

Seed orchards are usually designed to assure effective natural cross-pollination among trees. To make use of specific combining ability between orchard members artificial pollination may be practiced although natural crossing is, of course, much less expensive. If natural crossing is used it is again important to have as many different clones or unrelated progenies as possible to minimize inbreeding.

Once established, seed orchards are usually given intensive cultural treatment to promote flowering. Vigorous trees tend to flower at a younger age and more consistantly than low vigor trees. Therefore, all methods favoring vegetative growth should be used in orchard management including fertilization, irrigation, cultivation, crown pruning, and protection from animals or diseases. Ironically, the most damaging and troublesome factor to control in most tree seed orchards is the group of insects that attack the flowers, cones, and seed for which the orchard was established (Figure 12).

High resistance to greenbug is demonstrated by plants at left, the result of a cross between the varieties at center and right. (Photo courtesy of USDA, SEA, AR).

20

BREEDING APPROACHES IN WHEAT

E. H. Everson

Department of Crop and Soil Sciences, Michigan State University, East Lansing, Michigan

R. L. Gallun

Science Education Administration, U.S. Department of Agriculture, and Department of Entomology, Purdue University, West Lafayette, Indiana

1. WHEAT

1.1. Origin and Classification

There is archaeological evidence that wheat was among the first plants cultivated some 8,000 years ago and, along with barley, became a major source of energy for the new civilization in Asia Minor (Lelley, 1976). Domestication of these cereals led to rapid population increases, community settlement, and cultural development.

Wheat, barley, and rye belong to the tribe Triticeae which includes some of the most economically important genera of the Gramineae. Wheat belongs to the genus *Triticum* which, along with the genera *Secale, Aegilops, Agropyron,* (=*Eremopyrum*), and *Haynaldia,* is assigned to the subtribe Triticinae. Members of this subtribe readily hybridize, resulting in either a direct exchange of genetic material or amphiploidy.

Sakamura (1918), Sax (1922), Kihara (1919), and others have shown that the various species of *Triticum* form a polyploid series based on $x = 7$ with three levels of ploidy: diploids ($2n = 2x = 14$), tetraploids ($2n = 4x = 28$), and hexaploids ($2n = 6x = 42$).

The most primitive species of the diploid group, *Triticum monococcum* (wild and cultivated einkorns), is identified as the ancestor of the A genome (a chromosome set corresponding to the haploid set of the donor species). Wild einkorn, *T. monococcum* var, *boeoticum* is widely distributed in the fertile crescent and across the Anatolian Plateau into eastern Turkey and Greece. It is best adapted to soils of lava basalt and alluvial origin and seldom is found on sandy and limestone soils. The wild and cultivated einkorns are hulled and vary widely in vernalization requirement (Zohari, personal communication).

The tetraploid species, *Triticum turgidum* (emmer wheat), is believed to have arisen from a hybrid between *T. monococcum* and *Aegilops speltoides* or a close relative of the latter which was the donor of *B* genome. *T. dicoccoides* is a subspecies or variety of *T. turgidum*; wild and cultivated emmers belong to this group. Wild emmer lines are always found in primary (undisturbed) habitats and are distributed in a limited zone in northern Israel, Lebanon, northwestern Jordan, and southwestern Syria, as well as a few isolated places in western Iraq and southwestern Turkey. The most important cultivated wheats in this group are the durum varieties or the macaroni wheats, *T. turgidum* var. *durum*.

Most modern varieties of wheat, the bread and pastry wheats, belong to the hexaploid species, *Triticum aestivum*, which is widely distributed in an array of environments from 67 degrees north in Norway, Finland, and Russia to 45 degrees south in Argentina. In the tropics, cultivation has been restricted to higher elevations, but genotypes identified in recent years hold promise for the lowland tropics. Both winter and spring varieties are found in this species. The species arose through hybridization between *Triticum turgidum* var. *dicoccum* and *Aegilops squarrosa* = *T. tauschii. A. squarrosa* is the confirmed donor of the *D* genome. Its distribution only occasionally intersects with that of *T. turgidum* var. *dicoccum,* and where the populations do intersect, *A. squarrosa* is found only in secondary (disturbed) habitats. *A. squarrosa* is found in an area of spring and summer rainfall on a wide variety of soils in southeastern Russia, Afghanistan, northern Iran, and a few isolated areas in Iraq, Syria, and western Turkey. This species has a great variation in vernalization response and is found in areas in which extreme temperature, rainfall, and soil fertility prevail (Zohari, personal communication).

The reader is referred to Morris and Sears (1967), Feldman (1975), and Lelley (1976) for further discussions on the cytotaxonomic background of wheat and its relatives.

A study of the ranges of distribution of wheat and its relatives sheds light on some new approaches which must be taken in wheat breeding. Although tetraploid wheats, originating in the Middle East, are adapted to the mild Mediterranean climate with sparse winter rainfall, the addition of the *D* genome must have contributed to the adaptation of the hexaploid wheats to more severe continental climates. Since the distributions of the parents of *Triticum aestivum* rarely overlap, and when they do the parent species *T. turgidum* and *A. squarrosa* occur in different habitats (undisturbed and disturbed, respectively), the natural hybrid probably occurred only on rare occasions. This cross should be recreated with a variety of parents taken from many sites in their respective ranges of distribution. The hexaploid progeny would likely represent valuable resources in breeding for insect resistance as well as for many other desired characteristics.

Spring wheat is differentiated from winter wheat mainly by heading habit. Spring wheat sown in the spring will head, but spring-sown winter wheats usually remain in the rosette stage and fail to head. Spring habit is dominant and conditioned by one, two, or three genes. The dominant or recessive genes for spring or winter growth habit plus genes for date of heading and maturity, both of which depend on the date of seeding (day length × temperature response), have tended to isolate the cultivated hexaploid into two major groups. The spring and winter types have undoubtedly existed since the hexaploid wheats initially evolved. With this lengthy isolation, great gene pool differences have evolved for a multitude of factors between winter and spring varieties. Breeders and entomologists must examine the progeny of spring × winter wheat crosses for insect resistance and other valuable characteristics.

1.2.   Wheat Production

Wheat is grown over a wider area in the world than any other major crop. Its spring and winter habits, wide temperature tolerance, and ability to grow under a variety of soil types and rainfall paterns have endowed this crop with a wide range of adaptability. Wheat ranks next to rice in importance as a food crop and is the principal food for about one-third of the world's people (National Academy of Sciences, 1972). Ninety percent of the world's wheat production occurs in the northern hemisphere, with the USSR, Asia, and North America producing approximately 288 million metric tons, or about 70 percent of total world production (USDA-Agricultural Statistics, 1977). The major factors affecting wheat yields are climate (temperature, distribution, and amount of rainfall), soil type and fertility, plant pests, variety, and farming practices.

In the last three decades, breeders have improved spring wheat with a wide range of resistance to pathogens, shorter straw coupled with a fertilizer response (increased tillering), daylength insensitivity, and high temperature resistance. New semi-dwarf cultivars have extended spring wheat production into the tropics and dramatically raised yield ceilings on both rainfed and irrigated spring wheat. New agronomic practices were also developed to utilize the full genetic potential of these new cultivars. Shorter strawed (lodging resistant), high-yielding, winter wheat cultivars with more winter hardiness and drought, insect, and disease resistance have also been developed.

1.3. Wheat Germplasm Sources

Substantial collections of wheat germplasm covering a wide range of species are held by the U.S. Department of Agriculture at Beltsville, Maryland; the N. I. Vavilov Institue at Leningrad; the International Wheat and Maize Improvement Center (CIMMYT) in Mexico; and the FAO at Rome. There are also many national, individual, and special "working" collections throughout the world. These represent a small but substantial sample of the spectrum of genetic variation of the species, and they supply the wheat workers with the exotic germplasm needed in breeding programs for host plant resistance.

In 1974 the Consultative Group on International Agricultural Research set up the International Board for Plant Genetic Resources (IBPGR) to strengthen efforts already underway to conserve genetic variation. This group is charged with establishing a worldwide network of institutions, organizations, and programs to organize and document existing collections. Its charge is to identify general and specific needs for exploration, collection, evaluation, and conservation of plant genetic resources and to make this material available to plant breeding and other scientific programs as required.

2. INSECTS AND MITES INJURIOUS TO WHEAT

Wheat throughout the world is attacked by more than 100 percent species of insects and mites. In the United States alone, average annual losses estimated to be $42 million (Dahms, in Quisenberry and Reitz, 1967). The plant is attacked below ground by soil insects and above ground by chewing and sucking insects and mites that feed on the stem, leaves, and head.

The control measures for insect and mite pests of wheat have been mainly cultural, biological, and chemical. Major programs to develop resistance in wheat to insect pests have been started throughout the world, but as yet resistant wheat varieties have only been developed for protection against the Hessian fly, cereal leaf beetle, and wheat stem sawfly. Table 1 lists some of the important insect and mite pests of wheat that occur throughout the world.

Table 1. Important Insects and Mites that Attack Wheat

Acarina
  Petrobia latens, brown wheat mite
  Oligonychus pratensis, Banks grass mite
  Penthaleus major, winter grain mite
  Aceria tulipae, wheat curl mite

Orthoptera
  Melanoplus sanguinipes, migratory grasshopper
  M. differentialis, differential grasshopper
  M. bivittatus, two-striped grasshopper
  Ramburiella turcomana

Thysanoptera
  Haplothrips tritici
  Limothrips cerealium

Heteroptera
  Chlorochroa sayi, stinkbug
  C. uhleri, green grain bug
  Eurygaster integriceps, senn pest
  E. austriacus
  E. maurus
  Aelia rostrata
  A. acuminata
  A. furcula

Homoptera
  Schizaphis graminum, greenbug
  Rhopalosiphum padi, oat bird cherry aphid
  Macrosiphum avenae, English grain aphid
  Cuernavaca (Cavahyalopterus) noxius
  Endria inimica, leafhopper
  Psammotettix straitus, leafhopper
  Delphacodes pellucida, Calligypona pellucida, planthopper
  Macrosteles fascifrons, leafhopper

Lepidoptera
  Euxoa auxiliaris, army cutworm
  Agrotis orthogonia, pale western cutworm
  Apamea sordens, cutworm

Spodoptera frugiperda, fall armyworm
Papaipema nebris, stalk borer
Elasmopalpus lignosellus, lesser cornstalk borer
Syringopais (Scythrix) temperatella, cereal leaf miner

Coleoptera
  Oulema melanopus, cereal leaf beetle
  Marseulia dilativentris, leaf beetle
  Sphenophorus spp., billbugs
  Agriotes mancus, wheat wireworm
  Ctenicera aeripennis destructor, prairie grain wireworm
  Eleodes hispilabris, false wireworm
  E. opucus, fals wireworm
  E. tricostata, false wireworm
  E. suturalis, false wireworm
  Embaphion muricatum, false wireworm
  Phyllophaga lanceolata, white grub
  Phyllopertha nazarena, white grub
  Anisoplia austriaca, white grub
  A. leucaspis, white grub

Hymenoptera
  Cephus cinctus, wheat stem sawfly
  C. tabidus, black grain stem sawfly
  C. pygmaeus, European wheat stem sawfly
  Pachynematus sporax, Ross leaf-feeding sawfly
  Harmolita grandis, wheat strawworm
  H. tritici, wheat jointworm

Diptera
  Mayetiola destructor, Hessian fly
  Hylemya (Leptohylemyia) coarctata, wheat bulb fly
  Contarinia tritici, wheat blossom midge
  Sitodiplosis mosellana, wheat blossom midge
  Haplodiplosis equestris, gall midge
  Meromyza americana, wheat stem maggot
  M. saltatrix, wheat stem maggot
  Oscinella frit, frit fly

2.1. Mites

Mites damage plants by sucking sap from the leaves and head, causing the plant to show damage resembling that from drought. Severe infestations may kill the plant. The wheat curl mite, *Aceria tulipae* , is a vector of the virus causing wheat streak mosaic, an imporatnt disease of wheat in the Great Plains states. Both the winter gain mite, *Penthaleus major,* and the brown mite, *Petrobia latens,* are important pests in Iraq, Cyprus, and Northeast Africa. The Banks mite, *Oligonychus pratensis,* is a serious pest in the southwestern United States.

2.2. Grasshoppers

Grasshoppers cause greatest damage to wheat in the fall when they migrate into the edges of newly planted wheat fields and consume the plants. With the exception of *Ramburiella turcomana,* an economic pest in Iran, all species listed are sporadic pests of wheat in the western United States.

2.3. Thrips

Thrips cause some damage to wheat plants by removing chlorophyll from the leaves resulting in typical symptoms of streaked and blotched feeding. Thrip infestations have not been considered serious enough to apply control measures. The grain thrips *Limothrips cerealium* and *Haplothrips tritici* are found in Europe as well as in the United States.

2.4. Stinkbugs

The stinkbugs cause some damage to wheat in the United States, and considerable damage in Europe and the Middle East. The senn pest *Eurygaster integriceps* and its companion species of *Aelia* are of major importance in Afghanistan, Iran, Iraq, Lebanon, Syria, Turkey, Yugoslavia, and Greece. These pests suck sap from the stem, leaves, and kernels, resulting in wilting and stunting. Attacked kernels sometimes fail to fill, and even if they are filled the gluten is damaged, causing serious reduction in kneading and baking of bread. The senn pest has not yet entered the United States. If it should immigrate, it would be a serious threat to the bread industry. Resistance in wheats does occur, but no resistant varieties have been developed.

2.5.   Aphids, Leafhoppers, and Planthoppers

The aphids, leafhoppers and planthoppers are serious pests of wheat, both from feeding damage by the adults and nymphs and the virus diseases that they transmit. The greenbug, *Schizaphis graminum,* is widespread throughout the world and causes considerable damage to the wheat crop by toxins it injects into the plant while feeding. Along with the oat bird cherry aphid, *Rhopalosiphum padi,* and the English grain aphid, *Macrosiphum avenae,* it is a vector of the barley yellow dwarf virus of cereals in the United States, Canada, and Europe. Varieties are being developed for resistance to the greenbug.

The leafhopper, *Endria inimica,* and the planthopper, *Delphacodes pellucida,* transmit the virus causing the striate disease of wheat. *Psammotettix straitus* transmits the virus causing winter wheat striate.

2.6.   Cutworm and Armyworms

The cutworm and armyworms are lepidopterous pests of wheat. Cutworms consume plant parts both above and below the ground, whereas the armyworms feed only above the ground. The armyworm, *Pseudaletia unipunctata,* is found in North American east of the Rocky Mountains and in South America. *Pseudalatia* species are distributed throughout the world. The stalk borer, *Papaipema nebris,* and the lesser cornstalk borer, *Elasmopalpus ligonosellus,* are occasionally found tunneling within the culms of wheat but are not considered economically important in the United States. The cereal leaf miner, *Syringopais (Scythrix) temperatella,* attacks wheat and is a major pest in the Middle East, particularly Iran.

2.7.   Beetles

Of the numerous beetles that attack wheat, the cereal leaf beetle, *Oulema melanopus,* is considered the most injurious to wheat in the United States, Europe, and North Africa. It is most destructive in the larval stage, although some feeding is done by the adults. Leaves fed upon by the larvae have the upper parenchymous tissue removed, which results in reduced yields due to small and reduced numbers of kernels. A resistant variety has been released to protect the wheat crop in the eastern United States, and other resistant varieties are expected in the near future. Billbugs, *Sphenophorus* spp., and wireworms, *Eleodes* spp. and *Agriotes* spp., attack the wheat beneath the soil level. Damage caused by billbugs feeding below the base of the plant is seldom noticed, but losses in yield occur because of shrunken kernels. The wireworms and false wireworms feed on newly planted seed or germinating seed, causing reduced stands. *Agriotes* spp. are found in northeast Africa and southwest Asia as well as in the United States. White grubs, *Anisoplia* spp., damage wheat in the United States and also in Iran, Turkey, Syria, eastern Europe, and the USSR.

2.8. Sawflies

Sawflies are the major hymenopterous pests of wheat in the United States, Canada, Europe, North Africa, and Turkey. The larvae tunnel within the stem, feeding on the soft inner tissue. Sawfly-infested plants have reduced yields owing to lodging and shrunken kernels. The wheat stem sawfly, *Cephus pygmaeus*, is of great importance in Europe, North Africa, and Turkey. Resistant varieties have been developed in the United States, Canada, and the USSR to protect the crop from these insect pests. Other hymenopterous species attacking wheat are the *Harmolita* species and the leaf-feeding sawfly, *Pachynematus sporax*. None are considered serious pests of wheat. Some wheat breeding lines have been discovered that are resistant to the wheat jointworm, *Harmolita tritici*, but no varieties have been developed.

2.9. Flies

Numerous flies attack wheat but the Hessian fly, *Mayetiola destructor*, the wheat bulb fly, *Hylemya* (*Leptohylemyia*) *coarctata*, and the gout fly, *Chlorops pumilionis*, are considered the most injurious.

The Hessian fly is found in North America, Europe, North Africa, Turkey, and the USSR. The larvae damage the wheat plant during the fall and spring by stunting the seedlings resulting in winter kill or lodging the next summer, and reduced kernel size and weight. Nine biotypes of this insect have been isolated and determined on the ability of the larvae to survive on wheat plants having different genes for resistance. During 1974, approximately 30 resistant varieties were grown on more than 20 million acres (1974 National Wheat Survey, unpublished). The wheat bulb fly and the gout fly occur from Britain into Siberia and the Far East.

The wheat bulb fly maggots feed in the center of the tillers, cutting off the developing stem. The first sign of injury is a yellowish brown discoloration at the first node or in the stem above it.

The gout fly larvae eat through the leaf sheaths and feed in the top internode of the plant. This results in smaller heads and a reduction in the number, quality, and weight of kernels. Other dipterous pests causing damage to wheat but not considered serious are the wheat blossom midges, *Contarinia tritici* and *Sitodiplosis mosellana*, the saddle gall midge, *Haplodiplosis equestris*, a native of Europe, the wheat stem maggot, *Meromyza americana*, and the frit fly, *Oscinella frit.*

3. REARING AND EVALUATION TECHNIQUES

Success in breeding for resistance to insect and mite pests requires efficient rearing and evaluation techniques. Populations of insects must be maintained in large numbers either in the field or in the laboratory, preferably both, and methods should be developed to evaluate wheats for resistance both in the seedling and more mature stages. The genetics of both the plant and insect should be studied, and techniques to do this must be developed. Figure 1 is a diagram of a general program to follow in the development of plants resistant to insects and mites.

There are numerous measures to evaluate wheats for resistance. Those used to develop resistance to the Hessian fly, cereal leaf beetle, wheat stem sawfly, greenbug, and other major insect pests are based on death or adverse action to the insect and death or alteration of the plant. In the case of the Hessian fly, greenbug, and wheat stem sawfly, the mechanism of resistance which is utilized is antibiosis. In the case of the cereal leaf beetle, resistance is the result of the insect not preferring the plant for oviposition; no damage occurs from larval feeding because no eggs are laid.

Insects can be reared in the laboratory, both on artificial media or host plants. Artificial media lead to more refined studies for basic research, but naturally occurring hosts are better selectors for insect biotypes if they should occur. In the laboratory, insect rearing can be done by individual progenies from signle pair matings or by mass rearing. With individual progenies from single pair matings or by mass rearing. With individual use of wheats having different genes for resistance. When using mass rearing methods, homogeneous populations can be reared if wheats having specific genes for resistance are used as host plants. Homogeneous populations can also be derived from bulking progenies of individual matings of known biotypes. Individual progenies of specific biotypes can be used in basic research studies to determine the genetics of virulence and the nature of resistance. Homogeneous populations can be used to determine the genetics and nature of resistance in the plant.

Homogeneous laboratory populations of insects can also be used along with heterogeneous field populations to evaluate wheats for resistance. New

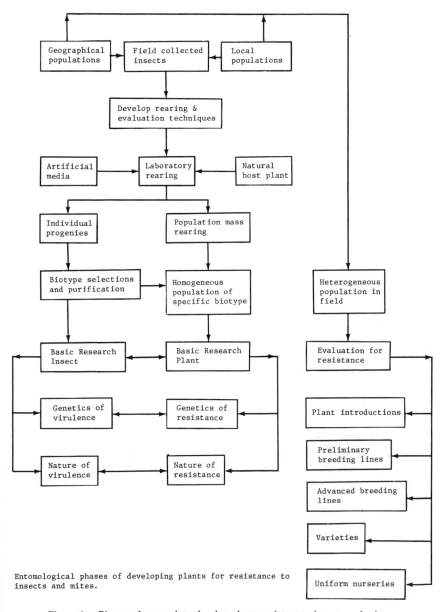

Entomological phases of developing plants for resistance to insects and mites.

Figure 1. Phases of research to develop plants resistant to insects and mites.

plant introductions, varieties, preliminary and advanced breeding lines, and uniform nurseries can be evaluated for resistance to different biotype populations if available. The heterogeneous field populations should be used to verify resistance under natural field conditions and corroborate those findings in the laboratory. The value of using laboratory populations as opposed to natural field populations is that more critical genetc and physiological research can be conducted when the purity of the insect population is established.

4. BREEDING TECHNIQUES AND METHODS

The wheat varieties grown commercially in the world are predominantly the common wheat species, *T. aestivum,* or the durum wheat species, *T. durum.* These species are self-pollinating (autogamous) and the varieties are generally highly inbred and true-breeding; consequently, selection within a variety seldom provides significant genetic advancement of a character such as insect resistance. Heritable (genetic) variability for selection purposes is usually obtained by crossing adapted varieties or breeders lines with unadapted, exotic, resistant varieties.

In a crossing program the anthers are removed by hand from the flowers of the female parent, which are later pollinated with pollen from the male parent. Plants in the following generations are allowed to self-pollinate naturally, producing a diverse population through genetic recombination in the second generation. Theoretically, 50 percent of the gene loci in every individual plant in the F_2 generation would be homozygous, and in each subsequent generation of self-pollination the heterozygosity would decline by 50 percent. By the fifth generation, approximately 93 percent of the gene loci of each plant in a population are homozygous. Thus with self-pollination the individuals in a population originating from a two-parent cross become increasingly more inbred and true-breeding with each subsequent generation.

4.1. Selecting for Resistance

The plant breeder is concerned with systems of selection within the segregating generations of crosses to identify the plants with desired new combinations of characters. He is usually interested in retaining most of the desired characteristics of the adapted parent, which already has approached the ideal agro-ecotype in essential features, and recombining only the desired characters (i.e., insect or disease resistance) with them. If resistance is dominant, as in the case of the Hessian fly, resistant plants can be identified in the F_2 and their progeny rated again in the F_3. However, most breeders select F_3 families with good agronomic type and subsequently rate and reselect those families on an individual plant basis for Hessian fly resistance. If resistance is recessive and controlled by more than one gene, a large part of the F_2 generation would be discarded if only the resistant plants were saved. For a recessive character, selection should begin in the F_3 generation. The F_3 lines considered for further progeny tests are studied on the basis of individual plants and are rated by their progenies' reaction in the F_4. Many gene combinations, in addition to those for insect resistance and susceptibility, form complexes of characters in the F_2 and subsequent generations. If the breeder concentrates on insect resistance in his selection, only the very rare individual will exhibit much potential in advanced generations owing to the lack of adaptability or agronomic worth.

4.2. Adaptability and Resistance

Most breeders have observed that a delicate genic balance for adaptability is easily lost by recombination in a two-parent cross between an adapted, insect susceptible variety and a nonadapted, insect resistant variety. Early attempts to retain adaptability included the use of large populations coupled with intense selection pressure for adaptation in early generations. Only the more adapted lines would be screened for insect resistance. Many breeders are now employing three-way and four-way crosses to maintain a workable level of stability or adaptability in their breeding populations. Table 2 illustrates the importance of the choice of parents in introducing exotic, nonadapted germplasm into breeding populations. Since adaptability is a complex trait controlled by many linked genes, the probability of selecting adapted, insect resistant lines from segregating populations, which will be released as varieties, increases measurably as the percentage of the adapted parent's genotype in the gene recombinations increases, especially since insect resistance is often dominant and simply inherited.

4.3. Breeding Strategies

MacKey (1963) discussed several long-term breeding strategies which would permit essential gene recombination but guarantee a more conservative trend when dealing with adaptability.

Long-term breeding strategies are presented in Figure 2. In all the diagrams, Parent A represents an adapted variety with a character deficiency

Table 2.   Some Possible Crosses Involving Two Adapted, Insect Susceptible Varieties (A and C) and Two Nonadapted Varieties (B and D) Each with a Different Dominant Gene for Insect Resistance, Illustrating Percent of Genetic Background with Adaptability Characteristics and Insect Resistance in Respective Populations

	Percent of Genetic Background		
Type of Cross	Adaptability		Insect Resistance
Two-way (2x)			
A/B	50		50
B/D	low		100
Three-way (3x)			
A/B//A (backcross)	75		25
A/B//C	75	(25 A)	25
		(50 C)	
A/B//D	25		75
B/D//A	50		50
Four-way (4x)			
A/B//C/D	50	(25 A)	50
		(25 C)	
A/B//A/D	50		50
A/B//A/C	75	(50 A)	25
		(25 C)	

such as insect susceptibility. Parents B, C, D, E, F, G, and H each contribute a different dominant gene for insect resistance with no linkage but little adaptability. In schemes 1 to 4 in Figure 2, we assume no selection by the breeder for adaptability or insect resistance. If selection pressure is applied for either character, the figures for percent adaptation and insect resistance would increase with each convergent cross.

Convergent Crossing with Maximum Recombination

Scheme 1 in Figure 2 illustrates convergent crossing to maximize recombinations. In the third year only 13 percent of the genes of the adaptive A parent are in the population. The breeding program will be of little value since resistant varieties with little adaptation or stability will be produced. Many diallel crossing programs use this approach to establish populations with great genetic diversity; however, with this approach agro-ecotypic stability deteriorates.

Crossing Scheme

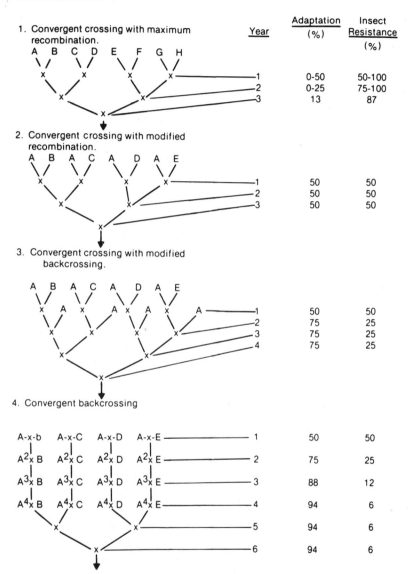

Figure 2. Long-term breeding strategies involving convergent recombination. Parent A represents an adapted variety with insect susceptibility. Each parent B, C, D, E, F, and G contributes a different dominant gene for insect resistance with no linkage and little adaptability.

Convergent Crossing with Modified Recombination

Scheme 2 illustrates convergent crossing to enhance transgressive segregation. This approach gives results very similar to a two-parent hybrid but results in greater diversity. Varieties with insect and/or disease resistance are developed but adaptation is impaired.

Convergent Crossing with Modified Backcrossing

Scheme 3 demonstrates convergent crossing using transgressive segregation and an incomplete backcross scheme to enhance adaptability. Several genes for insect resistance could be combined in an adapted genotype to produce a successful variety using this approach.

Convergent Backcrossing

Scheme 4 is convergent backcrossing, in which a number of backcross programs using the same adapted recurrent parent are converged through hybridization. This last scheme would be excellent where several genes for either insect or disease resistance are known, and the breeder wants to develope a variety that combines a number of resistances.

A variation of Scheme 4 would be to use different closely related, adapted parent varieties in place of parent A in some of the convergent backcross programs. With selection, both adaptability and insect resistance could be improved and maintained at levels above those shown in the schemes in Figure 2.

Many breeders are currently employing one or more of these approaches under such program titles as parent building, gene pool building, modified backcrossing, cyclic breeding, and multiple line breeding. All these breeding strategies are actually a type of convergent improvement used to retain certain desirable genic complexes (linkage groups) while adding specified, simply inherited genetic improvements such as insect and disease resistance and components of winter hardiness and drought resistance.

A different strategy must be employed where insect resistance is quantitatively inherited and a polygenic system is involved. A variety with a high level of insect resistance is crossed with the adapted, susceptible variety. A large number of F_2 plants are screened for resistance and their progeny rated again in the F_3. The 10 to 15 most resistant lines with plant types most closely resembling the susceptible adapted parent are selected and recrossed in all possible combinations. The bulk F_2 is again screened for resistance and their progeny rated again in the F_3. This cyclic recurrent crossing and selection continues until no further progress is made for adaptability and insect resistance. Should adaptability be unacceptable, the most resistant lines would be crossed with the adapted parent and the recurrent cycle repeated. In using recurrent selection, it is important to continue the cyclic crossing and selection in a closed population and to restrict the recrossing to 10 to 15 of the most superior F_2 plant lines.

4.4. Wheat Breeding Program

Figure 3 outlines a generalized wheat breeding program which can guide team research to integrate desired traits in highly productive varieties. Several essential components in this outline are common to all programs:

1. *Strong satellite research groups.*   Units which work on the entomology, plant pathology, plant physiology, cereal chemistry, and genetics of valuable characters. The satellite research groups are strong, independent, research units in the problem areas of the crop: insect resistance, disease resistance, winter hardiness, drought resistance, milling, baking, and nutritional quality. These groups maintain basic research in their respective disciplines studying the nature of the problems. As desirable traits are discovered, rapid, inexpensive, and (if possible) nondestructive screening tests are devised to identify that trait in individual plants. Germplasm collections are then screened to identify potential parents showing intense expression of the character.

Breeding for resistance to insects in wheat, like any other crop breeding program, involves procedures developed by experience as well as those adapted from research. Adequate insect populations and efficient evaluating techniques are prime requisites for breeding insect resistant wheats. For well-established programs, these techniques are already available. For programs with new insect pests, new techniques must be developed.

2. *Gene pool of exotic parents.*   Lines that have been screened for intense expression of a desirable trait, such as insect resistance, but are generally unadapted. Once the insect resistance is located in an exotic, generally nonadapted variety it becomes a component of the gene pool of exotic parents.

3. *Gene pool of adapted parents.*   Varieties that are systematically or frequently used in crosses to increase the gene frequency of adaptive characters so as to skew the population in the direction of agro-ecotypic adaptation.

4. *Parent crossing block.*   Genetic lines from both adapted and exotic gene pools, F_1, and recycled early-generation selections from the breeding program are recombined in two-, three-, and four-way crosses to fit long-term objectives.

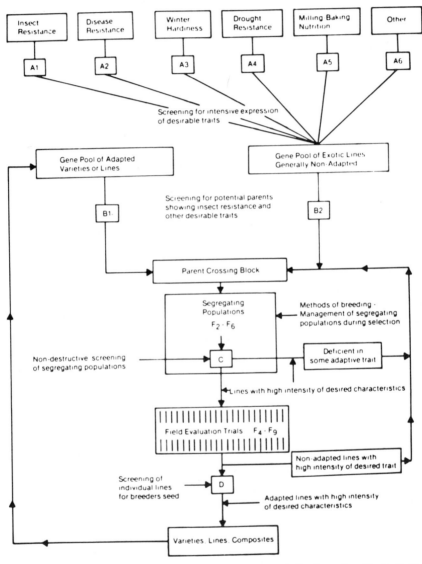

Figure 3. Generalized outline of a wheat breeding program indicating points (A, B, C, D) at which physiological and genetic tests for desirable characteristics are applied.

5. *Hybrid segregating populations.* The F_2 through F_6 generations are grown in spaced, bulk or headrow plantings, depending on the method of breeding (pedigree, bulk, backcross, etc.) used to manage the segregating populations during selection. Nondestructive screenings (C in Figure 2) of segregating populations are carried out on individual plants or their progeny (if the screening test is a destructive one) for any desired character (including yield) during any or all of these generations but usually from the F_3 onward. In any generation, a line deficient in some trait but with a high intensity of a desired character can be recycled to the crossing block to mate with a desirable adapted parent.

6. *Field evaluation trials and varietal release.* Selection is sequential through the advanced generations finishing with quality and field evaluation trials in the F_4 to F_8, depending on the program. High performing lines deficient in some character are recycled to the crossing block for further genetic recombination and improvement, whereas released varieties are recycled through the gene pool of adapted varieties to the crossing block.

Using a convergent crossing approach of this general description, adapted high-yielding wheat varieties with Hessian fly, sawfly, and cereal leaf beetle resistance have been developed.

4.5. Computerized Breeding

Computers are now used in some plant breeding programs to process performance data and, in the future, will be used for data storage and retrieval to aid in a multitude of plant breeding operations.

An example is the Michigan State University computer system for plant breeding developed for the wheat breeding program (Everson, 1973). This system facilitates the accumulation and storage of genetic and plant breeding data on parental lines, hybrid populations, and progeny through a system of character descriptors. Accession numbers, assigned to varieties and advanced selections, provide the computer code necessary for data storage and retrieval. When a cross is made between two parents, the hybrid is assigned a six-digit number which indicates the year the cross was made and the cross number (thus 770001 indicates the first cross made in 1977). In addition to the hybrid number, the accession numbers of the parents are also designated. Each time a certain hybrid population is grown, the hybrid number is used as the control number. The computer searches the memory bank for that number and retrieves the parental descriptions in the form of readily understandable character descriptors. The Hessian fly descriptors are as follows: 3H ('W38' resistance), 5H ('Ribiero' resistance) and 6H (durum resistance). Each time a generation field book is prepared by the computer program, the hybrid population is described by hybrid number, and the parents are listed with complete pedigree, generation, parentage, and character descriptors. By scanning the character descriptors, a breeder can determine what major genes have been identified in each parent and what variation the hybrid gene pool may contain for selection.

The computer can aid the plant breeder with many complex or repetitious tasks in record keeping, planning, and data analysis. Among other things, the MSU system is used in the analysis or field data, preparation of summaries, computation of varietal stability indexes, and even the preparation of labels for planting, harvesting, and quality tests.

4.6.   Utilization of Biotypes

Biotypes of Hessian fly and the greenbug have been utilized in a breeding program to develop wheats having different genes for resistance. In the case of the Hessian fly, there are nine biotypes, all differing in their genes for virulence. In the wheat plant there are eight genes for resistance. With the exception of the h_4 gene for resistance in 'Java,' resistance in wheat to Hessian fly is dominant. Virulence by Hessian fly biotypes is controlled by recessive genes. The Hessian fly-wheat relationship is comparable to the gene-for-gene system attributed to the wheat plant and wheat pathogens.

By maintaining pure line populations of Hessian fly biotypes, genetic sources of resistance in the world collection can be identified, distinctions can be made between different genetic sources without inheritance studies, and combinations of resistance genes can be determined.

For the greenbug, no resistance was found against Biotype C so resistance in wheat was developed by substituting a chromosome from rye into a common 21-chromosome wheat, using cytogentic techniques.

4.7.   Utilization of Morphological Characteristics

Morphological characters responsible for resistance can be used as a basis of selecting resistant plants in segregating populations. Gallun et al. (1966) reported that resistance to cereal leaf beetle in wheat is attributed to leaf pubescence. Ringlund and Everson (1968) verified the relationship between pubescent density and beetle resistance and determined that resistance was controlled by a number of partially dominant genes with additive effects. Hoxie et al. (1975) demonstrated that leaf hair length, as well as density, is responsible for nonpreference for oviposition. In the absence of suitable cereal leaf beetle populations, selections have been made on the basis of pubescence. It is important to mention here, however, that although this method is satisfactory for the selection of resistant plants, it does increase the risk of losing other sources of resistance that would normally show up in the presence of the insect.

5. INSECT RESISTANT WHEATS IN PEST MANAGEMENT

The use of wheats resistant to insect pests is a major control measure or is an adjunct to other control measures. The Hessian fly and wheat stem sawfly are controlled solely by resistant varieties, whereas the greenbug and cereal leaf beetle are also controlled by insecticides. To date, 42 wheat varieties have been released that are resistant to one or more biotypes of Hessian fly. Most of these varieties were grown in the United States in 1974, accounting for a total of about 20 million acres (National Wheat Survey, unpublished information).

In the same year five varieties resistant to the wheat stem sawfly were grown on 1–½ million acres in the United States, providing the sole protection against this insect pest.

No resistant wheat varieties have been released for the greenbug, but varieties are being developed. 'Downy' wheat, the first variety developed and released that has resistance to the cereal leaf beetle, was grown in Indiana, Michigan, and Ohio for the first time in 1977. It can also be used in conjunction with parasites and chemical control.

Breeders are turning to wild plants such as *Tripsacum,* a close relative of maize, to obtain desirable traits. Photo courtesy of CIMMYT.

21

FUTURE OPPORTUNITIES AND DIRECTIONS

Fowden G. Maxwell

Department of Entomology, Texas A&M University, College Station, Texas

The preceding chapters in this book have covered concepts and definitions, and have pointed to the successes in breeding plants for insect resistance with selected commodities over the past 25 years. However entomologists and plant breeders have barely scratched the surface in realizing the potential that breeding of plants resistant to insects offers for pest management. Opportunities for exploiting this method abound. With continued support and proper emphasis, achievements in the years ahead can contribute enormously to the elimination of world hunger and famine.

Crop productivity must double in the next 25 years to meet the food and fiber requirements of the rapidly expanding world population. In order for agriculture to meet this challenge, better varieties of crops carrying resistance to primary pests have to be developed. Much of this increased productivity must come from cropland in the food-deficient developing countries. Enhanced emphasis on food production by all nations of the world should stimulate the breeding of plants resistant to insect pests.

While opportunities are many for incorporating insect pest resistance into our most valuable crop plants we must be alert to the pitfalls inherent in many breeding programs. For example we must avoid, wherever possible, the dependence upon narrow genetic bases and extremely high levels of antibiosis that result in a rapid selection of biotypes causing loss of resistance.

Certain specific factors that will need to be considered include:

1. The capacity of insects to develop resistance to key insecticides. Many of our arthropod pests have acquired high levels of resistance to insecticides used extensively and intensively over time.

2. The increasing regulations that governmental agencies impose to limit the use of many insecticides in order to avoid environmental or health hazards.
3. The failure of industry to produce effective insecticides with new modes of action quickly enough to stay ahead of regulatory actions, and to compensate for loss due to insect resistance.
4. The increasing economic cost to industry for the production of insecticides with questionable life spans.
5. The costs and availability of the insecticides themselves, as well as the farmer's application expenses, which will increase as fossil energy becomes scarce.
6. The realization that even moderate levels of resistance, while perhaps inadequate as a prime defense, can be useful in an integrated control program.

The need for plant resistance and additional research will produce new techniques. For example, research in genetics on cell culture methods, on modification of DNA, and on induced genetic mutations is generating new information. If properly employed in the years ahead, they may allow for a wide crossing of species and for a great utilization of diverse resistant germplasm from exotic materials far removed from commercial types.

Research in insect nutrition and a mass rearing in the years ahead will contribute greatly to the development of fast and efficient screening and rearing techniques for evaluating germplasm. Technique development, especially in the use of chemical methodologies, has been one of the major restraints to progress in the field.

There has been little progress in determining the biochemical nature of resistance, even in the developed nations.

First, support, interest, and techniques have been lacking for exploring the basic causes of resistance. However, chemists and physiologists are now expressing interest in this question, and they are becoming important members of the resistance breeding teams. Adequate techniques for detecting, analyzing, and identifying chemical compounds from plants, often in minute quantities, have only recently been developed. New equipment and methods should greatly improve our capability to isolate and to identify those chemical substances in plants that contribute to resistance.

Second, crop resistance is a phenomenon involving not only physical and biochemical factors, but frequently complex interrelationships among the insect, the plant, and the environment. These complex relationships make it difficult, with existing techniques, to determine the major chemical causes for resistance. In studies of this basic chemical nature, elaborate and often very expensive tests must be devised to separate and identify the major factors that contribute to the total expression of resistance.

Third, we have not had the necessary basic behavioral and physiological information on insect pests. This is especially true for chemoreception and other physiological systems. These must be understood in great detail before we can develop adequate chemical and biological techniques that assess the response of insects to various plant-derived chemical fractions and compounds. This remains one of the serious limiting factors facing most teams attempting to determine the nature of resistance. As we look ahead, we must anticipate great progress in bringing new behavioral, electrophysiological, and electron microscopy techniques to bear on determining basic causes of resistance.

Fourth and finally, few people have thought it important in the past to investigate the chemical causes of resistance. Entomologists and plant breeders have ignored the subject for lack of interest, the complexity of the problems, and the lack of chemists and other trained personnel to cope with the problems. With new available techniques, chemists, physiologists, and workers in other basic disciplines will gain interest in the chemical causes of resistance. Health and environmental concerns require an understanding of the chemistry and toxicity of resistant factors in plants. This concern will become a powerful force in inspiring research into the basic mechanisms of resistance.

Chapter 6 of this book summarizes much of the current basic work on behavior, and explains the potential importance of determining the basic causes of resistance. The major advances being made in chemical technique make it possible not only to explore the bases of plant resistance, but also to explore the factors that make plants attractive or nonattractive to insects. While some research has been accomplished in this area, much needs to be done in the future. However breeding plants for lower or higher concentrations of biologically active chemical compounds is aided by an understanding of (1) the independent and collective effects of specific chemicals on the insect, and (2) the number of genes controlling the chemicals. Research conducted to date, with the preference mechanisms of resistance in plants, indicates that a blend of chemicals is usually involved, rather than a single specific compound. In these cases breeding becomes more difficult because each chemical is usually under different genetic control.

In summary, chemical research in developed countries on the bases of resistance in plants will increase in the years ahead. This will result from a need for a positive chemical approach to the field, the ability to meet this need with new technique developments, and the pressures of ever increasing regulatory and health issues.

In the less developed countries, expansion of the international systems for the identification of elite sources of resistance, and for the evaluation of promising materials from numerous locations and high volume centralized crossing programs will enhance opportunities for advancement of breeding plants for insect control. International linkages of centers and national programs will also help to alleviate professional isolation, lack of generalism, and scanty evaluation techniques that have plagued the field in the past. This expanding world network will insure a high degree of success in the years ahead.

In conclusion, breeding for insect resistance has a tremendous potential in the future. It offers one of the most effective means of controlling our major insect pests economically, especially when properly combined with other cultural, biological, and chemical tactics in a total pest management strategy. It offers a challenging field to students interested in basic insect-plant interactions, chemistry, ecology, physiology, insect behavior, and above all, plant breeding and economic entomology. Research in the field requires a team approach and a willingness by workers to share with others in research accomplishments. This concept of team research has contributed to the progress that breeding plants for insect control has enjoyed in the past; it will assure success in the future.

Part Three

AFTERWORD

BIBLIOGRAPHY

Aamodt, O. S. and J. S. Carlson, 1938. Grimm alfalfa flowers in spite of lygus injury. *Wis. Agric. Exp. Stn. Bull.* **440,** part II: 67.

Abdel-Malek, S. H., E. G. Heyne, and R. H. Painter, 1966. Resistance of F_1 wheat plants to green bug and Hesian fly. *J. Econ. Entomol.* **59:** 707–710.

Abernathy, C. O. and R. Thurston, 1969. Plant age in relation to the resistance of *Nicotiana* to the green peach aphid. *J. Econ. Entomol.* **62:** 1356–1359.

Adams, J. B. and M. E. Drew, 1969. Grain aphids in New Brunswick. IV. Effects of malathion and 2,4-D amine on aphid populations and on yield of oats and barley. *Can. J. Zool.* **47:** 423–426.

Adkisson, P. L. and J. C. Gaines, 1960. Pink bollworm control as related to the total cotton insect control program of Central Texas. *Tex. Agric. Exp. Stn. Misc. Publ.* **444:** 7 pp.

Adkisson, P. L., C. F. Bailey, and G. A. Niles, 1962. Cotton stocks screened for resistance to the pink bollworm, 1960–61. *Tex. Agric. Exp. Stn. Misc. Publ.* **606**.

Adkisson, P. L., 1972. The integrated control of cotton insect pests. *Proc. Tall Timbers Conf. Ecol. Animal Control Habitat Manage.* **4:** 1975–1988.

Agarwal, R. A., 1969. Morphological characteristics of sugarcane and insect resistance. *Entomol. Exp. Appl.* **12:** 767–776.

Agarwal, R. A., S. K. Banerjee, M. Singh, and K. N. Katiyar, 1976. Resistance to insects in cotton. II. To pink bollworm, *Pectinophora gossypiella* (Saunders). *Coton Fibres Trop.* **31:** 217–221.

Ajani, A., and J. H. Lonnquist. Selection for biological resistance to corn rootworm *Diabrotica, Can. J. Plant Sci.,* in press.

Alan, M. S., 1978. Resistance to biotype 3 of the brown planthopper, *Nilaparvata lugens* (Stål) in rice varieties. Ph.D. Thesis, Cornell University, Ithaca, N.Y., 108 pp.

Albrecht, H. R., and T. R. Chamberlain, 1941. Instability of resistance to aphids in some strains of alfalfa. *J. Econ. Entomol.* **34:** 551–554.

Albuquerque, M. de, 1976. *Cochonilha en mandioca na Amazonia.* Empresa Brasileira de Pesquisa Agropecuaria and Centro de Pesquisa Agropecuaria do Tropic Umido, Belem, Brazil, 10 pp.

All, J. N. and D. M. Benjamin, 1975. Influence of needle maturity on larval feeding preference and survival of *Neodiprion swainei* and *N. rugifrons* on jack pine, *Pinus banksiana. Ann. Entomol. Soc. Am.* **68:** 579–584.

All, J. N., C. W. Kuhn, R. W. Gallaher, M. O. Jellum, and R. S. Hussey, 1977. Influence of no-tillage-cropping, carbofuran, and hybrid resistance on dynamics of maize Chlorotic dwarf and maize dwarf mosaic diseases of corn. *J. Econ. Entomol.* **70:** 221–225.

Allan, R. E., E. G. Heyne, E. T. Jones, and C. O. Johnston, 1959. Genetic analyses of ten sources of Hessian fly resistance, their interrelationships and association with leaf rust reaction in wheat. *Kans. Agric. Exp. Stn. Tech. Bull.* **104.**

Allard, R. W., 1960. *Principles of Plant Breeding.* John Wiley & Sons, Inc., New York.

Allard, R. W., 1970. Population structure and sampling methods. In *Genetic Resources in Plants* O. H. Frankel and E. Bennet, Eds. F. A. Davis Co., Philadelphia, pp. 97–108.

Alston, F. H. and J. B. Briggs, 1968. Resistance to *Sappaphis devecta* (Wld.) in apple. *Euphytica* **17:** 468–472.

Alston, F. H. and J. B. Briggs, 1970. Inheritance of hypersensitivity to rosy apple aphid *Dysaphis plantaginea* in apple. *Can. J. Genet. Cytol.* **12:** 257–258.

Andres, F. L., 1975. Varietal resistance to the rice whorl maggot *Hydrellia philippina* Ferino. M.S. Thesis, University of the Philippines, Los Baños, 83 pp.

Andrew, R. H., and J. R. Carlson, Jr., 1976. Evaluation of sweet corn inbreds for resistance for European corn borer. *J. Am. Soc. Hortic. Sci.* **101:** 97–99.

Andries, J. A., J. E. Jones, L. W. Sloane, and J. G. Marshal, 1969. Effects of okra leaf shape on boll rot, yield and other important characters of Upland cotton, *Gossypium hirsutum* L. *Crop Sci.* **9:** 705–710.

Andries, J. A., J. E. Jones, L. W. Sloane, and J. G. Marshall, 1970. Effects of super okra leaf shape on boll rot, yield and other characters of Upland cotton, *Gossypium hirsutum* L. *Crop Sci.* **10:** 403–407.

Anglade, P., 1961. Influence sur le rendement de mais de l'infestation destiges par deuxieme generation de la sesame *Sesamia Monagroides* les Lep. (Noct.). *Ann. Epiphyt.* **12:** 327–372.

Anonymous, 1957. Spread of spotted alfalfa aphid in the United States. *U.S. Dep. Agric. ARS Econ. Insect Surv. Rep.,* April.

Anonymous, 1957a. The spotted aphid. *U.S. Dep. Agric. ARS* **22–39:** 8 pp.

Anonymous, 1966. *CIMMYT Annual Report.* International Maize and Wheat Improvement Center, Mexico D.F., Mexico.

Anonymous, 1968. Genetics and cytology of cotton, 1956–67. *South. Coop. Ser. Bull.* **139.**

Anonymous, 1972. *Genetic Vulnerability of Major Crops.* National Academy of Sciences, Washington, D.C., pp. 110–112.

Anonymous, 1973. *CIMMYT Annual Report on Maize Improvement.* International Maize and Wheat Improvement Center, Mexico D.F., Mexico.

Anonymous, 1974. The regional collection of *Gossypium germplasm. U.S. Dep. Agric. ARS* H-2.

Anonymous, 1976. *Maize Program. International Progeny Trials and Experimental Variety Trials. Preliminary and Supplementary Reports.* CIMMYT, Mexico D.F., Mexico.

Anonymous, 1977. *Cotton International,* 4th ann. ed., p. 260.

Anonymous, 1978. Variety trails of farm crops. *Univ. Minn. Agric. Exp. Stn. Misc. Rep.* **24:** 32 pp.

Anstey, T. H. and J. F. Moore, 1954. Inheritance of glossy foliage and cream petals in green sprouting broccoli. *J. Hered.* **45:** 39–41.

App, B. A. and G. R. Manglitz, 1972. Insects and related pests. Chapter 24 in *Alfalfa Science and Technology,* C. H. Hanson, Ed. American Society of Agronomy, Madison, Wis., pp. 527–554.

Applebaum, S. W., 1964. The action pattern and physiological role of *Tenebrio* larval amylase. *J. Insect Physiol.* **10:** 897–906.

Applebaum, S. W. and Y. Birk, 1972. Natural mechanisms of resistance to insects in legume seeds. In *Insect and Mite Nutrition,* J. G. Rodriquez, Ed. North-Holland, Amsterdam, pp. 629–636.

Arant, F. S. and C. M. Jones, 1951. Influence of lime and nitrogenous fertilizers on the population of greenbugs infesting oats. *J. Econ. Entomol.* **44:** 121–122.

Armbrust, E. J., C. E. White, and S. J. Roberts, 1970. Mating preference of eastern and western U.S. strains of the alfalfa weevil. *J. Econ. Entomol.* **63:** 674–675.

ARS Modeling Committee Report, 1977. Report of the ARS Modeling Committee set up by the administration of ARS. Appendix III. Concepts of using modeling as a research tool. File of Agriculture Research Models 1977. Data Systems Applications Division, U.S. Dept. of Agriculture. Mimeographed.

Aston, J. S. and B. Winfield, 1972. Insect section introduction. In *Cotton, Tech. Monogr.* **3:** 44–45. Ciba-Geigy Ltd., Basel, Switzerland.

Athwal, D. S., M. D. Pathak, E. H. Bacalangoo, and C. D. Pura, 1971. Genetics of resistance to brown planthoppers and green leafhoppers in *Oryza sativa* L. *Crop Sci.* **11:** 747–750.

Athwal, D. S. and M. D. Pathak, 1972. Genetics of resistance to rice insects. In *Rice Breeding,* IRRI, Los Baños, Philippines, pp. 375–386.

Auclair, J. L., J. B. Maltais, and J. J. Cartier, 1957. Factors in resistance of peas to the pea aphid, *Acyrthosiphon pisum* (Harr.). II. Amino Acids. *Can. Entomol.* **69:** 457–464.

Auclair, J. L., J. B. Maltais, and J. J. Cartier, 1957. Factors in resistance of peas to the pea aphid, *Acyrthosiphon pisum* (Harr.). (Homoptera: Aphididae). II. Amino Acids. *Can. Entomol.* **89:** 457–464.

Auclair, J. L., 1963. Aphid feeding and nutrition. *Ann. Rev. Entomol.* **8:** 439–490.

Auclair, J. L., 1969. Nutrition of plant-sucking insects on chemically defined diets. *Entomol. Exp. Appl.* **12:** 623–641.

Bahadur, R. R. and H. Robbins, 1950. The problem of the greater mean. *Ann. Math. Stat.* **21:** 469–487.

Bailey, J. C., F. G. Maxwell, and J. N. Jenkins, 1969. Boll weevil antibiosis studies with selected cotton lines utilizing egg-implantation techniques. *J. Econ. Entomol.* **60:** 1275–1280.

Baker, Whiteford L., 1972. Eastern forest insects. *U.S. Dept. Agric., For. Serv. Misc. Publ.* **1175,** 642 pp.

Ballard, W. W., 1951. Varietal differences in susceptibility to thrips injury in Upland cotton. *Agron. J.* **43:** 34–44.

Banks, C. J. and E. D. M. Macaulay, 1964. The feeding, growth and reproduction of *Aphis fabae* Scop. on *Vicia faba* under experimental conditions. *Ann. Appl. Biol.* **53:** 229–242.

Banks, D. J., 1976. Peanuts: germplasm resources. *Crop Sci.* **16:** 499:502.

Bardner, R. and K. E. Fletcher, 1974. Insect infestations and their effects on the growth and yield of field crops: a review. *Bull. Entomol. Res.* **64:** 141–160.

Barker, J. S. and O. E. Tauber, 1951. Fecundity of and plant injury by the pea aphid as influenced by nutritional changes in the garden pea. *J. Econ. Entomol.* **44:** 1010–1012.

Barnes, D. K. and R. H. Ratcliffe, 1967. Leaf disk method of testing alfalfa plants for resistance to feeding by adult alfalfa weevils. *J. Econ. Entomol.* **60:** 1561–1565.

Barnes, D. K., R. H. Ratcliffe, and C. H. Hanson, 1969a. Interrelationship of three laboratory screening procedures for breeding alfalfa resistance to the alfalfa weevil. *Crop Sci.* **9:** 77–79.

Barnes, D. K., B. L. Norwood, C. H. Hanson, R. H. Ratcliffe, and C. C. Blickenstaff, 1969b. A mass screening procedure for isolating alfalfa seedlings with resistance to alfalfa weevil. *Crop Sci.* **9:** 640–642.

Barnes, D. K., C. H. Hanson, R. H. Ratcliffe, T. H. Busbice, J. A. Schillinger, G. R. Buss, W. V. Campbell, R. W. Hemken, and C. C. Blickenstaff, 1970. The development and performance of Team alfalfa. *U.S. Dep. Agric. ARS Bull.* **34–115:** 41 pp.

Barnes, D. K., F. I. Frosheiser, E. L. Sorensen, J. H. Elgin, Jr., M. W. Nielson, W. F. Lehman, K. T. Leath, R. H. Ratcliffe, and R. J. Buker, 1974. Standard Tests to Characterize Pest Resistance in Alfalfa Varieties. *U.S. Dep. Agric. ARS* **NC-19:** 23 pp.

Barnes, D. K., E. T. Bingham, R. P. Murphy, O. J. Hunt, D. F. Beard, W. H. Skrdle, and L. R. Teuber, 1977. Alfalfa germplasm in the United States: Genetic vulnerability, use, improvement and maintenance. *U.S. Dep. Agric. ARS Tech. Bull.* **1571:** 21 pp.

Barnes, O. L., 1963. Resistance of Moapa alfalfa to the spotted alfalfa aphid in commercial-size fields in south-central Arizona. *J. Econ. Entomol.* **56**: 84–85.

Barrios, J. R., 1972. *Reaccion de 25 variedades de yuca, Manihot esculenta, al ataque de acaros. VII.* Jornadas Agronomicas, Caucagua, Venezuela. 8 pp.

Bath, J. E. and R. K. Chapman, 1967. Differential transmission of two pea enation mosaic virus isolates by the pea aphid, *Acyrthosiphon pisum* (Harris). *Virology* **33**: 503–507.

Batra, G. R. and D. S. Gupta, 1970. Screening of varieties of cotton for resistance to jassid. *Cotton Grow. Rev.* **47**: 285–291.

Bawden, F. D., 1964. *Plant Viruses and Virus Diseases,* 4th ed. The Ronald Press Co., New York, 361 pp.

Beasley, J. O., 1940. The origin of American tetroploid *Gossypium* species. *Am. Nat.* **64**: 285–286.

Beck, S. D. and J. H. Lilly, 1949. Report on European corn borer resistance investigations. *Iowa State Coll. J. Sci.* **23**: 249–259.

Beck, S. D., 1965. Resistance of plants to insect. *Ann. Rev. Entomol.* **10**: 207–232.

Beck, S. D., 1974. Theoretical aspects of host plant specificity in insects. In *Proc. Summer Inst. Biol. Control Plant Insects Dis.,* F. G. Maxwell and F. A. Harris, Eds.: 290–311. University Press of Mississippi, Jackson.

Beck, S. D. and J. C. Reese, 1976. Insect–plant interactions: nutrition and metabolism. *Rec. Adv. Phytochem.* **10**: 41–92.

Beck, S. D. and J. C. Reese, 1976. Insect–plant interactions: nutrition and metabolism. In *Recent Advances in Phytochemistry,* Vol. 10, J. W. Wallace and R. L. Mansell, Eds. Plenum Press, New York, pp. 41–92.

Bell, A. A. and R. D. Stipanovic, 1977. The chemical composition, biological activity and genetics of pigment glands in cotton. *Proc. Beltwide Cotton Prod. Res. Conf.:* 244–258. National Cotton Council, Memphis, Tenn.

Bellotti, A. C. and A. van Schoonhoven, 1977. World distribution, identification and control of cassava pests. *Proc. IV Symp. Int. Soc. Trop. Root Crops, Cali, Colombia*: 188–193.

Bellotti, A. C. and A. van Schoonhoven, 1978. Mite and insect pests of cassava. *Ann. Rev. Entomol.* **23**: 39–67.

Benepol, P. S. and C. V. Hall, 1967. The genetic basis of varietal resistance of *Cucurbita pepo* L. to squash bug Anasa tristis (DeGeer.) *Proc. Am. Soc. Hortic. Sci.* **90**: 301–303.

Bennett, F. O. and M. Yaseen, 1975. Investigation on the Cassava Mite *Mononychellus tanajoa* (Bondar) and its natural enemies in the Neotropics. Commonwealth Institute of Biological Control, Trinidad, West Indies, 12 pp.

Bergey's Manual, 1974. 8th ed., Wilkins & Wilkins, Baltimore, 1268 pp.

Berglund-Brücher, O. and H. Brücher, 1976. The South American wild bean (*Phaseolus aborigineus* Burk) as ancestor of the common bean. *Econ. Bot.* **30**: 257–272.

Bhalla, O. P. and A. G. Robinson, 1968. Effects of chemosterilants and growth regulators on the pea aphid fed an artificial diet. *J. Econ. Entomol.* **61:** 552–555.

Bird, J. B. and J. Mahler, 1951. America's oldest cotton fabrics. *Am. Fabrics* **20:** 73–78.

Bird, J. and K. Maramorosch, Eds., 1975. *Tropical Diseases of Legumes.* Academic Press, New York, 171 pp.

Bird, J. and K. Maramorosch, 1977. Viruses and virus diseases associated with whiteflies. *Adv. Virus Res.* **22:**55–110.

Bishara, S. I., A. Koura, and M. A. Elhalfway, 1972. Ovipositional preference of the granary and the rice weevils on the Egyptian rice varieties. *Bull. Soc. Entomol. Egypt* **56:** 145–150.

Blanchard, R. A. and J. E. Dudley, Jr., 1934. Alfalfa plant resistant to the pea aphid. *J. Econ. Entomol.* **27:** 262–264.

Blanchard, R. A., J. H. Bigger, and R. O. Snelling, 1941. Resistance of corn strains to the corn earworm. *J. Am. Soc. Agron.* **33:** 344–350.

Blanchard, R. A., A. F. Satterthwait, and R. O. Snelling, 1942. Manual infestation of corn strains as a method of determining differential earworm reactions. *J. Econ. Entomol.* **35:** 508–511.

Blickenstaff, C. C., D. D. Morey, and G. W. Burton, 1954. Effect of rates of nitrogen application on greenbug damage to oats, rye, and ryegrass. *Agron. J.* **46:** 338.

Blickenstaff, C. C., 1965. Partial intersterility of eastern and western U.S. strains of the alfalfa weevil. *Ann. Entomol. Soc. Am.* **58:** 523–526.

Blickenstaff, C. C., 1969. Mating competition between eastern and western strains of the alfalfa weevil, *Hypera postica. Ann. Entomol. Soc. Am.* **62:** 956–958.

Blum, A., 1968. Anatomical phenomena in seedlings of sorghum varieties resistant to the sorghum shoot fly (*Atherigona varia soccata*). *Crop Sci.* **8:** 388–390.

Blum, A., 1969. Oviposition preference by the sorghum shoot fly (*Atherigona varia soccata*) in progenies of susceptible × resistant sorghum crosses. *Crop Sci.* **9:** 695–696.

Bohn, G. W., A. N. Kishaba, J. A. Principe, and H. H. Toba, 1973. Tolerance to melon aphid in *Cucumis melo. J. Am. Soc. Hortic. Sci.* **98:** 37–40.

Boller, E. F. and R. J. Prokopy, 1976. Bionomics and management of Rhagoletis. *Ann. Rev. Entomol.* **21:** 223–246.

Bolton, J. L., 1962. *Alfalfa.* Interscience Publishers, Inc., New York, 474 pp.

Bolton, J. J., B. P. Goplen, and H. Baenzinger, 1972. World distribution and historical developments. Chapter 1 in *Alfalfa Science and Technology,* C. H. Hanson, Ed. American Society of Agronomy, Madison, Wis., pp. 1–34.

Borlaug, N. E., 1958. The use of multilineal or composite varieties to control airborne epidemic diseases of self-pollinated crop plants. *Proc. First Int. Wheat Genet. Symp.* 12–27. The Public Press Ltd., Winnipeg, Canada.

Bottger, G. T., E. T. Sheehan, and M. J. Lukefahr, 1964. Relation of ɔssypol content of cotton plants to insect resistance. *J. Econ. Entomol.* **57**: 283-285.

Bottrell, D. G., 1971. Entomological advances in sorghum production. *Tex. Agric. Exp. Stn. Prog. Rep.* **2940**: 28-40.

Bowden, J., 1965. Sorghum midge, *Contarinia sorghicola* (Coq.) and other causes of grain sorghum loss in Ghana. *Bull. Entomol. Res.* **56**: 169-189.

Bowling, C. C., 1963. Tests to determine varietal reaction to rice water weevil. *J. Econ. Entomol.* **56**: 893-894.

Bowling, C. C., 1973. Procedure for screening varieties for resistance to the rice water weevil. *J. Econ. Entomol.* **66**: 572-573.

Boyce, S. G. and J. J. Jokela, 1966. Report from the Central States Forest Tree Improvement Committee, In *Breeding Pest Resistant Trees,* H. D. Gerhold et al., Eds. Pergamon Press, London, pp. 69-70.

Branson, T. F., 1971. Resistance in the grass tribe Maydeae to larvae of the western corn rootworm. *Ann. Entomol. Soc. Am.* **64**: 861-863.

Branson, T. F., 1972. Resistance to the corn leaf aphid in the grass tribe Maydeae. *J. Econ. Entomol.* **65**: 195-196.

Branson, T. F., P. L. Guss, J. L. Krysan, and G. R. Sutter, 1975. Corn rootworms laboratory rearing and manipulation. *U.S. Dep. Agric. ARS* **NC-28**: 18 pp.

Brazzel, J. R. and D. F. Martin, 1956. Resistance of cotton to pink bollworm damage. *Texas Agric. Exp. Stn. Bull.* **843**.

Brett, C. H. and R. Bastida, 1963. Resistance of sweet corn varieties to the fall armyworm, *Laphygma frugiperda. J. Econ. Entomol.* **56**: 162-167.

Brewer, F. D., and D. F. Martin, 1976. Substitutes for agar in a wheat germ diet used to rear the corn earworm, *Heliothis Zea,* and the sugar cane borer *Diatraea saccharalis. Ann. Entomol. Soc. Am.* **69**: 255-256.

Briggs, J. B., 1965. The distribution, abundance, and genetic relationships of four strains of the rubus aphid (*Amphorophora rubi* Kalt.) in relation to raspberry breeding. *J. Hortic. Sci.* **40**: 109-117.

Brindley, T. A., and F. F. Dicke, 1963. Significant developments in European corn borer research. *Ann. Rev. Entomol.* **8**: 155-176.

Brindley, T. A., A. N. Sparks, W. B. Showers, and W. D. Guthrie, 1975. Recent research advances on the European corn borer in North America. *Ann. Rev. Entomol.* **20**: 221-239.

Brinholi, O., J. Nakagawa, D. A. S. Marcondes, and J. R. Machado, 1974. Estudo do Comportamento de alguns "cultivares" da mandioca ao ataque da brocha -dos- brotos (*Silba pendula*). *Rev. Agric.* **49**: 181-183.

Brown, L. G., J. W. Jones, J. D. Hesketh, J. D. Hartsog, F. D. Whisler, R. W. McClendon, F. A. Harris, D. W. Parvin, and H. N. Pitre, 1977. The use of simulation to predict cotton yield losses due to insect damage. *Proc. Beltwide Cotton Res. Conf.*: 131-135. National Cotton Council, Memphis, Tenn.

Browning, J. A., 1974. Relevance of knowledge about natural ecosystems to development of pest management programs for agro-ecosystems. *Proc. Am. Phytopathol. Soc.* **1:** 191-199.

Brues, C. T., 1946. *Insect Dietary.* Harvard University Press, Cambridge, Mass.

Buckley, B. R. and C. C. Burkhardt, 1962. Corn earworm damage and loss in grain sorghum. *J. Econ. Entomol.* **55:** 434-439.

Buckley, B. R. and C. C. Burkhardt, 1963. Notes on the biology and behavior of *Heliothis zea* (Boddie) on grain sorghum. *J. Kans. Entomol. Soc.* **36:** 127-132.

Buckton, G. B., 1899. Notes on two species of aphids. *Indian Mus. Notes* **4:** 277.

Burnett, J. H., 1975. *Mycogenetics.* John Wiley & Sons, New York, 375 pp.

Burton, R. L., and E. A. Harrell, 1966. Modification of a Lepidopterous larvae dispenser for a packing machine. *J. Econ. Entomol.* **59:** 1544-1545.

Burton, R. L., E. A. Harrell, H. C. Cox, and W. W. Hare, 1966. Devices to facilitate rearing of Lepidopterous larvae. *J. Econ. Entomol.* **59:** 594-596.

Burton, R. L., 1969. Mass rearing of the corn earworm in the laboratory. *U.S. Dep. Agric. ARS:* 33-134.

Burton, R. L., and W. D. Perkins, 1972. WSB, A new laboratory diet for the corn earworm and the fall armyworm. *J. Econ. Entomol.* **65:** 385-386.

Busbice, T. H., D. K. Barnes, C. H. Hanson, R. R. Hill, Jr., W. V. Campbell, C. C. Blickenstaff, and R. C. Newton, 1967. Field evaluation of alfalfa introductions for resistance to the alfalfa weevil *Hypera postica* (Gyllenhal). *U.S. Dep. Agric. ARS Bull.* **34-94:** 13 pp.

Busbice, T. H., R. R. Hill, Jr., and H. L. Carnahan, 1972. Genetics and breeding procedures. Chapter 13 in *Alfalfa Science and Technology.* C. H. Hanson, Ed. American Society of Agronomy, Madison, Wis., pp. 283-318.

Busbice, T. H., W. V. Campbell, L. V. Bunch, and R. Y. Gurgis, 1978. Breeding alfalfa cultivars resistant to the alfalfa weevil. *Euphytica:* **27(2):** 343-352.

Butler, G. D. and H. Muramoto, 1967. Banded-wing whitefly abundance and cotton leaf pubescence in Arizona. *J. Econ. Entomol.* **60:** 1176-1177.

Byrne, D. H. and E. L. Rittershausen, 1970. A technique for evaluation of alfalfa populations for resistance to alfalfa weevil larvae. *J. Econ. Entomol.* **63:** 652-653.

Byrne, H. D., 1969. The oviposition response of the alfalfa weevil *Hypera postica* (Gyllenhal). *Univ. Md. Agric. Exp. Stn. Bull.* **A160.**

Calderon, M., 1977. Effect of the nectariless character of cotton on the population dynamics of certain phytophagous and natural enemy insects. Ph.D. Dissertation. Entomology Department, Mississippi State University, State College.

Cardwell, R. M., W. B. Cartwright, and L. E. Compton, 1946. Inheritance of Hessian fly resistance derived from W38 and Durum P. I. 94587. *J. Am. Soc. Agron.* **38:** 398-409.

Callahan, P. S., 1957. Oviposition response of the corn earworm to differences in surface texture. *J. Kans. Entomol. Soc.* **30:** 59-63.

Callaham, R. Z., 1966. Nature of resistance of pines to bark beetles. In *Breeding Pest Resistant Trees*, H. D. Gerhold et al., Eds. Pergamon Press, London, pp. 197–201.

Cameron, J. W., L. D. Anderson, and H. H. Shorey, 1969. Corn earworm damage to corn with starch vs. sugary kernels on genetically uniform mother plants. *J. Econ. Entomol.* **62**: 986–988.

Campbell, W. V. and J. W. Dudley, 1965. Differences among *Medicago* species in resistance to oviposition by the alfalfa weevil *J. Econ. Entomol.* **58**: 245–248.

Camplis, J. V., E. V. Farias, and E. Salgado Sosa, 1976. Evaluation of cotton strains resistant to *Heliothis* spp. complex in Southern Tamaulipas. *Proc. Beltwide Beltwide Cotton Prod. Res. Conf.*: 80–90. National Cotton Council, Memphis, Tenn.

Cannon, W. N., Jr., and A. Ortega, 1966. Studies of *Ostrinia nubilalis* larvae (Lepidoptera: Pyraustidae) on corn plants supplied with various amounts of nitrogen and phosphorus. I. Survival. *Ann. Entomol. Soc. Am.* **59**: 631–638.

Carlson, J. R., and R. H. Andrew, 1976. Parameters for determination of damage to sweet corn genotypes from 1st and 2nd generation European corn borer. *Crop Sci.* **16**: 39–42.

Carnahan, H. L., R. N. Peaden, F. V. Lieberman, and R. K. Petersen, 1963. Differential reactions of alfalfa varieties and selections to the pea aphid. *Crop Sci.* **3**: 219–222.

Carter, J. J., 1966. Aphid responses to colors in artificial rearings. *Bull. Entomol. Soc. Am.* **12**: 378–380.

Carter, W., 1973. *Insects in Relation to Plant Disease*, 2nd ed. John Wiley & Sons, Inc., New York, 759 pp.

Cartier, J. J. and R. H. Painter, 1956. Differential reactions of two biotypes or the corn leaf aphid to resistant and susceptible varieties, hybrids and selections of sorghums. *J. Econ. Entomol.* **49**: 498–508.

Cartier, J. J., 1959. Recognition of three biotypes of the pea aphid from southern Quebec. *J. Econ. Entomol.* **52**: 293–294.

Cartier, J. J., 1963. Varietal resistance of peas to pea aphid biotypes under field and greenhouse conditions. *J. Econ. Entomol.* **56**: 205–213.

Cartier, J. J. and J. L. Auclair, 1964. Pea aphid behaviour: color preference on a chemical diet. *Can. Entomol.*, **96**: 1240–1243.

Cartier, J. J., A. Isaak, R. H. Painter, and E. L. Sorensen, 1965. Biotypes of peas aphid *Acyrthosiphon pisum* (Harris) in relation to alfalfa clones. *Can. Entomol.* **97**: 754–760.

Cate, J. R., Jr., D. G. Bottrell, and G. L. Teetes, 1973. Management of the greenbug on grain sorghum. 1. Testing foliar treatments of insecticides against greenbug and corn leaf aphid. *J. Econ. Entomol.* **66**: 945–951.

Cartwright, W. B. and G. A. Weibe, 1936. Inheritance of resistance to the Hessian fly in the wheat crosses Dawson × Poso and Dawson × Big Club. *J. Agric. Res.* **52**: 691–695.

Cartwright, W. B. and D. W. Lallue, 1944. Testing wheats in the greenhouse for Hessian fly resistance. *J. Econ. Entomol.* **37:** 385–387.

Cartwright, W. B., R. M. Caldwell, and L. E. Compton, 1946. Relation of temperature to the expression of resistance in wheats to Hessian fly. *J. Am. Soc. Agron.* **38:** 259–263.

Casagrande, R. A. and D. L. Haynes, 1976. The impact of pubescent wheat on the population dynamics of the cereal leaf beetle. *Environ. Entomol.* **5:** 153–159.

Caswell, H. and F. C. Reed, 1975. Indigestibility of C_{14} bundle sheath cells by the grasshopper, *Melanoplus confusus. Ann. Entomol. Soc. Am.* **68:** 686–688.

Centro Internacional de Agricultura Tropical, 1975. *Annual Report 1974.* Cali, Colombia, pp. 74–81.

Centro Internacional de Agricultura Tropical, 1976. *Annual Report 1975. Cassava Production Systems.* Cali, Columbia, 57 pp.

Centro Internacional de Agricultura Tropical, 1977. *Annual Report 1976. Cassava Production Systems.* Cali, Columbia, 76 pp.

Chada, H. L., 1959. Insectry technique for testing the resistance of small grains to the greenbug. *J. Econ. Entomol.* **52:** 276–279.

Chada, H. L., 1962. Reaction of sorghum varieties and hybrids to natural infestation by southwestern corn borer, Mangum, Oklahoma, 1961. *U.S. Dep. Agric. ARS Spec. Rep.* **W-160:** 8 pp.

Chada, H. L., I. M. Atkins, J. H. Gardenhire, and D. E. Weibel, 1961. Greenbug resistance studies with small grains. *Tex. Agric. Exp. Stn. Bull.* **B-982:** 18 pp.

Chambers, D. L., 1977. Quality control in mass rearing. *Ann. Rev. Entomol.* **22:** 289–308.

Chang, T. T. and E. A. Bardenas, 1965. The morphology and varietal characteristics of the rice plant. *Int. Rice Res. Inst. Tech. Bull.* **4:** 40 pp. Los Baños, Philippines.

Chang, S. H. and J. L. Brewbaker, 1975. Genetic resistance in corn to the corn leaf aphid. *Hawaii Agric. Exp. Stn. Misc. Publ.* **122:** 16.

Chang, T. T., 1976. The origin, evolution, cultivation, dissemination of Asian and African rices. *Euphytica* **25:** 425–441.

Chatterji, S. M., K. H. Siddiqui, V. P. S. Panwar, G. C. Sharma, and W. R. Young, 1968. Rearing of maize stem borer, *Chilo zonellus* (Swinhoe) on artificial diet. *Indian J. Entomol.* **30:** 8–12.

Chatterji, S. M., P. Sarup, K. K. Marwaha, V. P. S. Panwar, K. H. Siddiqui, and M. W. Bhamburkar, 1973. Studies on insect-plant relationship-comparative tolerance of some elite indigenous maize lines to *Chilo partellus* (Swinhoe) under artificial infestation. *Indian J. Entomol.* **35:** 156–159.

Chelliah, S. and A. Subramanian, 1972. Influence of nitrogen fertilization on the infestation by the gall midge, *Pachydiplosis oryzae* (Wood-Mason) Mani in certain rice varieties. *Indian J. Entomol.* **34:** 255–256.

Chen, T. H. and C. H. Liao, 1975. Corn stunt spiroplasma and proof of pathogenicity. *Science* **188**: 1015–1017.

Cheng, C. H. and M. D. Pathak, 1972. Resistance to Nephotettix virescens in rice varieties. *J. Econ. Entomol.* **65**: 1148–1153.

Cheng, C. H., 1977. The possible role of rice varieties in rice brown planthopper control. In *The Rice Brown Planthopper.* Food and Fertilizer Technology Center for the Asian and Pacific Region, Taiwan, Republic of China, pp. 214–229.

Chesnokov, P. G., 1962. *Methods of Investigating Plant Resistance to Pests.* Published for NSF and USDA by Israel Program for Scientific Translations, 107 pp.

Chiang, H. C., R. S. Raros, J. A. Mihm, and M. B. Windels, 1971. Artificially infesting corn with corn rootworms. *Minn. Sci.* **27**: 8, 9, 12.

Chiang, H. C., M. B. Windels, J. A. Mihm, D. E. Rasmussen, and L. K. French, 1975. Methods of mass production of corn rootworms in laboratory and artificial field infestation techniques. *Proc. N. Cent. Branch, Entomol. Soc. Am.* **30**: 37.

Chiang, H. C., 1973. Bionomics of the northern and western corn rootworm. *Ann. Rev. Entomol.* **18**: 47–72.

Chiang, H. C., 1978. Pest management in corn. *Ann. Rev. Entomol.* **23**: 101–123.

Chiang, M. S. and M. Hudson, 1973. Inheritance of resistance to the European corn borer in grain corn. *Can. J. Plant Sci.* **53**: 779–782.

Chiang, M. S. and M. Hudson, 1973a. Genetic studies of resistance to the European corn borer in grain corn. *Genetics* **74**, part 2.

Chiarappa, L., 1971. *Crop Loss Assessment Methods.* FAO Manual on the evaluation and prevention of losses by pests, disease, and weeds, FAO, Rome.

Chippendale, G. M., 1972. Composition of meridic diets for rearing plant feeding Lepidopteroud larvae. *Proc. N. Cent. Branch, Entomol. Soc. Am.* **27**: 114–121.

Chowdhuri, K. A. and G. M. Buth, 1971. Cotton seeds from the Neolithic in Egyptian Nubia and the origin of Old World cotton. *Linn. Soc. London, Biol. J.* **3**: 303–312.

Cibula, A. B., R. H. Davidson, F. W. Fisk, and J. B. LaPidus, 1967. Relationship of free amino acids of some Solanceous plants to trowth and development of *Leptinotarsa decemlineata* (Coleoptera: Chrysomelidae). *Ann. Entomol. Soc. Am.* **60**: 626–631.

Clark, T. B., 1977. Spiroplasma sp., a new pathogen in honey bees. *J. Invertebr. Pathol.* **29**: 112–113.

Cogburn, R. R., 1977. Susceptibility of varieties of stored rough rice to losses caused by storage insects. *J. Stored Prod. Res.* **13**: 29–34.

Cohen, M. and M. P. Russell, 1970. Some effects of rice varieties on the biology of the angoumois grain moth, *Sitotraga cerelella. Ann. Entomol. Soc. Am.* **63**: 930–931.

Commonwealth Institute of Entomology, 1957. Distribution maps of insect pests: *Aonidomytilus albus* (CK11). Ser. A, Map 81.

Comstock, R. E., H. F. Robinson, and P. H. Harvey, 1949. A breeding procedure designed to make maximum use of both general and specific combining ability. *Agron. J.* **41**: 360–367.

Connola, D. P. and K. Brinkafner, 1976. Large outdoor cage tests with eastern white pine being tested in field plots for white pine weevil resistance. *Proc. 23rd Northeast For. Tree Improv. Conf.*: 56–64.

Coon, B. F., 1959. Aphid populations on oats grown in various nutrient solutions. *J. Econ. Entomol.* **52**: 624–626.

Coppel, H. C. and J. W. Mertins, 1977. *Biological Insect Suppression,* Springer-Verlag, Berlin, 314 pp.

Costa, A. S., 1969. Whiteflies as virus vectors. In *Virus, Vectors, and Vegetation,* K. Maromorosch, Ed. John Wiley & Sons, Inc., New York, pp. 95–119.

Cowan, C. B. and M. J. Lukefahr, 1970. Characters of cotton plants that affect infestation of cotton fleahoppers. *Proc. Beltwide Cotton Prod. Res. Conf.:* 79–80. National Cotton Council, Memphis, Tenn.

Cowley, G. T. and E. P. Lichenstein, 1970. Growth inhibition of soil fungi by insecticides and annulment of inhibition by yeast extract or nitrogenous nutrients. *J. Gen. Microbiol.* **62**: 27–34.

Cram, W. T., 1965. Fecundity of the root weevils *Brachyrhinus sulcatus* and *Sciopithes obscurus* on strawberry at different conditions of host plant nutrition. *Can. J. Plant Sci.* **45**: 219–225.

Cramer, H. H., 1967. *Defensa vegetal y cosecha mundial.* Bayerischer Pflanzenschutz, Leverkusen, Germany, p. 179.

Cress, D. C. and H. L. Chada, 1971. Development of 1. a synthetic diet for the greenbug, *Schizaphis graminum*; 2. greenbug development as affected by zinc, iron, manganese and copper. *Ann. Entomol. Soc. Am.* **64**: 1240–1244.

Cross, W. H., M. J. Lukefahr, P. A. Fryxell, and H. R. Burke, 1975. Host plants of the boll weevil. *Environ. Entomol.* **4**: 19–26.

CRRI, 1954. *Annual Report for 1949–50 and 1950–51.* Central Rice Research Institute, Cuttack, 33 pp.

Cuthbert, F. P. and B. W. Davis, 1972. Factors contributing to cowpea curculio resistance in southern peas. *J. Econ. Entomol.* **65**: 778–781.

DaCosta, C. P. and C. M. Jones, 1971. Cucumber beetle resistance and mite susceptibility controlled by the bitter gene in *Cucumis salivus* L. *Science* **172**: 1145–1146.

Dadd, R. H. and T. E. Mittler, 1965. Studies on the artificial feeding of the aphid *Myzus persicae* (Sulzer). III. Some major nutritional requirements. *J. Insect Physiol.* **11**: 717–743.

Dadd, R. H. and D. L. Kreiger, 1968. Dietary amino acid requirements of the aphid, *Myzus persicae. J. Insect. Physiol.* **14**: 741–764.

Dadd, R. H., 1973. Insect nutrition: current developments and metabolic implications. *Ann. Rev. Entomol.* **18**: 381–420.

Dahms, R. G. and R. H. Painter, 1940. Rate of reproduction of the pea aphid on different alfalfa plants. *J. Econ. Entomol.* **33**: 482–485.

Dahms, R. G. and J. H. Martin, 1940. Resistance of F_1 sorghum hybrids to the chinch bug. *J. Am. Soc. Agron.* **32**: 141–147.

Dahms, R. G., 1943. Insect resistance in sorghums and cotton. *J. Am. Soc. Agron.* **35**: 704–715.

Dahms, R. G. and J. B. Sieglinger, 1954. Reaction of sorghum varieties to the chinch bug. *J. Econ. Entomol.* **47**: 536–537.

Dahms, R. G., T. H. Johnston, A. M. Schlehuber, and E. A. Wood, Jr., 1955. Reaction of small grain varieties and hybrids to greenbug attack. *Okla. Agric. Exp. Stn. Bull.* **T-55**: 61 pp.

Dahms, R. G., 1969. Theoretical effects of antibiosis on insect population dynamics. *U.S. Dep. Agric. ERD*: 5 pp.

Dahms, R. G., 1972. Techniques in the evaluation and development of host-plant resistance. *J. Environ. Qual.* **1**: 254–259.

Dahms, R. G., 1972a. The role of host plant resistance in integrated insect control. In *Control of Sorghum Shoot Fly,* M. G. Jotwani and W. R. Young, Eds. Oxford and IBH Publishing Co., New Delhi, pp. 152–167.

Daniels, N. E., 1957. Greenbug populations and their damage to winter wheat as affected by fertilizer applications. *J. Econ. Entomol.* **50**: 793–794.

Daniels, N. E. and K. B. Porter, 1958. Greenbug resistance studies in winter wheat. *J. Econ. Entomol.* **51**: 702–704.

Daniels, N. E., G. C. Wilson, and C. S. Clarke, 1968. Nitrogen and phosphorus content of greenbugs reared on fertilized wheat. *J. Econ. Entomol.* **61**: 1746–1747.

Das, Y. T., 1976a. Cross resistance to stem borers in rice varieties. *J. Econ. Entomol.* **69**: 41–46.

Das, Y. T., 1976b. Some factors of resistance to *Chilo suppressalis* in rice varieties. *Entomol. Exp. Appl.* **20**: 131–134.

David, W. A. L. and B. O. C. Gardiner, 1966. Mustard oil glucosides as feeding stimulants for *Peris brassicae* larvae in a semi-synthetic diet. *Entomol. Exp. Appl.* **9**: 247–255.

Davidson, A. G. and R. M. Prentice, 1967. Important forest insects and diseases of mutual concern to Canada, the United States and Mexico. *Can. Dep. For. Rural Dev. Publ.* **1180**: 248 pp.

Davis, D. D., J. V. Ellington, and J. C. Brown, 1973. Mortality factors affecting cotton insects: 1. Resistance of smooth and nectariless characters in Acala cotton to *Heliothis zea, Pectinophora gossypiella* and *Trichoplusia ni. J. Environ. Qual.* **2**: 530–535.

Davis, F. M., 1976. Production and handling of eggs of southwestern corn borer for host plant resistance studies. *Miss. Agric. For. Exp. Stn. Tech. Bull.* **74**.

Davis, F. M., G. E. Scott, and C. A. Henderson, 1973. Southwestern corn borer: preliminary screening of corn genotypes for resistance. *J. Econ. Entomol.* **66**: 503–506.

Davis, R. F., J. F. Worley, R. F. Whitcomb, F. Ishijima, and R. L. Steere, 1972. Helical filaments produced by a mycoplasma-like organism associated with corn stunt disease. *Science* **176**: 521–523.

Davis, R. L. and M. C. Wilson, 1953. Varietal tolerance of alfalfa to the potato leafhopper. *J. Econ. Entomol.* **46**: 242–245.

Day, P. R., 1974. *Genetics of Host-Parasite Interactions.* W. H. Freeman, San Francisco, 238 pp.

Del Valle, C. G. and J. C. Miller, 1963. Influence of husk length and tightness against corn earworm damage in sweet corn hybrids. *Am. Soc. Hortic. Sci. Proc.* **83**: 531–535.

Derr, R. F., D. D. Randall, and R. W. Kieckhefer, 1964. Feeding stimulant for western and northern corn rootworm adults. *J. Econ. Entomol.* **57**: 963–965.

Dethier, V. G., 1954. Evolution of feeding preference in phytophagous insects. *Evolution* **8**: 33–54.

Dethier, V. G., 1970. Chemical interactions between plants and insects. In *Chemical Ecology,* Sondheimer and Simeone, Eds., Academic Press, Inc., New York, pp. 83–102.

Devine, T. H., R. H. Ratcliffe, C. M. Rincker, D. K. Barnes, S. A. Ostazeski, T. H. Busbice, C. H. Hanson, J. A. Schillinger, G. R. Buss, and R. W. Cleveland, 1975. Registration of Arc Alfalfa. *Crop Sci.* **15**: 97.

Diaz, C. G., 1967. Some relationships of representative races of corn from the Latin American germplasm seed bank to intensity of infestation by the rice weevil *Sitophilus zeamais* Mots. (Coleoptera; Curculionidae). Ph.D. Thesis, Kansas State University, Manhattan.

Dicke, F. F., R. E. Atkins, and G. R. Pesho, 1963. Resistance of sorghum varieties and hybrids to the European corn borer (*Ostrinia nubilalis* Hbn.). *Iowa State J. Sci.* **37**: 247–247.

Dicke, F. F., 1975. The role of insects in some diseases of maize, *Iowa State J. Sci.* **49**: 553–558.

Dicke, F. F. and W. D. Guthrie, 1977. Proc. International Maize Symposium; Univ. of Illinois, Urbana, Ill., U.S.A., Genetics and Breeding. Genetics of Insect Resistance in Maize. John Wiley & Sons, Inc., New York, in press.

Dicker, G. H. L., 1940. The biology of *Rubus* aphids. *J. Pomol.* **18**: 1–33.

Dickson, J. G., 1956. *Diseases of Field Crops.* McGraw-Hill Book Co., Inc., New York, pp. 74–114.

Diener, T. O., 1971. A plant virus with properties of a free ribonucleic acid: potato spindle tuber virus. In *Comparative Virology,* K. Maramorosch, and E. Kurstak, Eds. Academic Press, Inc., New York, pp. 433–478.

Diener, T. O., 1977. Viroids. In *Insect and Plant Viruses: An Atlas,* K. Mara-morosch, Ed., Academic Press, Inc., New York, pp. 431–436.

Dilday, R. H. and T. N. Shaver, 1976a. Survey of the regional *Gossypium hirsutum* L. primitive race collection for flowerbud gossypol. *U.S. Dep. Agric. ARS* **S-80.**

Dilday, R. H. and T. N. Shaver, 1976b. Survey of the regional *Gossypium hirsutum* L. primitive race collection for flowerbud gosspypol and seasonal variation between years in gossypol percentage. *U.S. Dep. Agric. ARS* **S-146.**

Ditman, L. P. and J. L. Ditman, 1957. An apparatus for measuring corn earworm injury to sweet corn. *J. Econ. Entomol.* **50:** 371–372.

Djamin, A. and M. D. Pathak, 1967. The role of silica in resistance to Asiatic rice borer *Chilo suppressalis* (Walker) in rice varieties. *J. Econ. Entomol.* **60:** 347–351.

Dobie, P., 1974. The laboratory assessment of the inherent susceptibility of maize varieties to postharvest infestation by *Sitophilus zeamais* Motsch. (Coleoptera, Curculionidae). *J. Stored Prod. Res.* **10:** 183–197.

Dobson, R. C. and J. G. Watts, 1957. Spotted alfalfa aphic occurrences on seedling alfalfa as influenced by insecticides and varieties. *J. Econ. Entomol.* **50:** 132–135.

Dogger, J. R. and C. H. Hanson, 1963. Reaction of alfalfa varieties and strain to alfalfa weevil. *J. Econ. Entomol.,* **56:** 192–197.

Doggett, H., 1964. A note on the incidence of American boll worm *Heliothis armigera* (Hub.) (Noctuidae) in sorghum. *East Afr. Agric. For. J.* **29:** 348–349.

Doggett, H., 1970. *Sorghum.* Longmans, Green, London, 403 pp.

Doggett, H., K. J. Starks, and S. A. Eberhart, 1970. Breeding for resistance to the sorghum shoot fly. *Crop Sci.* **10:** 528–531.

Dolinka, B., Ed., 1975. *Report of the International Project on Ostrinia nubilalis, Phase II, Results.* Information Centre of the Ministry of Agriculture and Food, Budapest, Hungary, 137 pp.

Dolinka, B., H. C. Chiang, and D. Hadzistevic, Eds. 1973. Report of the international project on *Ostrinia nubilalis.* Phase I. Results 1969 and 1970. Information Center of Ministry of Agriculture and Food, Budapest, 168 pp.

Dorman, K. W., 1976. The genetics and breeding of southern pines. *U.S. Dep. Agric. Handb.* **471:** 407 pp.

Douwes, P., 1968. Host selection and host finding in the egg-laying female *Cidaria albulata* L. (Lep. Geometridae). *Opuscula Entomol.* **33:** 233–279.

Duke, W. B., R. D. Hagin, J. F. Hunt, and D. L. Linscott, 1975. Metal halide lamps for supplemental lighting in greenhouses: crop response and spectral distribution. *Agron. J.,* **67:** 49–53.

Dungan, G. H., C. M. Woodworth, A. L. Lang, J. H. Bigger, and R. O. Snelling, 1938. *Developments in Hybrids in Relation to the Illinois Farmer's Institute,* 51 pp.

Dunn, J. A. and D. P. H. Kempton, 1971. Differences in susceptibility to attack by *Brevicoryne brassicae* (L.) on Brussels sprouts. *Ann. Appl. Biol.* **68:** 121–134.

Dunn, J. A. and D. P. H. Kempton. 1976. Varietal differences in the susceptibility of Brussels sprouts to lepidopterous pests. *Ann. Appl. Biol.* **82:** 11–19.

Dunnam, E. W. and J. C. Clark, 1937. Thrips damage to cotton. *J. Econ. Entomol.* **30:** 855–857.

Dunnam, E. W. and J. C. Clark, 1939. The cotton aphid in relation to the pilosity of cotton leaves. *J. Econ. Entomol.* **31:** 633–666.

Dyck, V. A., E. A. Heinrichs, M. D. Pathak, and R. Feuer, 1976. Insect pest management in rice. Paper presented at the *Seventh Natl. Conf. Pest Control Council Philippines, Cagayan de Oro City*. Mimeographed, 29 pp.

Earnheart, A. T., Jr., 1973. Evaluation of cotton, *Gossypium hirsutum* L., race stocks for resistance to boll weevil attack M.S. Thesis, Mississippi State University, State College, Miss.

Eastop, V. F., 1977. Worldwide importance of aphids as virus vectors. In *Aphids as Virus Vectors,* K. F. Harris and K. Maramorosch, Eds. Academic Press, Inc., New York, pp. 3–61.

Eaton, S. V., 1952. Effects of potassium deficiency on growth and metabolism of sunflower plants. *Bot. Gaz.* **114:** 165–180.

Eberhart, S. A., M. N. Harrison, and F. Ogada, 1967. A comprehensive breeding system. *Der Zuchter.* **37:** 169.

Eberhart, S. A., 1972. Techniques and methods for more efficient population improvement in sorghum. In *Sorghum in the Seventies,* N. G. P. Rao and L. R. House, Eds. Oxford and IBH, New Delhi, pp. 197–213.

Eden, W. G., 1952. Effect of kernel characteristics and components of husk cover on rice weevil damage to corn. *J. Econ. Entomol.* **45:** 1084–1085.

Edminster, T. W., 1978. Concepts for using modeling as a research tool. *U.S. Dep. Agric. ARS Tech. Man.* **520.**

Eglinton, G. and R. J. Hamilton, 1963. The distribution of alkanes. In *Chemical Plant Taxonomy,* T. Swain, Ed., Academic Press, Inc., New York, pp. 187–217.

Ehler, L. E., 1974. A review of the spider-mite problem on sorghum and corn in West Texas. *Tex. Agric. Exp. Stn.* **B-1149,** 15 pp.

Ehrlich, P. R. and P. H. Raven, 1964. Butterflies and plants: a study in coevolution. *Evolution* **18:** 586–608.

Eichmeier, J. and G. Guyer, 1960. An evaluation of the rate of reproduction of the two-spotted spider mite reared on gibberlin-treated bean plants. *J. Econ. Entomol.* **53:** 661–664.

Ekman, N. V., N. A. Vilkova, and I. D. Shapior, 1973. Express method for determining resistance levels of cereals to *Eurygaster integriceps* Put. according to disintegration rates of caryopsis starch. *Proc. All Union Sci. Inst. Plant Prot.* **39:** 176–180.

Elias, L. A., 1970. Maize resistance to stalk borers in *Zeadiatrea* Box, and *Diatraea* Guilding (Lepidoptera: Pyralidae) at five localities in Mexico. Ph.D. Dissertation, Kansas State University, Manhattan, Kansas.

Elgin, J. H., Jr., R. R. Hill, Jr., and K. E. Zeiders, 1970. Comparison of four methods of multiple trait selection for five traits in alfalfa. *Crop Sci.* **10:** 190–193.

El-Tigani, M. El-Amin, 1962. De Finfluss der Mineraldungung der Pflanzen auf Entwicklung und Vermehrung von Blattausen. *Wiss. Z. Univ. Rostock* **11:** 307–324.

van Emden, J. F., 1964. Effect of (2-chloroethyl) trimethylammonium chloride on the rate of increase of the cabbage aphid *Brevicoryne brassicae* (L.). *Nature* **201:** 946–948.

van Emden, H. F. and C. H. Wearing, 1965. The role of the aphid host plant in delaying economic damage levels in crops. *Ann. Appl. Biol.* **56:** 323–324.

van Emden, H. F., 1966a. Plant resistance to insects induced by environment. *Sci. Hortic.* **18:** 91–102.

van Emden, H. F., 1966b. Studies on the relations of the insect and host plant. III. A comparison of the reproduction of *Brevicoryne brassicae* and *Myzus persicae* (Hemiptera: Aphididae) on Brussels sprout plants supplied with different rates of nitrogen and potassium. *Entomol. Exp. Appl.* **9:** 444–460.

van Emden, H. F. and M. A. Bashford, 1969. A comparison of the reproduction of *Brevicoryne brassicae* and *Myzus persicae* in relation to soluble nitrogen concentration and leaf age (leaf position) in the Brussels sprout plant. *Entomol. Exp. Appl.* **12:** 351–364.

van Emden, H. F., 1969a. Plant resistance to *Myzus persicae* induced by a plant growth regulator and measured by aphid relative growth rate. *Entomol. Exp. Appl.* **12:** 125–131.

van Emden, H. F., 1969b. Plant resistance to aphids induced by chemicals. *J. Sci. Food Agric.* **20:** 385–387.

van Emden, H. F. and M. A. Bashford, 1971. The performance of *Brevicoryne brassicae* and *Myzus persicae* in relation to plant age and leaf amino acids. *Entomol. Exp. Appl.* **14:** 349–360.

van Emden, H. F. and M. J. Way, 1973. Host plants in the population dynamics of insects. In *Insect/Plant Relationships,* H. F. van Emden, Ed. John Wiley & Sons, Inc., New York, pp. 181–199.

van Emden, H. F. and M. A. Bashford, 1976. The effect of leaf excision on the performance of *Myzus persicae* and *Brevicoryne brassicae* in relation to the nutrient treatment of the plants. *Physiol. Entomol.* **1:** 67–71.

Emery, W. T., 1946. Temporary immunity in alfalfa ordinarily susceptible to attack by the pea aphid. *J. Agric. Res.* **73:** 33–43.

Erickson, J. M. and P. Feeney, 1974. Sinigrin: A chemical barrier to the black swallowtail butterfly. *Ecology* **55:** 103–111.

Eschrich, W. 1970. Biochemistry and fine structure of phloem in relation to transport. *Ann. Rev. Plant Physiol.* **21**: 193–214.

Evans, A. D., 1938. Physiological relationships between insects and their host plant. I. The effect of the chemical composition of the plant on reproduction and the production of the winged forms in *Brevicoryne brassicae* L. (Aphidadae). *Ann. Appl. Biol.* **41**: 189–206.

Evans, H. J. and G. J. Sorger, 1966. Role of mineral elements with emphasis on the univalent cations. *Ann. Rev. Plant Physiol.* **17**: 47–76.

Everson, E. H., R. L. Gallun, J. A. Schillinger, Jr., D. H. Smith, and J. C. Craddock, 1966. Geographical distribution of resistance in Triticum to the cereal leaf beetle. *Mich. Agric. Exp. Stn. Quart. Bull.* **48**: 565–569.

Everson, E. H., 1973. *The Michigan State University Computer System for Plant Breeding.* Mimeographed, 41 pp.

FAO/IBPGR Task Force on the Safe Transfer of Genetic Material, 1977. *Plant Health and Quarantine in International Transfer of Genetic Resources.* FAO/ IBPGR, Rome, in press.

Feaster, C. V. and E. L. Turcotte, 1962. Genetic basis for varietal improvement of Pima cottons. U.S. Dep. Agric. ARS **34-31**.

Feeney, P., 1970. Seasonal changes in oak leaf tannings and nutrients as a cause of spring feeding by winter moth caterpillars. *Ecology* **51**: 565–581.

Feeny, P., 1975. Biochemical coevolution between plants and their insect herbivores. In *Coevolution of Animals and Plants,* L. E. Gilbert and P. H. Raven, Eds. University of Texas Press, Austin, pp. 3–19.

Feeny, P., 1976. Plant apparency and chemical defense. *Recent Adv. Phytochem.* **10**: 1–40.

Feldman, M., 1975. *Wheats, Evolution of Crop Plants.* Longmans, Green, London, pp. 120–128.

Fernando, H. E., 1972. Biology and laboratory culture of the rice gall midge and studies on varietal resistance. In *Rice Breeding.* International Rice Research Institute, Los Baños, Philippiness, pp. 343–351.

Fernando, H. E., 1975. Laboratory screening for all midge resistance in Sri Lanka. *Rice Entomol. Newsl.* **2**: 16.

Finlayson, D. G. and H. R. MacCarthy, 1965. The movement and persistence of insecticides in plant tissue. *Residue Rev.* **9**: 114–152.

Fitzgerald, P. J. and E. E. Ortman, 1964, Breeding for resistance to the western corn rootworm. *Proc. Ann. Corn Sorghum Res. Conf., 19th.:* 46–60.

Flor, H. H., 1942. Inheritance of pathogenicity in *Melampsora lini. Phytopathology* **32**: 653–659.

Flor, H. H., 1955. Host–parasite interaction in flax rust: its genetics and other implications. *Phytopathology* **45**: 680–685.

Flor, H. H., 1956. The complementary genetic systems in flax and flax rust. *Adv. Genet.* **8**: 29–54.

Flor, H. H., 1971. Current status of the gene-for-gene concept. *Ann. Rev. Phytopathol.* **9:** 275–295.

Food and Agriculture Organization, 1975. *Production Yearbook.* Rome.

Food and Agricultural Organization of the United Nations, 1975. *Mon. Bull. Agric. Stn.* **9:** 24.

Foster, D. G., G. L. Teetes, and J. W. Johnson, 1977a. Field evaluation of resistance in sorghums to Banks grass mite. *Crop Sci.* **16**, in press.

Foster, D. G., G. L. Teetes, J. W. Johnson, and C. R. Ward, 1977b. Resistance in sorghums to the Banks grass mite. *J. Econ. Entomol.* **70:** 259–262.

Foster, J. E., 1976. Current status of genetic control of Hessian fly populations with the dominant great plains race. *Proc. XV Int. Congr. Entomol.*: 157–163.

Fraenkel, G., 1959. The raison d'etre of secondary plant substances. *Science* **129:** 1466–1470.

Fraenkel, G., 1969. Evaluation of our thoughts on secondary plant substances. *Entomol. Exp. Appl.* **12:** 473–486.

Frankel, O. H. and E. Bennett, Eds., 1970. Genetic resources in plants—their exploration and conservation. *International Biological Programme,* Vol. 1. Blackwell Scientific Publications, Oxford.

Frankel, O. H. and J. G. Hawkes, Eds. 1975. Crop genetic resources for today and tomorrow. *International Biological Programme,* Vol. 2, Cambridge University Press, Cambridge.

Franzke, C. J. and M. E. Sanders, 1965. True-breeding colchicine-induced mutants from sorghum hybrids. *Am. J. Bot.* **52:** 211–221.

Frazer, B. D., 1972. Population of dynamics and recognition of biotypes in the pea aphid (Homoptera: Aphididae). *Can. Entomol.* **104:** 1729–1733.

Fryxell, P. A., 1968. A redefinition of the tribe Gossypieae. *Bot. Gaz.* **129:** 296–308.

Fryxell, P. A., 1969. A classification of *Gossypium* L. (Malvaceae). *Taxonomy* **18:** 585–591.

Fulton, J. P., H. A. Scott, and R. Bamez, 1975. Beetle transmission of legume viruses. In *Tropical Diseases of Legumes,* J. Bird and K. Maramorosch, Eds., pp. 123–131.

Furniss, M. M., 1972. Observations on resistance and susceptibility to Douglas-fir beetles. *Proc. 2nd For. Biol. Workshop, Corvallis, Or.*

Gahukar, R. T. and H. C. Chiang, 1976. Advances in European corn borer research. In *Report of the International Project on Ostrinia nubilalis, Phase III: Results,* B. Dolinka, Ed. Agricultural Research Institute of the Hungarian Academy of Sciences, Martonvasar, pp. 125–173.

Gall, A. and J. R. Dogger, 1967. Effect of 2,4-D on the wheat stem sawfly. *J. Econ. Entomol.* **60:** 75–77.

Gallun, R. L., 1965. The Hessian fly. *U.S. Dep. Agric. Leafl.* **533:** 8 pp.

Gallun, R. L., R. Ruppel, and E. H. Everson, 1966. Resistance of small grains to the cereal leaf beetle. *J. Econ. Entomol.* **59:** 827–829.

Gallun, R. L., R. T. Everly, and W. T. Tamazaki, 1967. Yield and milling quality of Monon wheat damage by feeding of cereal leaf beetle. *J. Econ. Entomol.* **60**: 356–359.

Gallun, R. L. and J. H. Hatchett, 1968. Interrelationship Between Races of Hessian Fly, *Mayetiola destructor* (Say) and Resistance in Wheat. *Proc. Third Int. Wheat Genet. Symp. Aust. Acad. Sci., Canberra:* 258–262.

Gallun, R. L. and J. H. Hatchett, 1969. Genetic evidence of elimination of chromosomes in the Hessian fly. *Ann. Entomol. Soc. Am.* **62**: 1095–1101.

Gallun, R. L., 1972. Genetic interrelationships between host plants and insects. *J. Environ. Qual.* **1**: 259–265.

Gallun, R. L., J. J. Roberts, R. E. Finny, and F. L. Patterson, 1973. Leaf Pubescence of field grown wheat: A deterrent to oviposition by the cereal leaf beetle. *J. Environ. Qual.* **2**: 333–334.

Gallun, R. L., K. J. Starks, and W. D. Guthrie, 1975. Plant resistance to insects attacking cereals. *Ann. Rev. Entomol.,* **20**: 337–357.

Gallun, R. L., 1977. The genetic basis of Hessian fly epidemics. *Ann. N.Y. Acad. Sci.,* **287**: 223–229.

Gallun, R. L., 1977. Genetic basis of Hessian fly epidemics. In *The Genetic Basis of Epidemics in Agriculture,* P. R. Day, Ed., *Ann. N.Y. Acad. Sci.* **287**: 1–400.

Gallun, R. L. and F. L. Patterson, 1977. Monosomic analyses of wheat for resistance to Hessian fly, *Mayetiola destructor* (Say). *J. Hered.* **68**: 223–226.

Gamez, R., 1972. Los virus del frijol en Centroamerica: II. Algunas propiedades y transmission por crisomelidos del virus del mosaico rugoso del frijol. (Bean viruses in Central America: II. Some properties and chrysomelid transmission of bean rugose mosaic virus.) *Turrialba* **22**: 249–257.

Garber, R. J. and S. S. Atwood, 1945. Natural crossing in sudangrass. *J. Am. Soc. Agron.* **37**: 365–369.

Gardenhire, J. H. and H. L. Chada, 1961. Inheritance of greenbug resistance in barley. *Crop Sci.* **1**: 349–359.

Gardenhire, J. H., 1965. Inheritance and linkage studies on greenbug resistance in barley (*Hordeium vulgare*). *Crop Sci.* **5**: 28–29.

Gardenhire, J. H., N. A. Tuleen, and K. W. Stewart, 1973. Trisomic analysis of greenbug resistance in barley, *Hordeum vulgare* L. *Crop Sci.* **13**: 684–685.

Gardner, C. O., 1972. Development of superior populations of sorghum and their role in breeding programs. In *Sorghum in the Seventies,* N. G. P. Rao and L. R. House, Eds. Oxford and IBH, New Delhi, pp. 180–196.

Garrett, P. W., 1972. Resistance of eastern white pine (*Pinus strobus* L.) to the white pine weevil (*Pissodes strobi* Peck.). *Silvae Genet.* **21**: 119–121.

Gates, C. T., 1964. The effect of water stress on plant growth. *J. Aust. Inst. Agric. Sci.* **30**: 3–22.

Gawaad, A. A. A. and A. S. Soliman, 1972. Studies on *Thrips tabaci* Lindman. IX.

Resistance of nineteen varieties of cotton to *Thrips tabaci* L. and *Aphis gossypii* G. Z. *Angew. Entomol.* **70:** 93–98.

Gawaad, A. A. A., F. H. El-Gayer, A. S. Soliman, and O. A. Zaghlool, 1973. Studies of *Thrips tabaci* Lindman. X. Mechanism of resistance to *Thrips tabaci* L. in cotton varieties. Z. *Angew. Entomol.* **73:** 251–255.

Gentry, H. S., 1969. Origin of the common bean *Pheseolus vulgaris. Econ. Bot.* **23:** 55–69.

Gentry, O. W., 1965. Crop insects in Northeast Africa—Southwest Asia. *U.S. Dep. Agric. Agric. Handb.* **273:** 210 pp.

George, B. W., E. S. Raun, D. C. Peters, and C. Mendoza, 1960. Artificial medium for rearing some Lepidopterous corn insects. *J. Econ. Entomol.* **53:** 318–319.

George Washington University, 1967, *Sorghum—A Biliography of the World Covering the Years 1930–1963.* Scarecrow, Metuchen, N.J., 301 pp.

Gerhold, H. D., R. E. McDermott, E. J. Schreiner, and J. A. Winieski, Eds., 1966. *Breeding Pest Resistant Trees.* Pergamon Press, London, 505 pp.

Gibson, R. W., 1971. Glandular hairs providing resistance to aphids in certain wild potato species. *Ann. Appl. Biol.* **68:** 113–119.

Gibson, R. W., 1972. the distribution of aphids on potato leaves in relation to vein size. *Entomol. Exp. Appl.* **15:** 213–223.

Gibson, R. W., 1974. Aphid-trapping glandular hairs on hybrids of *Solanum tuberosum* and *S. bertahultii. Potato Res.* **17:** 152–154.

Gibson, R. W., 1976. Glandular hairs on *Solanum polyadenium* lessen damage by the Colorado beetle. *Ann. Appl. Biol.* **82:** 147–150.

Gibson, R. W. and R. T. Plumb, 1977. Breeding plants for resistance to aphid infestation. In *Aphids as Virus Vectors,* K. F. Harris and K. Maramorosch, Eds. Academic Press, Inc., New York, pp. 473–500.

Gilbert, B. L., J. E. Baker, and D. M. Norris, 1967. Juglone (5-hydroxy-1,4-naphthoquinone) from *Carya ovata,* a deterrent to feeding by *Scolytus multistriatus. J. Insect Physiol.* **13:** 1453–1459.

Gilbert, L. E., 1971. Butterfly-plant coevolution: Has *Passiflora adenopoda* won the selectional race with Heliconiine butterflies? *Sci.* **172:** 585–586.

Giles, P. H. and F. Ashman, 1971. A study of preharvest infestation of maize by *S. oryzae* with varieties of maize and the extent of damage caused. *Ser. Cient. Inst. Invest. Agron. Angola* **12:** 26 pp.

Glover, C. and B. Melton, 1966. Inheritance patterns of spotted alfalfa aphid resistance in Zea plants. *N.M. Agric. Exp. Stn. Res. Rep.* **127:** 4 pp.

Glover, D. V. and E. H. Stanford, 1966. Tetrasonic inheritance of resistance in alfalfa to pea aphid. *Crop Sci.* **6:** 161–165.

Glover, D., D. F. Glover, and J. E. Jones, 1975. Boll weevil and bollworm damage as affected by Upland cotton strains with different cytoplasms. *Proc. Beltwide Cotton Prod. Res. Conf.:* 99–102. National Cotton Council, Memphis, Tenn.

Golding, F. D., 1936. *Bemisia nigeriensis* Corb., a vector of cassava mosaic in southern Nigeria. *Trop. Agric. Trinidad* **13**: 182–186.

Gonzales, J. C., 1974. Resistance to the rice leaf folder, *Cnaphalocrosis medinalis* Guenee, in rice varieties. M.S. Thesis, University of the Philippines, Los Baños, 104 pp.

Gothilf, S. and S. D. Beck, 1967. Larval feeding behaviour of the cabbage looper, *Trichoplusia ni. J. Insect Physiol.* **13**: 1039–1053.

Greany, P. D., H. R. Agee, A. K. Burditt, and D. L. Chambers, 1977. Field studies on color preferences of the Caribbean fruit fly, *Anastropha suspensa* (Diptera: Tephretidae). *Entomol. Exp. Appl.* **21**: 63–70.

Green, T. R. and C. A. Ryan, 1972. Wound-induced proteinase inhibitor in plant leaves: a possible defense mechanism against insects. *Science* **175**: 776–777.

Gulati, A. M. and A. N. Turner, 1928. A note on the early history of Cotton. *Indian Centr. Cotton Comm. Bull.* **17**: *Tech. Ser.* **12**.

Gupta, P. D. and A. J. Thorsteinson, 1960. Food plant relationships of the diamondback moth Plutella maculipennis (Curt.). II. Sensory regulation of oviposition of the adult female. *Entomol. Exp. Appl.* **3**: 305–314.

Guthrie, W. D., F. F. Dicke, and C. R. Neiswander, 1960. Leaf and sheath feeding resistance to the European corn borer in eight inbred lines of dent corn. *Ohio Agric. Exp. Stn. Res. Bull.* **860**: 38 pp.

Guthrie, W. D. and G. H. Stringfield, 1961. The recovery of genes controlling corn borer resistance in a backcrossing program. *J. Econ. Entomol.* **54**: 267–270.

Guthrie, W. D., E. S. Raun, F. F. Dicke, G. R. Pesho, and S. W. Carter, 1965. Laboratory production of European corn borer egg masses. *Iowa State J. Sci.* **40**: 65–83.

Guthrie, W. D., J. L. Huggans, and S. M. Chatterji, 1970. Sheath and collar feeding resistance to second brood European corn borer in six inbred lines of dent corn. *Iowa State J. Sci.* **44**: 297–311.

Guthrie, W. D., W. A. Russell, and C. W. Jennings, 1971. Resistance of maize to second-brood European corn borer. *Proc. Ann. Corn Sorghum Res. Conf., 26th;* 165–179.

Guthrie, W. D., 1974. Techniques, accomplishments and future potential of breeding for resistance to European corn borer in corn. In *Proc. Summer Inst. Biol. Contr. Plants Ins. Dis.,* F. G. Maxwell and F. A. Harris, Eds. 359–380. University of Mississippi Press, Jackson.

Guthrie, W. D., W. A. Russell, F. L. Neumann, G. L. Reed, and R. L. Grindeland, 1975. Yield losses in maize caused by different levels of infestation of second-brood European corn borers. *Iowa State J. Res.* **50**: 239–253.

Gutierrez, M. G., 1974. Maize germplasm preservation and utilization at CIMMYT. In *Proc. Symp. on World-Wide Maize Improvement in the 70's and the role of CIMMYT.* CIMMYT, Mexico D.F., Mexico.

Habgood, R. M., 1970. Designation of physiologic races of plant pathogens. *Nature* **227**: 1268–1269.

Hackerott, H. L., T. L. Harvey, E. L. Sorensen, and R. H. Painter, 1958. Varietal differences in survival of alfalfa seedlings infested with spotted alfalfa aphids. *Agron. J.* **50:** 139–141.

Hackerott, H. L. and T. L. Harvey, 1959. Effect of temperature on spotted alfalfa aphid reaction to resistance in alfalfa aphid reaction to resistance in alfalfa. *J. Econ. Entomol.* **52:** 949–953.

Hackerott, H. L., E. L. Sorensen, T. L. Harvey, E. E. Ortman, and R. L. Painter, 1963. Reaction of alfalfa varieties to pea aphids in the field and greenhouse. *Crop Sci.* **3:** 298–301.

Hackerott, H. L. and T. L. Harvey, 1971. Greenbug injury to resistant and susceptible sorghums in the field. *Crop Sci.* **11:** 641–643.

Hagen, A. F. and F. N. Anderson, 1967. Nutrient imbalance and leaf pubescence in corn as factors influencing leaf injury by the adult western corn rootworm. *J. Econ. Entomol.* **60:** 1071–1073.

Hahn, S. K., 1968. Resistance of barley (*Hordeum vulgare* L. Emend. Lam.) to cereal leaf beetle (*Oulema melanopus* L.). *Crop Sci.* **8:** 461–464.

Hahn, S. K., 1976. Personal communication. International Institute for Tropical Agriculture, Ibadan, Nigeria.

Hall, R. C., 1937. Growth and yield of shipmast on Long Island and its relative resistance to locust borer injury. *J. For.* **35:** 721–727.

Hamamura, Y., 1970. The substances that control the feeding behavior and growth of the silkmoth, *Bombyx mori.* In *Control of Insect Behavior by Natural Products,* D. L. Wood, R. M. Silverstein, and M. Nakajima. Eds., Academic Press, Inc., New York, pp. 55–80.

Handy, R. B., 1896. History and general statistics of cotton; the cotton plant. *U.S. Dep. Agric. Bull.* **33:** 17–66.

Hanny, B. W. and C. D. Elmore, 1974. Amino acid composition of cotton nectar. *J. Agric. Food Chem.* **22:** 476–478.

Hanover, J. W., 1975. Comparative physiology of eastern and western white pines: cleoresin composition and viscosity. *For. Sci.* **21:** 214–221.

Hanover, J. W., 1976. Physiology of tree resistance to insects. *Ann. Rev. Entomol.* **20:** 75–95.

Hanson, C. H., B. L. Norwood, C. C. Blickenstaff, and R. S. Vandenburgh, 1963. Recurrent phenotypic selection for resistance to potato leafhopper yellowing alfalfa. *Agron. Abstr.:* 80.

Hanson, C. H., 1972. *Alfalfa Science Technology.* American Society of Agronomy, Madison, Wis., 812 pp.

Hanson, C. H., T. H. Busbice, R. R. Hill, O. J. Hunt, and A. J. Oakes, 1972. Directed mass selection for developing multiple pest resistance and conserving germplasm in alfalfa. *J. Environ. Qual.* **1:** 106–111.

Hanson, W. D. and H. F. Robinson, 1963. Statistical genetics and plant breeding. *Natl. Acad. Sci., Natl. Res. Council Publ.* **982.**

Harlan, J. R. and J. M. deWet, 1971. Toward a rational classification of cultivated plants. *Taxonomy* **20:** 509–517.

Harlan, J. R., 1972. Genetics of disaster. *J. Environ. Qual.* 1(3): 212.

Harlan, J. R., 1972a. Genetics of disaster. *J. Environ. Qual.* **1:** 212–215.

Harlan, J. R., 1972b. Genetic resources in sorghum. In *Sorghum in the Seventies,* N. G. P. Rao and L. R. House, Eds., Oxford and IBH Publ. Co., New Delhi.

Harlan, J. R., 1975. Our vanishing genetic resources. *Science* **188:** 618–621.

Harlan, J. R., 1976. Genetic resources in wild relatives of crops. *Crop Sci.* **16:** 329–333.

Harlan, J. R., 1977. Sources of genetic defense. In *The Genetic Basis of Epidemics in Agriculture.* P. R. Day, Ed., *Ann. N.Y. Acad. Sci.* **287:** 1–400.

Harland, S. C., 1970. Gene pools in the New World tetraploid cottons. In *Genetic Resources in Plants—Their Exploration and Conservation,* O. H. Frankel and E. Bennett, Eds. F. A. Davis Co., Philadelphia.

Harpaz, I., 1955. Bionomics of *Therioaphis maculata* (Buckton) in Israel. *J. Econ. Entomol.,* **48:** 668–671.

Harrewijn, P., 1970. Reproduction of the aphid *Myzus persicae* related to mineral nutrition of potato plants. *Entomol. Exp. Appl.,* **13:** 307–319.

Harries, F. H., 1966. Reproduction and mortality of the two-spotted spider mite on fruit seedlings treated with chemicals. *J. Econ. Entomol.* **59:** 501–506.

Harries, H. C. and R. A. de Poerck, 1971. Coconut—classification of genetic resources for introduction and conservation. *Plant Genet. Resour. Newsl.* **25:** 26–27.

Harrington, C. D., 1941. Influence of aphid resistance in peas upon aphid development, reproduction and longevity. *J. Agric. Res.* **62:** 461–466.

Harrington, C. D., 1945. Biological races of the pea aphid. *J. Econ. Entomol.* **38:** 12–22.

Harris, H. B., G. W. Burton, and B. J. Johnson, 1965. Effects of gamma radiation on two varieties of *Sorghum vulgare* Pers., *Ga. Agric. Exp. Stn. Tech. Bull.* **40.**

Harris, K. F. and R. H. E. Bradley, 1973a. Importance of leafhairs in the transmission of tobacco mosaic virus by aphids. *Virology* **52:** 295–300.

Harris, K. F. and R. H. E. Bradley, 1973b. Tobacco mosaic virus: can aphids inoculate it into plants with their mouthparts? *Phytopathology* **63:** 1343–1345.

Harris, K. F., 1977. An ingestion-egestion hypothesis of noncirculative virus transmission. In *Aphids as Virus Vectors,* K. F. Harris and K. Maramorosch, Eds. Academic Press, Inc., New York, pp. 165–220.

Harris, K. F. and K. Maramorosch, Eds., 1977. *Aphids as Virus Vectors.* Academic Press, Inc., New York, 559 pp.

Harris, K. F., 1978. Aphid-borne viruses: ecological and environmental aspects. In *Viruses and Environment,* E. Kurstak and K. Maramorosch, Eds. Academic Press, New York, in press.

Harris, T. W., 1841. Harvest flies, etc. (Hemiptera, Homoptera). A report on the insects of Massachusetts injurious to vegetation **1841:** 186.

Harvey, T. L. and H. L. Hackerott, 1956. Apparent resistance to the spotted alfalfa aphid selected from seedlings of susceptible alfalfa varieties. *J. Econ. Entomol.* **49:** 289–291.

Harvey, T. L., H. L. Hackerott, and E. L. Sorensen, 1972. Pea aphid resistant alfalfa selected in the field. *J. Econ. Entomol.* **65:** 1661–1663.

Harwood, R. R., R. Y. Grandados, S. Jamornman, and R. G. Granados, 1973. Breeding for resistance to sorghum shoot fly in Thailand, in *Control of Sorghum Shoot Fly.* Oxford and IBH Publishers, New Delhi.

Hatchett, J. R. and R. L. Gallun, 1970. Genetics of the ability of the Hessian fly, *Mayetiola destructor,* to survive on wheats having different genes for resistance. *Ann. Entomol. Soc. Am.* **63:** 1400–1407.

Hatheway, W. H., 1957. Races of maize in Cuba. *Natl. Acad. Sci., Natl. Res. Council Pub.* **453:** 7.

Havens, J. N., 1972. Observations on the Hessian fly. *Soc. Agron. N.Y. Trans.* **1:** 89–107.

Hawkins, B. S., H. A. Peacock, and T. E. Steele, 1966. Thrips injury to Upland cotton (*Gossypium hirsutum* L.) varieties. *Crop Sci.* **6:** 256–258.

Hayes, W. P., 1922. Observations on insects attacking sorghum. *J. Econ. Entomol.* **51:** 349–356.

Hayes, H. K., F. H. Immer, and D. C. Smith, 1955. *Methods of Plant Breeding.* McGraw-Hill Book Co., Inc., New York, 551 pp.

Hedin, P. A., F. G. Maxwell, and J. N. Jenkins, 1974. Insect plant attractants, feeding stimulants, repellents, deterrents, and other related factors affecting insect behavior. In *Proc. Summer Inst. Biol. Control Plant Insects Dis.* F. G. Maxwell and F. A. Harris, Eds.: 494–527. University Press of Mississippi, Jackson.

Heftman, E., 1975. Functions of steroids in plants. *Phytochem.* **14:** 891–901.

Heilman, M. D., M. J. Lukefahr, L. N. Namken, and J. W. Norman, 1977. Field evaluation of a short season production system in the lower Rio Grande valley of Texas. *Proc. Beltwide Cotton Prod. Res. Conf.:* 80–83. National Cotton Council, Memphis, Tenn.

Heimburger, C., 1963. The breeding of white pine for resistance to weevil. *Proc. World Consult. For. Genet., Stockholm, FAO/For. Gen.* **63-6b**(5): 2 pp.

Heinrichs, E. A., 1979. Control of leafhopper and planthopper vectors of rice viruses. In *Leafhopper Vectors and Plant Disease Agents,* K. Maramorosch and F. K. Harris, Eds. Academic Press, Inc., New York, in press.

Henneberry, T. J., 1962. The effect of host-plant nitrogen supply and age of leaf tissue on the fecundity of the two-spotted spider mite. *J. Econ. Entomol.* **55:** 799–800.

Henneberry, T. J., 1963. Effect of host plant condition and fertilization on the two-spotted spider mite fecundity. *J. Econ. Entomol.* **56:** 503–505.

Heyn, C. C., 1963. The annual species of *Medicago*. Hebrew University, Jerusalem, 154 pp.

Hidaka, T., P. Vungsilabutr, and S. Rajamani, 1977. Geographical distribution of the rice gall midge *Orseolia oryzae* (Wood-Mason) (Diptera: Cecidomyiidae). *Appl. Entomol. Zool.* **12:** 4–8.

Hill, R. R., Jr. and R. C. Newton, 1972. A method for mass screening alfalfa for meadow spittlebug resistance in the greenhouse during the winter. *J. Econ. Entomol.* **65:** 621–623.

Hintz, S. D. and J. T. Schulz, 1969. The effect of selected herbicides on cereal aphids under greenhouse conditions. *Proc. North Cent. Branch Entomol. Soc. Am.* **24:** 114–117.

Holder, D. G., J. N. Jenkins, and F. G. Maxwell, 1968. Duplicate linkage of glandless and nectariless genes in Upland cotton, *Gossypium hirsutum* L. *Crop Sci.* **8:** 577–580.

Holmes, N. D. and L. K. Peterson, 1957. Effect of continuous rearing on Rescue wheat on survival of the wheat stem sawfly, *Cephus cinctus* Nort. (Hymenoptera: Cephidae). *Can. Entomol.* **89:** 363–365.

Holmes, N. D., R. I. Larson, L. K. Peterson, and M. D. MacDonald, 1960. Influence of periodic shading on the length and solidity of the internodes of Rescue wheat. *Can. J. Plant Sci.* **40:** 183–187.

Holmes, N. D. and L. K. Peterson, 1960. The influence of host on oviposition by the wheat stem sawfly, *Cephus cinctus* Nort. (Hymenoptera: Cephidae). *Can. J. Plant Sci.* **40:** 29–46.

Honeyborne, C. H. B., 1969. Performance of *Aphis fabae* and *Brevicoryne brassicae* on plants treated with growth regulators. *J. Sci. Food Agric.* **20:** 388–390.

Horber, E., 1972. Plant resistance to insects. *Agric. Sci. Rev.* **10:** 1–18.

Hormchong, T., 1967. Studies on the development of insect resistance sorghum varieties and hybrids. Ph.D. Thesis, Oklahoma State University, Stillwater, 87 pp.

Horowitz, S. and A. H. Marchioni, 1940. Herencia de la resistencia a la langosta en el maiz "Amargo." *An. Inst. Fito, Santa Catalina* **2:** 27–52.

House, H. L., 1976. Artificial diets for insects: A compilation of references with abstracts. *Can. Dept. Agric., Inf. Bull. Res. Inst.* **5:** 163 pp.

House, H. L., P. Singh, and W. W. Batsch, 1971. Artificial diets for insects: a compilation of references with abstracts. *Can. Dept. Agric. Inf. Bull. Res. Inst.* **7:** 156 pp.

Howe, W. L. 1949. Factors affecting the resistance of certain cucurbits to the squash borer. *J. Econ. Entomol.* **42:** 321–326.

Howe, W. L. and O. F. Smith, 1957. Resistance to the spotted alfalfa aphid in Lahontan alfalfa. *J. Econ. Entomol.* **50:** 320–324.

Howe, W. L. and G. R. Pesho, 1960a. Influence of plant age on the survival of alfalfa varieties differing in resistance to the spotted alfalfa aphid. *J. Econ. Entomol.* **53:** 142–144.

Howe, W. L. and G. R. Pesho, 1960b. Spotted alfalfa aphid resistance in mature growth of alfalfa varieties. *J. Econ. Entomol.* **53**: 234–238.

Howe, W. L. and G. R. Manglitz, 1961. Observations on the clover seed chalcid as a pest of alfalfa in eastern Nebraska. *Proc. North Cent. Branch Entomol. Soc. Am.* **16**: 49–51.

Howe, W. L., W. R. Kehr, and C. O. Calkins, 1965. Appraisal for combined pea aphid and spotted alfalfa aphid resistance in alfalfa. *Univ. Nebr. Agric. Exp. Stn. Res. Bull.* **221**: 31 pp.

Howitt, A. J. and R. H. Painter, 1956. Field and greenhouse studies regarding the sources and nature of sorghum (*Sorghum vulgare* Pers.) resistance to the corn leaf aphid, *Rhopalosiphum maidis* (Fitch). Kans. Agric. Exp. Stn. Bull. **82**: 38 pp.

Hoxie, R. P., S. G. Wellso, and J. A. Webster, 1975. Cereal leaf beetle response to wheat trichome length and density. *Environ. Entomol.* **4**: 365–370.

Hsu, Ting-wen, 1975. On the origin and phylogeny of cultivated barley with reference to the discovery of Ganze wild two-rowed barley *Hordeum spontaneum* C. Koch. *Acta Genet. Sinica* **2**: 137 (abstr.).

Huber, L. L. and G. H. Stringfield, 1940. Strain susceptibility to the European corn borer and the corn leaf aphid in maize. *Science* **92**: 172.

Huber, L. L. and G. H. Stringfield, 1942. Aphid infestation strains of corn as an index to their susceptibility to corn borer attack. *J. Agric. Res.* **64**: 283–291.

Hubert-Dahl, M. L., 1975. Anderung des Wirtswahlverhaltons dreier biotypen Von *Acyrthosiphon pisum* (Harris) nach anzueht auf versehiedenen virtspflanzen. *Beitr. Entomol.* **25**: 77–83.

Huettel, M. D., 1976. Monitoring the quality of laboratory-reared insects: a biological and behavioral perspective. *Environ. Entomol.* **5**(5): 807–814.

Hummel, K. and K. Staesche, 1962. Die Verbreitung der Haartypen in den naturlichen Verwandtschaftsgruppen. *Handbuch der Pflanzenanatomie*, Vol. 4, Part 5, *Histologie,* W. Zimmerman and P. G. Ozenda, Eds. Gebruder Borntraeger, Berlin, pp. 207–250.

Hunt, O. J., R. N. Peaden, H. L. Carahan, and F. V. Lieberman, 1966. Registration of Washoe alfalfa. *Crop Sci.* **6**: 610.

Hunt, O. J., H. Baenzinger, B. J. Hartman, D. H. Heinrichs, E. S. Horner, I. I. Kawaguchi, B. A. Melton, M. H. Schonhorst, and B. D. Thyr, 1978. Improved breeding lines of alfalfa. *U.S. Dep. Agric. Tech. Bull.*: in press.

Hunter, R. C., T. F. Leigh, C. Lincoln, B. A. Waddle, and L. A. Bariola, 1965. Evaluation of selected cross-section of cottons for resistance to the boll weevil. *Arkansas Agric. Exp. Stn. Bull.* **700**.

Hutchinson, J. B., 1938. The distribution of *Gossypium* and the evolution of the commercial cotton. *First Cotton Conf., Indian Cent. Cotton Comm., Bombay*.

Hutchinson, J. B., 1949. The dissemination of cotton in Africa. *Emp. Cott. Grow. Rev.* **26**: 256–270.

Hutchinson, J. B., 1951. Interspecific differentiation in *Gossypium hirsutum*. *Hered.* **5**: 161–193.

Hutchinson, J. B., 1962. The history and relationships of the world's cottons. *Endeavour* **21**(81): 5–15.

Ibrahim, M. A., 1954. Association tests between chromosomal interchanges in maize and resistance to the European corn borer. *Agron. J.* **46**: 293–298.

Ikeda, Toshiya, F. Matsumura, and D. M. Benjamin, 1977. Chemical basis for feeding adaptation of pine sawflies *Neodiprion rugifrons* and *Neodiprion swainei*. *Science* **197**: 497–499.

Inglett, G. E., Ed., 1970. *Corn: Culture, Processing, Products*. Avi Publishing, Westport, Conn.

Institute de Recherches Agronomiques Tropicales et des Cultures Vivieres, 1966. *Manioc. C.R. Anal. Trav. Relises 1965–1966*. 133 pp.

International Board for Plant Genetic Resources, 1976. Consultative group on international agricultural research—Priorities among crops and regions.

International Institute of Tropical Agriculture (IITA), 1977.*Annual Report for 1976*. Ibadan, Nigeria.

International Rice Research Institute, 1975. *Research Highlights for 1974*. Los Baños, Philippines, p. 90.

International Rice Research Institute, 1976. *Annual Report for 1975*. Los Baños, Philippines, pp. 101–110.

Isaak, A., E. L. Sorensen, and E. E. Ortman, 1963. Influence of temperature and humidity on resistance in alfalfa to the spotted alfalfa aphid and pea aphid. *J. Econ. Entomol.* **56**: 53–57.

Isaak, A., E. L. Sorensen, and R. H. Painter, 1975. Stability of resistance to pea aphid and spotted alfalfa aphid in several clones under various temperature regimes. *J. Econ. Entomol.* **58**: 140–143.

Isely, D. 1928. The relation of leaf color and leaf size to boll weevil infestation. *J. Econ. Entomol.* **21**: 553–559.

Isely, D., 1948. *Methods of Insect Control*, Part I, 3rd ed. Burgess Publishing Co., Minneapolis, Minn., 134 pp.

Ishii, S. and C. Hirano, 1963. Growth responses of larvae of the rice stem borer to rice plants treated with 2,4-D. *Entomol. Exp. Appl.* **6**: 257–262.

Isogai, A., C. Chang, S. Murakoshi, and A. Suzuki, 1973. Screening search for biologically active substances to insects in crude drug plants. *J. Agric. Chem. Soc. Japan* **47**: 443–447.

Israel, P., 1967. Varietal resistance to rice stem borers in India. In *The Major Insect Pests of the Rice Plant, Proc. Symp. Int. Rice Res. Inst., Los Baños, 1964*. Johns Hopkins Press, Baltimore, Md., pp. 391–403.

Israel, P., 1974. Current status of research and control of rice gall midge in India. *Plant Prot. News* **3**: 41–45.

James, W. C., C. S. Shih, L. C. Callbeck, and W. A. Hodgson, 1971. Evaluation of a method used to estimate yield loss of potatoes caused by late blight. *Phytopathology* **61**: 1471–1476.

James, W. C., 1973. Interplot interference in field experiments with late blight of potato (*Phytophthora infestans*). *Phytopathology* **63**: 1269–1275.

James, W. C., 1974. Assessment of plant diseases and losses. *Ann. Rev. Phytopathol.* **12**: 27–48.

James, W. C., 1976. Representational errors due to interplot interference in field experiments with late blight of potato. *Phytopathol.* **66**: 695–700.

Japan Cotton Trader's Association, 1976. *Japan Cotton Statistics and Related Data, 1976.* The Cotton. Economic Research Institute.

Jarvis, J. L. and W. R. Kehr, 1966. Population counts vs. nymphs per gram of plant material in determining degree of alfalfa resistance to the potato leafhopper. *J. Econ. Entomol.* **59**: 427–430.

Jenkins, J. N. and W. L. Parrott, 1971. Effectiveness of frego bract as a boll weevil resistance character in cotton. *Crop Sci.* **11**: 739–743.

Jenkins, J. N., W. L. Parrott, and J. C. McCarty, Jr., 1973. The role of a boll weevil resistant cotton in pest management research. *J. Environ. Qual.* **2**: 337–340.

Jenkins, J. N., F. G. Maxwell, and W. L. Parrott, 1964. A technique for measuring certain aspects of antibiosis in cotton to the boll weevil. *J. Econ. Entomol.* **57**: 679–781.

Jenkins, J. N., 1975. Application of modeling to cotton improvement. *Proc. Beltwide Cotton Prod. Res. Conf.*, pp. 161–164.

Jenkins, J. N., 1976. Boll Weevil resistant cottons. In *Boll Weevil Suppression, Management and Elimination Technology*. U.S. Dep. Agric. ARS **S-71**: 45–49.

Jenkins, J. N., J. C. McCarty, Jr., and W. L. Parrott, 1977. Inheritance of resistance to tarnished plant bugs in a cross of Stoneville 213 by Timok 811. *Proc. Beltwide Cotton Prod. Res. Conf.*: 97 (abstract). National Cotton Council, Memphis, Tenn.

Jenkins, M. T., 1940. The segregation of genes affecting yield of grain in maize. *Agron. J.* **32**: 55–63.

Jennings, C. W., W. A. Russell, and W. D. Guthrie, 1974. Genetics of resistance in maize to first- and second-brood European corn borer. *Crop Sci.* **14**: 394–498.

Jennings, P. R. and A. Pineda, 1970. Screening rice for resistance to the planthopper, *Sogatodes orizicola* (Muir.). *Crop Sci.* **10**: 687–689.

Jennings, P. R. and A. Pineda, 1970. *Sogatodes orizicola Resistance in Rice Varieties*. Centro Internacional Agricultura Tropical, Palmira, Colombia.

Jennings, P. R. and J. H. Cock, 1977. Centres of origin of crops and their productivity. *Econ. Bot.* **31**: 51–54.

Jensen, N. F. 1952. Intra-varietal diversification in cat breeding, *Agron. J.* **44**: 30–34.

Jensen, N. F., 1970. A diallel selective mating system for cereal breeding. *Crop Sci.* **10**: 629–635.

Jermy, T., F. E. Hanson, and V. G. Dethier, 1968. Induction of specific food preference in *Lepidopterous* larvae. *Entomol. Exp. Appl.* **11**: 211-230.

Jermy, T., 1976. *The Host-Plant in Relation to Insect Behavior and Reproduction*. Plenum Press, New York, 322 pp.

Jermy, T., 1976. Insect-host plant relationship—coevolution or sequential evolution? *Symp. Biol. Hung.* **16**: 109-113.

Jewett, D. M., F. Matsumura, and H. C. Coppel, 1976. Sex pheromone specificity in the pine sawflies: interchange of acid moieties in an ester. *Science* **192**: 51-53.

Johnson, B., 1953. The injurious effects of the hooked epidermal hairs of French beans (*Phaseolus vulgaris* L.) on *Aphis craccivora* Koch. *Bull Entomol. Res.* **44**: 779-788.

Johnson, H. B., 1975. Plant pubescence: An ecological perspective. *Bot. Rev.* **41**: 233-258.

Johnson, J. W., D. T. Rosenow, and G. L. Teetes, 1973. Resistance to the sorghum midge in converted exotic sorghum cultivars. *Crop Sci.* **13**: 754-755.

Johnson, J. W., G. L. Teetes, and C. A. Schaefer, 1976. Greenhouse and field technique for evaluating resistance of sorghum cultivars to the greenbug. *Southwestern Entomol.* **1**: 150-154.

Johnson, P. C. and R. E. Denton, 1975. Outbreaks of the western spruce budworm in the American northern Rocky Mountain area from 1922 through 1971. *U.S. Dep. Agric. For. Serv. Gen. Tech. Rep.* **INT-20**: 144 pp.

Johnson, T., 1961. Man-guided evolution in plant rusts. *Science* **133**: 357-362.

Johnson, W. H., 1926. *Cotton and its Production*. Macmillan and Co., Ltd., London.

Jones, A., P. D. Dukes, and F. P. Cuthbert, 1976. Mass selection in sweet potato; breeding for resistance to insects and diseases and for horticultural characteristics. *J. Am. Soc. Hortic.* **101**: 701-704.

Jones, B. F., E. L. Sorensen, and R. H. Painter, 1958. Tolerance of alfalfa clones to the spotted alfalfa aphid. *J. Econ. Entomol.* **61**: 1046-1050.

Jones, J. E., L. D. Newsom, and K. W. Tipton, 1964. Differences in boll weevil infestation among different biotypes of Upland cotton. *Proc. 16th Ann. Cotton Impr. Conf.*: 48-55. National Cotton Council, Memphis, Tenn.

Jones, J. E., 1972. Effect of morphological characters of cotton on insects and pathogens. *Proc. Beltwide Cotton Prod. Res. Conf.*: 88-92. National Cotton Council, Memphis, Tenn.

Jones, J. E., W. D. Caldwell, M. R. Milam, and D. F. Clower, 1976. Gumbo and Pronto—Two new open-canopy varieties of cotton. *La. Agric. Exp. Stn. Circ.*: 103.

Jones, J. W., 1975. A simulation model of boll weevil population dynamics as influenced by cotton crop status. Ph.D. Dissertation, Agricultural and Biological Engineering Department, North Carolina State University, Raleigh, N.C.

Jones, L. G., F. N. Briggs, and R. A. Blanchard, 1950. Inheritance of resistance to the pea aphid in alfalfa hybrids. *Hilgardia* **20:** 9–17.

Josephson, L. M., S. E. Bennett, and E. E. Burgess, 1966. Methods of artificially infesting corn with the corn earworm and factors influencing resistance. *J. Econ. Entomol.* **59:** 1322–1324.

Joshi, A. B. and S. B. Rao, 1959. The problem of breeding jassid resistant varieties of cotton in India. *Indian Cotton Grow. Rev.* **13:** 270–279.

Jotwani, M. G., S. P. Singh, and S. Chaudhari, 1971. Relative susceptibility of some sorghum lines to midge damage. Investigations on Insect Pests of Sorghum and Millets (1965–1970). *Div. Entomol. IARI, Final Tech. Rep.:* 123–130.

Jugenheimer, R. W., 1976. *Corn Improvement, Seed Production, and Uses.* John Wiley & Sons, Inc., New York.

Juliano, B. O., 1964. Hyroscopic equilibria of rough rice. *Cereal Chem.* **41:** 191–197.

Kajiura, M. and Y. Machida, 1961. Breeding experiment of new varieties of Japanese chestnut resistant to chestnut gall wasp (*Dryocosmus kuriphilus*). *Jap. J. Breed.* **11:** 73–76.

Kalode, M. B., 1971. Biochemical basis of resistance or susceptibility to brown planthopper and green leafhopper in some rice varieties. *IRRI Saturday Seminar.*

Kamel, S. A., 1965. Relationship between leaf hairiness and resistance to cotton leaf worm. *Emp. Cotton Grow. Rev.* **42:** 41–48.

Kamel, S. A. and F. Y. Elkassaby, 1965. Relative resistance of cotton varieties in Egypt to spider mites, leafhoppers and aphids. *J. Econ. Entomol.* **58:** 209–212.

Kaminkado, T., C. F. Change, S. Murakoshi, A. Sakurai, and S. Tamura, 1975. Isolation and structure elucidation of growth inhibitors on silkworm larvae from *Magnolia kobus*. *Agric. Biol. Chem.***39:** 833–886.

Kaukis, K. and O. J. Webster, 1956. Effects of thermal neutrons on dormant seeds of *Sorghum vulgare* Pers. *Agron. J.* **48:** 401–406.

Keaster, A. J., M. S. Zuber, and R. W. Straub, 1972. Status of resistance to corn earworm. *Proc. North Cent. Branch Entomol. Soc. Am.* **27:** 95–98.

Keep, E. and R. L. Knight, 1967. A new gene from *Rubus occidentali* L. for resistance to strains 1, 2, and 3 of the rubus aphid, *Amphorophora rubi* Kalt. *Euphytica* **16:** 309–314.

Keep, E., R. L. Knight, and J. H. Parker, 1969. Further data on resistance to the Rubus aphid, *Amphorophora rubi* (Ketb.). *Rep. E. Malling Res. Stn.:* 129–131.

Kehr, W. R., D. K. Barnes, E. L. Sorensen, W. H. Skrdle, C. H. Hanson, A. D. Miller, T. E. Thompson, I. T. Carlson, L. J. Elling, R. L. Taylor, M. D. Rumbaugh, E. T. Bingham, D. E. Brown, and M. K. Miller, 1975. Registration of alfalfa germplasm pools NC 83-1 and NC83-2 (Reg. Nos. GP45 and GP46). *Crop Sci.* **15:** 604–605.

Kennedy, J. S., K. P. Lamb, and C. O. Booth, 1958. Responses of *Aphis fabae* Scop. to water shortage in hostplants in pots. *Entomol. Exp. Appl.* **1:** 274–291.

Kennedy, J. S. and C. O. Booth, 1959. Responses of *Aphis fabae* Scop. to water shortage in hostplants in the field. *Entomol. Exp. Appl.* **2:** 1–11.

Kennedy, J. S., C. O. Booth, and W. J. S. Kershaw, 1961. Host findings by aphids in the field. III. Visual attraction. *Ann. Appl. Biol.* **49:** 1–21.

Kennedy, J. S., M. F. Day, and V. F. Eastop, 1962. *A Conspectus of Aphids as Vectors of Plant Viruses*. Commonwealth Institute of Entomology, London.

Kennedy, J. S., 1977. Olfactory responses to distant plants and other odor sources. In *Chemical Control of Insect Behavior*, H. H. Shorey and J. J. McKelvey, Eds. John Wiley & Sons, Inc., New York, pp. 67–91.

Khush, G. S. and H. M. Beachell, 1972. Breeding for disease and insect resistance at IRRI. In *Rice Breeding*. IRRI, Los Baños, Philippines, pp. 309–322.

Khush, G. S., 1977. Disease and insect resistance in rice. *Adv. Agron.*: **29:** 265–341.

Khush, G. S., 1977. Breeding for resistance in rice. *Ann. N.Y. Acad. Sci.* **28:** 296–308.

Khush, G. S., 1977b. Genetics of and breeding for resistance to brown planthopper. In *Brown Planthopper Symp.* IRRI, Los Baños, Philippines. Mimeographed.

Khush, G. S., M. D. Pathak, and G. S. Sidhu, 1977. Genetics and breeding for resistance to brown planthopper. In *Brown Planthopper Symp.*, IRRI, Los Baños, Philippines, April 18–22, 1977: 15 pp.

Kihara, H., 1919. Uber cytologische studien bei einigen Getreidearten. I. Species-Bastarde des Weizens und Weinzenroggen-Bastarde. *Bot. Mag. (Tokyo)* **33:** 17–38.

Kilian, L. and M. W. Nielson, 1971. Differential effects of temperature on the biological activity of four biotypes of the pea aphid. *J. Econ. Entomol.* **64:** 153–155.

Kim, M., H. Koh, T. Ichikawa, H. Fukami, and S. Ishii, 1975. Antifeedant of barnyard grass against the brown planthopper *Nilaparvata lugens* (Stol) (Homoptera: Delphacidae). *Appl. Entomol. Zool.* **10:** 116–122.

Kim, M., H. Koh, T. Obata, H. Fukami, and S. Ishii, 1976. Isolation and identification of trans-aconitic acid as the antifeedant in barnyard grass against the brown planthopper, *Nilaparvata lugens* (Stol) (Homoptera: Delphacidae). *Appl. Entomol. Zool.* **11:** 53–57.

Kindler, S. D. and R. Staples, 1969. Behavior of the spotted alfalfa aphid on resistant and susceptible alfalfas. *J. Econ. Entomol.* **62:** 474–479.

Kindler, S. D. and W. R. Kehr, 1970. Field tests of alfalfa selected for resistance to potato leafhopper in the greenhouse. *J. Econ. Entomol.* **63:** 1464–1467.

Kindler, S. D. and R. Staples, 1970a. Nutrients and the reaction of two alfalfa clones to the spotted alfalfa aphid. *J. Econ. Entomol.* **63:** 938–940.

Kindler, S. D. and R. Staples, 1970b. The influence of fluctuating and constant temperatures, photoperiod, and soil moisture on the resistance of alfalfa to the spotted alfalfa aphid. *J. Econ. Entomol.* **63:** 1198–1201.

Kindler, S. D. and W. R. Kehr, 1974. Evaluating potato leafhopper yellowing resistance, in *Standard Tests to Characterize Pest Resistance in Alfalfa Varieties. U.S. Dep. Agric. ARS* NC-19: 23 pp.

Kircher, H. W., W. B. Heed, J. S. Russell, and J. Grove, 1967. Senita cactus alkaloids: their significance to Sonoran Desert *Drosophila* ecology. *J. Insect Physiol.* **13**: 1869–1874.

Kircher, H. W., R. L. Misiorowski, and F. V. Lieberman, 1970. Resistance of alfalfa to the spotted alfalfa aphid. *J. Econ. Entomol.* **63**: 964–969.

Kirk, V. M. and A. Manwiller, 1964. Rating dent corn for resistance to rice weevils. *J. Econ. Entomol.* **57**: 850–852.

Kirk, V. M., 1965. Some flight habits of the rice weevil. *J. Econ. Entomol.* **58**: 155–156.

Kishaba, A. N. and G. R. Manglitz, 1965. Non-preference as a mechanism of sweetclover and alfalfa resistance to the sweetclover aphid and spotted alfalfa aphid. *J. Econ. Entomol.* **58**: 566–569.

Kishaba, A. N., G. W. Bohn, and E. H. Toba, 1971. Resistance to *Aphis gossypii* in muskmelon. *J. Econ. Entomol.* **64**: 935–937.

Klement, W. J. and N. M. Randolph, 1960. The evaluation of resistance of seedling alfalfa varieties and strains to the spotted alfalfa aphid, *Therioaphis maculata. J. Econ. Entomol.* **60**: 667–669.

Klun, J. A. and T. A. Brindley, 1966. Role of 6-methoxybenzoxazolinone in inbred resistance of host plant (maize) to first brood larvae of European corn borer. *J. Econ. Entomol.* **59**: 711–718.

Klun, J. A., C. L. Tipton, and T. A. Brindley, 1967. 2,4-dihydroxy-7-methoxy-1,4-benzoxazin-3-one (DIMBOA), an active agent in the resistance of maize to the European corn borer. *J. Econ. Entomol.* **60**: 1529–1533.

Klun, J. A. and J. F. Robinson, 1969. Concentration of two 1,4-benzoxazinones in dent corn at various stages of development of the plant and its relation to resistance of the host plant to the European corn borer. *J. Econ. Entomol.* **62**: 214–220.

Knapp, J. L., P. A. Hedin, and W. A. Douglas, 1965. Amino acids and reducing sugars in silks of corn resistant or susceptible to corn earworm. *Ann. Entomol. Soc. Am.* **58**: 401–402.

Knapp, J. L., F. G. Maxwell, and W. A. Douglas, 1967. Possible mechanisms of resistance of dent corn to the corn earworm. *J. Econ. Entomol.* **60**: 33–36.

Knerer, G. and C. E. Atwood, 1973. Diprionid sawflies: polymorphism and speciation. *Science* **179**: 1090–1099.

Knight, R. L., E. Keep, and J. B. Briggs, 1959. Genetics of resistance to *Amphorophora rubi* (Kalt.) in raspberry. I. The gene A_1 from Baumforrh. *Am. J. Genet.* **56**: 1–20.

Knight, R. L., E. Keep, and J. B. Briggs, 1960. Genetics of resistance to *Amphorophora rubi* (Kalt.) in the raspberry. II. The genes A_2–A_7 from the American variety, Chief. *J. Genet. Res. Cam.* **1**: 319–331.

Knight, R. L. and F. H. Alston, 1974. Pest resistance in fruit breeding. In *Biology of Pests and Disease Control*, D. Price-Jones and M. E. Solomon, Eds. Blackwell, Oxford, pp. 77-86.

Knipling, E. F., 1964. The potential role of the sterility method for insect population control with special reference to combining this method with conventional methods. *U.S. Dep. Agric. ARS* **33-98**: 54 pp.

Koehler, P. G., 1971. An association between the alfalfa weevil's larval growth response and adult feeding response to its host plant. *Ann. Entomol. Soc. Am.* **64**: 1230-1233.

Kogan, M., 1972. Intake and utilization of natural diets by the Mexican bean beetle, Epilachna varivestis. A multivariate analysis. In *Insect and Mite Nutrition: Significance and Implication in Ecology and Pest Management*, J. G. Rodriguez, Ed. North-Holland, Amsterdam, pp. 107-126.

Kogan, M., 1975. Plant resistance in pest management. In *Introduction to Insect Pest Management*, R. L. Metcalf, and W. Luckmann, Eds. John Wiley & Sons, Inc., New York, pp. 103-146.

Kogan, M., 1976. Resistance in soybean to insect pests. In *Expanding the Use of Soybeans*, R. M. Goodman, Ed. Proceedings of a conference for Asia and Oceania, Chiang Mai, Thailand. University of Illinois, Urbana, INTSOY.

Kogan, M., 1977. The role of chemical factors in insect/plant relationships. *Proc. Int. Cong. Entomol.* **15**: 211-227.

Kogan, M. and E. E. Ortman, 1978. Antixenosis—a new term proposed to replace Painter's "Nonpreference" modality of resistance. *ESA Bull.* **24**.

Kondo, F., A. H. McIntosh, S. B. Padhi, and K. Maramorosch, 1976. Electron microscopy of a new plant-pathogenic spiroplasma isolated from *Opuntia. 34th Ann. Proc. Electron Microsc. Soc. Am.*: 56-57.

Koyama, T. and J. Hirao, 1971. Varietal resistance of rice to the rice stem maggot. In *Symp. Trop. Agric. Res., July 19-24, 1971, Tokyo, Japan*: 251-266.

Krapovickas, A., 1969. The origin, variability and spread of the groundnut (*Arachis hypogaea*). In *The Domestication and Exploitation of Plants and Animals*, P. J. Ucko and G. W. Dimbleby, Eds. Aldine, Chicago, pp. 427-441.

Kring, J. B., 1972. Flight behavior of aphids. *Ann. Rev. Entomol.* **17**: 461-492.

Kronenberg, H. G. and H. J. De Fluiter, 1951. Resistance of respberries to the large raspberry aphid, *Amphorophora rubi* (Kalt.) Tijdschr. *Pl. Yiekt.* **57**: 114-123.

Kudagamage, C., 1977. Varietal resistance to the rice thrip, *Baliothrips biformis. Int. Rice Res. Newsl.* **5**: 11.

Kuhn, R. and A. Gouhe, 1947. Über die Bedeutung des Demissins fur die Resistenz von *S. demissum* gegen die Larven des Kartoffelkafers. *Z. Naturforsch.* **26**: 407-408.

Kuhn, R. and I. Low, 1955. Resistance factors against *Leptinotarsa decemlineata* (Say) isolated from the leaves of wild *Solanum* species. In *Origins of Resistance to Toxic Agents*, M. G. Sevag, R. D. Reid, and O. E. Reynolds, Eds. Academic Press, Inc., New York, pp. 122-132.

Kundu, G. G. and N. C. Pant, 1968. Studies on *Lipaphis erysimi* (Kalt.) with special reference to insect-plant relationship. III. Effect of age of plants on susceptibility. *Indian J. Entomol.* **30**: 169–172.

Lakshminarayana, A. and G. S. Khush, 1977. New genes for resistance to the brown planthopper in rice. *Crop Sci.* **17**: 96–100.

Laster, M. L. and W. R. Meredith, 1974a. Evaluating the response of cotton cultivars to tarnished plant bug injury. *J. Econ. Entomol.* **67**: 686–688.

Laster, M. L. and W. R. Meridith, 1974b. Influence of nectariless on insect pest populations. *Miss. Agric. For. Exp. Stn. Res. Highlights.*

Lee, J. A., 1968. Genetical studies concerning the distribution of trichomes on the leaves of *Gossypium hirsutum* L. *Genetics* **60**: 567–575.

Lee, J. A., 1971. Some problems in breeding smooth-leaved cottons. *Crop Sci.* **11**: 448–450.

Legge, J. B. B. and J. Palmer, 1968. Behavior of *Myzus persicae* (Sulz.) on tobacco. III. Choice by alatae of sites for feeding and larviposition on leaves and artificial substrates of differing nutrient status. *Bull. Entomol. Res.* **57**: 479–493.

Lehman, W. F., 1967. Alfalfa seed chalcid (*Bruchophagus roddi* Guss.). Infestation cycles in relation to alfalfa seed production and sampling for resistance in southern California. *Agron. J.* **59**: 403–406.

Lehman, W. F. and E. H. Stanford, 1971. Egyptian alfalfa weevil—breeding resistant alfalfa. *Calif. Agr.* **25**: 7–8.

Lehman, W. F. and E. H. Stanford, 1972. Progress in the development of alfalfa varieties with resistance to the Egyptian alfalfa weevil *Hypera brunneipennis* (Boh.). *Proc. 1972 Calif. Alfalfa Symp.*: 23–27.

Lehman, W. F. and E. H. Stanford, 1975. Possibilities for developing alfalfa varieties for resistance to the Egyptian alfalfa weevil *Hypera brunneipennis*. *Proc. Fifth Calif. Alfalfa Symp.*: 53–59.

Lehman, W. F., M. W. Nielson, V. L. Marble, and E. H. Stanford, 1977. CUF 101, a new variety of alfalfa is resistant to the blue alfalfa aphid. *Calif. Agric.* **31**: 3–4.

Lehman, W. F., M. W. Nielson, V. L. Marble, and E. H. Stanford, 1978. CUF 101 Alfalfa—a new variety of alfalfa with resistance to the blue alfalfa aphid. *Univ. Calif. Leafl.* **21009**: 3 pp.

Lehninger, A. L., 1975. *Biochemistry*, 2nd ed. Worth, New York.

Leigh, T. F. and A. H. Hyer, 1963. Spider mite resistant cotton. *Calif. Agric.* **17**: 6–7.

Leigh, T. F., A. H. Hyer, and R. E. Rice, 1972. Frego bract condition of cotton in relation to insect populations. *Environ. Entomol.* **1**: 390–391.

Lelley, J., 1976. *Wheat Breeding*. Publishing House of the Hungarian Acad. of Science, Budapest, 286 pp.

Leon, J., Ed. 1974. Handbook of plant introduction in tropical crops. *FAO Agric. Studies* **93**: 140 pp.

Leon, J., 1977. Origin, evolution, and early dispersal of root and tuber crops. *IV Symp. Inst. Soc. Trop. Root Crops, Cali, Colombia,* pp. 20–36.

LeRoux, E. J., 1954. Effects of various levels of nitrogen, phosphorus, and potassium in nutrient solution on the fecundity of the two-spotted spider mite, *Tetranychus bimaculatus* Harvey (Acarina: Tetranychidae) reared on cucumber. *Can. J. Agric. Sci.* **34:** 145–151.

Lesins, K. and C. B. Gillies, 1972. Taxonomy and cytogenetics of Medicago. Chapter 3 in *Alfalfa Science and Technology,* C. H. Hanson, Ed. American Society of Agronomy, Madison, Wis., pp. 53–86.

Leuschner, K., 1975. Major pests of cassava in Africa and preliminary guidelines for screening of resistance. *Proc. Interdiscipl. Workshop, IITA*: 55–56. Ibadan, Nigeria.

Levin, D. A., 1973. The role of trichomes in plant defense. *Quart. Rev. Biol.* **48:** 3–15.

Lewis, C. F. and T. R. Richmond, 1968. Cotton as a crop, in *Advances in Production and Utilization of Quality Cotton,* F. C. Elliot, M. Hoover, and W. K. Porter, Jr., Eds. Iowa State University Press, Ames.

Li, C. S. and S. F. Chiu, 1951. A study of the rice gall midge *Pachydiplosis oryzae* Wood-Mason. *Agric. Res. Taiwan* **2:** 1–13.

Libby, W. J., 1958. The backcross hybrid Jeffrey × (Jeffrey × Coulter) pine. *J. For.* **56:** 840–842.

Lichtenstein, E. P., F. M. Strong, and D. G. Morgan, 1962. Identification of 2-phenylethylisothiocyanate as an insecticide occurring naturally in the edible parts of turnips. *J. Agric. Food Chem.* **10:** 30–33.

Liener, I. E. and M. L. Kakade, 1969. Protease inhibitors. In *Toxic Constituents of Plant Foodstuffs,* I. E. Liener, Ed. Academic Press, Inc., New York, pp. 7–68.

Lincoln, C. and B. A. Waddle, 1966. Insect resistance of frego-type cotton. *Arkansas Farm Res.* **15:** 4–5.

Lindley, G., 1831. *A Guide to the Orchard and Kitchen Garden.* Landon, Longmans, Rees, Orme, Brown, and Green, 601 pp.

Lindquist, R. K., R. H. Painter, and E. L. Sorensen, 1967. Screening alfalfa seedlings for resistance to tarnished plant bug. *J. Econ. Entomol.* **60:** 1442–1445.

Ling, K. C., 1969. Nonpropagative leafhopper-borne viruses. In *Viruses Vectors, and Vegetation.* John Wiley & Sons, Inc., New York, pp. 255–277.

Ling, K. C. and E. R. Tiongco, 1978. Transmission of rice tungro virus at various temperatures: a transitory virus-vector interaction. In *Leafhopper Vectors and Plant Disease Agents,* K. Maramorosch and F. K. Harris, Eds. Academic Press, Inc., New York, in press.

Ling, L., 1974. Plant pest control on the international front. *Ann. Rev. Entomol.* **19:** 177–196.

Link, D. and C. J. Possetto, 1972. Relacao entre fissura na casca do arroz e infestacao de *Sitotroga cerealella* (Oliver, 1819) Lepidoptera: Gelechiidae). *Rev. Peruana Entomol. (An. Ier. Cong. Lationam. Entomol.).* **15:** 255–327.

Linser, H., K. H. Neumann, and H. El Damaty, 1965. Preliminary investigation of the action of (2-chloroethyl)-trimethyl ammonium chloride on the composition of the soluble N-fraction of young wheat plants. *Nature* **206:** 893–895.

Lipetz, J., 1970. Wound healing in higher plants. *Int. Rev. Cytol.* **27:** 1–28.

Lipke, H., G. S. Fraenkel, and I. E. Liener, 1954. Growth inhibitors: Effect of soybean inhibitors on growth of *Tribolium confusum. J. Agric. Food Chem.* **2:** 410–414.

Lippold, P. C., 1971. Rice pests, diseases, and varietal resistance trials in East Pakistan. Ford Foundation. Mimeographed paper, 52 pp.

Loegering, W. Q., 1951. Survival of races of wheat stem rust in mixtures. *Phytopathol.* **41:** 56–65.

Long, B. J., G. M. Dunn, J. S. Bowman, and D. G. Routley, 1977. Relationship of hybroxamic acid content in corn and resistance to the corn leaf aphid. *Crop Sci.* **17:** 55–57.

Loomis, R. S., S. D. Beck, and J. F. Stauffer, 1957. The European corn borer, *Pyrausta nubilalis* (Hubn.), and its principal host plant. V. A chemical study of host plant resistance. *Plant Physiol.* **32:** 379–385.

Lowe, C. C., V. L. Marble, and M. D. Rumbaugh, 1972. Adaptation, varieties, and usage. Chapter 18 in *Alfalfa Science and Technology,* C. H. Hanson, Ed. American Society of Agronomy, Madison, Wis., pp. 391–413.

Lowe, H. J. B., 1967. Interspecific differences in the biology of aphids (Homoptera: Aphididae) on leaves of *Vicia faba.* II. Growth and excretion. *Entomol. Exp. Appl.* **10:** 413–420.

Lowe, H. J. B., 1974. Testing sugar beet for aphid-resistance in the glasshouse: a method and some limiting factors. *Z. Angew. Entomol.* **76:** 311–321.

Lozano, J. C., 1977. Personal communication. Centro Internacional de Agricultura Tropical, Cali, Colombia.

Luckmann, W. H., A. M. Rhodes, and E. V. Wann, 1964. Silk balling and other factors associated with resistance of corn to corn earworm. *J. Econ. Entomol.* **57:** 778–779.

Luckmann, W. H., H. C. Chiang, E. E. Ortman, and M. P. Nichols, 1974. A bibliography of the northern corn rootworm *Diabrotica longcornis* (Say), and the western corn rootworm, *Diabrotica virgifera* LeConte (Coleoptera: Chrysomelidae) *Ill. Nat. Hist. Surv. Biol. Notes* **90.**

Luginbill, P., Jr., 1969. Developing resistant plants—the ideal method of controlling insects. *U.S. Dep. Agric. ARS. Prod. Res. Rep.* **111:** 14 pp.

Luginbill, P., Jr. and E. F. Knipling, 1969. Suppression of wheat stem sawfly with resistant wheat. *U.S. Dep. Agric. ARS Prod. Res. Rep.* **107:** 9 pp.

Lukefahr, M. J., D. F. Martin, and J. R. Meyer, 1956. Plant resistance to five Lepidoptera attacking cotton. *J. Econ. Entomol.* **58:** 516–518.

Lukefahr, M. J. and C. Rhyne, 1960. Effects of nectariless cottons on populations of Lepidopterous insects. *J. Econ. Entomol.* **53:** 242–244.

Lukefahr, M. J., D. F. Martin, and J. R. Meyer, 1965. Plant resistance to five Lepidoptera attacking cotton. *J. Econ. Entomol.* **586:** 516–518.

Lukefahr, M. J., C. B. Cowan, T. R. Pfrimmer, and L. W. Noble, 1966. Resistance of experimental cotton strain 1514 to the bollworm and cotton fleahopper. *J. Econ. Entomol.* **59:** 393–395.

Lukefahr, M. J., G. T. Bottger, and F. G. Maxwell, 1966. Utilization of gossypol as a source of insect resistance. *Proc. Ann. Cotton Dis. Council, Cotton Def. Phys. Conf., Cotton Impr. Conf.:* 215–222. National Cotton Council, Memphis, Tenn.

Lukefahr, M. J., C. B. Cowan, Jr., L. A. Bariola, and J. E. Houghtaling, 1968. Cotton strains resistant to the cotton fleahopper. *J. Econ. Entomol.* **61:** 661–664.

Lukefahr, M. J. and J. E. Houghtaling, 1969. Resistance of cotton strains with high gossypol content to *Heliothis* spp. *J. Econ. Entomol.* **62:** 588–591.

Lukefahr, M. J., C. B. Cowan, and J. E. Houghtaling, 1970. Field evaluations of improved cotton strains resistant to the cotton fleahopper. *J. Econ. Entomol.* **63:** 1101–1103.

Lukefahr, M. J., J. E. Houghtaling, and H. M. Graham, 1971. Suppression of *Heliothis* populations with glabrous cotton strains. *J. Econ. Entomol.* **64:** 486–488.

Lukefahr, M. J., T. N. Shaver, D. E. Cruhm, and J. E. Houghtaling, 1974. Location, transference and recovery of a *Heliothis* growth inhibition factor present in three *Gossypium hirsutum* race stocks. *Proc. Beltwide Cotton Prod. Res. Conf.:* 93–95. National Cotton Council, Memphis, Tenn.

Lukefahr, M. J. and J. E. Houghtaling, 1975. High gossypol cottons as a source of resistance to the cotton fleahopper. *Proc. Beltwide Cotton Prod. Res. Conf.:* 93–94. National Cotton Council, Memphis, Tenn.

Lukefahr, M. J., J. E. Houghtaling, and D. G. Cruhm, 1975. Suppression of *Heliothis* spp. with cottons containing combinations of resistant character. *J. Econ. Entomol.* **68:** 743–746.

Lukefahr, M. J., J. E. Jones, and J. E. Houghtaling, 1976. Fleahopper and leafhopper populations and agronimic evaluations of glabrous cottons from different genetic sources. *Proc. Beltwide Cotton Prod. Res. Conf.:* 84–86. National Cotton Council, Memphis, Tenn.

Lukefahr, M. J., 1977. Varietal resistance to cotton insects. *Proc. Beltwide Cotton Prod. Res. Conf.:* 236–237. National Cotton Council, Memphis, Tenn.

Lukefahr, M. J., R. D. Stipanovic, A. A. Bell, and J. R. Gray, 1977. Biological activity of new terpenoid compounds from *Gossypium hirsutum* against the tobacco budworm and pink bollworm. *Proc. Beltwide Cotton Prod. Res. Conf.:* 97–100. National Cotton Council, Memphis, Tenn.

Lupton, F. G. H., 1967. The use of resistance varieties in crop protection. *World Rev. Pest Control* **6:** 47–58.

Ma, W. C. and L. M. Schoonhoven, 1973. Tarsal contact chemosensory hairs of the large white butterfly *Pieris brassicae* and their possible role in oviposition behaviour. *Entomol. Exp. Appl.* **16**: 343–357.

MacKay, J., 1963. Autogamous plant breeding based on already high bred material. *Recent Plant Breeding Research, Svalof, 1946–1961.* Stockholm, pp. 73–88.

Major, R. T. 1967. The ginkgo, the most ancient living tree. *Science* **157**: 1270–1273.

Major, A. J. P., Jr., 1971. The interacting effects of okra leaf and frego bract on important agronomic characters in Upland cotton. M.S. Thesis, Louisiana State University, Baton Rouge.

Malcolm, D. R. 1953. Host relationship studies of *Lygus* in south-central Washington, *J. Econ. Entomol.* **46**: 485–488.

Maltais, J. B. and J. L. Auclair, 1957. Factors in resistance of peas to the pea aphid, *Acyrthosiphon pisum* (Harr.) (Homoptera: Aphididae). I. The sugar-nitrogen ratio. *Can. Entomol.* **89**: 365–370.

Mangelsdorf, P. C., 1974. *Corn: Its Origin, Evolution and Improvement.* Harvard University Press, Cambridge, Mass., pp. 19, 165–185.

Manglitz, G. R., C. O. Calkins, R. J. Walstrom, S. D. Hintz, S. D. Kindler, and L. L. Peters, 1966. Holocyclic strains of the spotted alfalfa aphid in Nebraska and adjacent states. *J. Econ. Entomol.* **59**: 636–639.

Manglitz, G. R. and H. J. Gorz, 1968. Inheritance of resistance in sweetclover to the sweetclover aphid. *J. Econ. Entomol.* **61**: 90–93.

Manglitz, G. R., H. J. Gorz, F. A. Haskins, W. R. Akeson, and G. L. Beland, 1976. Interactions between insects and chemical components of sweetclover. *J. Environ. Qual.* **5**: 347–352.

Manwan, I., 1975. Resistance of rice varieties to yellow borer, *Tryporyza incertulas* (Walker). Ph.D. Thesis, University of the Philippines, Los Baños, 185 pp.

Manwan, I., 1976. Resistance of rice varieties to yellow borer, *Tryporyza incertulas* (Walker). Unpublished Ph.D. Thesis, University of the Philippines, College of Agriculture, Los Baños.

Mamarorosch, K., 1958. Beneficial effect of virus-diseased plants on nonvector insects. *Tijdschr. Plantenziekten* **63**: 383–391.

Maramorosch, K., 1963. Arthropod transmission of plant viruses. *Ann. Rev. Entomol.* **8**: 369–414.

Maramorosch, K. and D. D. Jensen, 1963. Harmful and beneficial effects of plant viruses in insects. *Ann. Rev. Microbiol.* **17**: 495–530.

Maramorosch, K., Ed., 1969. *Viruses, Vectors, and Vegetation.* John Wiley & Sons, Inc., New York, 666 pp.

Maramorosch, K., 1975. Etiology of whitefly-borne diseases. In *Tropical Diseases of Legumes,* J. Bird and K. Maramorosch, Eds., Academic Press, Inc., New York, pp. 71–77.

Maramorosch, K., H. Hirumi, M. Kimura, and J. Bird, 1975. Mollicutes and rickettsia-like plant disease agents (Zoophytomicrobe in insects. *Ann. N.Y. Acad. Sci.* **266**: 276-292.

Maramorosch, K., 1976. Plant mycoplasma diseases. In *Encyclopedia of Plant Physiology,* New Series, Vol. 4, R. Heitefuss and P. H. Williams, Eds. Springer-Verlag, New York, pp. 150-171.

Maramorosch, K., 1977. In *Annual Report 1976-1977.* The Waksman Institute of Microbiology, Rutgers, The State University of New Jersey, pp. 23-24.

Maramorosch, K. and F. K. Harris, Eds., 1979. *Leafhopper Vectors and Plant Disease Agents.* Academic Press, Inc., New York, in press.

Marble, V. L., 1977. Alfalfa variety characteristics and adaptation in California. *Proc. Seventh Calif. Alfalfa Symp.*: 11-21.

Marke, M. E. and J. R. Meyer, 1963. Studies of resistance of cotton strains to the boll weevil. *J. Econ. Entomol.* **56**: 860-862.

Markkula, M. and K. Tittanen, 1969. Effect of fertilizers on the reproduction of *Tetranychus telarius* (L.), *Myzus persicae* (Sulz.) and *Acyrthosiphon pisum* (Harris). *Ann. Agric. Fenn.* **8**: 9-14.

Marston, A. R., 1930. Breeding corn for resistance to European corn borer. *J. Am. Soc. Agron.* **22**: 986-992.

Martin, J. H., 1970. History and classification of sorghum *Sorghum bicolor* (Linn.) Moench. In *Sorghum Production and Utilization,* J. S. Wall and W. M. Ross, Eds. Avi, Westport, Conn., pp. 1-27.

Martin, J. T. and B. E. Juniper, 1970. *The Cuticles of Plants.* St. Martin's Press, New York, p. 347.

Martin, H., 1973. *The scientific Principles of Crop Protection.* Edward Arnold, London.

Martin, G. A., C. A. Richard, and S. D. Hensley, 1975. Host resistance to *Diatraea saccharalis* (F.): Relationship of sugarcane internode hardness to larval damage. *Environ. Entomol.* **4**: 687-688.

Martin, T. J. and A. H. Ellingboe, 1976. Differences between compatible parasite/host genotypes involving the Pm4 locus of wheat and the corresponding genes in *Erysiphe graminis* f. sp. *tritici. Phytopathol.* **66**: 1435-1438.

Martinez, C. R. and G. S. Khush, 1974. *Crop Sci.* **14**: 167, Figure 2.

Mathes, R. and L. J. Charpentier, 1963. Some techniques and observations in studying the resistance of sugarcane varieties to the sugarcane borer in Louisiana. *Proc. Int. Soc. Sugarcane Technol.* **11**: 594-603.

Mattson, W. J. and N. D. Addy, 1975. Phytophagous insects as regulators of forest primary production. *Science* **190**: 515-521.

Maxwell, F. G. and R. H. Painter, 1959. Factors affecting rate of honeydew deposition by *Therioaphis maculata* (Buckton) and *Toxoptera graminum* (Rond.). *J. Econ. Entomol.* **52**: 368-373.

Maxwell, R. C. and R. F. Harwood, 1960. Increased reproduction of pea aphids on broad beans treated with 2,4-D. *J. Econ. Entomol.* **53**: 199-205.

Maxwell, F. G. and R. H. Painter, 1962. Auxin content of extracts of certain tolerant and susceptible host plants of *Toxoptera graminum, Macrosiphum pisi,* and *Therioaphis maculata* and relation to host plant resistance. *J. Econ. Entomol.* **55**: 46-56.

Maxwell, F. G., H. N. Lefever, and J. N. Jenkins, 1965. Blister beetles on glandless cotton. *J. Econ. Entomol.* **58**: 792-793.

Maxwell, F. G., J. N. Jenkins, W. L. Parrott, and W. T. Buford, 1969. Factors contributing to resistance and susceptibility of cotton and other hosts to the boll weevil, *Anthononus grandis. Entomol. Exp. App.* **12**: 801-810.

Maxwell, F. G., 1972. Host plant resistance to insects—nutritional and pest management relationships. In *Insect and Mite Nutrition,* Rodriguez, J. G., Ed. North-Holland, Amsterdam, pp. 599-609.

Maxwell, F. G., J. N,, Jenkins, and W. L. Parrott, 1972. Resistance of plants to insects. *Advan. Agron.* **24**: 187-265.

Mayers, W. F., 1868. *Notes and Queries on China and Japan* **2**(5): 73.

Mayo, Z. B., 1967. Resistance of native sorghums from India to the corn earworm and fall armyworm. Ph.D. Thesis, Oklahoma State University, Stillwater, 48 pp.

Mazakhin-Porshynakov, G. A., 1969. *Insect Vision.* Plenum Press, New York, p. 306.

McCarty, J. C., 1974. Evaluation of species and primitive races of cotton for boll weevil resistance and agronomic qualities. Ph.D. Dissertation, Mississippi State University, State College.

McColloch, J. W., 1921. The corn leaf aphid (*Aphis maidis* Fitch) in Kansas. *J. Econ. Entomol.* **14**: 89-94.

McDonald, G. I., 1976. Potential budworm resistance in Douglas-fir: and observational report. *U.S. Dep. Agric. For. Serv. Intermountain Stn. Off. Rep.* (dated Nov. 10, 1976).

McGaughey, W. H., 1973. Resistance to the lesser grain borer in "Dawn" and "Labelle" varieties of rice. *J. Econ. Entomol.* **66**: 10005.

McGaughey, W. H., 1974. Insect development in milled rice, effects of variety, degree of milling, parboiling and broken kernels. *J. Stored Food Res.* **10**: 81-86.

McKenzie, H., 1965. Inheritance of sawfly reaction and stem solidness in spring wheat crosses. *Can. J. Plant Sci.* **45**: 583-589.

McMillian, W. W. and K. J. Starks, 1967. Greenhouse and laboratory screening or sorghum lines for resistance to fall armyworm larvae. *J. Econ. Entomol.* **60**: 1462-1463.

McMillian, W. W., K. J. Starks, and M. C. Bowman, 1967. Resistance in corn to the corn earworm, *Heliothis zea,* and the fall armyworm, *Spodoptera frugiperda* (Lepidoptera: Noctuidae). Part I. Larval feeding responses to corn plant extracts. *Ann. Entomol. Soc. Am.* **60**: 871-873.

McMillian, W. W., N. W. Widstrom, and K. J. Starks, 1968. Rice weevil damage as affected by husk treatment with methods of artificially infesting field corn plots. *J. Econ. Entomol.* **61:** 918–921.

McMillian, W. W., B. R. Wiseman, and A. A. Sckul, 1970. Further studies on the responses of corn earworm larvae to extracts of corn silks and kernels. *Ann. Entomol. Soc. Am.* **63:** 371–378.

McMillian, W. W. and B. R. Wiseman, 1972. Separating egg masses of the fall armyworm. *J. Econ. Entomol.* **65:** 900–902.

McMillian, W. W. and B. R. Wiseman, 1972. A twentieth century look at the relationship between *Zea mays* L. and *Heliothis zea* (Boddic). *Fla. Agric. Exp. Stn. Monogr. Ser.* **2:** 131 pp.

McMurtry, J. A. and E. H. Stanford, 1960. Observations on the feeding habits of the spotted alfalfa aphid on resistant and susceptible alfalfa plants. *J. Econ. Entomol.* **53:** 714–717.

McMurtry, J. A., 1962. Resistance of alfalfa to spotted alfalfa aphid in relation to environmental factors. *Hilgardia* **32:** 501–539.

Melhus, I. E., R. H. Painter, and F. V. Smith, 1954. A search for resistance to the injury caused by species of *Diabrotica* in the corns of Guatemala. *Iowa State Coll. J. Sci.* **29:** 74–94.

Meredith, W. R., Jr., C. D. Ranney, M. L. Laster, and R. R. Bridge, 1973. Agronomic potential of nectariless cotton. *J. Environ. Qual.* **2:** 141–144.

Metcalf, C. L. and W. P. Flint (revised by R. L. Metcalf), 1962. *Destructive and Useful Insects, Their Habits and Control.* McGraw-Hill Book Company, Inc., New York, pp. 462–524.

Metcalf, R. L., 1971. Pesticides in the environment. In *The Chemistry and Biology of Pesticides,* Vol 1, R. White-Stevens, Ed. Marcel Dekker, Inc., New York.

Metcalf, R. L. and W. H. Luckmann, 1975. *Introduction to Insect Pest Management.* John Wiley & Sons, Inc., New York, 587 pp.

Meyer, J. R. and V. G. Meyer, 1961. Origin and inheritance of nectariless cotton. *Crop. Sci.* **1:** 167–169.

Meyer, V. G., 1973. Registration of sixteen germplasm lines of Upland cotton. *Crop Sci.* **13:** 778–779.

Meyer, V. G., 1974. Interspecific cotton breeding. *Econ. Bot.* **28:** 56–60.

Michel, E., 1963. Etude de la fecondite du puceron Myzus persicae en fonction de son alimentation sur tabac. I. Influence de la nutrition minerale de la plante. *Ann. Inst. Exp. Tab. Bergerac* **4:** 421–434.

Mihm, J. A., F. B. Peairs, and A. Ortega. New procedures for efficient mass production and artificial infestation with Lepidopterous pests of maize (unpublished).

Miles, D. W., 1968. Insect secretions in plants. *Ann. Rev. Phytopathol.* **6:** 137–164.

Mills, R. B., 1972. Host plant resistance applied to stored-product insect. *Proc. North Cent. Branch Entomol. Soc. Am.* **27:** 106–107.

Mills, R. B., 1976. Host resistance to stored-product insects—II. *Proc. Joint U.S.-Japan Semin. Stored Prod. Insects*: 77–87. Manhattan, Kansas.

Mishra, B. C., J. P. Kulshreshtha, J. K. Roy, and J. R. K. Rao, 1976. Shakti. A variety resistant to gall midge. *Indian Farming* **26**(4): 21, 23.

Mistric, W. J., Jr., 1968. Effects of nitrogen fertilization on cotton under boll weevil attack in North Carolina. *J. Econ. Entomol.* **61**: 282–283.

Mitchell, H. C., W. H. Cross, W. L. McGovern, and E. M. Dawson, 1973. Behavior of the boll weevil on frego bract cotton. *J. Econ. Entomol.* **66**: 677–680.

Mitchell, N. D., 1977. Differential host selection by *Pieris brassicae* L. (the large white butterfly) on *Brassica oleracea* subsp. oleracea (the wild cabbage). *Entomol. Exp. Appl.* **22**: in press.

Mitchell, R. G., N. E. Johnson, and K. H. Wright, 1974. Susceptibility of 10 spruce species and hybrids to the white pine weevil (Sitka spruce weevil) in the Pacific Northwest. *U.S. Dep. Agric. For. Serv. Res. Note* **PNW-225**: 8 pp.

Miyamoto, A. S. and Y. Miyamoto, 1966. Notes on aphid transmission of potato leafroll virus. *Sci. Rep. Hyogo Univ. Agric., Ser. Plant Prot.* **7**: 413–421.

Mode, C. J., 1958. A mathematical model for the coevoluation of obligate parasites and their hosts. *Evolution* **12**: 158–165.

Mohamed, A. H., M. M. El-Haddad, and G. S. Mallah, 1966. Inheritance of quantitative characters in *Zea mays*. VI. Genetic studies on the resistance to the European corn borer. *Can. J. Genet. Cytol.* **1**: 122–129.

Mohanty, H. K. and A. Roy, 1977. Parijat, a new upland rice variety from India. *Int. Rice Res. Newsl.* **2**(2): 2.

Morris, O. N., 1961. The development of clover mite, *Bryobia praetiosa* (Acarina: Tetranychidae) in relation to the nitrogen, phosphorus, and potassium nutrition of its plant host. *Ann. Entomol. Soc. Am.* **54**: 551–557.

Morris, R. F., 1967. Influence of parental food quality on the survival of *Hyphantria cunea. Can Entomol.* **99**: 24–33.

Morris, R. and E. R. Sears, 1967. The Cytogentics of Wheat and Its Relatives. Wheat and Wheat Improvement. *Agron. Monogr.* **13**: 19–87. American Society of Agronomy, Madison, Wis.

Mound, L. A., 1965. Effect of leaf hair on cotton whitefly population in the Sudan Gezira. *Emp. Cotton Grow. Rev.* **42**: 33–40.

Mulkern, G. B., 1969. Behavioral influence on food selection in grasshoppers (Orthoptera: Acrididae). *Entomol. Exp. Appl.* **12**: 509–523.

Muller, H. J., 1976. Ecological aspects of the joined evolution of plants and animals. *Nova Acta Leopoldina Suppl.* **7**: 433–440.

Muller, K. O., 1959. Hypersensitivity. In *Plant Pathology*, J. G. Horsfall, and A. E. Diamond, Eds., Vol. 1. Academic Press, New York, pp. 469–519.

Munakata, K. and D. Okamoto, 1967. Varietal resistance to rice stem borer in Japan. In *The Major Insect Pests of the Rice Plant. Proc. Symp. Int. Rice Res.*

Inst., Los Baños, Philippines, 1964. Johns Hopkins Press, Baltimore, pp. 419–430.

Naidu, B. S., P. S. Pal, and B. Jagannath, 1976. Yield performance of some gall midge resistant cultures in field trials in Karnataka, India. *Int. Rice Res. Newsl.* **1**: 16.

Naito, A., 1976a. Studies on the feeding habits of some leafhoppers attacking the forage crops. I. Comparison of the feeding habits of the adults. *Jap. J. Appl. Entomol. Zool.* **20**: 1–8.

Naito, A., 1976b. Studies on the feeding habits of some leafhoppers attacking the forage crops. II. Comparison of the feeding habits of the green rice leafhoppers in different developmental stages. *Jap. J. Appl. Entomol. Zool.* **20**: 51–54.

Nakano, K., G. Abe, N. Taketa, and C. Hirano, 1961. Silicon as an insect-resistance component of host plant, found in relation between the rice stem borer and the rice plant. *Jap. J. Appl. Entomol. Zool.* **5**: 17–27.

Nath, P. and C. V. Hall, 1963. Inheritance of resistance to the striped cucumber beetle in *Cucurbita pepo. Indian J. Genet. Plant Breed.* **23**: 342–345.

National Academy of Sciences, 1969. *Insect-Pest Management and Control,* Vol. 3. *Principles of Plant and Animal Pest Control.* National Academy of Sciences, Washington, D.C., 508 pp.

National Academy of Sciences, 1971. *Insect-Plant Interactions.* Report of a work conference. Washington, D.C., 93 pp.

National Academy of Sciences, 1972. *Genetic Vulnerability of Major Crops.* Washington, D.C., 300 pp.

National Extension Insect Pest-Management Workshop, 1972. Implementing practical pest management strategies. Purdue University, West Lafayette, Ind., 206 pp.

Nault, L. R. and W. B. Styer, 1972. Effects of sinigrin on host selection by aphids. *Entomol. Exp. Appl.* **15**: 423–437.

Nelson, R. R., 1972. Stabilizing racial populations of plant pathogens by use of resistance genes. *J. Environ. Qual.* **1**: 220–227.

Newton, R. C. and D. K. Barnes, 1965. Factors affecting resistance of selected alfalfa clones to the potato leafhopper. *J. Econ. Entomol.* **58**: 435–439.

Nichols, M. P., M. Kogan, and G. P. Waldbauer, 1974. *Ill. Nat. Hist. Surv. Biol. Notes* **85** Urbana, Ill.

Nielson, M. W. and W. E. Currie, 1959. Effect of alfalfa variety on the biology of the spotted alfalfa aphid in Arizona. *J. Econ. Entomol.* **52**: 1023–1024.

Nielson, M. W., 1962. A synonymical list of leafhopper vectors of plant viruses (Homoptera: Cicadellidae). *U.S. Dep. Agric. ARS* 33-74: 12 pp.

Nielson, M. W. and M. H. Schonhorst, 1965. Research on alfalfa seed chalcid resistance in alfalfa. *Prog. Agric. Ariz.* **27**: 20–21.

Nielson, M. W., 1967. Procedures for screening and testing alfalfa for resistance to the alfalfa seed chalcid. *U.S. Dep. Agric., ARS* **33–120**, 10 pp.

Nielson, M. W. and M. H. Schonhorst, 1967. Sources of alfalfa seed calcid resistance in alfalfa. *J. Econ. Entomol.* **60:** 1506–1511.

Nielson, M. W., W. F. Lehman, and V. L. Marble, 1970a. A new severe strain of the spotted alfalfa aphic in California. *J. Econ. Entomol.* **63:** 489–491.

Nielson, M. W., H. Don, M. H. Schonhorst, W. F. Lehman, and V. L. Marble, 1970b. Biotypes of the spotted alfalfa aphid in western United States. *J. Econ. Entomol.* **63:** 1822–1825.

Nielson, M. W., 1974. Evaluating spotted alfalfa aphid resistance. In *Standard Tests to Characterize Pest Resistance in Alfalfa Varieties. U.S. Dep. Agric.* **NC-19:** 19–20.

Nielson, M. W. and H. Don, 1974a. A new virulent biotype of the spotted alfalfa aphid in Arizona. *J. Econ. Entomol.* **67:** 64–66.

Nielson, M. W. and H. Don, 1974b. Probing behavior of biotypes of the spotted alfalfa aphid on resistant and susceptible alfalfa clones. *Entomol. Exp. Appl.* **17:** 477–486.

Nielson, M. W., H. Don, and J. Zaugg, 1974. Sources of resistance in alfalfa to *Lygus hesperus* Knight. *U.S. Dep. Agric. ARS* **WR-21:** 5 pp.

Nielson, M. W., W. F. Lehman, and R. T. Kodet, 1976. Resistance in alfalfa to *Acyrthosiphon kondoi. J. Econ. Entomol.* **69:** 471–472.

Nielson, M. W. and W. F. Lehman, 1977. Multiple aphid resistance in CUF-101 alfalfa. *J. Econ. Entomol.* **70:** 13–14.

Nielson, M. W., 1979. Taxonomic relationships of leafhopper vectors of plant pathogens. In *Leafhopper Vectors and Plant Disease Agents,* K. Maramorosch and K. F. Harris, Eds. Academic Press, Inc., New York, in press.

Nilakhe, S. S., 1976. Rice lines screened for resistance to the rice stink bug. *J. Econ. Entomol.* **69:** 703–705.

Noble, L. W., 1969. Fifty years of research on the pink bollworm in the United States. *U.S. Dep. Agric. Agric. Handb.* **357.**

Noble, W. B. and C. A. Suneson, 1943. Differentiation of two genetic factors for resistance to the Hessian fly in Dawson wheat. *J. Agric. Res.* **67:** 27–32.

Normanha, E. S. and A. S. Pereira, 1964. Cultura de mandioca. *Inst. Agron. Bol. (Campinas, Brazil)* **124:** 29 pp.

Normanha, E. S., 1970. General aspects of cassava root production in Brazil. In *2nd Int. Symp. Trop. Root Tuber Crops.*

Norris, D. M., 1970. Quinol stimulation and quinone deterrency of gustation by *Scolytus multistriatus. Ann Entomol. Soc. Am.* **63:** 476–478.

Norris, D. M., 1977. The role of repellents and deterrents in feeding of *Scolytus multistriatus.* In *The Chemical Basis for Plant Resistance to Pests,* American Chemical Society, Washington, D.C., in press.

Norris, D. M., J. M. Rozental, G. Samberg, and G. Singer, 1977. Protein sulfur dependent differences in the nerve receptors for repellent 1,4-naphthoquinones in two strains of *Periplaneta americana. Comp. Biochem. Physiol.* **57C:** 55–59.

Norwood, B. L., R. S. VandenBurgh, C. H. Hanson, and C. C. Blickenstaff, 1967a. Factors affecting resistance of field-planted alfalfa clones to alfalfa weevil. *Crop Sci.* **7**: 96–99.

Norwood, B. L., D. K. Barnes, R. S. VendenBurgh, C. H. Hanson, and C. C. Blickenstaff, 1967b. Influence of stem diameter on oviposition preference of the alfalfa weevil and its importance in breeding for resistance. *Crop Sci.* **7**: 428–430.

Nuorteva, P., 1952. Die Nahrungspflanzenwahl der Insekten in Lichte von Untersuchungen an Zikaden. *Ann. Akad. Sci. Fenn. A IV* **19**: 1–90.

Nyiira, Z. M., 1972. Report on investigation on Cassava Mite, *Mononychellus tanajoa* (Bondar). *Dep. Agric. Kawanda Res. Stn.* 14 pp.

Nyiira, Z. M., 1976. Advances in research on the economic significance of the green cassava mite, *Mononychellus tanajoa* (Bondar) in Uganda. *Int. Ech. Test. Cassave Germplasm Afr. Proc. Interdiscipl. Workshop, Ibadan, Nigeria*: 27–29.

Oka, I. N. and D. Pimentel, 1974. Corn susceptibility to corn leaf aphids and common corn smut after herbicide treatment. *Environ. Entomol.* **3**: 911–915.

Oka, I. N., 1976. Integrated control program on brown planthopper and yellow rice stem borer in Indonesia. Paper presented at the *Int. Rice Res. Conf. IRRI, Philippines.* Mimeographed, 12 pp.

O'Keefe, L. E., J. A. Callenbach, and K. L. Lebsock, 1960. Effect of culm solidness on the survival of the wheat stem sawfly. *J. Econ. Entomol.* **53**: 244–246.

Oksanen, H., V. Perttunen, and E. Kangas, 1970. Studies on the chemical factors involved in the olfactory orientation of *Blastophagus piniperda* (Coleoptera: Scolytidae). *Contrib. Boyce Thompson Inst.* **24**: 275–282.

Olembo, J. R., F. L. Patterson, and R. L. Gallun, 1966. Genetic analysis of the resistance to *Mayetiola destructor* (Say) in *Hordeum vulgara* L. *Crop Sci.* **6**: 563–566.

Oliver, B. F., F. G. Maxwell, and J. N. Jenkins, 1967. Measuring aspects of antibiosis in cotton lines to the bollworm. *J. Econ. Entomol.* **60**: 1459–1460.

Onate, B. T., 1965. Estimation of stem borer damage in rice fields. *Phil. Statist.* **14**: 201–221.

Orenski, S. W., 1964. Effects of a plant virus on survival, food acceptability and digestive enzymes of corn leafhoppers. *Ann. N.Y. Acad. Sci.* **118**: 374–386.

Ortega, A. and C. De Leon, 1974. Maize insects and diseases, in *Proc. Symp. on World-Wide Maize Improvement in the 70's and the Role of CIMMYT.* CIMMYT, Mexico D.F., Mexico.

Ortman, E. E. and R. H. Painter, 1960. Quantitative measurements of damage by the greenbug, *Toxoptera graminum,* to four wheat varieties. *J. Econ. Entomol.* **53**: 798–802.

Ortman, E. E., E. L. Sorensen, R. H. Painter, T. L. Harvey, and H. L. Hackerott, 1960. Selection and evaluation of pea aphid-resistant alfalfa plants. *J. Econ. Entomol.* **53**: 881–887.

Ortman, E. E. and P. J. Fitzgerald, 1964. Developments in corn rootworm research. *Proc. Ann. Hybrid Corn Ind. Res. Conf.* **19**: 1-8.

Ortman, E. E., D. C. Peters, and P. J. Fitzgerald, 1968. Vertical pull technique for evaluating tolerance of corn rootworm systems to northern and western corn rootworm. *J. Econ. Entomol.* **61**: 373-375.

Ortman, E. E. and E. D. Gerloff, 1970. Rootworm resistance: Problems in measuring and its relationship to performance. *Proc. Ann. Corn Sorghum Res. Conf.* **25**: 161-174.

Ortman, E. E., T. F. Branson, and E. D. Gerloff, 1974. Techniques, accomplishments, and future potential of host plant resistance to *Diabrotica,* In *Proc. Summer Inst. Biol. Control Plant Insects Dis.,* F. G. Maxwell and F. A. Harris, Eds.: 344-358. University Press of Mississippi, Jackson.

Ortman, E. E. and T. F. Branson, 1976. Growth pouches for studies of host plant resistance to larvae of corn rootworms. *J. Econ. Entomol.* **69**: 380-382.

Ossiannilsson, F. 1966. Insects in the epidemiology of plant viruses. *Ann. Rev. Entomol.* **11**: 213-232.

Owens, J. C., C. R. Ward, and G. L. Teetes, 1976. Current status of spider mites in corn and sorghum. *Proc. Ann. Corn Sorghum Res. Conf.,* **31**: 38-64.

Pablo, S., 1977. Varietal resistance to the white-backed planthopper, *Sogatella furcifera* (Horvath) in rice. Ph.D. Thesis, Indian Agricultural Research Institute, New Delhi, India (unpublished).

Painter, R. H., 1930. The biological strains of Hessian fly. *J. Econ. Entomol.* **23**: 322-329.

Painter, R. H. and C. O. Granfield, 1935. A preliminary report on resistance of alfalfa varieties to pea aphids, *Illinoia pisi* (Kalt.). *J. Am. Soc. Agron.* **27**: 671-674.

Painter, R. H., 1936. The food of insects and its relation to resistance of plants to insect attack. *Am. Nat.* **70**: 547-566.

Painter, R. H., 1941. The economic value and biological significance of plant resistance to insect attack. *J. Econ. Entomol.* **34**: 358-367.

Painter, R. H., 1951. *Insect Resistance in Crop Plants.* The Macmillan Co., New York, 520 pp.

Painter, R. H. and D. C. Peters, 1956. Screening wheat varieties and hybrids for resistance to the green bug. *J. Econ. Entomol.* **49**: 546-548.

Painter, R. H., 1958. Resistance of plants to insects. *Ann. Rev. Entomol.* **3**: 267-290.

Painter, R. H. and M. D. Pathak, 1962. The distinguishing features and significance of the four biotypes of the corn leaf aphid *Rhopalosiphum maidis* (Fitch). *Proc. XI Int. Congr. Entomol.* **11**: 110-115.

Painter, R. H., 1966. Lessons to be learned from past experience in breeding plants for insect resistance. In *Breeding Pest Resistant Trees.* H. D. Gerhold et al., Eds. Pergamon Press, London, pp. 349-355.

Painter, R. H., 1968. Crops that resist insects provide a way to increase world food supply. *Kans. State Agric. Exp. Stn. Bull.*: 520 pp.

Palmer, D. F., M. B. Windels, and H. C. Chiang, 1977. Artificial infestation of corn with western corn rootworm eggs in agarwater. *J. Econ. Entomol.* **70:** 277–278.

Pandya, P. S. and C. T. Patel, 1964. Possibilities of imparting resistance to pests in cotton by use of wild species of *Gossypium. Indian Cotton Grow. Rev.* **18:** 175–176.

Parisi, R. A., A. Ortega, and R. Reyna, 1973. El dano de *Diatraea saccharalis* Fabricius (Lepidoptera: Pyralidae) en relacion con densidad de plantas, nivel de fertilidad e hibridos de maiz, en Argentina. *Agrociencia* **13:** 43–63. Colegio de Postgraduados, Chapingo, Estado de Mexico, Mexico.

Parker, F. W. and N. M. Randolph, 1972. Mass rearing the chinch bug in the laboratory. *J. Econ. Entomol.* **65:** 894–895.

Parlevliet, J. E. and A. van Ommeren, 1975. Partial resistance of barley to leaf rust, *Puccinia hordei.* II. Relationship between field trials, micro-plot tests and latent period. *Euphytica* **24:** 293–303.

Parnell, F. R., 1925. Breeding jassid resistant cottons. *Emp. Cotton Grow. Rev.* **2:** 330–336.

Parnell, F. R., H. E. King, and D. F. Ruston, 1949. Jassid resistance and hairiness of the cotton plant. *Bull. Entomol. Res.* **39:** 539–575.

Parrott, W. L., J. N. Jenkins, and D. B. Smith, 1973. Frego bract cotton and normal bract cotton: How morphology affects control of boll weevils by insecticides. *J. Econ. Entomol.* **66:** 222–255.

Passlow, T., 1965. Bionomics of sorghum midge (*Contarinia sorghicola* Coq.) in Queensland with particular reference to diapause. *Queensl. J. Agric. Anim. Sci.* **22:** 150–167.

Patanakmjorn, S. and M. D. Pathak, 1967. Varietal resistance of the Asiatic rice borer, *Chilo suppressalis* (Lepidoptera: Crambidae), and its association with various plant characteristics. *Ann. Entomol. Soc. Am.* **60:** 287–292.

Patch, L. H. and L. L. Pierce, 1933. Laboratory production of clusters of European corn borer eggs for use in hand infestation of corn. *J. Econ. Entomol.* **26:** 196–204.

Patch, L. H., J. R. Holbert, and R. T. Everly, 1942. Strains of field corn resistant to the survival of the European corn borer. *U.S. Dep. Agric. Tech. Bull.* **823:** 22 pp.

Pathak, M. D. and R. H. Painter, 1958a. Differential amounts of material taken up by four biotypes of corn leaf aphids from resistant and susceptible sorghums. *Ann. Entomol. Soc. Am.* **51:** 250–254.

Pathak, M. D. and R. H. Painter, 1958b. Effect of the feeding of the four biotypes of corn leaf aphid, *Rhopalosiphum maidis* (Fitch) on susceptible white martin sorghum and spartan barley plants. *J. Kans. Entomol. Soc.* **31:** 93–100.

Pathak, M. D., 1969. Stem borer and leafhopper and planthopper resistance in rice varieties. *Entomol. Exp. Appl.* **12:** 789–800.

Pathak, M. D., C. H. Cheng, and M. E. Fortuno, 1969. Resistance to *Nephotettix impicticeps* and *Nilaparvata lugens* in rice varieties. *Nature* **223:** 502–504.

Pathak, M. D., 1970. Genetics of plants in pest management. In *Concepts of Pest Management,* R. L. Rabb and F. E. Guthrie, Eds. North Carolina State University, Raleigh, pp. 138–157.

Pathak, M. D., F. Andres, N. Galacgnac, and R. Raros, 1971. Resistance of rice varieties to the striped stem borer. *Int. Rice Res. Inst. Tech. Bull.* **11:** 69.

Pathak, M. D., 1972. Resistance to insect pests in rice varieties. In *Rice Breeding.* International Rice Research Institute, Los Baños, Philippines, pp. 325–341.

Pathak, M. D. and V. A. Dyck, 1973. Developing an integrated method of rice insect pest control. *PANS* **19**(4): 534–544.

Pathak, M. D., H. M. Beachell, and F. Andres, 1973. IR20, a pest- and disease-resistant high-yielding rice variety. *Int. Rice Comm. Newsl.* **22**(3): 1–8.

Pathak, M. D., 1975. Utilization of insect-plant interactions in pest control. In *Insects, Science and Society,* D. Pimentel, Ed. Academic Press, Inc., New York, pp. 121–148.

Pathak, M. D. and R. C. Saxena, 1976. Insect resistance in crop plants. *Curr. Adv. Plant Sci.* **8**(9): 1233–1252.

Pathak, M. D., 1977. Defense of the rice crop against insect pests, in *The Genetic Basis of Epidemics in Agriculture,* P. R. Day, Ed. *Ann. N.Y. Acad. Sci.* **287:** 287–295.

Pathak, M. D. and G. S. Khush, 1977. Studies on varietal resistance to brown planthopper at IRRI. In *Brown Planthopper Symposium.* IRRI, Los Baños, Philippines. Mimeographed, 48 pp.

Patterson, F. L. and R. L. Gallun, 1973. Inheritance of Seneca wheat to race E of Hessian fly. In *Proc. 4th Int. Wheat Genetic Symp. Mo. Agr. Exp. Stn. Columbia:* 445–449.

Pauley, S. S., S. H. Spurr, and F. W. Whitmore, 1955. Seed source trials of eastern white pine. *For Sci.* **1:** 244–256.

Payne, T. L., 1970. Electrophysiological investigations on response to pheromones in bark beetles. *Contrib. Boyce Thompson Inst.* **24:** 275–282.

Pedersen, M. W., D. K. Barnes, E. L. Sorensen, G. D. Griffin, M. W. Nielson, R. R. Hill, Jr., F. I. Frosheiser, R. M. Sonoda, C. H. Hanson, O. J. Hunt, R. N. Peaden, J. H. Elgin, Jr., T. E. Devine, M. J. Anderson, B. O. Goplen, L. J. Elling, and R. E. Howarth, 1976. Effects of low and high saponin selection in alfalfa on agronomic and pest resistance traits and the interrelationship of these traits. *Crop. Sci.* **19:** 193–199.

Penny, L. H. and F. F. Dicke, 1956. Inheritance of resistance in corn to leaf feeding by European corn borer. *Agron. J.* **48:** 200–203.

Penny, L. H. and F. F. Dicke, 1957. A single gene pair controlling segregation for European corn borer resistance. *Agron. J.* **49:** 193–196.

Penny, L. H. and F. F. Dicke, 1959. European corn borer damage in resistant and susceptible dent corn hybrids. *Agron. J.* **51:** 323–326.

Penny, L. H., G. E. Scott, and W. D. Guthrie, 1967. Recurrent selection for European corn borer resistance in maize. *Crop Sci.* **7:** 407–409.

Peregrine, W. T. H. and W. S. Catling, 1967. Studies on resistance in oats to the fruit fly. *Plant Pathol.* **16:** 170–175.

Person, C. O., D. J. Samborski, and R. Rohringer, 1952. The gene for gene concept. *Nature* **194:** 561–562.

Person, C. O., 1959. Gene-for-gene relationships in host: parasite systems. *Canad. J. Bot.* **37(5):** 1101–1130.

Person, H. L., 1931. Theory in explanation of the selection of certain trees by the western pine beetle. *J. For.* **29:** 696–699.

Pesho, G. R., F. V. Lieberman, and W. F. Lehman, 1960. A biotype of the spotted alfalfa aphid on alfalfa. *J. Econ. Entomol.* **53:** 146–150.

Peters, D. C., E. A. Wood, Jr., and K. J. Starks, 1975. Insecticide resistance in selections of the greenbug. *J. Econ. Entomol.* **68:** 339–340.

Peters, L. L., M. L. Fairchild, and M. S. Zuber, 1972. Effect of corn endosperm containing different levels of amylose on Angoumois grain moth biology. 3. Interrelationship of amylose levels and moisture content of diets. *J. Econ. Entomol.* **65:** 1168–1169.

Pierzchalski, T. and E. Werner, 1958. Changes in glycoalkaloid contents in leaves of cultivated and wild potatoes and their hybrids during growth and their influence on the Colorado potato beetle (*Leptinotarsa decemlineata* Say). Part I. Hodowla. *Rosl. Aklim. Nasienn.* **2:** 157–180 (in Polish).

Pillemer, E. A. and W. M. Tingey, 1976. Hooked trichomes: A physical plant barrier to a major agricultural pest. *Science* **193:** 482–484.

Pimentel, D., 1969. Animal populations, plant host resistance and environmental control. *Proc. Tall Timbers Conf. Ecol. Anim. Control Habitat Manage.* **1:** 19–28.

Pimentel, D. and A. C. Bellotti, 1976. Parasite host population systems and genetic stability. *Am. Nat.* **110:** 877–888.

Pirone, T. P. and K. F. Harris, 1977. Nonpersistent transmission of plant viruses by aphids. *Ann. Rev. Phytopathol.* **15:** 55–73.

Pirson, A., 1955. Functional aspects in mineral nutrition of green plants. *Ann. Rev. Plant. Physiol.* **6:** 71–114.

Platt, A. W., 1941. The influence of some environmental factors on the expression of the solid stem character in certain wheat varieties. *Sci. Agric.* **21:** 139–151.

Platt, W. W. and C. W. Farstad, 1946. The reaction of wheat varieties to wheat stem sawfly attack. *Sci. Agric.* **26:** 231–247.

Plumb, R. T., 1976. Barley yellow dwarf virus in aphids caught in suction traps. 1969–1973. *Ann. Appl. Biol.* **83:** 53–59.

Pollard, D. G. and J. H. Saunders, 1956. Jassid resistant Sakel and hairiness in relation to other cotton pests. *Emp. Cotton Grow. Rev.* **33:** 197–202.

Pongprasert, S., 1974. Resistance to the zigzag leafhopper, *Recilia dorsalis* (motschulsky) in rice varieties. M.S. Thesis, University of the Philippines, Los Baños, 76 pp.

Poole, C. F., 1936. New sweet corns resistance to earworm. *Proc. Am. Soc. Hortic. Sci.* **33**: 496–501.

Poole, R. T., 1965. The effect of 2,4-dichlorobenzyltributyphosphonium chloride and 2-chloroethyltrimethyl-ammonium chloride on growth, flowering and chemical composition of *Chrysanthemum morifolium* "Bluechip." *Diss. Abstr.* **25**: 6915–6916.

Poos, F. W. and F. F. Smith, 1931. Comparison of oviposition and nymphal development of *Empoasca fabae* (Harris) on different host plants. *J. Econ. Entomol.* **24**: 361–371.

Poos, F. W. and T. L. Bissell, 1953. The alfalfa weevil in Maryland. *J. Econ. Entomol.* **46**: 178–179.

Porter, K. B. and N. E. Daniels, 1963. Inheritance and heritability of greenbug resistance in a common wheat cross. *Crop Sci.* **3**: 116–118.

Prakasa Rao, P. S., 1975. Some methods of increasing field infestations of rice gall midge. *Rice Entomol. Newsl.* **2**: 16–17.

Pura, C. D., 1971. Resistance of the brown planthopper, *Nilaparvata legens* (Stal), in rice varieties. M.S. Thesis, University of the Philippines, Los Baños, 79 pp.

Quinby, J. R., N. W. Kramer, J. C. Stephens, K. A. Lahr, and R. E. Karper, 1958. Grain sorghum production in Texas. *Tex. Agric. Exp. Stn. B-912:* 35 pp.

Quinby, J. R. and K. F. Schertz, 1970. Sorghum genetics, breeding, and hybrid seed production. In *Sorghum Production and Utilization,* J. S. Wall and W. M. Ross, Eds., Avi, Westport, Conn., pp. 73–117.

Quinby, J. R., 1974. *Sorghum Improvement and the Genetics of Growth.* Texas A & M University Press, College Station, 108 pp.

Quiras, C. F., M. A. Stevens, C. M. Rick, and M. K. Kok-Yokomi, 1977. Resistance in tomato to the pink form of the potato aphid (*Macrosiphum euphorbiae* Thomas): The role of anatomy, epidermal hairs, and foliage composition. *J. Am. Soc. Hortic. Sci.* **102**: 166–171.

Quisenberry, K. S. and L. P. Reitz, 1967. Wheat and wheat improvement. *Agron. Ser.* **13**: XVI, 560 pp. American Society of Agronomy, Madison, Wis.

Radcliffe, E. G. and R. K. Chapman, 1965. Seasonal shifts in the relative resistance to insect attack of eight commercial cabbage varieties. *Ann. Entomol. Soc. Am.* **58**: 892–902.

Radcliffe, E. B. and R. K. Chapman, 1965. The relative resistance to insect attack of three cabbage varieties at different stages of plant maturity. *Ann. Entomol. Soc. Am.* **58**: 897–902.

Radcliffe, E. B. and R. K. Chapman, 1966. Varietal resistance to insect attack in various cruciferous crops. *J. Econ. Entomol.* **59**: 120–125.

Ramey, H. H., 1962. Genetics of plant pubescence in Upland cottons. *Crop. Sci.* **2**: 269.

Ramey, H. H., 1966. Historical review of cotton variety development. *Proc. Joint Mt. 26th Ann. Cotton Defol. Physiol. Conf. and 18th Ann. Cotton Improv. Conf.*: 310-226. National Cotton Council, Memphis, Tenn.

Rangdang, Y., 1971. Artificial media and rearing techniques for the corn stem borer, *Ostrinia salentialis. Proc. Seventh Inter-Asian Improv. Workshop, Philippines*: 116-119.

Rao, N. G. and L. R. House, Eds., 1972. *Sorghum in the Seventies.* Oxford and IBH, New Delhi, 636 pp.

Ratcliffe, R. H., 1974. Evaluating alfalfa weevil feeding resistance. In *Standard Tests to Characterize Pest Resistance in Alfalfa Varieties*, Barnes et al., Eds. *U.S. Dep. Agric. ARS* **NC-19**: 23 pp.

Raulston, J. R. and T. N. Shaver, 1970. A low agar casein-wheat germ diet for rearing tobacco budworms. *J. Econ. Entomol.* **63**: 1743-1744.

Reddy, M. S. and J. B. Weaver, Jr., 1975. Boll weevil nonpreference associated with several morphological characters in cotton. *Proc. Beltwide Cotton Prod. Res. Conf.*: 97. National Cotton Council, Memphis, Tenn.

Reddy, P. S. C., 1974. Effects of three leaf shape genotypes of *Gossypium hirsutum* L. and row types on plant microclimate, boll weevil survival, boll rot and important agronomic characters. Ph.D. Dissertation, Louisiana State University, Baton Rouge.

Reed, D. K. and P. L. Adkisson, 1961. Short-day cotton stocks as possible sources of host plant resistance to the pink bollworm. *J. Econ. Entomol.* **54**: 484-486.

Reed, G. L., T. A. Brindley, and W. B. Showers, 1972. Influence of resistant corn leaf tissue on the biology of the European corn borer. *Ann. Entomol. Soc. Am.* **65**: 658-662.

Reed, W., 1974. Selection of cotton varieties for resistance to insect pests in Uganda. *Cotton Grow. Rev.* **51**: 106-123.

Rees, C. J. C., 1969. Chemoreceptor specificity associated with choice of feeding site by the beetle, *Chrysolina brunsvicensis*, on its hostplant, *Hypericum hirsutum.* In *Insect and Host Plant*, J. de Wilde and L. M. Schoonhoven, Eds. North-Holland, Amsterdam, pp. 563-583.

Reinert, J. and Y. P. S. Bajaj, Eds., 1977. *Plant Cell, Tissue and Organ Culture.* Springer-Verlag, Berlin, 803 pp.

Reissig, W. H. and G. E. Wilde, 1971. Feeding responses of western corn rootworm on silks of fifteen genetic sources of corn. *J. Kans. Entomol. Soc.* **44**: 479-483.

Reitz, L. P. and K. L. Lebsock, 1972. Distribution of the varieties and classes of wheat in the United States in 1969. *U.S. Dep. Agric. ARS Stan Bull.* **475**: 70 pp.

Reitz, L. P., 1976. Improving germplasm resources. In *Agronomic Research for Food. Am. Soc. Agron. Spec. Publ.* **26**: 85-97.

Reitz, L. P. and W. G. Hamlin, 1978. Distribution of the wheat varieties and classes of wheat in the United States in 1974. *U.S. Dep. Agric. Stat. Bul.* **604**: 1-93.

Rezaul Karim, A. N. M., 1975. Resistance to the brown planthopper Nilaparvata lugens (Stal) in rice varieties. M.S. Thesis, University of the Philippines, Los Baños, 131 pp.

Rhine, J. J. and R. Staples, 1968. Effect of high amylose field corn on larval growth and survival of five species of stored-grain insects. *J. Econ. Entomol.* **61**: 280–282.

Rhodes, Am. M. and W. H. Luckhmann, 1967. Survival and reproduction of corn leaf aphid on twelve maize genotypes. *J. Econ. Entomol.* **67**: 527–530.

Rhoades, D. F. and R. G. Cates, 1976. Toward a general theory of plant anti-herbivore chemistry. *Recent Adv. Phytochem.* **10**: 168–213.

Ringlund, K. and E. H. Everson, 1968. Leaf pubescence in common wheat, *Triticum aestivum* L., and resistance to the cereal leaf beetle, *Oulema melanopus*. *Crop Sci.* **8**: 705–710.

Roberts, D. W. A. and C. Tyrrell, 1961. Sawfly resistance in wheat. IV. Some effects of light intensity on resistance. *Can. J. Plant Sci.* **41**: 457–465.

Robertson, D. S. and E. V. Walter, 1963. Genetic studies of earworm resistance in maize utilizing a series of chromosome-nine translocations. *J. Hered.* **54**: 267–272.

Robinson, A. G., 1960. Effect of maleic hydrazide and other plant growth regulators on the pea aphid. *Acyrthosiphon pisum* (Harris), caged on broad bean, *Vicia faba* L. *Can Entomol.* **92**: 494–499.

Robinson, A. G., 1961. Effects of amitrole, zytron, and other herbicides or plant growth regulators on the pea aphid, *Acyrthosiphon pisum* (Harris), caged on broad bean, *Vicia faba* L. *Can J. Plant Sci.* **41**: 413–417.

Robinson, R. A., 1971. Vertical resistance. *Rev. Pl. Pathol.* **50**: 233–239.

Robinson, R. A., 1973. Horizontal resistance. *Rev. Pl. Pathol.* **52**: 484–501.

Robinson, R. A., 1976. *The Pathosystem Concept. Plant Pathosystems.* Springer-Verlag, Berlin, 184 pp.

Robinson, R. A., 1977. Crop resistance may be our best crop protection. *World Crops Livest.* **29**: 104–106.

Robinson, R. A., 1977. Plant pathosystems. In *The Genetic Basis of Epidemics in Agriculture*, P. R. Day, Ed. *Ann. N.Y. Acad. Sci.* **287**: 1–400.

Robinson, T., 1974. Metabolism and function of alkaloids in plants. *Science* **184**: 430–435.

Rochow, W. F. and V. F. Eastop, 1966. Variation within *Ropalosiphym padi* and transmission of barley yellow dwarf virus by clones of four aphid species. *Virology* **30**: 286–296.

Rockefeller Foundation, 1973. *Sorghum—A Bibliography of the World Literature, 1964–1969*. Scarecrow, Metuchen, N.J. 393 pp.

Rodgers, R. R. and R. B. Mills, 1974. Reactions of sorghum varieties to maize weevil infestation under three relative humidities. *J. Econ. Entomol.* **67**: 692.

Rodriguez, J. G., H. H. Chen, and W. T. Smith, Jr., 1957. Effects of soil insecticides on beans, soybeans, and cotton and resulting effect on mite nutrition. *J. Econ. Entomol.* **50:** 587–593.

Rodriguez, J. G., 1960. Nutrition of the host and reaction to pests. *Am. Assoc. Adv. Sci.* **61:** 149–167.

Rodriguez, J. G., H. H. Chen, and W. T. Smith, Jr., 1960a. Effects of soil insecticides on apple trees and resulting effect on mite nutrition. *J. Econ. Entomol.* **53:** 487–490.

Rodriguez, J. G., D. E. Maynard, and W. T. Smith, Jr., 1960b. Effects of soil insecticides and absorbents on plant sugars and resulting effect on mite nutrition. *J. Econ. Entomol.* **53:** 491–495.

Rodriguez, J. G. and J. M. Campbell, 1961. Effects of gibberelin on nutrition of the mites, *Tetranychus telarius* and *Panonychus ulmi. J. Econ. Entomol.* **54:** 984–987.

Rodriguez, J. G., 1964. Nutritional studies in the Acarina. *Acarologia* **6:** 324–377.

Rodriguez, J. G., 1969. Dietetics and nutrition of *Tetranychus urticae* Koch. In *Proc. 2nd Int. Congr. Acarol.:* 469–475.

Rodriguez, J. G., C. E. Chaplin, L. P. Stoltz, and A. M. Lasheen, 1970. Studies on resistance of strawberries to mites. I. Effects of plant nitrogen. *J. Econ. Entomol.* **63:** 1855–1858.

Rodriguez, J. G., Ed., 1972. *Insect and Mite Nutrition.* North-Holland, Amsterdam, 702 pp.

Rodriguez-Rivera, R., 1972. Resistance to the white-backed planthopper *Sogatella furcifera* (Horvath) in rice varieties. M.S. Thesis, University of the Philippines, Los Baños, 60 pp.

Rogers, R. R. and R. B. Mills, 1974a. Evaluation of a world sorghum collection for resistance to the maize weevil. *Sitophilus zeamais* Motsch. (Coleoptera: Curculionidae). *J. Kans. Entomol. Soc.* **47:** 36–41.

Rogers, R. R. and R. B. Mills, 1974b. Reactions of sorghum varieties to maize weevil infestation under three relative humidities. *J. Econ. Entomol.* **67:** 692.

Rogers, R. R., W. A. Russell, and J. C. Ownes, 1976. Evaluation of a vertical pull technique in population improvement of maize for corn rootworm tolerance. *Crop Sci.* **16:** 591–594.

Rosenthal, G. A., D. L. Dahlman, and D. H. Janzen, 1976. A novel means of dealing with L-canavanine, a toxic metabolite. *Science* **192:** 256–258.

Rosenthal, G. A., D. H. Janzen, and D. L. Dahlman, 1977. Degradation and detoxification of canavanine by a specialized seed predator. *Science* **196:** 658–660.

Ross, E. S. and T. E. Moore, 1975. New species in the *Empoasca fabae* complex. *Ann. Entomol. Soc. Am.* **50:** 118–122.

Ross, W. M., 1973. Use of population breeding in sorghum—problems and progress. *Proc. 28th Corn Sorghum Res. Util. Conf. Am. Seed Trade Assoc.:* 30–42.

Rossetto, C. J. and N. V. Banzatto, 1967. Resistencia de variedades de sorgo a *Contarinia sorghicola* (Coquillet). VII. *Reunion Latino Americana de fitotecnia, Maracay, Venequela 17–23 Setiembre. Resume Trabajos Cient.*: 292–293.

Rossetto, C. J., and R. H. Painter, and D. Wilbur, 1973. Resistancia de variedades arroz en casca a *Sitophilus zeamais* Motchulsky. *Histochem. Latinoam.* **9:** 10–18.

Roth, E. R., 1970. Resistance: a literature review of important insects and diseases. U.S. Dep. Agric. For. Serv. Southeastern Area Rep. **28:** 59 pp.

Rout, G., B. Senapati, and T. Ahmed, 1976. Studies in relative susceptibility of some high yielding varieties of rice to the rice weevil, *Sitophilus oryzae* L. (Curculionidae: Coleptera). *Bull. Grain Technol.* **14:** 34–38.

Rudinsky, J. A., 1966. Host selection and invasion by the Douglas-fir beetle, *Dendroctonus pseudotsugae* Hopkins, in coastal Douglas-fir forests. *Can. Entomol.* **98:** 98–111.

Russell, G. E., 1969. Different forms of inherited resistance to virus yellows in sugar beet. *J. Int. Inst. Sugar Beet Res.* **4:** 131–144.

Russell, L. M., 1957. Distribution of legume-infesting *therioaphidine* aphids. *Plant Prot. Bull.* **5:** 78.

Russell, M. P., 1962. Effects of sorghum varieties on the lesser rice weevil, *Sitophilus oryzae* (L.) I. Oviposition, immature mortality, and size of adults. *Ann. Entomol. Soc. Am.* **55:** 678–685.

Russell, M. P., 1966. Effects of four sorghum varieties on the longevity of the lesser rice weevil, *Sitophilus oryzae* (L.). *J. Stored Prod. Res.* **2:** 75–79.

Russell, M. P., 1976. Resistance of commercial rice varieties to *Sitotroga cerealella* (Olivier) (Lepidoptera: Gelchiidae). *J. Stored Prod. Res.* **12:** 105–109.

Russell, M. P. and R. R. Cogburn, 1977. World collection rice varieties: resistance to seed penetration by *Sitotroga cerealella* (Olivier) (Lepidoptera: Gelchiidae). *J. Stored Prod. Res.* **13:** 103–106.

Russell, W. A., L. H. Penny, W. D. Guthrie, and F. F. Dicke, 1971. Registration of maize germplasm inbreds. *Crop Sci.* **11:** 140.

Russell, W. A., W. D. Guthrie, and R. L. Grindeland, 1974. Breeding for resistance in maize to first and second broods of the European corn borer. *Crop Sci.* **14:** 725–727.

Russell, W. A., 1975. Breeding and genetics in the control of insect pests. *Iowa State J. Res.* **49:** 527–551.

Ryan, C. A., 1973. Proteolytic enzymes and their inhibitors in plants. *Ann. Rev. Plant Physiol.* **24:** 173–196.

Saglio, P., M. L'Hospital, D. Lafleche, G. Dupont, J. M. Bove, J. G. Tully, and E. A. Freundt, 1973. *Spiroplasma citri* gen. and sp.n.: a mycoplasma-like organism associated with "stubborn" disease of citrus. *Int. J. Syst. Bacteriol.* **23:** 191–204.

Sajjan, S. S., 1972. Varietal resistance of rice to *Nephotettix nigropictus* (Stål). Report of the work done during the fellowship period, Nov. 18, 1971 to May 23, 1972 of International Rice Research Institute, Los Baños, Philippines, 20 pp.

Sakamura, T., 1918. Kurze Mitteilung uber die Chromosomenzahlen und die Verwandtschaftsverhaltnisse der *Triticum*-Arten. *Bot. Mag. (Tokyo)* **32:** 151–154.

Saksena, K. N., S. R. Singh, and W. H. Sill, Jr., 1964. Transmission of barley yellow dwarf virus by four biotypes of the corn leaf aphid *Rhopalosiphum maidis. J. Econ. Entomol.* **57:** 569–571.

Sanchez, F. F., H. A., Custodio, S. Bayubay, M. T. Madrid, Jr., E. D. Magallona, G. Bautista, V. A. Dyck, S. K. de Datta, and R. Feuer, 1976. 1976–1977 Interagency insect control guide for lowland transplanted rice in the Philippines. Unpublished report, National Food and Agriculture Council, Manila. Mimeographed, 36 pp.

Santamour, F. S., 1964. Is there genetic resistance to the white pine weevil in *Pinus strobus? Proc. 11th Northeast For. Tree Impr. Conf.*: 49–51.

Saringer, G., 1976. Oviposition behaviour of *Ceutorrhynchus maculaalba* Herbst. (Col.: Curculionidae). *Symp. Biol. Hung.* **16:** 241–245.

Sarup, P., B. K. Mukherjee, K. K. Marwaha, V. P. S. Panwar, K. H. Siddiqui, and N. N. Singh, 1974. Identification of a source of resistance to *Chilo partellus* (Swinhoe) in Colombia maize hybrid H207 and formulation of a suitable breeding procedure for its utilization. *Indian J. Entomol.* **36:** 1–5.

Sasamoto, K., 1957. Studies on the relation between the silica content in rice-plant and the insect pests. *Botyu-Kagaku (Inst. Insect Control)* **22:** 159–164.

Sasamoto, K., 1958. Studies on the relation between the silica content in rice plant and the insect pests. IV. On the injury of silicated rice plant caused by the rice stem borer and its feeding behavior. *Jap. J. Appl. Entomol. Zool.* **2:** 88–92.

Sastry, M. V. S. and P. S. Prakasa Rao, 1973. Inheritance of resistance to rice gall midge *Pachydiplosis oryzae* Wood-Mason. *Curr. Sci.* **42:** 652–653.

Satyanarayanaiah, K. and M. V. Reddi, 1972. Inheritance of resistance to insect gall-midge (*Pachydiplosis oryzae* Wood Mason) in rice. *Andhra Agric. J.* **19:** 1–8.

Sax, K., 1922. Sterility in wheat hybrids. II. Chromosome behavior in partially sterile hybrids. *Genetics* **7:** 513–552.

Saxena, R. C. and M. D. Pathak, 1977. Factors affecting resistance of rice varieties to the brown planthopper, *N. lugens.* Paper presented at the *8th Conf. Pest Control Council Philipp. Bacolod City, May 18–20, 1977,* 34 pp.

Scales, A. L. and J. Hacskaylo, 1974. Interaction of three cotton cultivars to infestations of the tarnished plant bug. *J. Econ. Entomol.* **67:** 602–604.

Schalk, J. M., S. D. Kindler, and G. D. Manglitz, 1969. Temperature and preference of the spotted alfalfa aphid for resistant and susceptible alfalfa plants. *J. Econ. Entomol.* **62:** 1000–1003.

Schertz, K. F. and L. E. Clark, 1967. Controlling dehiscence with plastic bags for hand crosses in sorghum. *Crop Sci.* **7:** 540–542.

Schillinger, J. A., F. C. Elliott, and R. F. Ruppel, 1964. A method for screening alfalfa plants for potato leafhopper resistance. *Mich. Agric. Exp. Stn. Bull.* **46:** 512–517.

Schillinger, J. A., 1966. Larval growth as a method of screening *Triticum* sp. for resistance to the cereal leaf beetle. *J. Econ. Entomol.* **59:** 1163–1166.

Schillinger, J. A., 1969. Three laboratory techniques for screening small grains for resistance to the cereal leaf beetle. *J. Econ. Entomol.* **62:** 360–363.

Schillinger, J. S. and R. L. Gallun, 1968. Leaf pubescence of wheat as a deterrent to the cereal leaf beetle, *Oulema melanopus. Ann. Entomol. Soc. Am.* **61:** 900–903.

Schlesinger, H. M., S. W. Applebaum, and Y. Birk, 1976. Effect of β-cyano-L-alanine on the water balance of *Locusta migratoria. J. Insect Physiol.* **22:** 1421–1425.

Schlosberg, J. and W. A. Baker, 1948. Tests of sweet corn lines for resistance to European corn borer larvae. *J. Agric. Res.* **77:** 137–159.

Schneider, F., 1944. Eine Ursache der raschen Blattausvermehrunj an Bohnen. *Fortschr. Ergebn. Gartenb.* **5:** 4.

Schonbörn, A. von, 1966. The breeding of insect-resistant forest trees in central and northwestern Europe. In *Breeding Pest Resistant Trees,* H. D. Gerhold et al., Eds. Pergamon Press, London, pp. 25–27.

Schoonhoven, A. V., C. E. Wassom, and E. Horber, 1972. Development of maize weevil on kernels of opaque-2 and floury-2, nearly isogenic corn inbred lines. *Crop Sci.* **12:** 862–863.

Schoonhoven, A. V., R. B. Mills, and E. Horber, 1974. Development of *Sitophilus zeamais* Motschulsky, in maize kernels and pellets made from maize kernel fractions. *J. Stored Prod. Res.* **10:** 73–80.

Schoonhoven, A. van, 1974. Resistance to thrips damage in cassava, *J. Econ. Entomol.* **67:** 728–730.

Schoonhoven, A. V., E. Horber, C. E. Wassom, and R. B. Mills, 1975. Selection for resistance to the maize weevil in kernels of maize. *Euphytica* **24:** 639–644.

Schoonhoven, A. V., E. Horber, and R. B. Mills, 1976. Conditions modifying expression of resistance of maize kernels to the maize weevil. *Environ. Entomol.* **5:** 163–168.

Schoonhoven, A. van, and J. Pena, 1976. Estimation of yield losses in cassava following attack from thrips. *J. Econ. Entomol.* **69:** 514–516.

Schoonhoven, L. M., 1967. Chemoreception of mustard oil glucosides in larvae of *Pieris brassicae. Proc. K. Ned. Akad. Wet. Ser. C* **70:** 556–568.

Schoonhoven, L. M., 1972. Plant recognition by lepidopterous larvae. *Symp. Roy. Entomol. Soc. London* **6:** 87–99.

Schoonhoven, L. M., 1977. On the individuality of insect feeding behaviour. *Proc. K. Ned. Akad. Wet.:* in press.

Schreiber, K., 1957. Naturally occurring plant resistance factors against the Colorado potato beetle (*Leptinotarsa decemlineata*) and their possible mode of action. *Zuchter.* **27**: 289–299.

Schroder, R. F. W. and A. L. Steinhauer, 1976. Studies of cross-mating among strains of the alfalfa weevil from the United States and western Europe. *Proc. Entomol. Soc. Washington* **78**: 1–5.

Schuster, D. J. and K. J. Starks, 1973. Greenbugs: components of host-plant resistance in sorghum. *J. Econ. Entomol.* **66**: 1131–1134.

Schuster, M. F., F. G. Maxwell, and J. N. Jenkins, 1972a. Resistance to the two-spotted spider mite in certain *Gossypium hirsutum* races, *Gossypium* species, and glandless counterpart cottons. *J. Econ. Entomol.* **65**: 1108–1110.

Schuster, M. F., F. G. Maxwell, and J. N. Jenkins, 1972b. Antibiosis to two-spotted spid·r mite in Upland and American Pima cotton. *J. Econ. Entomol.* **65**: 1110–1111.

Schuster, M. F., F. G. Maxwell, J. N. Jenkins, E. T. Cherry, W. L. Parrott, and D. G. Holder, 1973. Resistance to two-spotted spider mite in cotton. *Miss. Agric. For. Exp. Stn. Bull.* **802.**

Schuster, M. F. and F. G. Maxwell, 1974. The impact of nectariless cotton on plant bugs, bollworms and beneficial insects. *Proc. Beltwide Cotton Prod. Res. Conf.*: 86–87. National Cotton Council, Memphis, Tenn.

Schuster, M. F. and J. L. Frazier, 1976. Mechanisms of resistance to *Lygus* spp. in *Gossypium hirsutum* L. EUCARPIA/OILB Host Plant Resistance Proc., Wageningen, The Netherlands, pp. 129–135.

Schuster, M. F., D. G. Holder, E. T. Cherry, and F. G. Maxwell, 1976. Plant bugs and natural enemy insect populations on frego bract and smoothleaf cottons. *Miss. Agric. For. Exp. Stn. Tech. Bull.* **75**: 1–11.

Schuster, M. F. and F. G. Maxwell, 1976. Resistance to two-spotted spider mite in cotton. *Miss. Agric. For. Exp. Stn. Bull.* **821**: 13 pp.

Schwartze, C. D. and G. A. Huber, 1937. Aphid resistance in breeding mosaic escaping red raspberries. *Science* **86**: 158–159.

Scott, G. E., A. R. Hallauer, and F. F. Dicke, 1964. Types of gene action conditioning resistance to European corn borer leaf feeding. *Crop Sci.* **4**: 603–606.

Scott, G. E., F. F. Dicke, and L. H. Penny, 1965. Effects of first brood European corn borers on single crosses grown at different nitrogen and plant population levels. *Crop Sci.* **5**: 261–263.

Scott, G. E. and W. D. Guthrie, 1967. Reaction of permutations of maize double crosses to leaf feeding of European corn borers. *Crop Sci.* **7**: 233–235.

Scott, G. E., W. D. Guthrie, and G. R. Pesho, 1967. Effect of second-brood European corn borer infestation on 45 single-cross corn hybrids. *Crop Sci.* **7**: 229–230.

Scriber, J. M., W. M. Tingey, V. E. Gracen, and S. L. Sullivan, 1975. Leaf feeding resistance to the European corn borer in genotypes of tropical (Low Dimboa) and U.S. inbred (High Dimboa) maize. *J. Econ. Entomol.* **68**: 823–826.

Scriber, J. M., 1977. Limiting effects of low leaf-water content on the nitrogen utilization, energy budget, and larval growth of *Hyalophora cecropia* (Lepidoptera: Saturniidae). *Oecologia* **28**: 269–287.

Seaman, F., M. J. Lukefahr, and T. J. Mabry, 1977. The chemical basis of the natural resistance of *Gossypium hirsutum* L. to *Heliothis. Proc. Beltwide Cotton Prod. Res. Conf.*: 102–103. National Cotton Council, Memphis, Tenn.

Sebesta, E. E. (n.d.). Germplasm release of 'Amigo' greenbug resistant wheat. *Crop Sci.*: in press.

Secretariat IBPGR, 1977. *1976 Annual Report of International Board of Plant Genetic Resources.* FAO, Rome.

Seigler, D. and P. W. Price, 1976. Secondary compounds in plants: primary functions. *Am. Nat.* **110**: 101–105.

Self, L. S., E. F. Guthrie, and E. Hodgson, 1964. Adaptation of tobacco hornworms to the ingestion of nicotine. *J. Insect Physiol.* **10**: 907–914.

Selman, B. J. 1973. In *Viruses and Invertebrates,* A. J. Gibbs, Ed. American Elsevier, New York, pp. 157–177.

Selman, I. W. and V. Kandiah, 1971. Influence of gibberellic acid and nitrogen sprays on transmission of cabbage black ringspot virus by *Myzus persicae* (Sulz.). *Bull. Entomol. Res.* **60**: 359–365.

Sethi, R. L., P. L. Sethi, and T. R. Mehta, 1937. Inheritance of sheathed ear in rice. *Indian J. Agric. Sci.* **7**: 134–148.

Shade, R. E., T. E. Thompson, and W. R. Campbell, 1975. An alfalfa weevil larval resistance mechanism detected in *Medicago. J. Econ. Entomol.* **68**: 399–404.

Shands, R. G. and W. B. Cartwright, 1953. A fifth gene conditioning Hessian fly response in common wheat. *Agron. J.* **45**: 302–307.

Shao, Chi-chuan, Li Chang-sen, and Chiren Baschan, 1975. Origin and evolution of the cultivated barley—wild barley from the south-western part of China. *Acta Genet. Sinica* **2**: 128 (abstract).

Shapiro, I. D. and R. I. Bartoshko, 1973. Resistance of winter wheat varieties to larvae and newly emerged images of *Eurygaster integriceps* Put. *Proc. All Union Sci. Res. Inst. Plant Prot.* **37**: 41–48.

Sharma, R. K., V. M. Stern, and R. W. Hagemann, 1975. Blue alfalfa aphid damage and its chemical control in the Imperial Valley California. In *Proc. Fifth Calif. Alfalfa Symp.*: 29–30.

Sharma, V. K. and S. M. Chatterji, 1972. Further studies on the nature of antibiosis in maize (*Zea mays* Linn.) against the maize stem borer, *Chilo zonellus* (Swinhoe). *Indian J. Entomol.* **34**: 11–19.

Shastry, S. V. S., W. H. Freeman, D. V. Seshu, P. Israel, and J. K. Roy, 1972. Host plant resistance to rice gall midge. In *Rice Breeding.* IRRI, Los Baños, Philippines, pp. 353–365.

Shaver, T. N. and M. J. Lukefahr, 1971. A bioassay technique for detecting resistance of cotton strains to tobacco budworms. *J. Econ. Entomol.* **64**: 1274–1277.

Shaw, G. G. and C. H. A. Little, 1972. Effect of high urea fertilization of balsom fir trees of spruce budworm development. In *Insect and Mite Nutrition,* J. G. Rodriguez, Ed. North-Holland, Amsterdam, pp. 589–597.

Shea, K. R., 1971. Disease and insect activity in relation to intensive culture of forests. *Proc. 15th Congr. Int. Union For. Res. Organ.,* Gainesville, Fla., **3:** 14–20, 109–117.

Shikata, E. and K. Maramorosch, 1969. Electron microscopy of insect-borne viruses *in situ.* In *Viruses, Vectors, and Vegetation,* K. Maramorosch, Ed. John Wiley & Sons, Inc., New York, pp. 393–415.

Shimura, I., 1972. Breeding of chestnut varieties to chestnut gall wasp, *Dryocosmus kuriphilus* Yasumatsu. *Jap. Agric. Res. Quart.* **6:** 224–230.

Shindy, W. and R. J. Weaver, 1967. Plant regulators alter translocation of photosynthetic products. *Nature* **214:** 1024–1025.

Shinji, J. O. and T. Kondo, 1938. Aphidae of Manchoukuo with the description of two new species. *Kontyu* **12:** 55–69.

Shorey, H. H. and R. L. Hale, 1965. Mass rearing of the larvae of nine noctuid species of a simple artificial medium, *J. Econ. Entomol.* **58:** 522–524.

Showers, W. B. and G. S. Reed, 1972. Methods of infesting corn with the European corn borer. *Iowa State J. Sci.* **46:** 429–434.

Shurtleff, M. C., Ed., 1973. *A Compendium of Corn Diseases.* The American Phytopathological Society, Inc., St. Paul, Minn.

Sifuentes, J. A., 1964. Reaction of field corn parents and hybrids to attack by the corn earworm (*Heliothis zea* Boddie). Ph.D. Dissertation, Kansas State University, Manhattan.

Sifuentes, J. A. and R. H. Painter, 1964. Inheritance of resistance to western corn rootworm adults in field corn. *J. Econ. Entomol.* **57:** 475–477.

Simpson, G. W. and W. A. Shands, 1949. Progress in some important insect and disease problems of Irish potato production in Maine. *Bull. Maine Agric. Exp. Stn.* **470.**

Singh, A. and D. K. Butani, 1963. Insect pests of cotton: I. Incidence of pink bollworm (*Pectinophora gossypiella* Saunders) in the Punjab. *Indian Cotton Grow. Rev.* **17:** 343–348.

Singh, B. B., H. H. Hadley, and R. L. Bernard, 1971. Morphology of pubescence in soybeans and its relationship to plant vigor. *Crop Sci.* **11:** 13–16.

Singh, D. N. and F. S. McCain, 1963. Relationship of some nutritional properties of the corn kernel to weevil infestation. *Crop Sci.* **3:** 259–261.

Singh, I. D. and J. B. Weaver, Jr., 1972. Studies on the heritability of gossypol in leaves and flower buds of *Gossypium. Crop Sci.* **12:** 294–297.

Singh, J., 1967. Studies on breeding in maize for resistance to top shoot borer (*Chilo zonellus* Swin.). Ph.D. Thesis, P. G. School, Indian Agricultural Research Institute, New Delhi.

Singh, K., N. S. Agarwal, and G. K. Girish, 1972. The oviposition and development of *Sitophilus oryzae* (L.) (Coleoptera: Curculionidae) in different maize hybrids and corn parts. *Indian J. Entomol.* **34:** 148–154.

Singh, P., 1970. Host-plant nutrition and composition: effects on agricultural pests. *Can. Dep. Agric. Res. Inst. Inf. Bull.* **6:** 1–102.

Singh, P., 1972. Bibliography of artificial diets for insects and mites. *N. Z. Dep. Sci. Ind. Res. Wellington Bull.* **209:** 75 pp.

Singh, R., 1953. Inheritance in maize of reaction to the European corn borer. *Indian J. Genet. Plant Breed.* **13:** 18047.

Singh, S. R. and R. H. Painter, 1965. Reactions of four biotypes of corn leaf aphid, *Rhopalosiphum maidis* (Fitch) to differences in host plant nutrition. *Proc. 12th Int. Cong. Entomol.* **1964:** 543.

Singh, S. R. and D. P. Sharma, 1971. Behavior of stem fly, *Atherigona varia soccata* Rond. under different environmental conditions and its reaction on different sorghum varieties. *Proc. 13th Int. Congr. Entomol.* (*1968*) **2:** 388.

Singh, S. R. and S. Soenardi, 1973. Insect pests of Java, Indonesia. *Int. Rice Comm. Newsl.* **22:** 22–25.

Singh, S. R., 1975. A proposal for integrated control of cowpea insect pests. In *Proc. I.I.T.A. Collab. Meet. Grain Legume Improv.*: 41–43.

Sinskaya, E. N., 1950. Flora of cultivated plants of the U.S.S.R. XIII. Part Transl. 1961. Israel Program of Scientific Translations, Jerusalem, 425 pp.

Siwi, B. H. and G. S. Kush, 1977. New genes for resistance to the green leafhopper in rice. *Crop Sci.* **17:** 17–20.

Smith, B. D., 1969. Spectra of activity of plant growth retardants against various parasites of one host species. *J. Sci. Food Agric.* **20:** 398–400.

Smith, C. N., Ed., 1966. *Insect Colonization and Mass Production.* Academic Press, Inc., New York, 618 pp.

Smith, D. H., Jr., T. Ninam, E. Rothke, and C. E. Cress, 1971. Weight gain of cereal leaf beetle larvae on normal and induced leaf pubescence. *Crop Sci.* **11:** 639–641.

Smith, D. H. and J. A. Webster, 1973. Resistance to cereal leaf beetle in Hope substitution lines. In *Proc. 4th Int. Genet. Symp., Columbia, Mo.*: 761–764.

Smith, D. S., 1960. Effects of changing the phosphorus content of the food plant on the migratory grasshopper, *Melanoplus bilituratus* (Walker) (Orthoptera: Acrididae). *Can. Entomol.* **92:** 103–107.

Smith, F. F., 1965. Plant virus–vector relationships. *Adv. Virsus Res.* **9:** 61–96.

Smith, H. S., 1941. Racial segregation in insect populations and its significance in applied entomology. *J. Econ. Entomol.* **34:** 1–13.

Smith, K. M., 1965. Plant virus-vector relationships. *Adv. Virus Res.* **11:** 61–96.

Smith, O. F., R. N. Peaden, and R. K. Petersen, 1958. Moapa alfalfa. *Nev. Agric. Exp. Stn. Circ.* **15:** 15 pp.

Smith, O. F. and R. N. Peaden, 1960. A method of testing alfalfa plants for resistance to the pea aphid. *Agron. J.* **52:** 609–610.

Smith, R. F., J. E. Swift, and J. Dibble, 1956. Rapid spread of alfalfa pest. *Calif. Agr.* **10:** 5.

Smith, R. H., 1960. Resistance of pines to the pine reproduction weevil. *J. Econ. Entomol.* **53:** 1044–1048.

Smith, R. H., 1966a. Resin quality as a factor in the resistance of pines to bark beetles. In *Breeding Pest Resistant Trees*, H. D. Gerhold et al., Eds. Pergamon Press, London, pp. 189–196.

Smith, R. H., 1966b. Research and development of pines resistant to the pine reproduction weevil, *Cylindrocopturus eatoni* Buch. In *Breeding Pest Resistant Trees*, in H. Gerhold et al., Eds. Pergamon Press, London, pp. 363–365.

Smith, R. H., 1975. Formula for describing effect of insect and host tree factors on resistance to western pine beetle attack. *J. Econ. Entomol.* **68:** 841–844.

Smith, R. K., D. A. Miller, and E. J. Armbrust, 1974. Effect of boron on alfalfa weevil oviposition. *J. Econ. Entomol.* **67:** 130.

Smith, R. L., R. L. Wilson, and F. D. Wilson, 1975. Resistance of cotton plant hairs to mobility of first instars of the pink bollworm. *J. Econ. Entomol.* **68:** 679–683.

Snelling, R. O., R. H. Painter, J. H. Parker, and W. M. Osborn, 1937. Resistance of sorghums to the chinch bug. *U.S. Dep. Agric. Tech. Bull.* **585:** 56 pp.

Snelling, R. O., R. A. Blanchard, and J. H. Bigger, 1940. Resistance of corn strains to the leaf aphid, *Aphis maidis* Fitch. *Agron. J.* **32:** 371–381.

Snelling, R. O., 1941. Resistance of plants to insect attack. *Bot. Res.* **7:** 543–586.

Snelling, R. O., 1941. Resistance of plants to insect attack. *Bot. Rev.* **7(10):** 543–586.

Søegaard, B., 1964. Breeding for resistance to insect attack in forest trees. *Unasylva* **18:** 82–88.

Soehardjan, M. Rudhendi, and J. Leeuwangh, 1975. Varietal screening for resistance to the rice gall midge. In *Agricultural Cooperation, Indonesia–The Netherlands Research Reports 1968–1974. Section II—Technical Contributions Republic of Indonesia.* Ministry of Agriculture, Jakarta, pp. 155–159.

Saojitno, J. and G. van Vreden, 1975. Varietal screening for resistance to the armyworm, *Spodoptera mauritia*, in rice. In *Agricultural Cooperation, Indonesia–The Netherlands Research Reports 1968–1974. Section II—Technical Contributions Republic of Indonesia.* Ministry of Agriculture, Jakarta, p. 166.

Sogawa, K. and M. D. Pathak, 1970. Mechanisms of brown planthopper resistance to Mudgo rice variety of rice (Hemiptera: Delphacidae) *Appl. Entomol. Zool.* **5:** 145–158.

Sogawa, K., 1971. Feeding behaviors of the brown planthopper and varietal resistance of rice to this insect. In *Symp. Rice Insects. Proc. Symp. Trop. Agric. Res., 19–24 July, 1971:* 195–200.

Sohi, S. S. and K. G. Swenson, 1964. Pea aphid biotypes differing in bean yellow mosaic virus transmission. *Entomol. Exp. Appl.* **7:** 9–14.

Soles, R. L. and H. D. Gerhold, 1968. Caged white pine seedlings attacked by white pine weevil, *Pissodes strobi* Peck, at five population densities. *Ann. Entomol. Soc. Am.* **61**: 1468–1473.

Sorensen, E. L., R. H. Painter, H. L. Hackerott, and T. L. Harvey, 1969. Registration of Kanza Alfalfa (Reg. No. 41). *Crop Sci.* **9**: 847.

Sorensen, E. L., M. C. Wilson, and G. R. Manglitz, 1972. Breeding for insect resistance. In *Alfalfa Science and Technology*, C. H. Hanson, Ed. American Society of Agronomy, Madison, Wis. pp. 371–390.

Sorensen, E. L., 1974. Evaluating pea aphid resistance. In *Standard Tests to Characterize Pest Resistance in Alfalfa Varieties*, Barnes, et al., Eds., *U.S. Dept. Agric.* **NC-19**: 18–19.

Sosa, O., Jr. and J. E. Foster, 1976. Temperature and the expression of resistance in wheat to the Hessian fly. *Environ. Entomol.* **5**: 333–336.

Soto, P. E. and K. Laxminarayana, 1971. A method for rearing the sorghum shoot fly. *J. Econ. Entomol.* **64**: 553.

Soto, P. E., 1972. Mass rearing of the sorghum shoot fly and screening for host plant resistance under greenhouse conditions. In *Control of Sorghum Shoot Fly*, M. G. Jotwani and W. R. Young, Eds. Oxford and IBH, New Delhi, 324 pp.

Soto, P. E., 1974. Ovipositional preference and antibiosis in relation to resistance to a sorghum shoot fly. *J. Econ. Entomol.* **67**: 256–257.

Soto, P. E. and Z. Siddiqi, 1976. Screening for resistance to African rice insect pests. Paper presented at the *WARDA Varietal Improv. Semin. Bouake, Ivory Coast, Sept. 13–17, 1976*: 13 pp.

Sozonov, A. P., 1973. Anatomy of caryopses of wheat varieties showing different resitance to *Eurygaster integriceps* Put. *Proc. All Union Sci. Res. Inst. Plant Prot.* **39**: 76–81.

Sparks, A. N. and E. A. Harrell, 1976. Corn earworm rearing mechanization. *U.S. Dep. Agric. ARS Tech. Bull.* **1554**.

Sprague, G. F., Ed., 1977. *Corn and Corn Improvement.* Monograph 18. American Society of Agronomy, Madison, Wisconsin, 774 pp.

Sprague, G. F. and R. G. Dahms, 1972. Development of crop resistance to insects. *J. Environ. Qual.* **1**: 28–34.

Sprott, J. M., R. D. Lacewell, G. A. Niles, J. K. Walker, and J. R. Gannaway, 1976. Agronomic economic, energy and environment implications of short-season, narrow-row cotton production. *Tex. Agric. Exp. Stn. Misc. Publ.* **1250C**, 23 pp.

Stadelbacher, E. A. and A. L. Scales, 1973. Technique for determining oviposition preference of the bollworm and tobacco budworm for varieties and experimental stocks of cotton. *J. Econ. Entomol.* **66**: 418–421.

Stakman, E. C., 1968. The need for international testing centers for crop pests and pathogens. *Bull. Entomol. Soc. Am.* **14**: 124–128.

Stanford, E. H., 1952. Transfer of Disease Resistance to Standard Varieties. *Proc. Sixth Int. Grassland Cong.*: 1585–1589.

Staniland, L. N., 1924. The immunity of apple stocks from attacks of the wooly aphis (*Eriosoma lanigerum* Hausmann). Part II. The causes of the relative resistance of the stocks. *Bull. Entomol. Res.* **15**: 157–170.

Starks, K. J. and W. W. McMillian, 1967. Resistance in corn to the corn earworm and fall armyworm. Part II. Types of field resistance to the corn earworm. *J. Econ. Entomol.* **60**: 920–923.

Starks, K. J., M. C. Bowman, and W. W. McMillian, 1967. Resistance in corn to the corn earworm, *Heliothis zea,* and the fall armyworm, *Spodoptera frugiperda* (Lepidoptera: Noctuidae). Part III. Use of plant parts of inbred corn lines by the larvae. *Ann. Entomol. Soc. Am.* **60**: 873–874.

Starks, K. J., S. A. Eberhart, and H. Doggett, 1970. Recovery from shoot fly attack in a sorghum diallel. *Crop Sci.* **10**: 519–522.

Starks, K. J. and H. Doggett, 1970. Resistance to a spotted stem borer in sorghum and maize. *J. Econ. Entomol.* **63**: 1790–1795.

Starks, K. J., 1970. Increasing infestations of the sorghum shoot fly in experimental plots. *J. Econ. Entomol.* **63**: 1715–1716.

Starks, K. J., D. E. Weibel, and J. W. Johnson, 1972. Sorghum resistance to green bug. *Sorghum Newsl.* **15**: 130.

Starks, K. J., R. Muniappan, and R. D. Eikenbary, 1972. Interaction between plant resistance and parasitism against the greenbug on barley and sorghum. *Ann. Entomol. Soc. Am.* **65**: 650–655.

Starks, K. J., E. A. Wood, Jr., and G. L. Teetes, 1973. Effects of temperature on the preference of two greenbug biotypes for sorghum selections. *Environ. Entomol.* **2**: 351–354.

Starks, K. J. and R. L. Burton, 1977. Greenbugs: determining biotypes, culturing, and screening for plant resistance. *U.S. Dep. Agric. ARS Tech. Bull.* **1556**: 12 pp.

Steiner, K., 1974. Genetic differences in resistance of Scotch pine to eastern pineshoot borer. *Great Lakes Entomol.* **7**: 103–107.

Stephens, J. C. and J. R. Quinby, 1933. Bulk emasculation of sorghum flowers. *J. Am. Soc. Agron.* **25**: 233–234.

Stephens, J. C., F. R. Miller, and D. T. Rosenow, 1967. Conversion of alien sorghums to early combine genotypes. *Crop Sci.* **7**: 396.

Stephens, S. G., 1957. Sources of resistance of cotton strains to the boll weevil and their possible utilization. *J. Econ. Entomol.* **50**: 415–418.

Stephens, S. G., 1959. Laboratory studies on feeding and oviposition preferences of *Anthonomus grandis* Boh. *J. Econ. Entomol.* **52**: 390–396.

Stephens, S. G. and H. S. Lee, 1961. Further studies on the feeding and oviposition preferences of the boll weevil (*Anthonomus grandis*). *J. Econ. Entomol.* **54**: 1085–1090.

Stern, V. M., R. F. Smith, R. van den Bosch, and K. S. Hagen, 1959. The integrated control concept. *Hilgardia* **29**: 81–101.

Stern, V. M., P. L. Adkisson, O. G. Beingolea, and G. A. Viktorov, 1977. Cultural controls. In *Theory and Practice of Biological Control,* C. F. Huffaker, Ed. Academic Press, Inc. New York, pp. 593–613.

Stevens, R. A. and R. B. Mills, 1973. Comparison of technique for screening sorghum grain varieties for resistance to rice weevil. *J. Econ. Entomol.* **66**: 1222–1223.

Storms, J. J. H., 1969. Observations on the relationship between mineral nutrition of apple rootstocks in gravel culture and the reproduction rate of *Tetranychus urticae* (Acarina: Tetranychidae). *Entomol. Exp. Appl.* **12**: 297–311.

Straub, R. W. and M. L. Fairchild, 1970. Laboratory studies of resistance in corn to the corn earworm. *J. Econ. Entomol.* **63**: 1901–1903.

Strickberger, M. W., 1968. *Genetics.* Macmillan Co., New York, p. 868.

Strong, F. E., 1962. The reaction of some alfalfas to seed chalcid infestation. *J. Econ. Entomol.* **55**: 1004–1005.

Sullivan, S. L., V. E. Gracen, and A. Ortega, 1974. Resistance of exotic maize varieties to the European corn borer, *Ostrinia nubilalis* (Hubner). *Environ. Entomol.* **3**: 718–720.

Summers, C. G. and W. F. Lehman, 1976. Evaluation of non-dormant alfalfa cultivars for resistance to the Egyptian alfalfa weevil. *J. Econ. Entomol.* **69**: 29–34.

Suneson, C. A. and W. B. Noble, 1950. Further differentiation of genetic factors in wheat for resistance to the Hessian fly. *U.S. Dep. Agric. Tech. Bull.* **1004.**

Sutherland, O. R. W., 1969. The role of the host plant in the production of winged forms by two strains of the pea aphid. *Acyrthosiphon pisum. J. Insect Physiol.* **15**: 2179–2201.

Sylvester, E. S. (1956). Beet yellows virus transmission by the green peach aphid. *J. Econ. Entomol.* **49**: 789–800.

Sylvester, E. S. and J. Richardson, 1969. Additional evidence of multiplication of the sowthistle yellow vein virus in an aphid vector—Serial passage. *Virology* **37**: 26–31.

Tahori, A. S., A. H . Halevy, and G. Zeidler, 1965. Effect of some plant growth retardants on the oleander aphid, *Aphis nerii* (Boyer). *J. Sci. Food Agric.* **16**: 568–569.

Tanton, M. T., 1962. The effect of leaf "toughness" on the feeding of larvae of the mustard beetle *Phaedon cochleariae* Fab. *Entomol. Exp. Appl.* **5**: 74–78.

Tauber, M. J., B. Shalucha, and R. W. Langhans, 1971. Succinic acid-2,2-dimethylhydrazide (SADH) prevents whitefly population increase. *Hortsci.* **6**: 458.

Teetes, G. L. and J. W. Johnson, 1973. Damage assessment of the greenbug on grain sorghum. *J. Econ. Entomol.* **66**: 1181–1186.

Teetes, G. L. and J. W. Johnson, 1974. Assessment of damage by the greenbug in grain sorghum hybrids of different maturities. *J. Econ. Entomol.* **67**: 514–516.

Teetes, G. L., C. A. Schaefer, and J. W. Johnson, 1974. Resistance in sorghums to the greenbug; laboratory determination of mechanisms of resistance. *J. Econ. Entomol.* **67**: 393–396.

Teetes, G. L., C. A. Schaefer, J. R. Gipson, R. C. McIntyre, and E. E. Latham, 1975. Greenbug resistance to organophosphorous insecticides on the Texas High Plains. *J. Econ. Entomol.* **69**: 214–216.

Teetes, G. L., J. W. Johnson, and D. T. Rosenow, 1975a. Response of improved resistant sorghum hybrids to natural and artificial greenbug populations. *J. Econ. Entomol.* **68**: 546–548.

Teetes, G. L., E. G. Lopez, and C. A. Schaefer, 1975b. Seasonal abundance of the greenbug and its natural enemies in grain sorghum in the Texas High Plains. *Tex. Agric. Exp. Stn.* **B-1162**: 4 pp.

Teetes, G. L., J. W. Johnson, and D. T. Rosenow, 1976. Screening for corn leaf aphid resistant sorghums. *Sorghum Newsl.* **19**: 128.

Thielges, B. A., 1968. Altered polyphenol metabolism in the foliage of *Pinus sylvestris* associated with European pine sawfly attack. *Can. J. Bot.* **46**: 724–725.

Thielges, B. A. and R. L. Campbell, 1972. Selection and breeding to avoid the eastern spruce gall aphid. *Am. Christmas Tree J.* **1972** (May): 3–6.

Thomas, J. G., E. L. Sorensen, and R. H. Painter, 1966. Attached vs. excised trifoliates for evaluation of resistance in alfalfa to the spotted alfalfa aphid. *J. Econ. Entomol.* **59**: 444–448.

Thomas, J. G. and E. L. Sorensen, 1971. Effect of excision duration on spotted alfalfa aphid resistance in alfalfa cuttings. *J. Econ. Entomol.* **64**: 700–704.

Thompson, K. F., 1963. Resistance to the cabbage aphid (*Brevicoryna brassicae*) in Brassica plants. *Nature* **198**: 209.

Thurston, R. and J. A. Webster, 1962. Toxicity of *Nicotiana gossei* Domin to *Myzus persicae* (Sulzer). *Entomol. Exp. Appl.* **5**: 233–238.

Thurston, R., J. C. Parr, and W. T. Smith, 1966a. The phylogeny of *Nicotiana* and resistance to insects. *Fourth Int. Tobacco Sci. Congr. Proc., Natl. Tobacco Board Greece, Athens*: 424–430.

Thurston, R., W. T. Smith, and B. Cooper, 1966b. Alkaloid secretion by trichomes of *Nicotiana* species and resistance to aphids *Entomol. Exp. Appl.* **9**: 428–432.

Thurston, R., 1970. Toxicity of trichome excudates of *Nicotiana* and *Petunia* species to tobacco hornworm larvae. *J. Econ. Entomol.* **63**: 272–274.

Tidke, P. M. and P. V. Sane, 1962. Jassid resistance and morphology of cotton leaf. *Indian Cotton Grow. Rev.* **16**: 324–327.

Tija, B. and D. B. Houston, 1975. Phenolic constituents of Norway spruce resistant or susceptible to the eastern spruce gall aphid. *For. Sci.* **22**: 180–184.

Tingey, W. M., T. F. Leigh, and A. H. Hyer, 1973. Three methods of screening cotton for ovipositional nonpreference by lygus bugs. *J. Econ. Entomol.* **66:** 1312–1314.

Tingey, W. M. and Nielson, M. W., 1974. Alfalfa seed chalcid: nonpreference resistance in alfalfa. *J. Econ. Entomol.* **67:** 219–221.

Tingey, W. M. and T. F. Leigh, 1974. Height preference of lygus bugs for oviposition on caged cotton plants. *Environ. Entomol.* **3:** 350–351.

Tingey, W. M., V. E. Gracen, and J. M. Scriber, 1975. European corn borer—resistance maize genotypes. *N.Y. Food Life Sci. Quart.* **8** (3): 3–7.

Tingey, W. M., T. F. Leigh, and A. H. Hyer, 1975. *Lygus hesperus:* Growth, survival and egg laying resistance of cotton genotypes. *J. Econ. Entomol.* **68:** 28–30.

Tingey, W. M. and M. W. Nielson, 1975. Developmental biology of the alfalfa seed chalcid on resistant and susceptible alfalfa clones. *J. Econ. Entomol.* **68:** 167–168.

Tingey, W. M. and E. A. Pillemer, 1977. *Lygus* bugs: Crop resistance and physiological nature of feeding injury. *Entomol. Soc. Am. Bull.* **23:** 277–287.

Titus, E. G., 1910. The alfalfa leaf weevil. *Utah Agric. Exp. Stn. Bull.* **110:** 1–72.

Toda, R., 1974. Forest tree breeding in the world. *Gov. For. Exp. Stn. Jap. Tokyo:* 205 pp.

Todd, G. W., A. Getahun, and D. C. Cress, 1971. Resistance in barley to the greenbug, *Schizaphis graminum.* 1. Toxicity of phenolic and flavonoid compounds and related substances. *Ann. Entomol. Soc. Am.* **64:** 718–722.

Torres, C., C. Soza, and A. Ortega, 1973. Comportamiento ce variedades e hibridos de maiz frente al ataque de *Diatraea saccharalis* Fabricius (Lepidoptera: Pyralidae) en Argentina. *Agrociencia* **13:** 31–41. Colegio de Postgraduados, Chapingo, Estado de Mexico, Mexico.

Uichanco, L. B., 1921. The rice bug *Leptocorisa acuta* Thunberg in the Philippines. *Phil. Agric. Res.* **14:** 87–125.

Ullstrup, A. J., 1974. Corn diseases in the United States and their control. *U.S. Dep. Agric. Agric. Handb.* **199.**

Umanah, E. E. and R. W. Hartman, 1973. Chromosome numbers and karyotypes of some Manihot species. *J. Am. Soc. Hortic. Sci.* **98**(3): 272–274.

Universidad Federal de Bahia, 1973. *Projecto mandioca.* Cruz das Almas, Bahia, Brazil, 115 pp.

Uphof, J. C. Th., 1962. Plant hairs. *In Handbuch der Pflantzenanatomie.* Vol. 4, · Part 5, *Histologie,* W. Zimmerman and P. G. Ozenda, Eds. Gebruder Borntraeger, Berlin, pp. 1–206.

U.S. Forest Service, 1977. Spruce budworm infestation continues. *For. Serv. News* **N-40** (11-15-77).

U.S. Department of Agriculture, 1976. *Agric. Stat. 1976.* Wheat. pp. 1–11.

VanDenburgh, R. S., B. L. Norwood, C. C. Blickenstaff, and C. H. Hanson, 1966.

Factors affecting resistance of alfalfa clones to adult feeding and oviposition of the alfalfa in the laboratory. *J. Econ. Entomol.* **59**: 1193–1198.

Van der Graaff, N. A., 1977. Personal communication. *Inst. Agric. Res. PMB* **192**: Kimma, Ethiopia.

Vanderlip, R. L., 1972. How a sorghum plant develops. *Coop. Ext. Serv. Kans. State Univ.* **C-477**: 19 pp.

Van der Plank, J. E., 1963. *Plant Diseases: Epidemics and Control.* Academic Press, Inc., New York, 349 pp.

Van der Plank, J. E., 1968. *Disease Resistance in Plants.* Academic Press, Inc., New York, 206 pp.

Van der Plank, J. E., 1975. *Principles of Plant Infection.* Academic Press, Inc., New York, 216 pp.

Vander Schaaf, P., D. A. Wilbur, and R. H. Painter, 1969. Resistance of corn to laboratory infestation of the larger rice weevil, *Sitophilus zeamais. J. Econ. Entomol.* **62**: 352–355.

Vanderzant, E. S., 1974. Development, significance and application of artificial diets for insects. *Ann. Rev. Entomol.* **19**: 139–160.

Van Duyn, J. W., S. G. Turnipseed, and J. D. Maxwell, 1972. Resistance in soybeans the Mexican bean beetle: II. Reactions of the beetle to resistant plants. *Crop Sci.* **12**: 561–562.

van Emden, H. F., Ed. 1973. *Insect/Plant Relationships.* Symp. Roy. Entomol. Soc., London 6. 215 pp.

Varis, A. L., 1958. On the susceptibility of the different varieties of big-leafed turnip to damage caused by cabbage maggots (*Hylemyia* ssp.). *J. Sci. Agric. Soc. Fin.* **30**: 271–275.

Vashistha, R. N. and B. Choudhury, 1975. Inheritance of resistance to red pumpkin beetle in muskmelon. *Sabarao J.* **6**: 95–97.

Velusamy, R., I. P. Janaki, A. Subramanian, and J. Chandra, 1975. Varietal resistance of rice to insect pests. *Rice Entomol. Newsl.* **3**: 13–16.

Venkata Rae, G. and K. Muralidharan, 1977. Outbreak of rice hispa in Nellore district, Andhra Pradesh, India. *Int. Rice Res. Newsl.* **5**: 18–19.

Venkataswamy, T., 1968. Evolution of high yielding varieties of paddy with special reference to gall midge and other insects. Paper presented at the *All India Rice Res. Workers' Conf., Hyderabad, India*: 8 pp.

Verschaffelt, E., 1910. The cause determining the selection of food in some herbiverous insects. *Proc. K. Akad. Wet. Amst. Sect. Sci.* **13**: 536–542.

Viajante, V. and R. M. Herrera, 1976. Resistance of IR2070-414-3-9 to the rice whorl maggot. *Int. Rice Res. Newsl.* **1**: 16.

Vilkova, N. A. and I. D. Shapiro, 1973. An indirect express method for determining resistance levels of cereals to *Eurygaster integriceps* Put. according to sorption rates of a vital stain by salivary glands of this insect. *Proc. All Union Sci. Res. Inst. Plant Prot.* **39**: 181–184.

Villacis, J., C. Sosa, and A. Ortega, 1972. Comportamiento de *Sitotroga cerealella* Oliver (Lepidoptera: Gelechiidae) y de *Sitophilus* zeamais Mots. (Coleoptera: Curculionidae) en diez tipos de maiz con caracteristicas contrastantes. *Agrociencia, Ser. D.* **9**: 3–16. Colegio de Postgraduados, Chapingo, Estado de Mexico, Mexico.

Virtanen, A. I., A. Aura, and T. Ettala, 1957. Formation of benzoxazolinone in rye seedlings. *Suom. Kem.* B30: 246.

Vité, J. P. and D. L. Wood, 1961. A study of the applicability of the measurement of oleoresin exudation pressure in determining susceptibility of second-growth ponderosa pine to bark beetle infestations. *Contrib. Boyce Thompson Inst.* **21**: 67–78.

Waghray, R. N. and S. R. Singh, 1965. Effect of N,P, and K on the fecundity of the ground nut aphid, *Aphis craccivora* Koch. *Indian J. Entomol.* **27**: 331–334.

Walker, D. W., A. Alemany, V. Qunitana, F. Padovani, and K. S. Hagen, 1966. Improved xenic diets for rearing the sugar cane borer in Puerto Rico. *J. Econ. Entomol.* **59**: 1–4.

Walker, J. K., Jr. and G. A. Niles, 1971. Population dynamics of the boll weevil and modified cotton types: Implications for pest management. *Tex. Agric. Exp. Stn. Bull.* **1109**: 14 pp.

Walker, J. K., Jr. and G. A. Niles, 1973. Studies of the effects of bollworms and cotton fleahoppers on yield reductions in different cotton genotypes. *Proc. Beltwide Cotton Prod. Res. Conf.*: 100–102. National Cotton Council, Memphis, Tenn.

Walker, J. K., G. A. Niles, J. R. Gannaway, J. V. Robinson, C. B. Cowan, and M. J. Lukefahr. 1974. Cotton fleahopper damage to cotton genotypes. *J. Econ. Entomol.* **67**: 537–542.

Walker, J. K., J. R. Gannaway, and G. A. Niles, 1977. Age distribution of cotton bolls and damage from the boll weevil. *J. Econ. Entomol.* **70**: 5–8.

Walker-Simons, M. and C. A. Ryan, 1977. Wound-induced accumulation of trypsin inhibitor activites in plant leaves. *Plant Physiol.* **59**: 437–439.

Wall, J. W. and W. M. Ross, Eds., 1970. *Sorghum Production and Utilization.* Avi, Westport, Conn. 702 pp.

Wallace, L. E. and F. H. McNeal, 1966. Stem sawflies of economic importance in grain crops in the United States. *U.S. Dep. Agric. ARS Tech. Bull.* **1350**: 60 pp.

Wallace, L. E., F. H. McNeal, and M. A. Berg, 1973. Minimum stem solidness required in wheat for resistance to the wheat stem sawfly. *J. Econ. Entomol.* **66**: 1121–1123.

Walter, E. V. and A. M. Brunson, 1940. Differential susceptibility of corn hybrids to *Aphis maidis. J. Econ. Entomol.* **33**: 623–628.

Walter, E. V., 1962. Sources of earworm resistance for sweet corn. *Proc. Am. Soc. Hortic. Sci.* **80**: 485–487.

Walter, E. V., 1965. Northern corn rootworm resistance in sweet corn. *J. Econ. Entomol.* **58**: 1076–1078.

Walters, H. J., 1969. Bettle transmission of plant viruses. *Adv. Virus Res.* **15**: 339–363.

Walters, H. J. and D. G. Henry, 1970. Bean leaf beetle as a vector of cowpea strain of southern bean mosaic virus. *Phytopathol.* **60**: 177–178.

Ward, C. R. and F. M. Tan, 1977. Organophosphate resistance in the Banks grass mite. *J. Econ. Entomol.* **70**: 250–252.

Ware, J. O., 1936. Plant breeding and the cotton industry. *U.S. Dep. Agric. Yearb.*: 657–744.

Warren, L. O., 1956. Behaviour of Angoumois grain moth on several strains of corn at two moisture levels. *J. Econ. Entomol.* **49**: 316–319.

Watson, M. A. and F. M. Roberts, 1939. A comparative study of the transmission of *Hyocyamus* virus 3, potato virus Y, and cucumber virus 1 by the vectors *Myzus persicae* (Sulz.), *M. circumflexus* (Buckton) and *Macrosiphum gei* (Koch). *Proc. Roy. Soc. London, Ser. B.* **127**: 543–576.

Watson, T. F., 1964. Influence of host plant condition on population increase of *Tetranychus telarius* (Linnaeus) (Acarina: Tetranychidae). *Hilgardia* **35**: 273–322.

Watt, G., 1907. *The Wild and Cultivated Cotton Plants of the World.* Longmans, Green and Co., London.

Way, M. J. and G. Murdie, 1965. An example of varietal variations in resistance of Brussels sprouts. *Ann. Appl. Biol.* **56**: 326–328.

Way, M. J., 1976. Entomology and the world food situation. *Bull. Entomol. Soc. Am.* **22**(2): 125–129.

Wearing, C. H. and H. F. van Emden, 1967. Studies on the relations of insect and host plant. I. Effects of water stress in hostplants in pots on infestation by *Aphis fabae* Scop., *Myzus persicae* (Sulz.), and *Brevicoryne brassicae* (L.). *Nature* **213**: 1051–1052.

Wearing, C. H., 1972a. Responses of *Myzus persicae* and *Brevicoryne brassicae* to leaf age and water stress in Brussels sprouts grown in pots. *Entomol. Exp. Appl.* **15**: 61–80.

Wearing, C. H., 1972b. Selection of Brussels sprouts of different water status by apterous and alate *Myzus persicae* and *Brevicoryne brassicae* in relation to the age of leaves. *Entomol. Exp. Appl.* **15**: 139–154.

Weatherley, D. E., A. J. Peel, and G. D. Hill, 1959. The physiology of the sieve tube. Preliminary experiments using aphid mouthparts. *J. Exp. Bot.* **10**: 1–16.

Weaver, J. B., Jr., 1971. Methods of utilizing male sterility in cotton. *Proc. Beltwide Cotton Prod. Res. Conf.*: 60–61. National Cotton Council, Memphis, Tenn.

Weaver, J. B., 1974. Boll weevil nonpreference and bee activity in cytoplasmic male-sterile cotton. *Proc. Beltwide Cotton Prod. Res. Conf.*: 95. National Cotton Council, Memphis, Tenn.

Weaver, J. B., Jr., 1977. Present status of fertility restoration in cytoplasmic male-sterile Upland cotton. *Proc. Beltwide Cotton Prod. Res. Conf.,* National Cotton Council, Memphis, Tenn. pp. 95–97.

Webster, F. M., 1912. Preliminary report on the alfalfa weevil. *U.S. Dept. Agric. Bur. Entomol. Bull.* **112.**

Webster, J. A., S. H. Gage, and D. H. Smith, Jr., 1973. Suppression of the cereal leaf beetle with resistant wheat. *Environ. Entomol.* **2:** 1089–1091.

Webster, J. A., 1975. Association of plant hairs and insect resistance. An annotated bibliography. *U.S. Dep. Agric. ARS Misc. Publ.* **1297:** 1–18.

Webster, J. A., 1977. The cereal leaf beetle in North America: breeding for resistance in small grains. In *The Genetic Basis of Epidemics in Agriculture,* P. R. Day, Ed. *Ann. N.Y. Acad. Sci.* **287:** 1–400.

Webster, T. A., E. L. Sorensen, and R. H. Painter, 1968a. Temperature, plant-growth stage, and insect-population effects on seedling survival of resistant and susceptible alfalfa infected with potato leafhopper. *J. Econ. Entomol.* **61:** 142–145.

Webster, T. A., E. L. Sorensen, and R. H. Painter, 1968b. Resistance of alfalfa varieties to the potato leafhopper: Seedling survival and field damage after infestation. *Crop Sci.* **8:** 15–17.

Weerawooth, K., W. Yaklai, and S. Chantarasaard, 1977. Some gall midge resistant rice hybrids in northeast Thailand. *Int. Rice Res. Newl.* **3:** 4.

Weibel, D. E., K. J. Starks, E. A. Wood, and R. D. Morrison, 1972. Sorghum cultivars and progenies rated for resistance to greenbugs. *Crop Sci.* **12:** 334–336.

Weissenberg, K. van, 1976. Indirect selection for improvement of desired traits. In *Modern Methods in Forest Genetics,* J. P. Miksche, Ed., Springer-Verlag, Berlin, pp. 217–228.

Wellso, S., 1973. Cereal leaf beetle: Larval feeding, orientation, development, and survival on four small grain cultivars in the laboratory. *Ann. Entomol. Soc. Am.* **66:** 1201–1208.

Westwood, M. N. and P. H. Westigard, 1969. Degree of resistance among pear species to the woolly pear aphid, *Eriosoma pyricola. J. Am. Soc. Hortic. Sci.* **94:** 91–93.

White, C. E., E. J. Armbrust, and J. Ashley, 1972. Cross-mating studies of eastern and western strains of alfalfa weevil. *J. Econ. Entomol.* **65:** 85–89.

Whittaker, R. H., 1970. The biochemical ecology of higher plants. In *Chemical Ecology,* Sondheimer and Simeone, Eds. Academic Press, Inc., New York, pp. 43–70.

Whittaker, R. H. and P. Feeny, 1971. Allelochemics: Chemical interactions between species. *Science* **171:** 757–770.

Widstrom, N. W., 1967. An evaluation of methods for measuring corn earworm injury: *J. Econ. Entomol.* **60:** 791–794.

Widstrom, N. W. and J. B. Davis, 1967. Analysis of two diallel sets of sweet corn inbreds for corn earworm injury. *Crop Sci.* **7**: 50–52.

Widstrom, N. W. and J. H. Hamm, 1969. Combining abilities and relative dominance among maize inbreds for resistance to earworm injury. *Crop Sci.* **9**: 216–219.

Widstrom, N. W. and R. L. Burton, 1970. Artificial infestation of corn with suspensions of corn earworm eggs. *J. Econ. Entomol.* **63**: 443–446.

Widstrom, N. W., W. J. Wiser, and L. F. Bauman, 1970. Recurrent selection in maize for earworm resistance. *Crop Sci.* **10**: 674–676.

Widstrom, N. W., B. R. Wiseman, and W. W. McMillian, 1972a. Genetic parameters for earworm injury in maize populations with Latin American germplasm. *Crop Sci.* **12**: 358–359.

Widstrom, N. W., B. R. Wiseman, and W. W. McMillian, 1972b. Resistance among some maize inbreds and single crosses to fall armyworm injury. *Crop Sci.* **12**: 290–292.

Widstrom, N. W. and W. W. McMillian, 1973. Genetic effects conditioning resistance to earworm in maize. *Crop Sci.* **13**: 459–461.

Widstrom, N. W. and B. R. Wiseman, 1973. Locating major genes for resistance to the corn earworm in maize inbreds. *J. Hered.* **64**: 83–86.

Widstrom, N. W., W. D. Hansen, and L. M. Redlinger, 1975. Inheritance of maize weevil resistance in maize. *Crop Sci.* **15**: 467–470.

Wiklund, C., 1974. Ovipositon preferences in Papilio machaon in relation to the host plants of the larvae. *Entomol. Exp. Appl.* **17**: 189–198.

de Wilde, J., W. Bongers, and A. H. Schooneveld, 1969. Effects of host plant age on phytophagous insects. *Entomol. Exp. App.* **12**: 714–720.

Wilde, G. and H. Feese, 1973. A new corn leaf aphid biotype and its effect on some cereal and small grains. *J. Econ. Entomol.* **66**: 570–571.

Wilde, G. and C. Ohiagu, 1976. Relation of corn leaf aphid to sorghum yield. *J. Econ. Entomol.* **69**: 195–196.

Wilkinson, R. C., 1978. Personal communication. U.S. Forest Service Northeastern Forest Experiment Station, Durham, N.H.

Williams, J. T., C. H. Lamoureux, and N. Wulijarni-Soetjipto, 1975. South East Asian plant genetic resources. International Board for Plant Genetic Resources, Bogor SEAMEO, Biotrop et al.

Williamson, D. D. and R. F. Whitcomb, 1974. Helical, wall-free prokaryotes in Drosophila, leafhoppers, and plants. *INSERM* (*Inst. Natl. Sante Rech. Med.*) **33**: 283–290.

Williamson, D. L. and R. F. Whitcomb, 1975. Plant mycoplasmas: a cultivable spiroplasma causes corn stunt disease. *Science* **188**: 1018–1020.

Wilson, F. D. and J. A. Lee, 1971. Genetic relationship between tobacco budworm feeding response and gland number in cotton seedlings. *Crop Sci.* **11**: 419–421.

Wilson, L. F., 1976. Entomological problems of forest crops grown under intensive culture. *Iowa State J. Res.* **50**: 277–286.

Wilson, M. C. and R. L. Davis, 1953. Varietal tolerance to alfalfa to the meadow spittlebug. *J. Econ. Entomol.* **46**: 238–241.

Wilson, M. C. and R. L. Davis, 1958. Development of alfalfa having resistance to the meadow spittlebug. *J. Econ. Entomol.* **51**: 219–222.

Wilson, M. C. and R. L. Davis, 1966. Host plant resistance research on *Philaenus spumarius* (L.) in alfalfa. *Ninth Int. Grassland Congr. Proc.* **2**: 1283–1286.

Wilson, R. L. and D. C. Peters, 1973. Plant introductions of *Zea mays* as sources of corn rootworm tolerance. *J. Econ. Entomol.* **66**: 101–104.

Wilson, R. L. and F. D. Wilson, 1974. Laboratory diets for screening cotton for resistance to pink bollworm. *Cotton Grow. Rev.* **51**: 302–308.

Wilson, R. L. and F. D. Wilson, 1975. A laboratory evaluation of primitive cotton (*Gossypium hirsutum* L.) races for pink bollworm resistance. *U.S. Dep. Agric. ARS* **W-30.**

Wilson, R. L. and F. D. Wilson, 1975a. Comparison of an X-ray and a greenboll technique for screening cotton for resistance to pink bollworm. *J. Econ. Entomol.* **68**: 636–638.

Wilson, R. L. and F. D. Wilson, 1975b. Effects of pilose, pubescent and smooth cottons on the cotton leafperforator. *Crop Sci.* **15**: 807–809.

Wilson, R. L. and F. D. Wilson, 1976. Nectariless and glabrous cottons: effect on pink bollworm in Arizona. *J. Econ. Entomol.* **69**: 623–624.

Wiseman, B. R., C. E. Wassom, and R. H. Painter, 1967a. An unusual feeding habit to measure differences in damage to 81 Latin American lines by the fall armyworm. *Agron. J.* **59**: 279–281.

Wiseman, B. R., R. H. Painter, and C. E. Wassom, 1966. Detecting corn seedling differences in the greenhouse by visual classification of damage by the fall armyworm. *J. Econ. Entomol.* **59**: 1211–1214.

Wiseman, B. R., 1967. Resistance of corn *Zea mays* L. and related plant species to the fall army worm, *Spodoptera frugiperda* (J. E. Smith). Ph.D. Dissertation, Kansas State University, Manhattan.

Wiseman, B. R., R. H. Painter, and C. E. Wassom, 1967b. Preference of first-instar fall armyworm larvae for corn compared with *Tripsacum dactyloides*. *J. Econ. Entomol.* **60**: 1738–1742.

Wiseman, B. R., W. W. McMillian, and N. W. Widstrom, 1970. Husk and kernel resistance among maize hybrids to an insect complex. *J. Econ. Entomol.* **63**: 1260–1262.

Wiseman, B. R., W. W. McMillian, and N. W. Widstrom, 1972. Tolerance as a mechanism of resistance in corn to the corn earworm. *J. Econ. Entomol.* **65**: 835–837.

Wiseman, B. R., D. B. Leuck, and W. W. McMillian, 1973. Effect of crop fertilizer on feeding of larvae of fall armyworm on excised leaf sections of corn foliage. *J. Ga. Entomol. Soc.* **8:** 136–141.

Wiseman, B. R., W. W. McMillian, and N. W. Widstrom, 1973. Registration of SGIRL-MR-1 sorghum germ plasm. *Crop Sci.* **13:** 398.

Wiseman, B. R., N. W. Widstrom, and W. W. McMillian, 1974a. Methods of application and numbers of eggs of the corn earworm required to infest ears of corn artificially. *J. Econ. Entomol.* **67:** 74–76.

Wiseman, B. R., W. W. McMillian, and N. W. Widstrom, 1974b. Techniques, accomplishments, and future potential of breeding for resistance in corn to the corn earworm, fall armyworm and maize weevil; and in sorghum to the sorghum midge. In *Proc. Summer Inst. Biol. Control Plant Insects Dis.*, F. G. Maxwell and F. M. Harris, Eds.: 381–393. University Press of Mississippi, Jackson.

Wiseman, B. R., N. W. Widstrom, W. W. McMillian, and W. D. Perkins, 1976. Greenhouse evaluations of leaf feeding resistance in corn to corn earworm (Lep. Noct.). *J. Ga. Entomol. Soc.* **11:** 63–67.

Wiseman, B. R., N. W. Widstrom, and W. W. McMillian, 1977. Ear characteristics and mechanisms of resistance among selected corns to corn earworm. *Fla. Entomol.* **60:** 97–103.

Witz, J. A., 1973. Integration of Systems Science Methodology and Scientific Research. *Agric. Sci. Rev. U.S. Dep. Agric. 2nd Quart.:* 37–48.

Wongsiri, T., P. Vungsilabutr, and T. Hidaka, 1971. Study on the ecology of the rice gall midge in Thailand. In *Symp. Rice Insects. Proc. Symp. Trop. Agric. Res., 19–24 July, 1971, Tokyo, Japan:* 267–290.

Wood, D. L., 1973. Selection and colonization of ponderosa pine bark beetles. In *Insect/Plant Relationships*, H. F. van Emden, Ed. John Wiley & Sons, Inc., New York, pp. 101–117.

Wood, E. A., Jr., 1961. Description and results of a new greenhouse technique for evaluating tolerance of small grains to the greenbug. *J. Econ. Entomol.* **54:** 303–305.

Wood, E. A., H. L. Chada, D. E. Weibel, and F. F. Davies, 1969. A sorghum variety highly tolerant to the greenbug, *Schizaphis graminum* (Rond). *Okla. Agric. Res. Prog. Rep.* **P-614:** 10 pp.

Wood, E. A., Jr. and K. J. Starks, 1972. Effect of temperature and host plant interaction on the biology of three biotypes of the greenbug. *Environ. Entomol.* **1:** 230–234.

Worthing, C. R., 1969. Use of growth retardants on chrysanthemums: effect on pest populations. *J. Sci. Food Agric.* **20:** 394–397.

Wright, J. W. and W. J. Gabriel, 1959. Possibilities of breeding weevil-resistant white pine strains. *U.S. Dep. Agric. For. Serv. Stn. Pap.* **NE 115.**

Wright, J. W. and L. F. Wilson, 1972. Genetic differences in Scotch pine resistance to pine root collar weevil. *Mich. State Univ. Agric. Exp. Stn. Res. Rep.* **159:** 5 pp.

Wright J. W., 1976. *Introduction to Forest Genetics.* Academic Press, New York, 463 pp.

Yamamoto, R. and R. Y. Jenkins, 1972. Host plant preferences of tobacco hornworm moths. In *Insect and Mite Nutrition,* J. G. Rodriguez, Ed., North-Holland, Amsterdam, pp. 567–574.

Yen, D. E., 1974. The sweet potato and Oceania. *Bernice P. Bishop Mus. Bull.* **236:** 1–389.

York, J. O. and W. H. Whitcomb, 1966. Progress in breeding for southwestern corn borer resistance. *Arkansas Farm Res.* **15:** 5.

Yoshida, S., Y. Onishi, and K. Kitagishi, 1959. The chemical nature of silicon in rice plants. *Soil Plant Food (Tokyo)* **5:** 23–27.

Young, W. R. and G. L. Teetes, 1977. Sorghum entomology. *Ann. Rev. Entomol.* **22:** 193–218.

Zohren, E., 1968. Laboruntersuchungen zu Massenanzucht, Lebenweise, Eiablage und Eiablageverhalten de Kohlfliege, *Chortophila brassica* Bouche (Diptera, Anthonoymiidae). *Z. Angew. Entomol.* **62:** 139–188.

Zuber, M. S., G. S. Musick, and M. L. Fairchild, 1971a. A method of evaluating corn strains for tolerance to the western corn rootworm. *J. Econ. Entomol.* **64:** 1514–1518.

Zuber, M. S., M. L. Fairchild, A. J. Keaster, V. L. Fergason, G. F. Krause, E. Hilderbrand, and P. J Loesch, Jr., 1971b. Evaluation of 10 generations of mass selection for corn earworm resistance. *Crop Sci.* **11:** 16–18.

GLOSSARY

Additive Resistance. Resistance governed by more than one gene, each of which can be expressed independently, but which is reinforced by the expression of each of the additional genes.

Adult Plant Resistance. The generally horizontal resistance of adult plants, also known as age resistance or mature plant resistance.

Aglycone. An organic compound, usually a phenol or an alcohol, combined with the sugar portion of a glycoside and obtainable by hydrolysis.

Agroecosystem. Cultivated plants and the environmental factors (including human intervention) that act upon or influence them under field conditions.

Allele. Either of a pair of alternative Mendelian characters.

Allelochemic. As defined by Whittaker (1970), a nonnutritional chemical produced by an individual of one species that affects the growth, health, behavior, or population biology of another species.

Allomone. An allelochemic that induces a response in an individual of another species (e.g., an insect) that is beneficial to the emitting organism. Many allomones are essentially defensive, that is, toxic or repugnant to potential attackers. However a scent that attracts bees and therefore facilitates pollination is also an allomone. (See also *Kairomone.*)

Allotetraploid. Amphidiploid, that is, having a complete diploid chromosome set from each parent strain as a result of chromosome doubling in the first hybrid generations.

Anemotaxis. The tendency of certain insects to orient themselves in relation to wind direction.

Anthesis. The action or period of the opening of a flower.

Antibiosis. A toxic or other direct detrimental effect of one organism upon another. (See also *Antixenosis.*)

Antifeedant. A substance that deters feeding by an insect but does not kill the insect directly.

Antixenosis. Term proposed by Kogan and Ortman to replace *Non-preference*; intended to parallel *Antibiosis.*

Auxin. Any organic substance that in very low concentrations is able to induce growth of plant shoots along the longitudinal axis.

Avirulent Gene. A gene that does not contribute to parasitic ability.

Backcross. Mating the result of a cross with its parent to reinforce or increase the gene frequency of a desired characteristic.

Barbadense. *Gossypium barbadense,* Sea Island, Egyptian, Pima, or long-staple cottons.

Benzophenone. A ketone $C_6H_5COC_6H_5$ having a roselike odor and often used in perfumery.

Benzoxazolinones. A group of chemical compounds occurring in maize and having an allomonic effect on the European corn borer.

Bioassay. Measurement of the effect on an organism of a given stress, such as a toxic substance, heat, cold, or drought.

Biotype. In entomology, an individual or a population that is distinguished from the rest of its species by criteria other than morphology, for example, a difference in parasite ability.

Biotype Specific Resistance. Resistance that is effective only for a given biotype of the pest species. Such resistance is often, but not necessarily, vertical.

Breakdown of Resistance. The inability of a variety to maintain resistance when attacked by a newly selected insect biotype that has a gene for virulence at every locus corresponding to a gene for resistance in the host.

Byssinosis. A disease frequently found in textile mill workers, caused by inhalation of airborne particles of fiber; also known as brown lung disease.

Chemical Ecology. The chemistry and biochemistry of ecological situations involving interactions among organisms and between organisms and their environment.

Cinnamic Acid. An acide $C_6H_5CH:CHOOH$ found especially in cinnamon oil and sterax.

Circulative. Term used by Kennedy et al. (1962) and by Harris (1977, 1978) to refer to disease agents that are persistent in the vector.

Cleistogamous. Adjective for a plant in which pollination and fertilization may occur in an unopened flower bud; with reference to sorghums, those with long papery glumes through which the anthers fail to extrude during anthesis.

Coefficient of Selection. A measure of the relative change in gene frequency between generations as a result of differential selection.

Colchicine. A poisonous yellow crystalline alkaloid $C_{22}H_{25}NO_6$ extracted from the meadow saffron (*Colchicum autumnale*) that induces polyploidy when applied to mitotic cells.

Complementary Resistance. Resistance governed jointly by two or more genes, one of them alone being ineffective.

Coumarin. A substance that deters feeding by certain insect pests. Derived from glucosides in the plant, it is formed when tissues are disrupted by the feeding insect.

Culm. A monocotyledonous stem.

Cytogenetic Breeding. Breeding techniques that involve the manipulation or alteration of genetic material in cells, for example, exposure to radiation or application of chemicals such as colchicine.

Dalton. A unit of mass equal to a molecular weight of one (i.e., to one-twelfth the mass of an atom of carbon 12). Daltons are used in preference to molecular weight when the structure being described is not a molecule.

Dehisce. To split open along a natural line, especially when associated with discharge of contents, as in the opening of seedpods or the freeing of pollen by the anther.

Dent. A type of maize having kernels that contain both soft and hard starch and thus become indented at maturity.

Diallel (or Diallele) Cross. The intercrossing of parents in all combinations of two; all possible crosses among individuals in a group.

Diapause. A period of dormancy.

DIMBOA. Acronym for 2,4-dihydroxy-7-methoxyl-1,4-benzoxazin-3-one, a naturally occurring compound that confers resistance to certain pests in corn.

Dimeric. Consisting of two parts.

Directional Selection. The adjustment, through natural selection, of a pest population to a change in host resistance or of a host population to a change in the parasitic ability of its pests; used by Van der Plank (1968) to denote what is referred to by other authors (see chapter by MacKenzie) as *Stabilizing Selection.*

Domestication. The process of breeding for a given desirable characteristic found in the wild so as to increase, enhance, and stabilize its occurrence in cultivated plants; used by Robinson.

Economic Injury Level. The degree of crop damage at which economic losses become significant.

Environmental Biotypes. Biotypes distinguished by their degree of sensitivity to environmental factors (e.g., moisture, temperature) rather than to host resistance factors.

Epistasis.   The suppression of the effect of a gene by a nonallelic gene.

Escapes.   Individual susceptible plants that show no infestation damage because pests did not happen to attack or feed on them.

Esodemic.   An infestation, or phase thereof, in which the host plant is attacked by a parasite hatched or otherwise originating on the plant itself.

Exodemic.   An infestation, or phase thereof, in which the host plant is attacked by a parasite originating on another plant.

Field Resistance.   Resistance observed under field conditions as distinguished from resistance observed in the laboratory or greenhouse.

Flavonoid.   Any of a number of compounds, including plant pigments, related to or similar in chemical structure to flavone $C_{15}H_{10}O_2$.

Flint.   A maize (*Zea mays indurata*) having usually rounded kernels with a hard outer layer.

Full-Sib.   Having both parents in common.

Fundatrice.   A vivaparous parthenogenetic female aphid produced on a primary host plant from an overwintering fertilized egg.

Gene-for-Gene Resistance.   See *Vertical Resistance.*

General Resistance.   Analogous to *Horizontal Resistance,* biotype non-specific resistance.

Genetic Sanitation.   *Vertical Resistance,* as used by MacKenzie.

Genic.   Of or relating to a gene or genes.

Genome.   One haploid set of chromosomes with their constituent genes.

Glabrous.   Hairless or bald; the opposite of pubescent.

Glume.   In grasses, either of two empty bracts at the base of a spikelet.

Gossypol.   A pigment $C_{30}H_{30}O$ that occurs naturally in cotton plants and is toxic to some insect pests.

Habgood Name.   A number representing a combination of genes in the mathematical system of nomenclature devised by Habgood (1970).

Half-Sib.   Having one parent in common.

Homozygosity.   The condition of state of having pairs of genes that are identical for one or more allelomorphic characters.

Horizontal Resistance.   A resistance trait that does not involve a gene-for-gene relationship, that is, any resistance that is not *Vertical Resistance.*

Host Resistance.   The relative genetic ability of a race or variety to produce a larger or higher quality crop or to produce more offspring compared with other races or varieties at the same level of infestation.

Inbreeding Depression.   Loss of vigor and increased mortality in successive generations due to inbreeding.

Indirect Selection. Selection for a trait other than the one desired to improve, based upon the existence of a genetic correlation between the two traits.

Interallelic. Expression governed by alleles of more than one pair.

Interference. Distortion of test results by an extraneous factor or factors. Specifically, increased levels of infestation on resistant test plants due to their proximity to susceptible plants.

Intraallelic. Expression governed by one of a single pair of alleles.

Introgression. The entry or introduction of a gene from one gene complex into another.

Isoprenoid. Of or related to the hydrocarbon isoprene C_5H_8.

Juglone. A reddish yellow crystalline compound $C_{10}H_5O_2(OH)$.

Juvenile Resistance. The characteristically vertical resistance of plants in the seedling stage; also known as seedling resistance.

Kairomone. A chemical or mixture of chemicals emitted by an organism (e.g., a plant) that induces a response in an individual of another species (e.g., an insect) that is beneficial to the receiving organism. An example of a kairomone is a plant scent that makes the plant more easily identifiable to an insect pest. (See also *Allomone*.)

Key Pests. Serious, perennially occurring persistent species that dominate control practices and that in the absence of human intervention commonly attain population densities that exceed economic injury levels.

Land Variety or Landrace. A variety developed locally by indigenous people, presumably without benefit of scientific knowledge of plant breeding.

L-Arginine. A common amino acid of proteins: $H_2N—C(=NH)—NH—CH_2—CH_2—CH_2—CH(NH_2)CO_2H$.

L-Canavanine. A nonprotein amino acid closely related to arginine and highly toxic to most insects: $H_2N—C(=NH)—NH—O—CH_2—CH_2—CH_2—CH(NH_2)CO_2H$.

Leaf-Disk Test. A test in which resistance is measured by allowing pests to feed on disks cut from leaves of test plants. Resistance is based on the amount of leaf uneaten after a given test period.

Lemma. In grasses, the lower bract that with the palea encloses the flower.

Lignin. A complex amorphous substance that, in association with cellulose, causes the thickening of plant cell walls and thereby forms wood or woody parts.

Lodging. The condition of a plant that has been beaten to the ground or otherwise damaged so that it cannot stand upright.

Lyophilized.   Freeze-dried.

Meiosis.   The stage of the reproductive process in which the pairs of chromosomes within the nucleus of a cell separate so that only one chromosome from each pair enters each daughter cell.

Mines.   Deep holes or tunnels in plant parts caused by burrowing insects or their larvae.

Model.   A description or representation of a system, entity, or state of affairs. The use of model in a scientific context implies (1) that the thing being represented cannot be directly observed or, in some cases, directly manipulated, and (2) that the model itself is in some degree hypothetical and subject to validation. A model can consist of statements in ordinary language, but models of complex systems often utilize graphic or mathematical symbols. A model may also be made up three-dimensional shapes, with or without moving mechanical, hydraulic, or aerodynamic components. An advantage of mathematical models is that much of the labor of testing them can be performed by computers. (For a detailed taxonomy of models, see Chapter 10.)

Monocarpellary.   Having or consisting of a single carpel.

Monoecious.   Having male and female flower parts in the same plant.

Monogenic.   Relating to or controlled by a single gene.

Monophagous.   Feeding on or utilizing a single kind of food.

Multigenic.   Relating to or controlled by more than one gene.

Multiline.   A crop or field grown from a mixture of seed of several resistant lines, thereby confronting insect pests with a mixture of host genetypes.

Nonpreference.   The behavior of insects when they avoid or exhibit negative reactions to a plant of a host species; also used to describe resistance trait that induces such behavior. (See also *Antixenosis.*)

Occasional Pests.   A pest generally under natural control, that exceeds the economic injury level only sporadically or in localized areas.

Olfactometer.   An instrument for measuring the sensitivity of the sense of smell.

Oligogenic.   Governed by a few genes.

Oligophagous.   Feeding on or utilizing a few kinds of food.

Palea.   The upper bract that with the lemma encloses the flower in grasses.

Parenchyma.   The plant tissue consisting of cells capable of photosynthesis or storage, as distinguished from supportive tissue.

Pathodeme.   A host population in which all individuals have a given resistance in common.

Pathosystem. An ecological subsystem defined by the phenomenon of parasitism. A plant pathosystem may include one or more host plant species along with the various parasites—insects, fungi, bacteria, and others—that utilize the hosts. Herbivorous birds and mammals, however, are generally not classified as parasites.

Pathotype. A parasite (i.e., pest) population in which all individuals have a given parasitic ability in common.

Pedicellate. Having a slender stalk or peduncle that supports a larger fruiting or spore-bearing organ.

Peduncle. See *Pedicellate*.

Persistent. With reference to a disease agent, one that remains in a virulent state in the vector's system for an extended period, often for the lifetime of the vector.

Person Differential Interaction. A mathematical technique for demonstrating the existence of a gene-for-gene relationship between host and pest.

Phytoalexin. A substance in plants that, in combination with antibodies, causes destruction of bacteria and other antigens.

Pilosity. Being covered with short, usually soft hair; pubescence.

Pleiotropic. Having multiple phenotypic expressions.

Ploidy. The degree of replication of chromosomes or genomes (e.g., haploid, diploid, tetraploid, etc.).

Polygenic. Governed by several or many genes. (See also *Monogenic* and *Oligogenic*.)

Polyphagous. Feeding on or utilizing many kinds of foods.

Propagule. A propagable shoot.

Proprioceptor. A receptor that responds to stimuli arising within the organism.

Pseudoresistance. Apparent resistance in potentially susceptible host plants resulting from chance, from transitory (nonheritable) traits, or from environmental conditions.

Pulling Resistance. Resistance to being pulled out of the ground or uprooted by pulling.

Raceme. An influorescence, usually conical in form, consisting of an elongated axis bearing flowers on short stems.

Rogue. To weed out or cull diseased or defective individuals from a crop field or from a stand of timber.

Sclerenchyma. A protective or supporting plant tissue made up of cells with thickened, lignified walls.

Secondary Pests. Pests that do not normally occur in economically important numbers, but which can exceed the economic injury level as a result of changes in cultural practices or crop varieties, or the injudicious use of insecticides applied for a key pest.

Segregating Population. A population, generally the progeny of a cross, in which genetic differences are detectable, thus permitting identification of individuals having a desired trait and their selection for further breeding.

Selection Pressure. Exposure to stresses, such as pests, that tend to induce natural selection in the exposed population. Negative selection pressure is used by Robinson to denote the absence of or interference with selection pressure.

Self. To induce pollination with pollen from the same plant.

Shikimic Acid. A crystalline acid $C_6H_6(OH)_3COOH$ formed in plants as a precursor in the biosynthesis of aromatic amino acids and of lignin.

Square. An unopened cotton flower with its enclosing bracts.

Stabilizing Selection. Natural selection tending to maintain an established equilibrium between host and pest populations.

Staling. Decrease and eventual cessation of growth of a fungus believed to be due in whole or in part to the effects of the metabolic products of the fungus itself.

Stover. Grain stalks after removal of the ears, usually cured and used as feed.

Strength and Weakness. Used with reference to vertical resistance, these terms do not describe the intrinsic effectiveness of the resistance trait but rather the relative rarity of the matching pest pathotype. A "strong" vertical resistance will tend to last longer because of the scarcity of opposition.

Stylet-borne. A term proposed by Kennedy et al. (1962) to refer to disease agents that are not persistent in the vector.

Subeconomic. See *Economic Injury Level*.

Sympatric. Occurring in the same area without loss of identity through interbreeding.

Synthetic. A line or variety produced by combining genes unlikely to occur in nature; lines having a high general combining ability.

Terpene. A class of hydrocarbons characterized by $(C_5H_8)_n$ and occurring in plant oils and resins.

Tetrasomic. Having one or a few chromosomes tetraploid in otherwise diploid nuclei.

Tiller. A sprout or stalk, especially one from the base of a plant or from the axils of its lower leaves.

Tolerance. Ability of a plant to withstand infestation and to support insect populations that would severely damage susceptible plants.

Top Cross. A cross between an inbred line and a random bred variety or strain.

Turgor Pressure. The pressure exerted against the wall of a plant cell by the fluid contents of the cell.

Vernalization. To treat seeds, bulbs, or seedlings of a plant to shorten its vegative period and thereby induce earlier flowering and fruiting.

Vertical Resistance. In theory, resistance governed by one or more genes in the host plant, each of which corresponds to a matching gene for parasitic ability in the pest species; sometimes called gene-for-gene resistance. (See *Horizontal Resistance*.)

Vertifolia Effects. The decline of horizontal resistance due to the absence of selection pressure, during breeding for vertical resistance.

Water stress. Drought.

LIST OF INSECT AND
MITE SPECIES

627

SCIENTIFIC NAME	COMMON NAME	PAGE

SCIENTIFIC NAME	COMMON NAME	PAGE
Cucujidae		
Oryzaephilus surinamensis		
(Linnaeus)	Sawtoothed grain beetle	450
Curculionidae		
Anacentrinus deplanatus		
Casey	Sugarcane rootstock weevil	463
Anthonomus grandis Boheman	Cotton boll weevil	13, 26, 33, 35, 42, 48,
		52, 59, 60, 74, 75,
		222-229, 239-240,
		247, 249, 251, 260,
		352-355, 363-364,
		358
Ceutorrhynchus maculaalba		
Herbst	Weevil	123
Chalcodermus aeneus Boheman	Cowpea curculio	43
Coelosternus sp.	Weevil	317, 322
Curculio elephas (Gyllenhal)	Chestnut weevil	48
Cylindrocopturus eatoni		
Buchanan	Pine reproduction weevil	497
Cylindrocopturus furnissi		
Buchanan	Douglas fir twig weevil	261
Graphognathus spp.	Whitefringed beetles	283
Hylobius radicis Buchanan	Pine root collar weevil	497
Hypera brunneipennis		
(Boheman)	Egyptian alfalfa weevil	283, 298-299
Hypera postica (Gyllenhal)	Alfalfa weevil	283, 296-298, 299, 304,
		306-307, 308, 310
Hypera punctata (Fabricius)	Clover leaf weevil	283
Lissorhoptrus oryzophilus		
Kuschel	Rice water weevil	425, 449-450
Otiorhynchus ligustici		
(Linnaeus)	Alfalfa snout beetle	283
Pissodes strobi (Peck)	White pine weevil, Sitka	497
	spruce weevil, Englemann	
	spruce weevil	
Sitona sp.	Weevil	283
Sitona hispidula (Fabricius)	Clover root curculio	283
Sitophilus spp.	Maize weevils	376, 379, 386, 394-395
Sitophilus oryzae (Linnaeus)	Rice weevil	95, 402, 425, 450, 451,
		465, 480
Sitophilus zeamais		
Motschulsky	Maize weevil	95, 402, 425, 450, 451,
		465, 480
Sphenophorus spp.	Billbugs	518, 520
Elateridae		
Aeolus spp.	Click beetles, wireworm	462
Agriotes spp.	Click beetles	521

SCIENTIFIC NAME	COMMON NAME	PAGE
Agriotes mancus (Say)	Wheat wireworm	518
Conoderus spp.	Click beetles	462
Ctenicera aeripennis destructor (Brown)	Prairie grain wireworm	518
Meloidae		
Epicauta spp.	Blister beetles	25, 26, 33, 40
Scarabaeidae		
Anisoplia spp.	White grubs	521
Anisoplia austriaca (Herbst)	White grub	518
Anisoplia leucaspis Laporte	White grub	518
Holotrichia consanguinea (Blanchard)	White grub	462
Leucopholis sp.	White grub	317
Phyllopertha nazarena Marseul	White grub	518
Phyllophaga sp.	White grub	317
Phyllophaga crinata (Burmeister)	May beetle, white grub	462
Phyllophaga lanceolata (Say)	May beetle, white grub	518
Schizonycha sp.	White grub	462
Scolytidae		
Blastophagus piniperda (Linnaeus)	Scolytid	33
Dendroctonus spp.	Bark beetle	489, 496
Dendroctonus brevicomis LeConte	Western pine beetle	33
Dendroctonus frontalis Zimmerman	Southern pine beetle	33
Dendroctonus rufipennis (Kirby)	Spruce beetle	492
Hylurgopinus rufipes (Eichhoff)	Native elm bark beetle	139, 491
Ips spp.	Bark beetles	496
Scolytus multistriatus (Marsham)	Smaller European elm bark beetle	25, 26, 29, 34-36, 139, 491
Scolytus quadrispinosus Say	Hickory bark beetle	25
Tenebrionidae		
Eleodes hispilabris (Say)	False wireworm	518
Eleodes spp.	False wireworms	462, 520
Eleodes opacus (Say)	Plains false wireworm	518
Eleodes suturalis (Say)	False wireworm	518
Eleodes tricostata (Say)	False wireworm	518
Embaphion muricatum (Say)	False wireworm	518
Tenebrio molitor Linnaeus	Yellow mealworm	38
Tribolium castaneum (Herbst)	Red flour beetle	38, 450, 465, 480
Tribolium confusum du Val	Confused flour beetle	38, 465

COLLEMBOLA
Sminthuridae

SCIENTIFIC NAME	COMMON NAME	PAGE
Deuterosminthurus yumanensis (Wray)	Springtail	50
DIPTERA		
Anthomyiidae		
Atherigona varia soccata Rondani	Sorghum shoot fly	43, 48, 71-72, 83, 97, 460, 462, 472-473, 479
Hylemya brassicae (Bouche)	Cabbage maggot, cabbage root fly	120
Hylemya coarctata Fallén	Wheat bulb fly	518, 521
Hylemya floralis (Fallén)	Turnip maggot	42
Hylemya genitalis Schnabl	Spring fly	49
Cecidomyiidae		
Contarinia sorghicola (Coquillett)	Sorghum midge	460, 464, 474-475, 479, 483-484
Contarinia tritici (Kirby)	Wheat blossom midge	518, 522
Dasineura sp.	Midge, seed midge	283
Haplodiplosis equestris (Wagner)	Saddle gall midge	518, 522
Jatrophobia brasiliensis Rubsaamen	Gall midge	317
Mayetiola destructor (Say)	Hessian fly	4, 6, 12, 49, 72, 73, 78-81, 91-92, 109, 134, 185, 199, 200, 211, 236, 258, 517, 518, 521, 525, 532, 533
Orseolia oryzae (Wood-Mason)	Rice gall midge	82, 242, 424, 446-448, 452
Pachydiplosis oryzae Wood-Mason	Rice stem gall midge	71, 97, 99, 446
Sitodiplosis mosellan (Gehin)	Wheat midge	518, 522
Chloropidae		
Chlorops oryzae Matsumura	Rice stem maggot	425, 448-449
Chlorops pumilionis (Bjerkander)	Gout fly	518, 521
Oscinella frit (Linnaeus)	Frit fly	42, 48, 49, 418, 518, 522
Meromyza americana Fitch	Wheat stem maggot	518, 522
Meromyza saltatrix (Linnaeus)	Wheat stem maggot	518
Diopsidae		
Diopsis thoracica Westwood	Stalk-eyed borer	425, 449
Drosophilidae		
Drosophila melanogaster Meigen	Vinegar fly	40

SCIENTIFIC NAME	COMMON NAME	PAGE
Drosophila pachea Patterson and Wheeler	Vinegar fly or pomace fly	37
Ephydridae		
Hydrellia phillippina Ferino	Rice whorl maggot	242, 425, 448
Lonchaeidae		
Lonchaea sp.		316, 323
Silba pendula (Bezzi)	Cassava shoot fly	316, 323
Tephritidae		
Anastrepha manihoti Lima	Fruit fly	316
Anastrepha pickeli Lima	Cassava fruit fly	316
Anastrepha suspensa (Loew)	Caribbean fruit fly	122
HEMIPTERA		
Alydidae		
Leptocorisa spp.	Rice bugs	450
Leptocorisa varicornis (Fabricius)	Rice bug	424
Anthocoridae		
Orius insidiosus Say	Flower bug	60
Coreidae		
Anasa tristis (DeGeer)	Squash bug	77, 83
Lygaeidae		
Blissus leucopterus leucopterus (Say)	Chinch bug	19, 72, 260, 464, 475, 478-481
Nysius raphanus Howard	False chinch bug	464
Miridae		
Calocoris angustatus Lethierry	Jowar earhead bug	464
Calocoris norvegicus (Gmelin)	Potato capsid	282
Lygus spp.	Lygus bugs	301, 358, 366-367
Lygus desertinus Knight	Lygus bug	302
Lygus elisus Van Duzee	Pale legume bug	302
Lygus hesperus Knight	Western lygus bug	59, 107, 282, 302, 349, 353
Lygus lineolaris (de Beauvois)	Tarnished plant bug	222, 223, 228, 282, 302, 349, 353, 365
Pseudatomoscelis seriatus (Reuter)	Cotton fleahopper	49, 53, 247, 349, 358, 367
Pentatomidae		
Aelia spp.	Stinkbugs	518, 519
Chlorochroa sayi Stål	Say stink bug	282, 518
Chlorochroa uhleri (Stål)	Green grain bug	518
Eurygaster spp.	Senn pests	518
Eurygaster integriceps Puton	Senn pest	518, 519, 533
Oebalus pugnax (Fabricius)	Rice stink bug	424, 450, 460, 480

SCIENTIFIC NAME	COMMON NAME	PAGE
Pyrrhocoridae		
Dysdercus superstitiosus (Fabricius)	Pyrrhocorid bug	464
Pyrrhocoris apterus (Linnaeus)	Pyrrhocorid bug	282
Tingidae		
Vatiga manihotae (Drake)	Lacebug	317, 321
		143
HOMOPTERA		
Aleyrodidae		
Aleurodes brassicae (Walker)	Cabbage whitefly	58
Aleurolobus barodensis (Maskell)	Sugarcane whitefly	49
Aleurotrachelus sp.	Whitefly	316, 321, 322, 329-330
Bemisia tabaci (Gennadius)	Cotton whitefly, sweet potato whitefly	48, 50, 151, 316, 319, 322, 357
Trialeurodes abutilonea Haldeman	Bandedwing whitefly	48, 49, 356
Trialeurodes vaporariorum (Westwood)	Greenhouse whitefly	50
Aphididae		146-149, 258
Acyrthosiphon kondoi Shinji	Blue alfalfa aphid	282, 295-296, 308, 309-310
Acyrthosiphon pisum (Harris)	Pea aphid	12, 20, 42, 50, 77, 82, 97, 101, 102, 104, 105, 107, 109, 131, 136, 155, 237, 259, 282, 294-295, 303, 306, 308, 309
Amphorophora agathonica Hottes		149
Amphorophora rubi (Kaltenbach)	Rubus aphid	58, 76, 78, 82, 261
Aphis craccivora Koch	Cowpea aphid	49, 282
Aphis fabae Scopoli	Bean aphid	49, 54, 97, 103, 105, 107-108, 147, 148
Aphis gossypii Glover	Melon aphid, cotton aphid	48, 76, 356
Aphis nerii (Boyer)	Oleander aphid	103
Aphis varians Patch	Aphid	103
Aulacorthum solani (Kattenbach)	Foxglove aphid	108
Brevicoryne brassicae (Linnaeus)	Cabbage aphid	42, 58, 82, 97, 99, 103, 109, 131, 147
Cuernavaca noxius (Mordvilko)	Aphid	518
Dysaphis devecta (Walker)	Rosy leaf curling aphid of apple	75, 155
Dysaphis plantaginea (Passerini)	Rosy apple aphid	75

SCIENTIFIC NAME	COMMON NAME	PAGE
Erisoma lanigerum (Hausmann)	Wooly apple aphid	4, 73-76, 147, 149, 155, 236
Erisoma pyricola Baker and Davidson	Wooly pear aphid	259, 260
Longuiunguis sacchari (Zehntner)	Sugarcane aphid	462
Macrosiphum avenae (Fabricius)	English grain aphid	101, 518, 520
Macrosiphum euphoribiae (Thomas)	Potato aphid	44, 45, 50, 55
Myzus persicae (Sulzer)	Green peach aphid	40, 50, 55, 94, 97, 99, 103, 104, 106, 108, 109, 145, 146-147, 148, 182, 261
Rhopalosiphum maidis (Fitch)	Corn leaf aphid	10, 71, 82, 101, 148, 154, 376, 378, 401-402, 462, 479, 485
Rhopalosiphum padi (Linnaeus)	Oat bird cherry aphid	154, 518, 520
Schizaphis graminum (Rondani)	Greenbug	8, 35, 67, 71, 72, 82, 91, 97, 148, 259, 261, 460, 462, 469-472, 479, 481-483, 518, 520, 532
Sipha flava (Forbes)	Yellow sugarcane aphid	462
Therioaphis maculata (Buckton)	Spotted alfalfa aphid	10, 12, 20, 67, 77, 82, 90, 92, 96, 106-107, 108, 109, 136, 147, 148, 236, 259, 282, 287, 291, 292-294, 302-303, 304-305, 307-309
Therioaphis riehmi (Borner)	Sweetclover aphid	77
Cercopidae		143
Philaenus spumarius (Linnaeus)	Meadow spittlebug	259, 300-301, 308, 310
Cicadellidae		
Aceratagallia curvata Oman	Western alfalfa leafhopper	282
Amrasca spp.	Leafhoppers	49
Cicadulina spp.	Leafhoppers	165, 376
Dalbulus spp.	Leafhoppers	376
Dalbulus maidis (Delong and Wolcott)	American corn leafhopper	153, 154, 415
Empoasca spp.	Leafhoppers	49, 59, 345, 348-349, 358
Empoasca devastans Distant	Cotton leafhopper	348
Empoasca fabae (Harris)	Potato leafhopper	12, 46, 49, 50, 259, 282, 299-300, 308, 311
Empoasca facialis (Jacobi)	Cotton leafhopper	235, 348

SCIENTIFIC NAME	COMMON NAME	PAGE
Pseudococcidae		
Phenacoccus gossypii Townsend and Cockerell	Mexican mealybug	317, 321
Pseudococcus spp.	Mealybugs	317, 331
HYMENOPTERA		
Cephidae		
Cephus cinctus Norton	Wheat stem sawfly	6, 44, 73, 93, 100-101, 237, 260, 517, 518
Cephus pygmaeus (Linnaeus)	European wheat stem sawfly	518, 521
Cephus tabidus (Fabricius)	Black grain stem sawfly	518
Cynipidae		
Dryocosmus kuriphilus Yasumatsu	Chestnut gall wasp chalcid	82, 235, 496
Diprionidae		
Diprion spp.	Conifer sawflies	498
Neodiprion spp.	Conifer sawflies	28, 495, 498-504
Neodiprion abietis (Harris)	Balsam fir sawfly	498-504
Neodiprion rugifrons Middleton	Conifer sawfly	118, 504
Neodiprion swainei Middleton	Swaine jack pine sawfly	118, 504
Eurytomidae		
Bruchophagus roddi (Gussakovsky)	Alfalfa seed chalcid	259, 283, 301, 304, 311-312
Harmolita spp.	Jointworms	521
Harmolita grandis (Riley)	Wheat strawworm	518
Harmolita tritici (Fitch)	Wheat jointworm	518, 521
Formicidae		143
Acromyrmex sp.	Ant	317
Atta sp.	Leafcutting ant	317
Tenthredinidae		
Pachynematus sporax Ross	Ross leaf-feeding sawfly	518, 521
ISOPTERA		
Rhinotermitidae		
Coptotermes spp.	Termites	317
Coptotermes paradoxis (Sjostedt)	Termite	317
LEPIDOPTERA		
Bombycidae		
Bombyx mori (Linnaeus)	Silkworm	35
Gelechiidae		
Pectinophora gossypiella (Saunders)	Pink bollworm	35, 43, 49, 59, 60,

SCIENTIFIC NAME	COMMON NAME	PAGE
Pseudoplusia includens (Walker)	Soybean looper	12
Scotogramma trifolii (Rottenberg)	Clover cutworm	12
Sesamia spp.	Pink sorghum borers	379, 463
Sesamia calamistis Hampson		376, 398
Sesamia inferens (Walker)	Pink rice borer	418, 429
Spodoptera spp.	Leaf feeders	392-395
Spodoptera exempta (Walker)	Nutgrass armyworm	376, 463
Spodoptera exigua (Hübner)	Beet armyworm	463
Spodoptera frugiperda (Smith)	Fall armyworm	74, 83, 98, 376, 383, 385, 386, 400, 406, 463, 479, 518
Spodoptera littoralis (Boisduval)	Egyptian cotton leaf worm	49, 52
Spodoptera mauritia (Boisduval)	Paddy cutworm	450
Trichoplusia ni (Hübner)	Cabbage looper	49, 59, 114
Nolidae		
Celama sorghiella (Riley)	Sorghum webworm	464
Olethreutidae		
Cryptoplebia leucotreta (Meyrick)	False codling moth	351
Eucosma gloriola Heinrich	Eastern pineshoot borer	497
Leguminivora glycinivorella Matsmura	Soybean pod borer	48
Papilionidae		
Papilio machaon Linnaeus	Swallowtail butterfly	127-128
Papilio polyxenes Fabricius	Black swallowtail butterfly	131
Pieridae		
Colias croceus (Fourcroy)	Clouded yellow butterfly	283
Colias eurytheme Boisduval	Alfalfa caterpillar	283
Pieris brassicae Linnaeus	Large white butterfly	125, 126
Pieris rapae (Linnaeus)	Imported cabbageworm	42
Pyralidae		
Chilo partellus (Swinhoe)	Asian maize borer, stem borer	376, 379, 460, 463, 479
Chilo suppressalis (Walker)	Asiatic rice borer or striped rice borer	43, 48, 57, 60, 101, 242, 260, 424, 429-437
Chilotraea polychrysa (Meyrick)	Dark-headed rice borer	429
Chilo zonellus (Swinhoe)	Maize stem borer or maize and jowar borer	394-395, 418
Cnaphalocrosis medinalis Guenée	Rice leaf folder	425, 449
Corcyra cephalonica (Stainton)	Rice moth	450

SCIENTIFIC NAME	COMMON NAME	PAGE
Cryptoblabes adoceta Turner	Head caterpillar	464
Diatraea spp.	Borers	376, 394-395, 389-399, 415
Diatraea grandiosella Dyar	Southwestern corn borer	383, 385, 398-399, 463, 479
Diatraea lineolata Walker	Neotropical corn borer	379, 385
Diatraea saccharalis (Fabricius)	Sugarcane borer	44, 59, 156, 231, 379, 383, 385, 463, 479
Dichocrocis punctiferalis Guenée	Yellow peach moth	464
Elasmopalpus lingnosellus (Zeller)	Lesser cornstalk borer	463, 518, 520
Eldana saccharina Walker	Sorghum maize borer	376, 463
Etiella zinckenella (Treitschke)	Lima bean pod borer	49
Loxostege commixtalis (Walker)	Alfalfa webworm	283
Ostrinia furnacalis (Guenée)		376
Ostrinia nubilalis (Hübner)	European corn borer	7, 8, 13, 19, 25, 73, 83, 94, 106, 116-117, 120, 236-237, 258, 259, 376, 379, 381, 383, 394-395, 396, 397-398, 404, 405, 418, 463, 479
Plodia interpunctella (Hübner)	Indian meal moth	465
Scirpophaga nivella (Fabricius)	Sugarcane top borer	49
Stenachroia elongella Hampson	Webworm	464
Tryporyza incertulas (Walker)	Yellow stem borer or rice stem borer	97, 260, 424, 429, 437-438
Tryporyza innotata (Walker)	White rice borer	429
Saturniidae		
Hyalophora cecropia (Linnaeus)	Cecropia moth	99
Scythridae		
Syringopais temperatella (Lederer)	Cereal leaf miner	518, 520
Sesiidae		
Melittia cucurbitae (Harris)	Squash vine borer	44
Sphingidae		
Erinnyis ello (Linnaeus)	Cassava hornworm	316
Manduca sexta (Linnaeus)	Tobacco hornworm	37, 39, 50, 57, 123, 125, 127
Tortricidae		
Choristoneura fumiferana (Clemens)	Spruce budworm	491, 504
Yponomeutidae		
Plutella xylostella (Linnaeus)	Diamondback moth	123

SCIENTIFIC NAME	COMMON NAME	PAGE
NEUROPTERA		
Chrysopidae		
Chrysopa spp.	Lacewings	60
ORTHOPTERA		
Acrididae		
Locusta migratoria (Linnaeus)	Migratory locust	118
Locusta migratoria migra-		
torioides (Reiche and		
Fairmaire)	African migratory locust	39
Melanoplus bilituratus (Walker)	Migratory grasshopper	97
Melanoplus bivittatus (Say)	Two-striped grasshopper	518
Melanoplus confusus Scudder	Grasshopper	43
Melanoplus differentialis		
(Thomas)	Differential grasshopper	518
Melanoplus femurrubrum		
(DeGeer)	Redlegged grasshopper	48
Melanoplus sanguinipes		
(Fabricius)	Migratory grasshopper	518
Ramburiella turcomana (Fischer-		
Waldheim)	Grasshopper	518, 519
Schistocerca gregaria (Forskal)	Desert locust	118
Zonocerus elegans (Thunberg)	Grasshopper	317
Zonocerus variegatus (Linnaeus)	Variegated grasshopper	317
Blattidae		
Periplaneta americana		
(Linnaeus)	American cockroach	29
THYSANOPTERA		
Phlaeothripidae		
Haplothrips tritici Kurdjumov	Thrips	518, 519
Thripidae		
Baliothrips biformis (Bagnall)	Rice thrips	425, 450
Caliothrips masculinus (Hood)	Thrips	316
Corynothrips stenopterus		
Williams	Thrips	316, 321
Frankliniella spp.	Thrips	321, 356
Frankliniella occidentalis		
(Pergande)	Western flower thrips	282
Frankliniella williamsi Hood	Thrips	316
Limothrips cerealium (Haliday)	Grain thrips	518, 519
Sericothrips occipitalis Hood	Thrips	99
Sericothrips variabilis (Beach)	Soybean thrips	48
Thrips spp.	Thrips	75, 326, 356, 358
Thrips tabaci Lindeman	Onion thrips	49

LIST OF VASCULAR
PLANT SPECIES

SCIENTIFIC NAME	COMMON NAME	PAGE
Amaranthaceae	Amaranth family	
Achyranthes L.		32
Apocynaceae	Dogbane family	
Nerium oleander L.	Oleander	103
Bromeliaceae	Pineapple family	
Ananas comosus Merr.	Pineapple	152
Byblidaceae		
Byblis Salisb.		59
Cactaceae	Cactus family	
Lophocereus schottii	Cina,	37
(Engelm.) Britt. & Rose	Cabesa de Viejo	
Opuntia tuna (L.) Mill.	Prickly pear	141
Chenopodiaceae	Goosefoot family	
Beta vulgaris L.	Sugar beet	94, 97, 148, 164, 174, 269-270
Spinaca oleracea L.	Spinach	151
Compositae	Composite family	48
Callistephus chinensis Nees	China aster	153
Chrysanthemum L.	Chrysanthemum	97, 98, 103, 104
Helianthus annus L.	Sunflower	262, 272
Lactuca sativa L.	Lettuce, garden lettuce	151
Taraxacum officinale Weber.	Dandelion	127
Convolvulaceae	Morning-glory family	
Ipomoea batatas Poir.	Sweet potato	181, 256, 268
Cruciferae	Mustard family	40, 131
Brassica juncea (L.) Coss.	Mustard	366
Brassica oleracea L.	Cultivated cabbage	42, 125, 139
Brassica oleracea L. var.		
acephala DC	Narrow-stem kale, kale	43, 58
Brassica oleracea L. var.		
gemmifera Zenker	Brussels sprouts	42, 43, 97, 99, 103, 109, 147

SCIENTIFIC NAME	COMMON NAME	PAGE
Brassica oleracea L. var. *italica* Plenck.	Sprouting broccoli	58
Brassica rapa L.	Turnip	40, 42, 43, 105
Cucurbitaceae	Gourd family	33, 48
Cucumis melo L.	Muskmelon	76-77
Cucurbita L.	Gourd, squash, pumpkin	44
Cucurbita maxima Duchesne	Squash	67, 76-77
Cucurbita sativus L.	Cucumber	97, 98
Cyperacea	Sedge family	57
Dioscoreaceae	Yam family	
Dioscorea alata L.	White yam	272
Dioscorea cayensis Lam.*	Guinea yam, yellow yam	272
Dioscorea esculenta (Lour.) Burk.	Yam, potato, potato yam	256, 272, 372
Droseraceae	Sundew family	
Drosera L.	Sundew	59
Drosophyllum Link.		59
Euphorbiaceae	Spurge family	
Croton L.	Croton	367
Manihot esculenta Crantz.	Cassava, mandioca, yuca, manioc, tapioca	151, 169, 176, 179, 181, 262, 269, 312-335, 372
Ricinus communis L.	Castor bean	48
Fagaceae	Beech family	
Castanea Mill.	Chestnut	48, 235, 496
Castanea crenata Sieb. & Zucc.	Japanese chestnut	496
C. mollissima Blume	Chinese chestnut	496
Quercus L.	Oak	118
Quercus macrocarpa Michx.	Mossy-cup oak	35
Gingkoales, Gingkoaceae	Gingko family	
Gingko biloba L.	Gingko, maidenhair tree	495
Gramineae	Grass family	48, 57, 422, 514
Aegilops L.	Goat-grass	514
Aegilops squarrosa L.†		515
Aegilops speltoides Tausch		515
Agropyron Gaertn.	Wheatgrass	514
Andropogon gerardi Vitman	Beardgrass	43
Avena L.	Oat	42, 268
Avena sativa L.	Cultivated oat	48, 97, 263, 268
Echinocloa crusgalli (L.) Beav.	Barnyard grass	439, 444
Eleusine coracana (L.) Gaertn.	African millet, ragi millet, finger millet	268-269
Eremopyrum Jaub. & Spach.‡		514

D. rotundata Poir.
†T. tauschii
‡*Agropyron* Gaertn.

SCIENTIFIC NAME	COMMON NAME	PAGE
Euchlaena mexicana Schrad.	Teosinte	153
Haynaldia Schur		514
Hordeum vulgare L.	Barley	69, 73, 101, 148, 260, 263, 266, 372
Oryza glaberrima Steud.	African rice	258, 266, 422, 426
Oryza sativa L.	Asian cultivated rice	43, 48, 57-58, 59-60, 67, 70-71, 82, 84-85, 97, 99, 101, 150-151, 176, 202, 203, 205, 235, 242, 258, 260, 262, 265-266, 372, 420-455
Panicum miliaceum L.	Broomcorn millet, proso millet	269
Panicum virgatum L.	Switchgrass	43
Paspalum scrorbiculatum L.	Kodo millet	269
Saccharum officinarum L.	Sugarcane	169, 171, 174, 179, 181, 256
*Schizachyrium scoparium** Nash.	Little bluestem	43
Secale cereale L.	Rye	72, 261, 263, 270, 514
Setaria italica (L.) Beauv.	Italian or foxtail millet	269
Sorghum Moench.	Sorghums	260-261
Sorghum vulgare Pers.[†]	Sorghum	43, 47, 48, 60, 71-72, 91, 95, 109, 139, 257, 267-268, 362, 457-458
Sorghum vulgare var. *Sudanense* Hitchc.	Sudan grass	466
Tripsacum L.	Gamagrass	261, 402, 534
Tripsacum dactyloides L.	Eastern gamagrass	400, 401
Tripsacum floridanum 'Porter ex. Vasey'	Florida gamagrass	402
Triticum aestivum L.	Common wheat	4, 6, 8, 12, 38, 44, 49, 51, 53, 67, 69, 72-73, 78, 80, 91, 93, 97, 100-101, 109, 134, 175, 176, 185, 199-200, 206-207, 208, 210, 211, 212, 237, 260, 261, 263-265, 372
Triticum dicoccoides Koern.		515

*Andropogon scoparius Michx.
†*S. bicolor* (L.) Moench.

Aegilops squarrosa L.
[†]*T. dicoccum* Schrank
[‡]*T. durum* Desf.

SCIENTIFIC NAME	COMMON NAME	PAGE
Medicago intertexta All.		298
Medicago laciniata Mill.		298
Medicago marina L.		298
Medicago minima (L.) Desr.		298
Medicago pironae Vis.		281
Medicago polymorpha L.		298
Medicago prostrata Jacq.		281
Medicago rhodopea Velen		281
Medicago rugosa Lam.		298
Medicago sativa L.	Cultivated alfalfa	10, 12, 20, 49, 90, 92, 96, 106-107, 108, 109, 147, 148, 236, 259, 276, 279-311
Medicago scutellata (L.) Willd.	Snail clover	298
Medicago tianschanica Vassilcz.		
var. *agropyretorum*		311
Medicago truncatula Gaertn.		298
*Medicago varia**		281
Melilotus Mill.	Sweet clover	309
Melilotus alba Desv.	White sweet clover	40, 77, 306
Melilotus officinalis (L.) Lam	Yellow sweet clover	40, 77
Phaseolus L.	Bean	49, 53, 104, 176, 262, 271
Phaseolus limensis Macfad.	Lima bean	38, 97
Phaseolus vulgaris L.	Dry bean, snap bean	101, 102, 104-105, 146
Pisum sativum L.	Pea	42, 105, 107, 237, 262, 372
Robinia pseudoacacia L.	Black locust	496
Trifolium L.	Clover	47, 309
Trifolium pratense L.	Red clover	306
Vicia faba L.	Broadbean	103, 104-105, 147, 295, 306
Vicia sativa L.	Common vetch	39
Vigna sinensis Endl.	Cowpea	43, 99
Lentibulariaceae	Bladderwort family	
Penguicula L.	Butterwort	59
Liliaceae†	Lily family	
Agave L.	Sisal	169
Linaceae	Flax family	
Linum usitatissimum L.	Common flax	199-200
Magnoliaceae	Magnolia family	
Magnolia kobus DC	Kobus magnolia	34, 35
Malvaceae	Mallow family	48

M. media Martyn
†Alternatively grouped under the Agavaceae

SCIENTIFIC NAME	COMMON NAME	PAGE
Gossypium L.	Cotton	11-12, 13, 25, 26, 35, 43, 48, 52-53, 74-75, 97, 102, 107, 151, 221-229, 235-236, 239-241, 242, 244-251, 270, 336-339
Gossypium anomalum Wawra & Peyr.		343, 354
Gossypium arboreum L.	Asiatic tree cotton, Indian tree cotton, deo cotton, nurma cotton	340, 342, 343, 344, 347, 348, 350, 354, 359
Gossypium areysianum Defl.		343
Gossypium aridum Skovsted.		343
Gossypium armourianum Kearney		343, 348, 350, 359
Gossypium australe F. Muell.		343, 356
Gossypium barbadense L.	Gallini cotton, Sea Island cotton, American cotton, Barbados cotton, black seed cotton	340, 341, 343, 344, 348, 350, 352, 356, 359, 362
Gossypium barbosanum L.		343
Gossypium bickii Prokh.		343
Gossypium capitis-viridis Mauer		343
Gossypium costulatum Tod.		343
Gossypium cunninghamii Tod.		343
Gossypium darwinii Watt		343
Gossypium davidsonii Kell.		343
Gossypium gossypoides Standl.		343
Gossypium harknessii Brandg.		343, 354
Gossypium herbaceum L.	Levant cotton, Bamia cotton	340, 341, 342, 343, 344, 347, 348, 350, 354
Gossypium hirsutum L.	American upland cotton	33, 59, 60, 139, 260, 341-346, 348, 350, 352, 355, 356, 357, 359, 362, 365
Gossypium incanum (Schwartz) Hillcoat		343
Gossypium klotzschianum Anderss.		343
Gossypium laxum Phill.		343
Gossypium lobatum Gentry		343, 356
Gossypium longicalyx Hutch & Lee		343

SCIENTIFIC NAME	COMMON NAME	PAGE
Gossypium nelsonii Fryx.		343
Gossypium pilosum Fryx.		343
Gossypium populifolium F. Muell.		343
Gossypium pulchellum (Gardn.) Fryx.		343
Gossypium raimondii Ulbrich.		343, 348
Gossypium robinsonii F. Muell.		343
Gossypium somalense Gurke		343, 350
Gossypium stocksii Mast. ex Hook.		343
Gossypium sturtianum (Der.) Fryx.		343
Gossypium thurberi Tod.		343, 350
Gossypium tomentosum Nutt. ex Seem.		342, 343, 348, 357, 360, 362
Gossypium trilobum (Moc. & Sess. ex DC) Skov. emend. Kearney		343, 350
Gossypium triphyllum Hochr.		343
Musaceae	Banana family	
Musa sapientum L.	Common eating banana	169, 272
Palmae	Palm family	57
Cocos nucifera L.	Coconut palm	139, 271
Papaveraceae	Poppy family	
Papaver L.		123
Passifloraceae	Passion flower family	49
Passiflora adenopoda Moc. & Sess. ex DC		53
Pinacea	Pine family	498
Abies balsamea (L.) Mill	Balsam fir, fir-balsam	32, 97, 498, 500-501
Abies concolor (Gord.) Engelm.	Silver fir, white fir	498, 499
Picea abies (L.) Karst	Norway spruce	495-496, 497
Picea engelmanni (Parry) Engelm.	Engelmann spruce	492
Picea glauca (Moench.) Voss	White spruce	498
Picea sitchensis Trautv. & Meg.	Sitka spruce	497
Pinus L.	Pine	139, 261, 497
Pinus banksiana Lamb.	Jackpine	28, 118, 499, 504
Pinus coulteri D. Don.	Coulter pine	497
Pinus jeffreyi Murr.	Jeffrey pine	497
Pinus ponderosa Laws.	Western yellow pine, bull pine ponderosa pine	33
Pinus silvestris L.	Scotch pine, scotch fir	33, 497

SCIENTIFIC NAME	COMMON NAME	PAGE
Pinus strobus L.	Eastern white pine, white pine	497
Pinus taeda L.	Loblolly pine	33
Pseudotsuga menziesii (Mirb.) Franco	Douglas fir	497, 504
Polypodiaceae	Fern family	
Pteridium aquilinum (L.) Kuhn	Brake fern, pasture fern	32
Rosaceae	Rose family	50
Crataegus L.	Hawthorn	75
Fragaria L.	Strawberry	50, 97
Prunus L.	Cherry	97
Prunus persicola (L.) Batsch.	Peach	97
Prunus serotina Ehrh.	Black or wild cherry	99
Pyrus americana (Marsch.) DC	Mountain ash	75
Pyrus communis L.	Common pear	75, 260
Pyrus malus L.	Apple	4, 75-76, 102, 104, 139, 147, 149, 155, 236, 262
Rubus idaeus L.	Raspberry	58, 76, 78, 149, 261
Rubus phoenicolasius Maxim.	Wineberry	58
Rubiaceae	Madder family	
Coffea L.	Coffee	139, 181
Rutaceae	Rue family	
Citrus sinensis Osbeck.	Orange	181, 262
Salicaceae	Willow family	
Populus L.	Poplars, willows	489
Saxifragaceae	Saxifrage family	
Ribes americanum Mill.	American wild black currant	103
Solancecae	Nightshade family	37, 50, 54
Capsicum frutescens L.	Green pepper	120
Lycopersicum esculentum Mill.	Cultivated tomato	38, 97, 125, 152-153, 176, 262
Lycopersicum hirsutum H. B. & K.	Wild tomato relative	44, 45
Nicotiana tabacum L.	Tobacco	37, 57, 97, 106, 125, 146, 151, 175, 261
Petunia Juss.	Petunia	57
Solanum (Tourn.) L.		37
Solanum berthaultii Hawkes	Wild potato	55, 146
Solanum chacoense Bitter.		94
Solanum demissum Lindl.		94, 172
Solanum polyadenium Greenm. or Greenm?	Wild potato	55, 56
Solanum tarijense Hawkes	Wild potato	55

SCIENTIFIC NAME	COMMON NAME	PAGE
Solanum tuberosum L.	Cultivated potato	38, 94, 97, 120, 125, 146, 147, 148, 151, 172, 175, 179, 181, 209, 256, 262, 263, 266-267
Sterculiaceae	Sterculia family	
Theobroma cacao L.	Cocoa	139, 152, 181
Theaceae	Tea family	
Camellia sinensis (L.) Kuntze	Common tea, Chinese tea, Japanese tea	181
Ulmaceae	Elm family	
Ulmus L.	Elm	34, 35, 75, 139
Umbelliferae	Parsley family	131
Angelica L.	Angelica	127
Apium graveolens L.	Celery	131
Daucus L.	Carrot	151
Peucedanum L.		127
Peucedanum officinale L.	Sulfur-root	73
Vitaceae	Vine or grape family	50
Vitis L.	Grape	4, 181, 235

INDEX

651

RESOURCE RECOVERY AND RECYCLING
A. F. M. Barton

QUANTITATIVE TOXICOLOGY
V. A. Filov, A. A. Golubev, E. I. Liublina, and N. A. Tolokontsev

ATMOSPHERIC MOTION AND AIR POLLUTION
Richard A. Dobbins

INDUSTRIAL POLLUTION CONTROL—Volume I: Agro-Industries
E. Joe Middlebrooks

BREEDING PLANTS RESISTANT TO INSECTS
Fowden G. Maxwell and Peter Jennings, Editors